Water Wells and Boreholes

Water Wells and Boreholes

Second Edition

BRUCE MISSTEAR

Trinity College Dublin, Ireland

DAVID BANKS

Holymoor Consultancy Ltd and University of Glasgow, UK

LEWIS CLARK

(Deceased) – formerly of Clark Consult Ltd, Henley on Thames, UK

WILEY Blackwell

This edition first published 2017 © 2017 by John Wiley & Sons Ltd

Registered Office
John Wiley & Sons Ltd, The Atrium, Southern Gate, Chichester, West Sussex, PO19 8SQ, UK

Editorial Offices
9600 Garsington Road, Oxford, OX4 2DQ, UK
The Atrium, Southern Gate, Chichester, West Sussex, PO19 8SQ, UK
111 River Street, Hoboken, NJ 07030-5774, USA

For details of our global editorial offices, for customer services and for information about how to apply for permission to reuse the copyright material in this book please see our website at www.wiley.com/wiley-blackwell.

The right of Bruce Misstear, David Banks and Lewis Clark to be identified as the author of the editorial material in this work has been asserted in accordance with the UK Copyright, Designs and Patents Act 1988.

Library of Congress Cataloging-in-Publication data applied for:

9781118951705

A catalogue record for this book is available from the British Library.

Wiley also publishes its books in a variety of electronic formats. Some content that appears in print may not be available in electronic books.

Cover image: Courtesy of the author
Cover design: Wiley

Set in 10/12pt Times by SPi Global, Pondicherry, India

10 9 8 7 6 5 4 3 2 1

Contents

Preface to Second Edition

For this second edition we have retained the structure and emphasis of the original book: the text follows a life-cycle approach - from choosing a suitable well site, through the processes of designing, constructing, testing and sampling the well, to monitoring, maintenance and, if required, rehabilitating or finally abandoning the well. The target audience for this new edition continues to be students, professionals in hydrogeology and engineering and aid workers and other practitioners involved in well projects.

This second edition contains many updates on new well guidelines and standards published since the first edition. We also provide additional text on several topics, for example: the siting and construction of wells for economically-disadvantaged communities; specialist well designs for applications such as heating, cooling and aquifer recharge; drilling techniques such as sonic drilling and dual rotary that are becoming increasingly popular in the water well industry; new techniques in downhole geophysical logging; methods for analysing pumping test data under "non-ideal" conditions; and sampling wells for stable isotopes and dissolved gases.

Whilst we include some additional guidance on health and safety issues, we would again like to stress, as we did in the first edition, that the book is not intended to be a manual. The reader should always consult the relevant regulations and guidance within their own country on these and other issues relating to water well projects.

We hope readers will enjoy this new edition and find it useful in their studies and workplace.

Bruce Misstear and David Banks
July 2016

Legal disclaimer

Although the authors and the publisher have used their best efforts to ensure the accuracy of the material contained in this book, complete accuracy cannot be guaranteed. Neither the authors nor the publisher accept any responsibility for loss or damage occasioned, or claim to have been occasioned, in part or in full, as a consequence of any person acting, or refraining from acting, as a result of matter contained within this publication. For well construction projects, the services of experienced and competent professionals should always be sought.

Preface to First Edition

The *Field Guide to Water Wells and Boreholes,* published by Lewis Clark in 1988, was a practical guide to designing and constructing wells and boreholes. It was primarily intended to be of use to field workers involved in implementing groundwater projects (it was written as one of the Geological Society of London Professional Handbook Series). This new book aims to update and expand the content of the *Field Guide.* It maintains the practical emphasis, but it has also been written with students in mind. The target readership includes:

- final-year undergraduate students in geology and civil engineering;
- graduate students in hydrogeology, groundwater engineering, civil engineering and environmental sciences;
- research students who are involved in using data from wells as part of their research;
- professionals in hydrogeology, water engineering, environmental engineering and geotechnical engineering;
- aid workers and others involved in well projects.

With its wider target audience, the new book has a broader scope than the *Field Guide.* Although it remains a practical guide, the book introduces additional theoretical detail on matters relating to the siting, design, construction, operation and maintenance of water wells and boreholes. Only a basic level of mathematical ability is assumed in the reader: the book includes a number of simple equations for the analysis of groundwater flow and well design problems which can be solved manually using a hand-calculator. Although the use of computer software is helpful for the longer and more repetitive computations, the authors are keen to promote a basic understanding of the issues, and do not support indiscriminate use of computer software without an appreciation of the basics.

The main focus of the book is on water wells that are used for drinking, industry, agriculture or other supply purpose, although other types of wells and boreholes are also covered, including boreholes for monitoring groundwater level and groundwater quality. Just as the potential car buyer looks for a certain combination of performance, reliability, durability, cost (including running cost) and personal and environmental safety in his or her new vehicle, the potential water well owner requires that:

- the well (or group of wells) should have sufficient yield to meet the demand;
- the water quality should be fit for the particular purpose;
- the well should be reliable, requiring little maintenance (although, as with a vehicle, some regular programme of maintenance will be required);
- the well should be durable, with a design life suited to its purpose.
- the construction and operating costs should not be excessive;
- the well should not impact unacceptably on neighbouring wells or on the environment, and therefore should not violate local water resources, planning or environmental legislation.

These principles underpin the guidance given throughout this text. The book follows a 'life-cycle' approach to water wells, from identifying a suitable well site through to the successful implementation, operation and maintenance of the well, to its eventual decommissioning. The structure of the book is illustrated in the figure below.

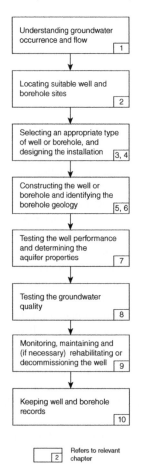

The book is not a driller's manual: it does not describe drilling procedures in detail; nor does it deal in detail with issues such as drilling permits, abstraction licences, or health and safety procedures in constructing and operating wells: readers should always consult local country guidance and regulations on these issues.

Lewis Clark (1937–2004): An Appreciation

Lewis Clark died in July 2004, when the first edition of this book was at an early stage of drafting. Lewis was an inspiration to many hydrogeologists in Britain and further afield: his co-authors would like to dedicate this new edition of the book to him, and to include this short appreciation of his work.

Following a PhD from the University of Leeds in 1963 (on the subject of metamorphic geology), Lewis first became involved in hydrogeology whilst working for the Geological Survey of Uganda in the 1960s. In 1968 he joined the Hunting consultancy group where he worked on hydrogeological projects in many developing countries, including Sudan, Thailand and Saudi Arabia. He was part of the talented Hunting Technical Services and Sir M Macdonald and Partners team (which also included Wiktor Bakiewicz, Roy Stoner and the late Don Milne) that worked on a major groundwater supply project for the Saudi Arabian capital Riyadh in the early 1970s, a project which involved the design and construction of a well field with more than 50 large capacity wells tapping a deep sandstone aquifer. This and subsequent experience in the design, drilling and testing of wells led Lewis to publish his *Field Guide to Water Wells and Boreholes* in the 1980s. He also published a significant and much-quoted paper on step drawdown tests in the *Quarterly Journal of Engineering Geology* in 1977 (Clark, 1977).

In 1976 Lewis Clark joined the Water Research Centre (now WRc plc) and he soon became involved in applied research in groundwater quality and pollution, which is perhaps the work for which he is best remembered in Britain. He studied the origins and transport of organic contaminants including chlorinated solvents and pesticides, and the resulting research publications were always insightful and useful. In 1993 he was appointed Visiting Professor in Hydrogeology at University College London. He retired from WRc and set up his own consultancy, Clark Consult, in 1997. During that year, his contribution to hydrogeology was recognized by his peers in the award of the prestigious Whitaker medal by the Geological Society of London. He continued to work as a consultant hydrogeologist up until his death, making several visits for UN agencies to groundwater projects in Africa and central Asia.

Acknowledgements

Many people contributed directly or indirectly to the completion of this book. Individual chapters in the first edition were reviewed by Paul Ashley, John Barker, Charles Jones, Atle Dagestad, Mike Jones, Nick Robins, Vin Robinson, Stuart Smith, Geoff Wright and Paul Younger. We are also indebted to Aonghus McNabola for his patience in drafting several of the original figures in the book. Many individuals and organizations were involved in making available their own illustrations, and these are acknowledged in the relevant figure captions. We would especially like to thank the following for their help in sourcing figures and photos: Asgeir Bårdsen, Kim Beesley, Aidan Briody, Rolv Dahl, Bjørn Frengstad, Jeff Meehan, Laurence Gill, Peter O'Connor, David Roberts, Jan Steiner Rønning, Henrik Schiellerup, Svein Stoveland and Alan Waters. Bruce Misstear would like to acknowledge his colleagues in the School of Engineering at Trinity College Dublin, and also the contribution of the University of New South Wales in Sydney where he spent a sabbatical working on the first edition. David Banks wishes to thank the University of Glasgow and his colleagues in the School of Engineering at that venerable institution for their support. Others who helped in the preparation of the book, or provided inspiration to its authors, include: Ian Acworth, Wiktor Bakiewicz, David Ball, Sarah Beeson, Donal Daly, the late Eugene Daly, James Dodds, Jane Dottridge, Robin Farbridge, Robin Hazell, Peter Howsam, Paul Johnston, John Lloyd, the late Don Milne, David Misstear, Karen Misstear, Gillian Misstear, Steve Parsons, Alan Rendell, Peter Rippon, Roy Stoner, John Tellam, Jan van Wonderen and Paul Younger.

1

Introduction

1.1 Wells and boreholes

Water wells in some form or other have existed for almost as long a time as people have occupied this planet. The earliest wells were probably simple constructions around springs and seeps, or shallow excavations in dry river beds, but such wells have not left any traces for archaeologists. One of the oldest well discoveries is in Cyprus, dating from 7000 to 9000 BC (Fagan, 2011), whilst the earliest well remains in China have been dated at around 3700 BC (Zhou *et al.*, 2011). Since the first millennium BC, horizontal wells or *qanats* have been widely used for water supply and irrigation in the Middle East and western Asia, notably Iran, and continue to be used today (Figure 1.1). In Europe, the development of many towns and cities in the middle ages and on through the industrial period was aided considerably by the abstraction of relatively pure water supplies from wells and springs (Figure 1.2). In the nineteenth century, new drilling technology was used to construct deep wells to exploit artesian (flowing) aquifers (see Section 1.2 for explanations of aquifer terminology), including the Grenelle well in the Paris basin, which was drilled between 1833 and

1841, and reached a depth of 548 m (Margat *et al.*, 2013). The first mechanically-drilled well in the United States dates from 1823, whereas the first drilled well in the Great Artesian Basin of Australia was constructed in 1878 (Margat and van der Gun, 2013).

Wells continue to have an important role in society today. Some 2 billion people obtain their drinking water supplies directly from drilled or hand-dug wells (UNICEF and WHO, 2012). A further 4 billion people have access to piped water or public taps, a proportion of which will be sourced from groundwater, so it is likely that more than 3 billion people worldwide rely on water wells for their drinking water. Over half the public water supplies in European Union countries come from groundwater, ranging from between 20% and 30% of drinking water supplied in Spain and the United Kingdom, to nearly 100% in Austria, Lithuania and Denmark (Hiscock *et al.*, 2002).

The largest use of groundwater worldwide is for irrigation (70%), with India, China and the United States the leading countries in terms of total groundwater withdrawals (Margat and van der Gun, 2013). The last 30 years have witnessed a huge increase in the use of wells for agricultural

Water Wells and Boreholes, Second Edition. Bruce Misstear, David Banks and Lewis Clark.
© 2017 John Wiley & Sons Ltd. Published 2017 by John Wiley & Sons Ltd.

Figure 1.1   *Open section of falaj (qanat) running through a town in northern Oman. Here, the channel is divided into three, with two of the channels then rejoining (at the bottom of the picture), in order to produce a two-thirds: one third split in the flow downstream. This Falaj al Khatmeen is included on the UNESCO list of World Heritage Sites. Photo by Bruce Misstear*

irrigation, especially in Asia (Figure 1.3): in China 54% of irrigation water is supplied from groundwater while this proportion rises to 89% in India and 94% in Pakistan. In the United States, groundwater pumping increased by 144% between 1950 and 1980, with 71% of the annual withdrawal of 111.7 km^3 in 2010 being used for irrigated agriculture (Margat and van der Gun, 2013). According to the National Ground Water Association, 44% of the population of the United States depends on

groundwater for its drinking water and there are about 500 000 new private wells constructed each year for domestic supplies.

Other uses of wells are many and diverse and include livestock watering (Figure 1.4), industrial supplies, geothermal energy or ground-source heating/cooling (Figure 1.5), construction dewatering, brine mining, water injection to oil reservoirs, aquifer clean up, river support and artificial recharge of aquifers. Wells and boreholes are also

Figure 1.2   *Hand-dug well in Brittany, France. Photo by Bruce Misstear*

used extensively for monitoring water levels and groundwater quality.

Wells have long had a religious significance in many societies. In India, the Holy Vedic Scriptures dating back to 8000 BC contain references to wells (Limaye, 2013). In the Bible and Koran, wells and springs feature prominently, sometimes as places for meeting and talking and often as metaphors for paradise. Holy wells remain an important feature of local culture throughout the Celtic lands in western Europe, for example, where there may be as many as 3000 holy wells in Ireland alone (Logan, 1980; Robins and Misstear, 2000). Many of these wells are still visited regularly and votive offerings such as rags, statues and coins are common (see Box 3.7 in Chapter 3).

Water wells have also been a source of conflict since Biblical times:

> *But when Isaac's servants dug in the valley and found there a well of springing water, the herdsmen of Gerar quarrelled with Isaac's herdsmen, saying "This water is ours".*
>
> Genesis 26:19-20

Figure 1.3 A dual purpose irrigation and drainage well in the Indus valley, Pakistan. In this 'scavenger well' the outlet pipe in the foreground of the picture is discharging fresh groundwater from the upper part of the well, whereas the pipe to the right is discharging saline water from the lower section of the well, thus preventing the saline water from moving upwards and contaminating the good quality water. The good quality water is used for irrigation whilst the saline water is diverted to the drainage system. Photo by Bruce Misstear

They remain so today. A major point of contention in the Middle East is the control of the groundwater resources in the region (Shuval and Dweik, 2007; Younger, 2012).

Water wells come in many forms, orientations and sizes. Traditionally most water wells were excavated by hand as shallow, large diameter, shafts; nowadays, the majority are constructed from relatively small diameter boreholes drilled by machine, sometimes to great depths. Water wells are typically vertical but can be horizontal (infiltration gallery), a combination of vertical and horizontal well (radial collector well), or occasionally inclined (Figure 1.6). The water may be abstracted by hand-operated or motorized pumps, or it may flow to the surface naturally under positive upward pressure (artesian well; Figure 1.7) or by gravity drainage (*qanat* or *falaj*). This book deals mainly with drilled wells (often called boreholes), since readers are likely to encounter these most often, but other types of wells are also covered.

Water well terminology is not standard throughout the world, and different names are commonly applied to identical constructions. The terms used in this book are explained in Box 1.1. Further details of the different types of wells and boreholes, and their component parts, are included in Chapter 3.

Figure 1.4 *Drilled well fitted with a windmill pump used for livestock watering, New South Wales, Australia. Photo by Bruce Misstear*

1.2 Groundwater occurrence

The remainder of this chapter provides the non-specialist reader with a brief introduction to the occurrence of groundwater and the principles of groundwater flow, including radial flow to water wells. For a more comprehensive coverage of these topics the reader is referred to standard hydrogeology texts including Freeze and Cherry (1979), Fetter (2001), Todd and Mays (2005), and Hiscock and Bense (2014).

1.2.1 Aquifers, aquicludes and aquitards

Figure 1.8 illustrates some of the basic terminology used to describe groundwater and aquifers. While some authorities define *groundwater* as any water occurring in the subsurface – that is, water occurring in both the *unsaturated* and the *saturated* zones – we follow the tradition of defining groundwater as that portion of water in the subsurface that occurs in the saturated zone. A geological formation that is able to store and

Figure 1.5   *Drilling rig being set up for constructing a well in a gravel aquifer used as a source of geothermal energy, Dublin, Ireland. Photo by Bruce Misstear*

transmit groundwater in useful quantities is called an *aquifer*. Aquifer is thus a relative term, since a low permeability geological formation that would not be considered as an aquifer capable of meeting public water supply or irrigation water demands, may be able to supply 'useful quantities' of groundwater to a village or domestic well in regions where water is otherwise scarce. In this context, one can argue, for example, that

low-permeability mudstones in parts of Africa are hugely valuable aquifers (MacDonald, 2003).

Aquifers are often described according to their water level or pressure head conditions (see Boxes 1.2 and 1.3 for explanations of groundwater head). An aquifer is said to be *unconfined* where its upper boundary consists of a free groundwater surface at which the pressure equals atmospheric. This free surface is known as the *water table* and

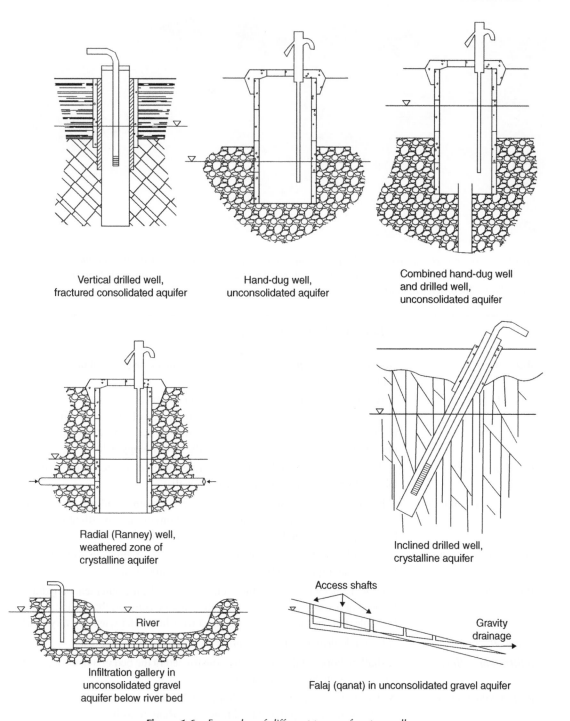

Figure 1.6   *Examples of different types of water well*

Figure 1.7 *Flowing artesian well, northern Myanmar. The well was drilled into a strongly confined sandstone aquifer. Children are enjoying the 'swimming pool' created by the discharge until such time as the well is capped. Photo by Bruce Misstear*

Box 1.1   Well and borehole terminology

Water well	Any hole excavated in the ground that can be used to obtain a water supply
Drilled well	A water well constructed by drilling. Synonyms are tubewell or, simply, borehole. As drilled wells are the main focus of this book they will be referred to as wells for simplicity. Other types of water well will be distinguished, where necessary, using the terminology below
Hand-dug well	A large-diameter, usually shallow, water well constructed by manual labour. Synonyms are dug well or open well
Exploratory borehole	A borehole drilled for the specific purpose of obtaining information about the subsurface geology or groundwater. Synonyms are investigation borehole, exploration borehole or pilot borehole
Observation borehole	A borehole constructed to obtain information on variations in groundwater level or water quality. Also known as observation well
Piezometer	A small diameter borehole or tube constructed for the measurement of hydraulic head at a specific depth in an aquifer. In a piezometer, the section of the borehole (the screened section) in contact with the aquifer is usually very short
Test well	A borehole drilled to test an aquifer by means of pumping tests
Infiltration gallery	A shallow horizontal well usually constructed in the bed of a river or along a river bank in an alluvial aquifer
Radial collector well	A large diameter well with horizontal boreholes extending radially outwards into the aquifer. Also known as a Ranney well
Qanat	An infiltration gallery in which the water flows to the point of abstraction under gravity. There are many synonyms, including *falaj* (Oman), *karez* (Afghanistan) and *kariz* (Azerbaijan)

Box 1.2 What is groundwater head?

There is a common misconception that water always flows from areas of high pressure to areas of low pressure, but it does not. Consider two points, A and B, in the tank of water illustrated in Figure B1.2(i). The pressures (P) at points A and B are given by:

$$P = H\rho g$$

where H is the height of the column of water above the point (dimension [L]), ρ is the density of the water ([M][L]$^{-3}$=c.1000 kg m^{-3}) and g the acceleration due to gravity ([L] [T]$^{-2}$=9.81 m s^{-2}).

Thus, at point A, the water pressure is 14,715 N m^{-2}, and at point B it is 53,955 N m^{-2}. But water does not flow from B to A - the water in the tank is static. Clearly we need a more sophisticated concept. In fact, we can use the concept of *potential energy*: groundwater always flows from areas of high potential energy to low potential energy. *Groundwater head (h)* is a measure of the potential energy of a unit mass of groundwater at any particular point. This is the sum of potential energy due to elevation and that due to pressure.

$$\text{Potential energy} = \frac{P}{\rho} + zg \ \text{(in J kg}^{-1}\text{)}$$

To obtain head (in metres), we divide by g (a constant):

$$h = \frac{P}{\rho g} + z$$

where z is the elevation above an arbitrary datum [L]. Returning to the tank of water example, the heads at A and B, relative to the base of the tank, are:

$$h_A = \frac{14,715}{1000 \times 9.81} + 5 = 6.5\,m$$

$$h_B = \frac{53,955}{1000 \times 9.81} + 1 = 6.5\,m$$

In other words, they are identical and there is no tendency to flow between the two points. Note that we can compare heads in different locations relative to an arbitrary datum *only* if the density is constant (i.e., 1 m in elevation is equivalent in energy terms to the pressure exerted by a 1 m column of fluid). If we are considering groundwater systems of variable salinity (and density), it is easy to get into difficulties by applying simplistic concepts of head.

In an unconfined aquifer, the elevation of the water table represents groundwater head at that point in the aquifer. While it is often assumed that the water table represents the boundary between unsaturated and saturated aquifer material, this is not quite true, as there is a thin capillary fringe of saturated material above the water table. Strictly speaking, the water table is the surface at which the pressure is equal to atmospheric (i.e., the water pressure is zero).

For confined aquifers, we can imagine contours joining all locations of equal head. These contours then define a surface which is called the *piezometric surface* or *potentiometric surface*. The slope of this surface defines the hydraulic gradient, which in turn controls the direction of groundwater flow. Water will rise in a borehole sunk into the confined aquifer to a level corresponding to the potentiomentric surface.

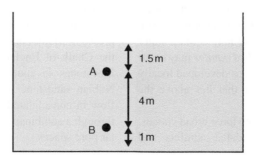

Figure B1.2(i) Sketch of a water tank showing two points where pressure and head can be calculated

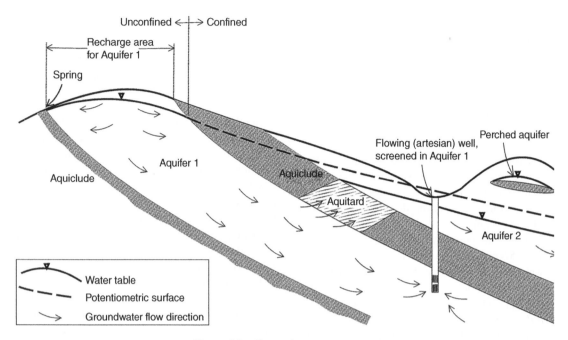

Figure 1.8 *Groundwater occurrence*

unconfined aquifers are sometimes known as *water-table aquifers*. An aquifer is said to be *confined* when it is fully saturated and its *potentiometric surface* (hydraulic head) lies in an overlying, low-permeability confining layer. Very low permeability layers bounding aquifers are often known as *aquicludes*. However, no formation is truly impermeable and many low permeability formations can transmit quantities of groundwater that may be significant on a regional scale: thus, the term *aquitard* is often preferred for such formations. Where an aquitard allows some leakage of water to or from an aquifer, the aquifer is often said to be *semi-confined* or *leaky*. In a system of aquifers separated by aquitards or aquicludes, each aquifer may have a different hydraulic head, as depicted in Figure 1.8, and may contain water of a different quality. A *perched aquifer* may occur where a shallow water table has developed locally on a low permeability layer that lies above the regional water table.

Aquifers can be divided into three broad classes: crystalline aquifers, consolidated aquifers and unconsolidated aquifers. Crystalline aquifers are typified by the igneous and metamorphic rocks that underlie large areas of the world. They include the ancient granites and gneisses that form the 'basement complex' of sub-Saharan Africa and the younger volcanic rocks of the Deccan traps in southern India. Groundwater flow in crystalline aquifers takes place through discrete fractures, rather than through intergranular pore spaces.

Consolidated aquifers are composed of lithified (but not metamorphosed) sedimentary rocks such as sandstones and limestones (the term consolidated is used here in its general meaning of any sediment that has been solidified into a rock, rather than in the geotechnical engineering sense of a fine-grained cohesive soil that has been compressed). Major consolidated aquifers are found in the Chalk of England and France, the Floridan limestones in southeast United States and the Nubian sandstone in north Africa. Groundwater flow in consolidated aquifers tends to take place through a combination of fractures and intergranular pore spaces.

Box 1.3   Groundwater head as a three-dimensional concept

The distribution of groundwater head in an aquifer can be imagined as a *three dimensional scalar field*. Each point in the scalar field has a unique value of groundwater head $h(x,y,z)$. Points of equal head can be joined by groundwater head contours. Groundwater flow has a tendency to follow the maximum gradient of head; in other words, the groundwater flow vector (Q) is proportional to $-\text{grad}(h)$. In vector-speak:

$$Q \propto -\nabla h$$

Thus, if we construct groundwater head contours in a porous medium aquifer, the groundwater flow lines will be perpendicular to the head contours (in fractured aquifers, groundwater flow *may* not be perpendicular to the regional head contours, as the groundwater is constrained to flow along fracture pathways which may not exist parallel to the head gradient).

Figure 1.8 implies that artesian boreholes can occur in confined aquifers where the potentiometric surface is higher than ground level. However, artesian boreholes *can* also occur in unconfined aquifers. Consider the two aquifer sections below. Figure B1.3(i) shows a relatively high permeability aquifer. The water-table gradient is shallow and groundwater flow is predominantly horizontal. Thus, the head contours are approximately vertical and the head at any depth in the aquifer at a given horizontal (x,y) coordinate is approximately equal to the elevation of the water table. Hence wells exhibit similar static water levels, irrespective of depth [wells A and B in Figure B1.3(i)]. Groundwater flow thus approximately follows the gradient of the water table.

Consider, then, the second drawing [Figure B1.3(ii)], of groundwater flow in a low permeability aquifer in an area of high topography. Here, head is truly three-dimensional, varying with elevation (z) as well as horizontally (x,y). Head contours are complex and *not* necessarily vertical. Groundwater flow has upwards and downwards components. Typically, in recharge areas, head decreases with increasing depth, and groundwater flow has a downward component. A deep-drilled well here (well C) will have a lower static water level than a shallow one (well D). In discharge areas, head increases with increasing depth and groundwater flow has an upward component. A deep-drilled well here (well E) will

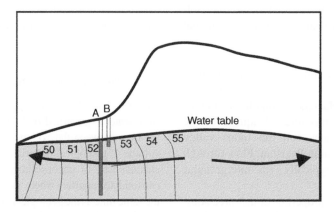

Figure B1.3(i)   *Cross section through a relatively permeable aquifer. The water table gradient is flat. Contours on piezometric head (numbered contours, in m above sea level) are approximately vertical. Wells A and B have similar static water levels irrespective of depth*

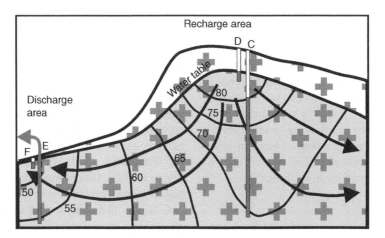

Figure B1.3(ii)   *Cross section through a relatively low permeability aquifer, such as granite. The water table gradient reflects topography. Contours on piezometric head (numbered contours, in m above sea level) are strongly three-dimensional. Pairs of wells (C, D and E, F) have differing static water levels depending on well depth. Deep wells may even be artesian (overflowing) in discharge areas (well E)*

have a static water level higher than a shallow one (well F). In extreme cases, deep wells in discharge areas in *unconfined* aquifers may even have artesian heads, and overflow at the ground surface [as shown by well E in Figure B1.3(ii)].

Aquifers with strongly three-dimensional head distributions will typically either have a strong topography or have relatively low permeability (or both). Erosionally resistant crystalline bedrock aquifers are typically of this type. Note that a two-dimensional network of observation boreholes with long well screens may be adequate to characterize the head distribution in aquifers of the type illustrated in Figure B1.3(i), but are inadequate to characterize three-dimensional head distributions of the type in Figure B1.3(ii). For the latter type, a 3-D network of piezometers to varying depths is required. Each piezometer will have a very short open section, and will give a reading of head (*h*) at a specific point (*x,y,z*).

Unconsolidated aquifers are typically formed of relatively young sediments laid down by water, wind or glaciers. Notable examples include the High Plains alluvial aquifer of the mid-west United States and the Indus valley alluvial aquifer system in Pakistan. Flow through such sediments is typically via intergranular pore spaces.

The main hydraulic properties of the three aquifer classes are described in the following sections. The three-fold aquifer classification also forms the basis of the general introduction to drilled well design given in Chapter 3.

1.2.2   Porosity and aquifer storage

Porosity. The ability of a geological formation to store water is governed by its porosity (*n*), which is the ratio between the volume of voids and the total volume of geological material. *Primary porosity* is a characteristic of unconsolidated aquifers and some consolidated aquifers where the voids were formed at the same time as the geological material. In crystalline aquifers and in consolidated aquifers where the original pores have been infilled with cement, porosity results from openings formed at a later time due to fracturing and weathering. This is known as *secondary porosity* and typically comprises tectonic fractures and dissolution fissures. Secondary porosity is usually much smaller than primary porosity. In *karst* limestone aquifers, secondary porosity can develop into conduit-like or even extensive cavern flow systems because of dissolution of soluble calcium carbonate minerals along the fractures [Figure 1.9(a)]. Groundwater flow rates of several hundred metres per hour can occur, comparable to surface water velocities (Banks *et al.* 1995; Coxon and Drew, 2000), and springs issuing from karstic aquifers can provide substantial water supplies [Figure 1.9(b)].

Porosity values for a range of geological formations are given in Table 1.1. Figure 1.10 illustrates different types of porosity. Sometimes, active groundwater flow only occurs through a portion of an aquifer's total porosity (some of the pores may be "blind" or too small to permit efficient flow). This porosity is often referred to as the *effective porosity* (n_e).

Aquifer storativity or coefficient of storage. While porosity gives an indication of the amount of water that can be held by a geological formation it does

(a)

Figure 1.9 *(a) Entrance to large limestone cave in Kras (karst) area of Slovenia; (b) major karst limestone spring near the city of Dubrovnik, Croatia. Photos by Bruce Misstear*

(b)

Figure 1.9   *(Continued)*

not indicate how much it will release. The amount of water that an aquifer will readily take up or release is determined by its *storativity* or *coefficient of storage*. Aquifer storativity is defined as the volume of water that an aquifer will absorb or release per unit surface area, for a unit change in head. It is a dimensionless quantity. Aquifer storativity has two facets (Figure 1.11): unconfined storage (*specific yield, S_y*) and confined storage (*specific storage S_s or elastic storage*).

The *specific yield* of an unconfined aquifer is the volume of water that will drain from it by gravity alone, per unit area, when the water table falls by one unit. The quantity is dimensionless. The water that is unable to drain and which is retained in the pores is termed the *specific retention (S_r)*. Specific yield and specific retention together equal the porosity. Fine-grained materials such as clays and silts have a high specific retention. Because of this, and because of their low permeability, they do

Table 1.1 *Typical values for hydraulic properties of geological formations*

Lithology	Dominant porosity type	Porosity (%)	Specific yield (%)	Hydraulic conductivity (m day^{-1})
Unconsolidated sediments				
Clay	P (S)	30–60	1–10	10^{-7}–10^{-3}
Silt	P	35–50	5–30	10^{-3}–1
Sand	P	25–50	10–30	1–100
Gravel	P	20–40	10–25	50–1000
Consolidated sediments				
Shale	S	<1–10	0.5–5	10^{-7}–10^{-3}
Sandstone	P/S	5–30	5–25	10^{-4}–10
Limestone	S/P	1–20	0.5–15	10^{-4}–1000
Crystalline rocks				
Granite	S	<1–2[a]	<1–2	10^{-8}–1
Basalt	S/P	<1–50	<1–30	10^{-8}–1000
Schist	S	<1–2[a]	<1–2	10^{-8}–10^{-1}
Weathered Crystalline Rocks				
Clayey Saprolite	S*	[b]		10^{-2}–10^{-3}
Sandy Saprolite	S*	2–5		10^{-1}–10
Saprock	P/S*	<2		1–100

P, primary porosity; S, secondary porosity (fractures, vesicles, fissures); S*, intergranular secondary porosity due to weathering and disaggregation of crystals.
[a] The typical kinematic (effective) porosity of crystalline rock aquifers may be <0.05% (Olofsson, 2002).
[b] Effective (not total) porosity of clayey saprolite 0.1–2 % (Rebouças 1993).
Main sources: Freeze and Cherry (1979), Heath (1983), Open University (1995), Robins (1990), Rebouças (1993), US EPA (1994), Todd and Mays (2005).

not normally form good aquifers. Examples of specific yield values for different geological formations are included in Table 1.1.

The *specific storage* of a confined aquifer is the volume of water released from storage per unit volume of aquifer per unit fall in head. The aquifer remains saturated: this storage is related to the elastic deformation (compressibility) of water and of the aquifer fabric.

$$S_s = \frac{\Delta Q}{V \Delta h} \qquad (1.1)$$

where V is the volume of confined aquifer, ΔQ the amount of water released to or from storage and Δh the change in head. S_s has units of [L]$^{-1}$; for example m^{-1}.

The *storativity* of a confined aquifer (i.e., the volume of water released per unit change in head per unit *area* [A]) is given by the product of S_s and aquifer thickness (b), and is dimensionless:

$$S = \frac{\Delta Q}{A \Delta h} = S_s b \qquad (1.2)$$

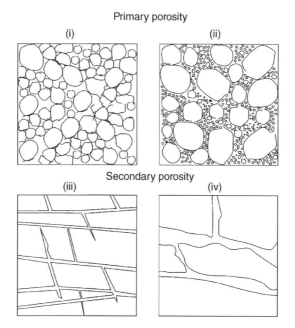

Primary porosity

(i)   (ii)

Secondary porosity

(iii)   (iv)

Figure 1.10 *Types of porosity: (i) primary porosity, well sorted unconsolidated formation; (ii) primary porosity, poorly sorted unconsolidated formation; (iii) secondary porosity, consolidated formation; (iv) secondary porosity, carbonate formation, illustrating enlargement of fractures by chemical dissolution*

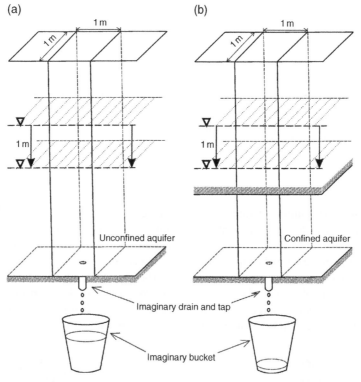

(a)   (b)

1 m   1 m

1 m   1 m

1 m   1 m

Unconfined aquifer   Confined aquifer

Imaginary drain and tap

Imaginary bucket

Figure 1.11 *Schematic diagrams illustrating concepts of (a) specific yield and (b) confined storage. In the case of specific yield (a), a relatively large amount of water drains into our hypothetical bucket because there is a partial emptying of the aquifer pores when the water table is lowered. With confined storage (b), the aquifer remains saturated after the potentiometric surface is lowered, and only a small amount of water is released into our bucket (in response to compression of the aquifer matrix and, to a lesser degree, expansion of the water)*

Box 1.4   Calculations involving specific yield and coefficient of storage

An unconfined sand and gravel aquifer has a porosity of 0.32 and a specific retention of 0.06. If the water table is lowered by an average of 4 m over an area of 5 ha due to pumping, estimate the volume of water removed.

First, we need to calculate the specific yield. This equals the porosity minus the specific retention:

$$S_y = 0.32 - 0.06 = 0.26$$

Now we can calculate the volume of water removed. This equals the fall in water table over the area of aquifer affected, multiplied by the specific yield:

$$4\,\text{m} \times 50\,000\,\text{m}^2 \times 0.26 = 52\,000\,\text{m}^3$$

The same aquifer has a coefficient of storage of 0.0002 where it is confined. Estimate the volume removed if the potentiometric surface also fell by 4 m over an area of 5 ha (assuming the aquifer remains confined).

The volume removed is:

$$4\,\text{m} \times 50\,000\,\text{m}^2 \times 0.0002 = 40\,\text{m}^3$$

These simple calculations illustrate that coefficient of storage in a confined aquifer is much smaller than specific yield in an unconfined aquifer.

The elastic storativity of a confined aquifer is usually two or three orders of magnitude smaller than the specific yield of an unconfined aquifer of a similar lithology (Box 1.4).

It should also be noted that an unconfined aquifer subject to a change in water table will release water both from drainable storage (specific yield S_y) and elastic storage (S_s). However, as specific yield is much greater in magnitude, we will often (but not always) neglect the latter term. Further discussion can be found in Chapter 7 on well testing.

1.3   Groundwater flow

Groundwater under natural conditions flows from areas of recharge, normally the aquifer's outcrop area, to points of discharge at springs, rivers or in the sea. The driving force of groundwater flow is the *hydraulic gradient*—the difference in head between the recharge and discharge areas, divided by the length of the flow path. Hydraulic gradients vary vertically as well as laterally along the flow path: in the recharge area, the vertical component of the hydraulic gradient will be downwards whereas in the discharge area the gradient and therefore flow direction will be upwards (Box 1.3).

1.3.1   Darcy's equation

The flow of water through the saturated zone of an aquifer can be represented by the Darcy equation:

$$Q = -AKi \qquad (1.3)$$

where Q is the groundwater flow rate (dimension $[L]^3 [T]^{-1}$), A the cross-sectional area of flow ($[L]^2$), K the hydraulic conductivity ($[L][T]^{-1}$) and i the hydraulic gradient in the direction of flow (dimensionless). This most fundamental equation of groundwater flow is empirical: it is based on Darcy's experimental observations of flow through sand filters in the 1850s (Box 1.5). The negative sign indicates that flow takes place in the direction of negative (i.e., decreasing) hydraulic gradient, although in subsequent equations in this chapter it will be omitted as we are usually only concerned with the magnitude of flow (the direction being obvious).

The flow rate per unit cross-sectional area of saturated aquifer is given by the Darcy velocity or flux (v_D), also known as the specific discharge:

$$v_D = \frac{Q}{A} = Ki \qquad (1.4)$$

The cross-sectional area of aquifer includes the solid material as well as the pores. To obtain an

Box 1.5   Henry Darcy (1803–1858)

Henry Philibert Gaspard Darcy was born in Dijon, France on June 10th 1803 (Brown, 2002). He was not a hydrogeologist (although he did assist on the development of Dijon's Saint Michel well, and also worked on the Blaizy tunnel, where he would have been able to observe water seepage), as that science had not yet been formally invented. He was a water engineer, educated at Paris's *L'Ecole Polytechnique* and, subsequently *L'Ecole des Ponts et Chaussées* (School of Roads and Bridges). Darcy was a practical, empirical researcher, rather than a pure theoretician. Most of his life he worked with public water supply for his home city of Dijon [Figure B1.5(i)], but was also employed later by the cities of Paris (as Chief Director for Water and Pavements) and Brussels (as a consultant). He developed *inter*

alia formulae for water velocity in various types of open channels, and formulae for estimating water flow in pipes (Darcy, 1857). Some of this work was published posthumously (Darcy and Bazin, 1865) by his protégé and collaborator, Henri Emile Bazin (1829–1917).

In 1855, Darcy's health deteriorated and he returned to Dijon to work and experiment on further hydraulic issues that presumably had long interested him. During 1855 and 1856, Darcy and his friend Charles Ritter empirically studied the flow of water through columns of sand in the laboratory. Ostensibly, this was to improve the design of the sand filters used for purification of surface water supplies (and still widely used today). Brown (2002) argues, however, that Darcy would also clearly have appreciated the importance of such

Figure B1.5(i)   *The fountains and water feature in the Square Henry Darcy in Dijon, France, created in honour of the great French water engineer in recognition of his work in bringing about the first potable water supply reservoir for the town. Photo by Bruce Misstear*

studies for understanding groundwater flow. In 1856 Darcy published a report on the water supply of Dijon city, and a technical appendix to this report contained the results of his experimentation [Darcy (1856); an English translation of *Les fontaines publiques de la Ville de Dijon* was published in 2004 – see Bobeck (2006)]. The famous appendix contended that the flow of water (Q) through a sand filter was proportional to the area (A) of the filter and the difference in water head across the filter, and that it was inversely proportional to the filter's thickness (L). In other words (Brown 2002):

$$Q = KA\frac{(\varsigma_1 + z_1) - (\varsigma_2 + z_2)}{L}$$

where ς and z are pressure head and elevation at locations 1 and 2 on the flow path, respectively, and K is a coefficient of proportionality (hydraulic conductivity). Note that total head $h = \varsigma + z$. This is what we know today as Darcy's law [also expressed in the main text as Equation (1.3)]. Henry Darcy died of pneumonia, while on a trip to Paris, on January 3rd 1858 (Tarbé de St-Hardouin, 1884). As noted by Simmons (2008), only seven years elapsed from the publication of Darcy's law and its first application to a groundwater flow problem, when Jules Dupuit used it to develop his equation for radial flow to a water well [see Equation (1.26) and Box 7.4 in Chapter 7].

estimate of the flow velocity through the pores it is necessary to divide the Darcy velocity by the effective porosity, n_e. This gives the linear seepage velocity v_s:

$$v_s = \frac{Ki}{n_e} \quad (1.5)$$

This equation represents only the *average* linear seepage velocity. The actual velocity of a water particle is affected by dispersion, which depends mainly on the tortuosity of the flow path. This is especially important in contaminant transport problems, as the first arrival of a contaminant at a well may be much faster than the *average linear velocity* predicted by the above equation (Box 1.6).

The terms *hydraulic conductivity* and *coefficient of permeability* are often used interchangeably, especially in engineering texts. It is important to note, however, that some engineers and most petroleum geologists also use a quantity called *intrinsic permeability* (k). The term *hydraulic conductivity* assumes that the fluid under consideration is water (in our case groundwater). The *intrinsic permeability* of the porous medium is independent of the properties of the fluid involved,

is a characteristic of the porous medium alone and is related to the hydraulic conductivity by the equation:

$$k = \frac{K\mu}{\rho g} = \frac{K\upsilon}{g} \quad (1.6)$$

where k is the intrinsic permeability (dimension $[L]^2$), K the hydraulic conductivity ($[L][T]^{-1}$), υ the fluid kinematic viscosity ($[L]^2[T]^{-1}$), μ the dynamic viscosity ($[M][L]^{-1}[T]^{-1}$), ρ the fluid density ($[M][L]^{-3}$) and g the gravitational acceleration ($[L][T]^{-2}$). Intrinsic permeability is a particularly useful parameter for the petroleum industry when dealing with multi-phase fluids with different kinematic viscosities. In hydrogeology, the kinematic fluid viscosity does not vary much over the normal temperature and density range of most groundwaters, and so hydraulic conductivity is the parameter of permeability most commonly used.

Typical hydraulic conductivity values for a range of geological formations are given in Table 1.1. (Metric units are used throughout this book for hydraulic conductivity and other parameters, but conversion tables to Imperial units are included in Appendix 1). The hydraulic

Box 1.6   The use and misuse of Darcy's Law

Consider the multilayered aquifer system illustrated in Figure B1.6(i) below, subject to a hydraulic gradient of 0.01. Let us suppose we have carried out some kind of pumping test and determined that the transmissivity (T) of the sequence is $300\,m^2\,day^{-1}$. We can thus calculate that the (average) hydraulic conductivity (K) of the sequence is $300\,m^2\,day^{-1}/9\,m = 33\,m\,day^{-1}$.

The groundwater flux (Q) through the entire thickness (b) of the aquifer is given by:

$Q = Ti = Kbi = 33\,m\,day^{-1} \times 0.01 \times 9\,m = 3\,m^3$ day^{-1} per m aquifer width

Now, let us consider a contamination incident, such that a volume of polluted groundwater starts migrating in the aquifer. We wish to find out how long it will take to travel 100 m to a protected spring area. If we make the reasonable assumption that the effective porosity (n_e) is 0.10, and apply Equation (1.5) to calculate linear flow velocity (v_s):

$$v_s = \frac{Ki}{n_e} = 33\,m\,day^{-1} \times 0.01/0.10 = 3.3\,m\,day^{-1}$$

We thus calculate that it will take 30 days for the pollution to migrate 100 m. We will return to our office and relax a little, imagining that we have about a month to try and come up with a remediation scheme. However, the telephone rings after only 12 days to tell us that the pollution has already arrived at the spring.

We have made the mistake of calculating the *average* linear flow velocity and assuming that the aquifer was homogeneous. It is not: the water in the coarse sand will be travelling at $80\,m\,day^{-1} \times 0.01/0.10 = 8\,m\,day^{-1}$, while that in the medium sand will only be travelling at $1\,m\,day^{-1}$.

Darcy's Law is very robust when considering problems of bulk groundwater flux. We do not need to know too much about the detailed aquifer structure. However, when considering problems involving actual groundwater and contaminant flow velocity, it is very easy to make mistakes. Not only is the result very sensitive to the value of n_e selected, we also need to know how hydraulic conductivity and porosity are distributed throughout the aquifer.

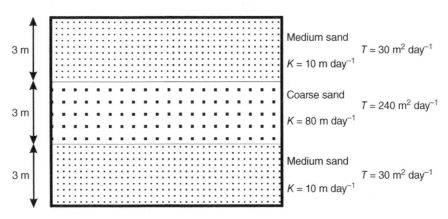

Figure B1.6(i)   Multilayered aquifer system

conductivity of an intergranular aquifer depends on the grain size and sorting of the aquifer material and the degree of cementation. In fractured/fissured aquifers the intensity of fracturing and the aperture, continuity and connectivity of individual fractures control the hydraulic conductivity.

The Darcy equation only applies where the flow is laminar. In laminar flow the water particles move along streamlines that are approximately parallel to each other. This is normally the situation with groundwater flow in intergranular aquifers, where the flow velocities are very small. However, higher flow velocities can occur in fissured aquifers (notably where the fissures have been enlarged through karstification) and near wells in both fissured and intergranular aquifers. Flow in these situations may become turbulent, whereby the water particles move erratically in speed and direction. A dimensionless ratio, known as the Reynolds number (Appendix 2), can be used to indicate whether the flow is likely to be laminar or non-laminar:

$$R = \frac{v_D d}{\upsilon} \qquad (1.7)$$

where R is the Reynolds number (dimensionless), v_D the Darcy velocity (dimension $[L][T]^{-1}$), d the average pore diameter ($[L]$) and υ the fluid kinematic viscosity ($[L]^2[T]^{-1}$). Studies quoted in the standard hydrogeological literature indicate that Darcy's law is valid when the Reynolds number is less than some value between 1 and 10 (Freeze and Cherry, 1979; Fetter, 2001; Todd and Mays, 2005). In most natural groundwater flow situations the Reynolds number is less than 1.

Transmissivity is the rate at which water can pass through the full thickness of aquifer per unit width under a unit hydraulic gradient. The transmissivity in a uniform aquifer is the hydraulic conductivity multiplied by the saturated aquifer thickness. However, as uniform aquifers are uncommon in nature, the transmissivity (T) is usually derived by summing, over the entire aquifer, the transmissivities of individual horizons ($i=1$ to n), where the transmissivity of each horizon is given by the product of the horizon's hydraulic conductivity (K_i) and its thickness (b_i):

$$T = \sum_{i=1}^{n} K_i b_i \qquad (1.8)$$

An aquifer having the same properties in all directions from a point is referred to as *isotropic*. If the properties of the aquifer are also the same at all locations the aquifer is said to be *homogeneous*. Sedimentary aquifers can be relatively homogeneous but they are rarely isotropic. This is because the hydraulic conductivity along the direction of the bedding planes is usually greater than that at right angles to the bedding. Crystalline aquifers, and consolidated aquifers with secondary porosity are both heterogeneous and anisotropic.

The nature of hydraulic conductivity of fractured rock aquifers is considered in more detail in Box 1.7, whilst Box 1.8 compares groundwater flow velocity in fractured and porous aquifers.

1.3.2 General equations of groundwater flow

Groundwater flow in a confined aquifer. The general equations for groundwater flow in porous media are based on the Darcy equation and on the principles of conservation of energy and mass. For transient flow in a confined aquifer, the general equation of groundwater flow is:

$$K_x \frac{\partial^2 h}{\partial x^2} + K_y \frac{\partial^2 h}{\partial y^2} + K_z \frac{\partial^2 h}{\partial z^2} = \frac{S}{b} \frac{\partial h}{\partial t} = S_s \frac{\partial h}{\partial t} \qquad (1.9)$$

where K_x, K_y and K_z are the hydraulic conductivities in the principal directions x, y and z; h is the hydraulic head, S is the dimensionless aquifer coefficient of storage, S_s the specific storage coefficient, b the aquifer thickness and t the time. In a homogeneous and isotropic aquifer where $K_x = K_y = K_z$, the equation becomes:

$$\frac{\partial^2 h}{\partial x^2} + \frac{\partial^2 h}{\partial y^2} + \frac{\partial^2 h}{\partial z^2} = \frac{S}{Kb} \frac{\partial h}{\partial t} = \frac{S}{T} \frac{\partial h}{\partial t} \qquad (1.10)$$

The quantity T/S (or K/S_s) is called the *hydraulic diffusivity* ($[L]^2[T]^{-1}$). For steady state flow, head

Box 1.7   The hydraulic conductivity of fractured aquifers

Imagine a single horizontal fracture in an impermeable rock mass. If the fracture's aperture is b_a, and if its sides are smooth, planar and parallel, then its *fracture transmissivity* (T_f) is given by (Snow, 1969; Walsh, 1981):

$$T_f = \frac{\rho g b_a^3}{12\mu} \approx 629\,000 b_a^3$$

where T_f is in m^2 s^{-1} and b_a is in m, ρ is the density of water (c. 1000 kg m^{-3}), g the acceleration due to gravity (9.81 m s^{-2}) and μ is the dynamic viscosity of water (c. 0.0013 kg s^{-1} m^{-1}). We see that transmissivity (the ability of the fracture to transmit water) is proportional to the cube of the aperture. Thus, one ideal plane-parallel fracture of aperture 1 mm is hydraulically equivalent to 1000 fractures of aperture 0.1 mm. The implication of this in a real aquifer is that the bulk of the groundwater is transported through fractures of large aperture. As fracture apertures in natural geological media are typically approximately log-normally distributed (Long *et al.*, 1982), these will be relatively few and far between. In real boreholes in crystalline rocks, the entire well yield typically comes from only a few major fractures.

From the above equation, and changing to units of m^2 day^{-1}, we can see that an idealized fracture of aperture 0.1 mm thus has a transmissivity of 0.05 m^2 day^{-1}, while one of aperture 0.5 mm has a T_f of 7 m^2 day^{-1}.

The *hydraulic conductivity* of fractured rocks can be defined as the total transmissivity within a given interval divided by the thickness of that interval (B). Thus, for the total interval of 5 m in Figure B1.7(i), the total transmissivity of the two fractures is 7.05 m^2 day^{-1}, and the hydraulic conductivity is around 1.4 m day^{-1} (equivalent to that, say, for a fine sand). However, if we now consider only the 1 m interval containing the larger fracture, we would calculate a hydraulic conductivity of 7 m day^{-1}. The calculated hydraulic conductivity thus depends heavily on the interval of measurement and is said to be *scale-dependent*. This scale dependence can be significantly reduced by choosing a large enough interval, which can be referred to as a *representative element*.

In reality, fractures are not smooth, planar or parallel. Rather, flow within a fracture plane may be canalized; indeed, in some limestone aquifers, the flow features penetrated by wells are distinctly cylindrical and pipe-like in appearance [see Figure 6.26(b) in Chapter 6]. The Frenchman, Jean Louis Poiseuille (1799–1869), conducted experiments on laminar (non-turbulent) fluid flow in cylindrical tubes and found that the rate of flow (Q) was proportional to the

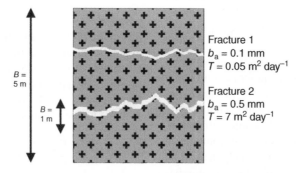

Figure B1.7(i)   *Relationship between fracture aperture and transmissivity*

fourth power of the pipe's diameter. More formally:

$$Q = \frac{\pi r^4 \Delta P}{8\mu\Delta l} \text{ or, in terms of head, } Q = \frac{\pi\rho g r^4 \Delta h}{8\mu\Delta l}$$

where Δl is the length of the tube (dimension [L]), r the radius of the tube ([L]), ΔP the difference in pressure between the two ends of the tube ([M][L]$^{-1}$[T]$^{-2}$), Δh the head difference between the two ends of the pipe ([L]), μ the dynamic viscosity of the fluid ([M][L]$^{-1}$[T]$^{-1}$), ρ the fluid density ([M][L]$^{-3}$), and g is the acceleration due to gravity ([L][T]$^{-2}$).

Rearranging in terms of head loss per unit length of pipe:

$$\frac{\Delta h}{\Delta l} = \frac{8v\mu}{\rho g r^2}$$

where v is average flow velocity $= Q/\pi r^2$. This is, in fact, identical to the Darcy-Weisbach equation [named after Henry Darcy - see Box 1.5 - and the Saxonian Julius Weisbach (1806–1871)]:

$$\frac{\Delta h}{\Delta l} = \frac{fv^2}{4rg}$$

where f is a friction loss factor. For laminar flow $f = \frac{64}{R}$, where R is the Reynolds Number, named after Osborne Reynolds (1842–1912) of the University of Manchester. For circular pipes, R is given by:

$$R = \frac{2r\rho v}{\mu} = \frac{2\rho Q}{\pi r \mu}$$

If R is low (<2300, for circular pipes), flow is typically laminar. When R exceeds this figure, flow gradually becomes turbulent and the Poiseuille equation is not valid. For more information on flow in pipes and Reynolds numbers, the reader is referred to Appendix 2.

While these concepts and equations help our understanding of flow in fissures, they can also be useful in well design, for example, in estimating head losses during flow within the well casing (Section 4.5).

Box 1.8 Groundwater flow velocity in fractured and porous aquifers

Consider the two 5 m thick aquifers shown in Figure B1.8(i) below. One is a crystalline granite containing a single ideal plane parallel fracture of aperture 0.5 mm. We saw, from Box 1.7, that its transmissivity would be around 7 m^2 day^{-1}, and thus the bulk hydraulic conductivity of the aquifer is 7/5 m day^{-1} = 1.4 m day^{-1}. The second aquifer is a homogeneous fine sand aquifer with a hydraulic conductivity also equal to 1.4 m day^{-1}.

The hydraulic conductivities and transmissivities of the aquifers are the same and Darcy's Law [Equation (1.3)] thus states that they will transmit the same flow of groundwater (Q) under a hydraulic gradient of 0.01, namely:

$$Q = KBi = 1.4 \text{ m day}^{-1} \times 0.01 \times 5 \text{ m}$$
$$= 0.07 \text{ m}^3 \text{ day}^{-1} \text{per m of aquifer width}$$

However, if we are investigating a contamination incident and wish to know the velocity (v_s) at which the contamination is flowing towards a well, then Equation (1.5) should be used:

$$v_s = \frac{Ki}{n_e}$$

For the sand, the effective porosity (n_e) might be, say, 0.17, resulting in a derived v_s of 0.08 m day^{-1}. This is the average linear velocity. Dispersion effects will mean that some contaminant travels somewhat faster than this and some slightly slower.

In the fractured rock aquifer, the effective porosity is probably no more than 0.0005 m/5 m = 0.0001, yielding a transport velocity of some 140 m day^{-1}.

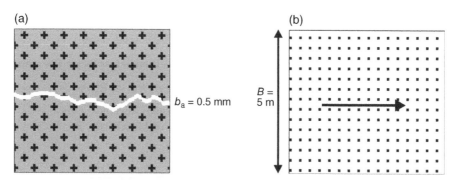

Figure B1.8(i)   *Groundwater flow in (a) fractured granite and (b) porous sand aquifers*

From this, we can learn two valuable lessons:

i. Groundwater flows much faster in fractured and fissured aquifers than in equivalent porous aquifers.

ii. Groundwater flow velocity depends very heavily on the value selected for effective porosity (and in many aquifers, especially a low permeability one, this is very difficult to derive).

in the aquifer does not change with time $\left(\dfrac{\partial h}{\partial t}=0\right)$ and the equation reduces to:

$$\frac{\partial^2 h}{\partial x^2}+\frac{\partial^2 h}{\partial y^2}+\frac{\partial^2 h}{\partial z^2}=0 \qquad (1.11)$$

This is known as the Laplace equation. The assumptions underlying this equation are:

• the aquifer is confined;
• the aquifer is homogeneous and isotropic;
• the fluid is incompressible (a density term can be introduced into the equation for compressible fluids);
• groundwater flow is in steady state;
• all the flow comes from water stored within the aquifer (that is, there is no leakage into the aquifer from overlying or underlying layers).

The solution to Equation (1.11) describes hydraulic head in terms of the x, y and z coordinates. The solution to Equation (1.10) describes the hydraulic head at any point in the three-dimensional flow system at any time t. These equations are often reduced to two dimensions—and sometimes to one dimension—to facilitate their solution by graphical, analytical or numerical methods.

Groundwater flow in an unconfined aquifer. Whereas Darcy's equation (1.3) can be applied to simple one-dimensional flow problems in a confined aquifer (under steady-state conditions), the problem is more complex for the situation of the unconfined aquifer in Figure 1.12 because the flow is not horizontal; indeed, the water table represents a flow line whose shape is both governed by, and plays a role in governing, flow in the remainder of the aquifer (Todd and Mays, 2005). Darcy's Law strictly speaking states that flow (q) is proportional to hydraulic gradient along the direction of flow; that is, $q \propto \dfrac{dh}{ds}$, where s is a distance coordinate along a flow line. If flow is horizontal then $\dfrac{dh}{dx}=\dfrac{dh}{ds}$, where x is the horizontal distance coordinate. In an unconfined aquifer this is not strictly the case. A solution to this problem was proposed by Dupuit (1863; see Box 7.4 in Chapter 7) and developed by Forchheimer (1930). The Dupuit-Forchheimer solution allows this two-dimensional flow problem to be reduced to one dimension by assuming (a) that flow is horizontal and uniform throughout the vertical section, and (b) that $q \propto \dfrac{dh}{dx}=\dfrac{dh}{ds}$. Applying Darcy's equation (1.3)

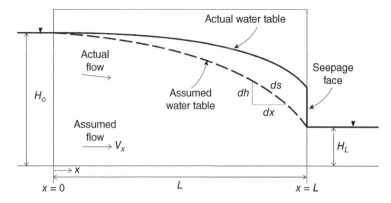

Figure 1.12 *Dupuit-Forchheimer assumptions for groundwater flow in an unconfined aquifer*

and the Dupuit-Forchheimer assumptions, the flow per unit width (q) through the vertical section of aquifer in Figure 1.12 is given by:

$$q = -KH\frac{dH}{dx} \qquad (1.12)$$

where K is the hydraulic conductivity, H the elevation of the water table (hydraulic head) *relative to the impervious base of the aquifer* and $\dfrac{dH}{dx}$ the hydraulic gradient (along a horizontal axis). Integrating gives:

$$qx = -\frac{KH^2}{2} + c \qquad (1.13)$$

where c is the coefficient of integration. For the boundary conditions $H=H_0$ at $x=0$, and $H=H_L$ at $x=L$, then:

$$q = \frac{K}{2L}\left(H_0^2 - H_L^2\right) \qquad (1.14)$$

This is known as the Dupuit equation or the Dupuit-Forchheimer discharge equation. In Figure 1.12, we see that the hydraulic gradient increases downstream. This is because the saturated thickness (and hence the transmissivity) of the aquifer decreases as the water table falls. As the hydraulic gradient steepens, the Dupuit-Forchheimer assumptions become increasingly violated and the calculated water table departs increasingly from the actual water table. Indeed, in actuality a *seepage* face develops at the downstream end of the aquifer

block. Nevertheless, the Dupuit-Forchheimer assumptions are useful in a variety of situations: they can be applied to the estimation of recharge (Section 2.6.3) and to radial flow to a well in an unconfined aquifer (Chapter 7).

If the thickness of an unconfined aquifer is great relative to variations in the water table level and if water table gradients are relatively low, we can assume that transmissivity does not vary greatly nor depend on water table level. It thus becomes possible to apply equations derived for confined aquifers to unconfined aquifer situations.

1.3.3   Radial flow to wells

The natural flow conditions in an aquifer are disturbed when a well is pumped. The action of pumping water from the well lowers the level of groundwater and creates a hydraulic head difference between the water in the well and that in the aquifer. This head difference causes water to flow into the well and so lowers the hydraulic head in the aquifer around the well. The effects of pumping spread radially through the aquifer. The lowering of the water table or potentiometric surface forms a *cone of depression* around the pumping well. This cone can be seen and measured in observation wells (Figure 1.13).

Radial flow to a well in a confined aquifer. If it is assumed that flow is horizontal, Equations (1.10) and (1.11) can be reduced to the following

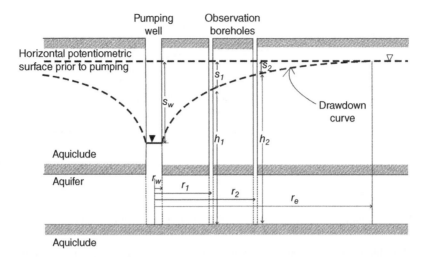

Figure 1.13 *Cone of depression of potentiometric surface around a pumping well*

expressions for two-dimensional flow in a confined aquifer:

$$\frac{\partial^2 h}{\partial x^2} + \frac{\partial^2 h}{\partial y^2} = \frac{S}{Kb}\frac{\partial h}{\partial t} = \frac{S}{T}\frac{\partial h}{\partial t} \qquad (1.15)$$

$$\frac{\partial^2 h}{\partial x^2} + \frac{\partial^2 h}{\partial y^2} = 0 \text{ for steady-state conditions} \qquad (1.16)$$

In analysing groundwater flow to water wells we must convert Equations (1.15) and (1.16) into radial coordinates, with $r = \sqrt{x^2 + y^2}$:

$$\frac{\partial^2 h}{\partial r^2} + \frac{1}{r}\frac{\partial h}{\partial r} = \frac{S}{Kb}\frac{\partial h}{\partial t} = \frac{S}{T}\frac{\partial h}{\partial t} \qquad (1.17)$$

and

$$\frac{\partial^2 h}{\partial r^2} + \frac{1}{r}\frac{\partial h}{\partial r} = 0 \quad \text{ for steady-state flow} \qquad (1.18)$$

The first solution to Equation (1.17) for *transient* (i.e., non-steady-state) flow to a well in a confined aquifer was proposed by Theis (1935). The Theis solution enables the estimation of aquifer transmissivity and storage coefficient from a well pumping test by a curve-matching technique. This and other methods for analysing pumping test data are described in Chapter 7.

Simple equilibrium equations for steady-state radial flow to a well have applications in well siting and design (Chapters 2 to 4), and so are introduced here rather than in Chapter 7. Thiem (1906; see Box 7.4 in Chapter 7) developed a solution for radial flow to a well in a confined aquifer by applying Darcy's equation to a cylindrical flow section. Different boundary conditions are illustrated in Figure 1.13. For the case of two observation wells located near a pumping well, the Thiem equation can be written as:

$$h_2 - h_1 = \frac{Q}{2\pi T}\ln\frac{r_2}{r_1} \qquad (1.19)$$

where h_1 and h_2 are the hydraulic heads in the observation wells located at distances r_1 and r_2 from the pumping well, respectively; Q is the discharge rate in the pumping well and T the transmissivity.

The hydraulic head values can be replaced by the drawdown values ($h_2 - h_1 = s_1 - s_2$) to give:

$$s_1 - s_2 = \frac{Q}{2\pi T}\ln\frac{r_2}{r_1} \qquad (1.20)$$

For the case where s_2 is zero—which occurs at a distance from the pumping well known as the *radius of influence*—and where the first observation well is replaced by measurements in the pumping well itself, Equation (1.20) becomes:

$$s_w = \frac{Q}{2\pi T}\ln\frac{r_e}{r_w} \qquad (1.21)$$

where s_w is the drawdown in the pumping well, r_e the radius of influence of the pumping well and r_w

the well radius. It should be added that the "radius of influence" at equilibrium is a somewhat flawed concept, since theory dictates that the cone of depression will continue to expand with continued pumping unless there is a source of recharge to the aquifer (this inconsistency provided much of the motivation for C.V. Theis's work—see Box 7.5 in Chapter 7).

Rearranging in terms of transmissivity, Equation (1.21) becomes:

$$T = \frac{Q}{2\pi s_w} \ln \frac{r_e}{r_w} \qquad (1.22)$$

Equations (1.19) to (1.22) apply to a single pumping well that penetrates the entire thickness of an infinite, homogeneous and isotropic aquifer of uniform thickness, which is confined, and where the potentiometric surface is horizontal prior to pumping. These underlying assumptions are never met fully in reality, but they are more likely to be at least partly satisfied in extensive unconsolidated and consolidated aquifers characterized by primary porosity than in consolidated or crystalline aquifers with fracture-porosity. The equilibrium equations can be applied in the case of several wells in a well field in order to calculate the interference drawdowns between the wells, using the principle of superposition (Section 2.9).

A number of simplifications of the Thiem equilibrium equation have been proposed (Misstear, 2001), including that by Logan (1964). In Equation (1.22) the ratio r_e/r_w cannot be determined accurately during a pumping test unless we have data from several observation boreholes. Although this ratio may vary significantly, the log term is relatively insensitive to these variations. Logan proposed a value of 3.32 as typical for the $\log_{10}$ ratio (equal to 7.65 for the ln ratio), and thus reduced Equation (1.22) to the following approximation:

$$T = \frac{1.22Q}{s_w} \qquad (1.23)$$

Similar approximations have been proposed elsewhere. For example, the relationship:

$$T = \frac{1.32Q}{s_w} \qquad (1.24)$$

was obtained from a large number of well pumping tests in alluvial aquifers in the Indus valley of Pakistan (Bakiewicz *et al.*, 1985). This type of equilibrium approximation equation can be used to calculate the required length of screen when designing a well in a thick, uniform aquifer (Section 3.1.4).

Radial flow to a well in an unconfined aquifer. The equations for a confined aquifer can be applied approximately also to an unconfined aquifer, provided that the aquifer thickness is relatively large compared to the amount of drawdown (i.e., that transmissivity does not vary significantly with drawdown). If this is not the case, we can derive a modified, unconfined variant of the Thiem equation by supposing a well pumping at rate Q, and then imagining that this induces a flow Q (at steady state) through an imagined cylinder at a distance r around the well. Applying Darcy's Law and the Dupuit assumptions:

$$Q = -2\pi K H \frac{dH}{dr} \qquad (1.25)$$

where H is the height of the water table (head) relative to the impermeable base of the aquifer. Integration between two radii r_1 and r_2 yields:

$$Q = \pi K \frac{\left(H_2^2 - H_1^2\right)}{\ln\left(r_2 / r_1\right)} \qquad (1.26)$$

This is sometimes known as the Dupuit equation (Kruseman *et al.*, 1990). Note that, if the aquifer thickness H is very much greater than the drawdown:

$$Q = \pi K \frac{\left(H_2^2 - H_1^2\right)}{\ln\left(r_2 / r_1\right)}$$

$$= \pi K \frac{\left(H_2 - H_1\right)\left(H_2 + H_1\right)}{\ln\left(r_2 / r_1\right)} \approx 2\pi K H \frac{\left(H_2 - H_1\right)}{\ln\left(r_2 / r_1\right)} \qquad (1.27)$$

which is identical to Thiem's equation for confined aquifers [i.e., to Equation (1.19) when expressed in terms of Q].

2

Groundwater Investigations for Locating Well Sites

This chapter describes a systematic programme of desk and field studies for locating suitable sites for water wells. The main focus is on groundwater investigations for production wells – for drinking water, irrigation or other supply purposes – but we can follow a similar programme for locating other types of wells, adapting our approach as necessary to suit the particular purpose of these wells (aquifer clean-up, artificial recharge, ground source heat pumps, groundwater monitoring, or whatever).

The objectives of the groundwater investigations should be to find locations where wells can be designed and constructed to supply the required demand of water, of a quality suitable for the intended use, at reasonable cost and with least impact to either fellow groundwater users or to the aqueous environment. Environmental impacts to be avoided include (i) significant reductions in groundwater flow to ecologically important wetlands, spring areas or baseflow-supported rivers, (ii) saline intrusion in coastal aquifers and (iii) ground subsidence caused by large drawdowns in unconsolidated, compressible aquifers or by dewatering organic-rich subsoils or sediments.

Figure 2.1 is a flow diagram illustrating the sequence of groundwater investigations for locating well sites and planning a well scheme. The figure also indicates the sections of Chapter 2 where each of the investigation tasks is described. In some cases, the only tasks that will be required for selecting the well site are a desk study followed by a field reconnaissance. This might be the case with a small groundwater development in an extensive, homogeneous aquifer, where the groundwater resources and groundwater quality are clearly understood. In most cases, however, we will need to follow the steps in Figure 2.1 until it is clear that sufficient data are available to enable the investigation to proceed to the final planning stage for the well scheme. Some of the individual investigation tasks may be broken down into different phases so that expenditure on investigations is optimized against the results obtained. For example, an exploration drilling programme that follows on from a geophysical survey may lead to further geophysical surveys, which in turn enables the selection of a suitable site for a test well.

Examples of possible well locations for a variety of aquifer situations are shown in Figure 2.2. These are based on rather simple conceptual models; in reality, the location of a suitable source of groundwater supply is more complicated, since it requires information on aquifer characteristics, recharge,

Water Wells and Boreholes, Second Edition. Bruce Misstear, David Banks and Lewis Clark.
© 2017 John Wiley & Sons Ltd. Published 2017 by John Wiley & Sons Ltd.

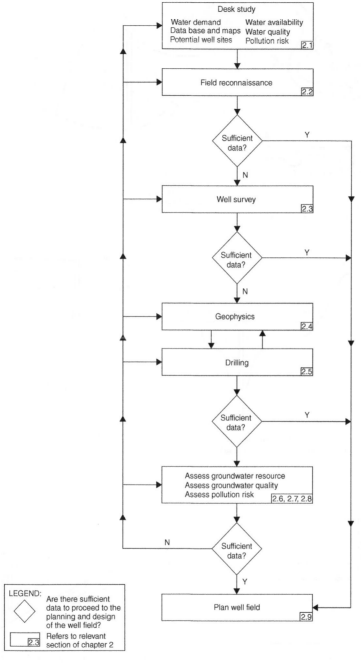

Figure 2.1   *Flow diagram showing the sequence of tasks in a groundwater investigation programme. The number in the bottom right-hand corner of each box refers to the relevant section number in this chapter*

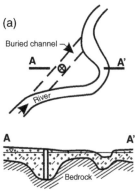

(a)

Buried channel

A   A'

River

A   A'

Bedrock

Well in buried channel may encounter
thicker gravel aquifer, and will not have
the same risk of flooding as a well site
along the modern river channel.

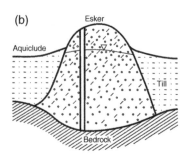

(b)

Esker

Aquiclude

Till

Bedrock

In glaciated terrain, esker gravels can provide
good well sites. The esker can be identified
on aerial photographs as a long ridge with well-drained
soils surrounded by poorly-drained
clayey soils formed on tills.

(c)

Granite
inselberg

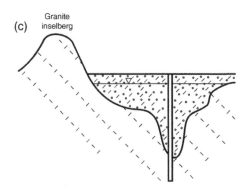

Well site in fractured and weathered crystalline rock aquifer.
The zone of deep weathering corresponds to a vertical fault.

(d)

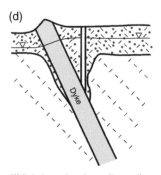

Dyke

Well site in weathered crystalline aquifer
where groundwater flow is impeded by
cross-cutting dyke. The water table is
therefore raised on the upgradient side
of the dyke.

(e)

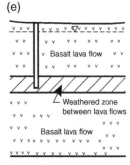

Basalt lava flow

Weathered zone
between lava flows

Basalt lava flow

The aquifer characteristics of
volcanic rocks are very variable.
Good sources of groundwater
can sometimes be found in the
weathered zone between
individual lava flows.

(f)

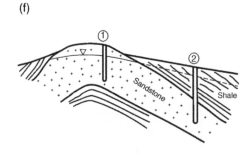

①   ②

Sandstone   Shale

Two well sites in a sandstone aquifer. Site ① may encounter
better aquifer characteristics owing to greater fracturing near
the centre of the anticline. However, site ① is more
vulnerable to pollution than the second well site ②, located
downdip where the aquifer is confined.

Figure 2.2   *Examples of well sites in a variety of hydrogeological situations*

other abstractions, interactions between ground-water and surface water sources, water quality, vulnerability to future pollution hazards and so forth. The amount of money that should be spent on a groundwater investigation will depend on the size and nature of the proposed groundwater abstraction scheme, the complexity of the hydrogeology, the existing information, and the success rate of the exploration techniques that may be used.

2.1 Desk studies

The first task should be to establish the aims of the groundwater investigation by addressing questions such as:

- How much water is required?
- What water quality criteria apply – potable, livestock, irrigation, other?

- How many wells are likely to be needed?
- How much hydrogeological information is already available?
- What are the data gaps to be filled by additional investigations?
- What are the social, environmental, community and land ownership criteria that will influence the siting and operation of the well scheme?

Potential data requirements and data sources for the desk study are summarized in Table 2.1. In most industrialized countries a large amount of the necessary geological and hydrogeological information is already available, and so the collection and analysis of existing data can sometimes eliminate the need for a detailed field exploration programme. Hydrogeological data in many developing countries are sparse and more difficult to obtain. Nevertheless, an energetic and persistent investigator can usually find useful data. Even if written

Table 2.1 Desk study: data requirements for a groundwater investigation

Data requirements	Main data sources
Topographic maps	National state mapping agency
Bedrock geology maps	Geological survey
Soils, subsoils and land use maps	Geological survey, agriculture ministry
Hydrogeology and groundwater vulnerability maps	Geological survey, water ministry, environmental agency
Geology, hydrogeology, site investigation and other relevant reports	Geological survey, water ministry, environmental agency, local authorities, consulting firms, non-governmental organizations
Aerial photographs, satellite imagery	National mapping agencies plus international agencies for distribution of Landsat, SPOT and other satellite imagery (e.g. NASA), Google Earth
Well and borehole records	Geological survey, water ministry, environmental agency, consulting firms, drilling firms, non-governmental organizations
Water level and water quality monitoring data	Geological survey, water ministry, environmental agency, environmental health office, local authorities
Existing groundwater abstractions	Geological survey, ministry of water, rural development, irrigation or public works (depending on national government structure), environmental agency
River flow records	Ministry of water, hydrometric agency, environmental agency, consulting firms, electricity (hydropower) authorities
Climate data, including rainfall and evapotranspiration	Meteorological office, hydrometric agency, water ministry, environmental agency

records are poor, useful information can be obtained by talking to practitioners in the country: drillers, aid workers and personnel from the local water ministry and geological survey.

The information derived from all available sources needs to be entered into a project database and plotted as a series of base maps of the area, showing:

- topography and surface drainage;
- soils and land use;
- geology, including distribution of superficial deposits and bedrock formations, locations of faults;
- hydrogeology, including depth and thickness of target aquifers, aquifer boundaries, recharge/discharge areas, locations of springs and existing wells, well yields, piezometry, groundwater quality, river flow gauging stations.

For the larger investigation programmes it is advantageous to use a geographical information system (GIS) for storing and retrieving the data, including the preparation of maps and sections. Care needs to be taken with the accuracy of the data entry (correct coordinates, datum, etc.), and the quality of the input data should be assessed (reliability of borehole logs, etc.). It should also be borne in mind that well yield and specific capacity information stored in databases may not be a truly representative subset of the total hydrogeological regime: dry boreholes are often not included in the database and pumping test information tends to be from the more productive wells (Sander, 1999). More information on the use of well databases can be found in Chapter 10.

Remote sensing data (aerial photographs, radar data, satellite imagery) are useful for drawing inferences on groundwater conditions: for example, information on soil cover, topography, drainage patterns and vegetation gleaned from interpretation of satellite imagery can help identify aquifer recharge and discharge areas. Satellites contain sensors that record reflected and emitted radiation in the visible and infrared parts of the electromagnetic spectrum. Earth materials exhibit different spectral reflectances: vegetation, for example, has a higher reflectance than dry rock in

the near infrared wavelengths of the spectrum. Whilst satellite imagery is available at very high spatial resolution (down to less than 1 m) from missions such as QuickBird and IKONOS, these data can be relatively expensive to purchase, and so the most widely used satellite imagery in groundwater exploration is from medium resolution, low-cost systems such as Landsat, SPOT, IRS and ASTER (Sander, 2007). The U.S. Landsat and the European SPOT missions include several generations of satellites with different sensors and spatial and spectral resolutions. For example, Landsat-7 (with Enhanced Thematic Mapper) produces images at a spatial resolution of 15 m to 60 m, depending on the waveband, whereas the SPOT-5 satellite provides a spatial resolution ranging from 2.5 m to 20 m (Meijerink *et al.*, 2007). Data from some of the satellite systems, including ASTER (spatial resolution 15 m to 90 m), can be used to produce digital elevation models (DEMs), which can be very useful in analysing topographical features such as those associated with geological 'lineaments'.

The use of remote sensing is perhaps especially relevant when choosing well sites in crystalline aquifers and in some consolidated aquifers. In these situations, the best potential well sites may (but not always) relate to 'lineaments' that can be observed on the remote sensing imagery or on DEMs derived from the satellite or radar data. These lineaments may correspond to fracture zones, faults or to other hydrogeological features of significance, including lithological or hydraulic boundaries (Figure 2.3). Figure 2.4 shows an example of a good positive correlation between well yield and proximity to lineaments mapped on satellite imagery (SPOT) in Botswana and the value of global positioning systems (GPS) for locating sites on the ground (Sander, 1999). Figure 2.5 is an aerial photograph from Norway, where the main fracture patterns in the crystalline bedrock can be clearly seen.

Not all fractures represent zones of enhanced permeability – some fracture zones may be infilled with low permeability clays derived from hydrothermal alteration or weathering material

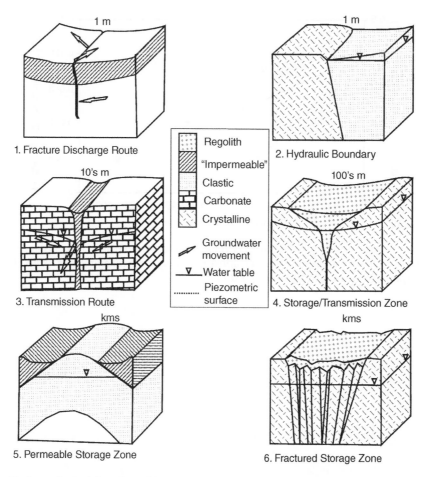

Figure 2.3 *Hydrogeological function of lineaments. From Waters et al. (1990), Remote Sensing Reviews, 4:223–264. Reproduced by permission of Taylor & Francis Ltd, www.tandf.co.uk/journals*

(Misstear *et al.* 1980). Therefore, the detection of a lineament on an aerial photograph or satellite image should always be followed up by ground investigations such as field reconnaissance, geophysical surveys or exploratory drilling. However, even field reconnaissance and geophysics may not adequately distinguish between clay-filled and transmissive fracture zones, as shown by case studies from Hvaler in Norway (Banks *et al.*, 1992a, 1993c; Banks and Robins, 2002).

The outcomes of the desk study should be:

- a *conceptual model* of the hydrogeology of the study area, indicating the main aquifers and their boundaries, the recharge and discharge areas, etc;

- a project database, often incorporated within a GIS;

- maps, including a hydrogeological map showing lineaments or other relevant features identified on remote sensing data;

- initial proposals for field investigations (Sections 2.2, 2.3, 2.4 and 2.5);

- initial estimate of groundwater resources available for the new wells (Section 2.6);

- initial assessment of groundwater quality and potential pollution risks (Sections 2.7 and 2.8).

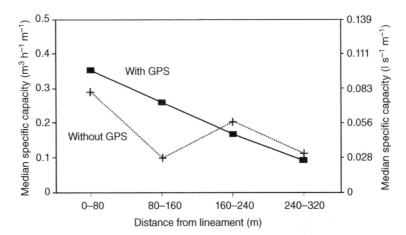

Figure 2.4 *Example relationship between well performance and proximity to lineaments in Botswana. From Sander (1999), in 'Water resources of hard rock aquifers in arid and semi-arid zones' edited by J.W. Lloyd. Reproduced by permission of J.W. Lloyd and UNESCO*

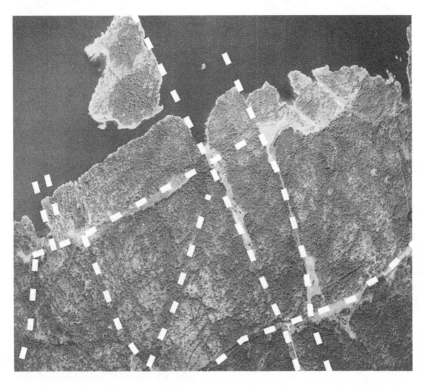

Figure 2.5 *Fracture mapping in Norway. (For scale, length of coastline = approximately 2.7 km). Aerial photo reproduced by permission of Blom Geomatics AS*

The hydrogeological conceptual model developed during the desk study will be refined and updated as more detailed information becomes available throughout the groundwater investigation programme. Further guidance on developing conceptual models is provided in Rushton (2003), MacDonald *et al.* (2005) and American Society for Testing and Materials (2014b).

The health and safety of persons engaged on the hydrogeological investigations, and the general public, should also be considered at the desk study stage, and a health and safety plan prepared (Appendix 3). The plan should identify the potential risks associated with activities such as the well survey, geophysical surveys and the drilling investigations, and indicate how these risks are to be managed. The preparation of this health and safety plan will be aided by the information collected during the initial field reconnaissance.

2.2   Field reconnaissance

While much information can be gleaned from the desk studies it is *always* useful to carry out a reconnaissance of the project area prior to planning the detailed exploration programme. The field reconnaissance enables us to:

1. Develop a better understanding of the hydrogeology of an area and hence improve the conceptual model. The aquifer and superficial materials can be examined at outcrop or in quarries (Figure 2.6); the geomorphology, topography and drainage patterns can be assessed with respect to surface water and groundwater linkages, including aquifer recharge and discharge areas; existing wells and boreholes in the area can be identified.

2. Talk to local people about their existing groundwater supplies, to find out whether there are

Figure 2.6   *Examination of a quarry exposure of a fractured sandstone formation, County Monaghan, Ireland. Photo by Bruce Misstear*

any problems relating to well yields, water quality or reliability. If the local people are the potential beneficiaries of the new well, then it will be essential to involve the community in site selection and scheme development (Figure 2.7; Box 2.1).

3. Assess the practicalities involved in carrying out geophysical surveys: for example, are there power lines present that would inhibit electromagnetic techniques, or are there significant noise levels that might affect shallow seismic surveys? (Section 2.4).

4. Look for potential sources of pollution and estimate the groundwater vulnerability (Section 2.7).

5. Identify and examine potential sites for exploratory boreholes. Although borehole sites will be selected primarily on hydrogeological criteria, the field reconnaissance should also consider issues such as land availability and community acceptance, access for a drilling

rig, potential flooding problems (Figure 2.8), health and safety issues, and borehole security with respect to vandalism.

2.3 Well survey

A systematic survey of existing wells in the study area often follows on from the field reconnaissance. The information collected is added to the project database or GIS. An essential requirement is to be able to record the locations of the wells accurately on the project maps. Global positioning systems (GPS) are very useful for mapping the coordinates of a well and the elevation of a suitable datum such as the wellhead (Figure 2.4). However, it is important to bear in mind that the accuracy of elevation readings from many current GPS systems is limited. Also, the highly accurate lateral GPS coordinates may not correspond with the accuracy of older map sheets in many countries,

Figure 2.7 *A well siting committee in southern Oman. Photo by Bruce Misstear*

Box 2.1   Involving local people in choosing well sites

Where local people are to be the beneficiaries of a new water well we must take account of their views when choosing the well site. Many new wells in developing countries have failed because the local people were not properly consulted about the new water supply and its location. Unfortunately, it was not uncommon practice in the past for the hydrogeologist or engineer to arrive in the village and select the well sites purely on technical criteria, with little or no attempt at engaging with the local community to find out about their opinions. Whilst technical considerations such as potential well yield and groundwater quality are, of course, fundamental to a successful well supply, cultural and socio-economic issues are also critical when seeking to establish a sustainable groundwater supply:

a. Most of the water collection in sub-Saharan Africa and Asia is carried out by women and children [Figure B2.1(i)], so their needs must be taken into account when choosing the well site [Inter-agency Standing Committee (IASC), 2006]. The local communities are usually run by men, and it may prove difficult to engage in dialogue with the women – but we must persist until the views of the women have been ascertained.

b. The potential security of the women and children at the potential well site should be considered (IASC, 2005; Asaba *et al.*, 2015). There may be increased risks associated with isolated sites at particular locations. Villagers may be able to provide information on areas to be avoided.

Figure B2.1(i)   *Water collection in many developing countries is by women and children. In this example from rural Uganda, the boy is using a bicycle to help transport the water but the women and girls have to carry the water by hand. Photo by Bruce Misstear*

c. The most powerful people in the village may wish to have the wells located close to their own houses. To overcome such challenges, wells are sometimes sited amongst the poorest sections of the community, where the needs may be greatest (Carter *et al.*, 2010). It is also important to be aware that some villages may contain more than one ethnic group, and this may have a bearing on where certain people wish to see their water well situated.

d. In developing countries, most of the improved hand-dug wells – and many drilled wells – are fitted with hand pumps (Section 3.7), and therefore the potential walking distance for the people collecting the water must be taken into account. If the time taken to walk to a well exceeds half an hour, then water consumption will drop off significantly (Cairncross and Feachem, 1993), so the well should not be sited too far from the village. On the other hand, there may be hydrogeological considerations that make it impractical or undesirable to locate a well within the village itself. For example, many villages are situated on high ground whilst the most

accessible groundwater sources are normally in the valleys. Again, there may be a greater contamination risk if wells are located at the heart of a village (pollution risk assessment and prevention is discussed in Section 2.8).

e. The new rural water supply wells may also be used for livestock and other purposes, and the siting of the well should therefore take account of the different potential water demands.

f. It is important to establish that the potential beneficiaries are clear about their economic stake in the project and the ongoing financial obligations that a new well may imply.

By talking to local people about their needs and wishes, the hydrogeologist or engineer can learn a lot about the existing water supplies and their problems, and therefore what is most likely to succeed in terms of the new well supplies. He/she must try and involve the community in the decision-making about their new well sites. This process may be challenging and the reader is referred to MacDonald *et al.* (2005) for further guidance on *how to* work with local communities on site selection.

leading to apparent discrepancies. The type of information that should be gathered during a well survey is included in Box 2.2. The survey questions will need to be modified for each project, since each survey will have its own particular objectives and the information potentially available varies from region to region. In general terms, a well survey provides the hydrogeologist with information on:

- groundwater levels (at rest and during pumping);
- well construction;
- present status and use of the well;
- pumping rates, hours pumped, groundwater use;
- groundwater quality.

The data are combined with the data from the desk study and the exploration drilling phase

(Section 2.5) to establish the piezometry and groundwater quality in the area. A number of the existing wells may be incorporated in a monitoring programme to investigate seasonal fluctuations in water level and water quality, and any longer-term trends.

We can gather information on well construction by inspecting the wells, measuring their depth and diameter, and by talking to the owners. With unlined hand-dug wells, it may be possible to observe the shallow geology from the wellhead (the hydrogeologist should never descend into the well, as this can be extremely dangerous). Also, the spoil from recently excavated wells may be available near the wellhead for inspection (Figure 2.9). It is not as straightforward to collect information in the field on well

Figure 2.8 *The remains of a well located in a wadi in northern Oman. The original wadi surface is preserved as a remnant mound around the top of the well casing. From the height of the hydrogeologist standing nearby, we can see that the wadi floor has been lowered by about 2 m due to erosion from the occasional, but violent, floods that occurred in the 20 year period between well construction and when this photo was taken. Photo by Bruce Misstear*

construction and geology from drilled wells. However, a down-hole TV camera can be employed to inspect a selection of existing drilled wells, giving useful information on the construction and condition of the well and, where it is unlined in the aquifer zone, on inflow horizons (see Section 6.4). Where possible, the well data should be correlated with the original construction data and borehole logs; these may be held in an existing well database at the national geological survey or other organization.

The information on groundwater usage is important for assessing potential yields for new wells in the area. This information will also be included in the groundwater balance for the area for determining the resource that is available for development (Section 2.6).

A limited number of groundwater quality parameters can be assessed rapidly using field testing equipment and water samples can be collected from a selection of wells for full laboratory analysis (Section 2.7 and Chapter 8).

Box 2.2   Well survey information

A well survey proforma should be designed to meet the specific needs of each project, but the information to be collected may include the following:

General	Well number: local survey number or the number assigned in a national well database, if applicable
	Locality where well is located, description of well location, grid reference, location map
	Date and time of visit, name of investigator
	Name of well owner (public authority, private household, industry)
Well details	Well type (drilled, hand-dug, radial collector)
	Water use (drinking water supply, industrial, irrigation, monitoring)
	Abstraction permit number, if applicable
	Is construction drawing available? (Include copy with survey form)
	Is borehole log available? (Include copy with survey form)
	Depth of well, method of measurement
	Diameter at surface, type of lining, type of well cover
	When constructed? By whom?
	Description of reference point for measurements (e.g., top of casing) and elevation of reference point above ground level
	Aquifer type
	Depth to water level (static, pumping), corresponding pumping rate and method of measurement
Pumping details	Type and make of pump, pump setting depth, rising main diameter
	Normal pumping periods (hours per day, months per year)
	Pumping rate during visit, method of measurement
	Are pumping test data available? (Include summary with survey sheet) If not, should the well be tested?
Installed monitoring equipment	Water level monitoring (transducer, float recorder, dipping tube?)
	Flow measurement (flowmeter?)
	Water quality (sampling tap?)
Water quality	Type of sample (pumped, bailed); treated or untreated water
	Sample appearance (turbid, clear)
	Any field filtration or chemical conservation of sample?
	Site measurements: temperature, pH, electrical conductivity, redox potential, dissolved oxygen, alkalinity (and other parameters)
	Sample collected for laboratory analysis?
Health and safety issues (see Appendix 3)	Physical dangers around wellhead
	Access to wellhead
	Risk of degassing of carbon dioxide, methane, radon
	Risk of working in enclosed space
Other information	Has the well been geophysically logged or inspected by CCTV? Should it be?

Figure 2.9 *Examining a hand-dug well in northern Oman. The well is fitted with a belt-drive turbine pump. Note the excavated spoil material piled up behind the well, which can give a useful indication of the well geology. Photo by Bruce Misstear*

The well survey may be combined with an *environmental survey* or *water features survey* (see Section 7.2.1), designed to identify any other water users, springs, streams, lakes and so on, which might be adversely affected by either the drilling operation or subsequent groundwater abstraction. Such surveys are important for the preparation of an environmental impact assessment.

2.4 Geophysical surveys

Geophysical surveys can provide useful data on geology, aquifer geometry and water quality. Geophysical surveys are sometimes undertaken without proper planning, but rather in the hope that they will show something useful. This approach is rarely productive. A better approach is to:

1. Identify the nature of the physical problem to be investigated.
2. Select the geophysical method or methods appropriate for that problem.
3. Plan the investigation programme.

Geophysical surveys do not lead to a unique geological model; more than one interpretation of the data is possible. Borehole control is essential to reduce this ambiguity. Therefore, geophysical surveys should be carried out in conjunction with exploratory boreholes rather than as a replacement for a drilling programme (Figure 2.10). The combined use of geophysics and drilling can produce results more cheaply than relying on drilling alone, since the number of exploratory boreholes can be reduced.

The main geophysical methods used in groundwater exploration are summarized in Table 2.2.

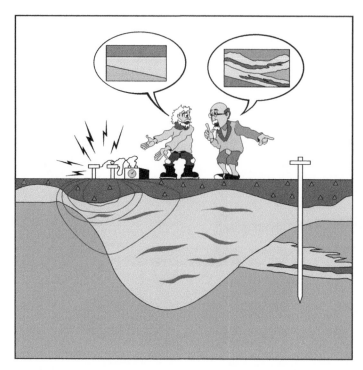

Figure 2.10   *Cartoon highlighting that geophysical surveys and exploratory drilling give the best results when carried out together. With kind permission from Springer Science+Business Media: Hydrogeology Journal, 'Hydrogeological mapping as a basis for establishing site-specific groundwater protection zones in Denmark', 12, 2004, 550–562, Thomsen R, Søndergaard VH and Sørensen KI*

Different methods provide data on different geophysical properties of the ground and, therefore, the best survey results are usually achieved by using more than one method. The electrical resistivity and electromagnetic techniques that are most commonly used in groundwater investigations are described in Sections 2.4.1 and 2.4.2 below, whilst further information on the ground penetrating radar (Georadar) method is included in Box 2.3.

2.4.1   Electrical resistivity

The electrical resistivity of the ground depends mainly on the properties of the geological material, the degree of saturation and the nature of the fluid. Dry soils and crystalline bedrock have high electrical resistivities, saturated sands and consolidated aquifers have lower resistivities, while clayey materials or strata containing salt water

have very low resistivities. Electrical resistivity methods are, therefore, useful for investigating both aquifer geometry and groundwater quality. The methods can also be used to estimate aquifer properties such as transmissivity and hydraulic conductivity (Heigold *et al.*, 1979; MacDonald *et al.*, 1999; Soupios *et al.*, 2007; Utom *et al.*, 2012; Attwa *et al.*, 2014).

Electrical resistivity is measured by passing an electrical current into the ground between two electrodes and measuring the potential difference between two other electrodes. The resistance is calculated using Ohm's law. This resistance is multiplied by a geometric factor relating to the electrode configuration to calculate the electrical resistivity of the subsurface (usually expressed in units of ohm m). There are many different electrode configurations or arrays. Those most common in

Table 2.2 *Geophysical methods used in groundwater exploration*

Principle	Method	Main applications in groundwater exploration
Electrical resistivity	Vertical electrical soundings (VES)	Depth to bedrock; thickness of superficial deposits; depth to water table; depth of weathering in crystalline rock aquifers; depth to saline water interface in coastal aquifers; aquifer properties
	Electrical resistivity profiling (constant electrode separation traversing)	Location of buried valleys; detection of vertical/near vertical fracture zones; depth of weathering in crystalline rock aquifers; location of contaminant plumes
	Electrical imaging (tomography)	2-D and 3-D imaging combines many of the applications of VES and resistivity profiling. Time-lapse (or 4-D) imaging can monitor water movement in the subsurface
Electromagnetics (EM)	Ground conductivity profiling (frequency-domain EM)	Similar applications to resistivity profiling
	Time-domain EM (TDEM)	Similar applications to VES, but often used for greater depths of investigation
	Very low frequency (VLF)	Mainly for location of vertical/near vertical fracture zones; also to determine depth to bedrock; depth to water table
	Surface nuclear magnetic resonance (SNMR)	Aquifer geometry; aquifer properties
Magnetometry	Total magnetic field anomaly	Location of igneous dykes. Location of fracture zones
Seismic	Seismic refraction	Depth to bedrock; thickness of superficial deposits; depth to water table; depth of weathering in crystalline rock aquifers; location of fracture zones
Ground penetrating radar (Georadar)		Thickness of sand and gravel aquifers; depth to bedrock; depth to water table; location of sub-horizontal fractures or of cavities in karst limestones
Gravity	Gravity and microgravity surveys	Geometry of extensive sedimentary aquifers; location of buried valleys; location of cavities in karst limestones (microgravity)

groundwater exploration are the Wenner and the Schlumberger arrays (Figure 2.11). A modification of the Wenner configuration, known as the Offset Wenner, employs a five-electrode array and is used to reduce the effects of near-surface lateral inhomogeneities on the results (Barker, 1981). There are three main types of electrical resistivity survey: vertical electrical sounding, electrical profiling and electrical imaging.

Vertical electrical sounding. Vertical electrical sounding (VES) involves expanding the electrode array about a central point (Figures 2.12 and 2.13). VES can be carried out using any of the electrode configurations shown in Figure 2.11. A resistivity measurement (the *apparent resistivity*) is taken at each increase in the electrode separation. The depth of investigation increases with the electrode separation. The maximum effective depth of

Box 2.3   Georadar

Ground penetrating radar (or Georadar) is conceptually simple, consisting of a downward orientated radar sender and a radar receiver some short distance behind it. A pulse of high frequency radio signal (often 10–1000 MHz) is reflected from interfaces between materials with differing electrical properties. In practice, this means that reflectors can be boundaries between different lithologies, between overburden and bedrock [Figure B2.3(i)] or sub-horizontal fractures. Under good conditions (e.g., homogeneous sand), the water table can also be detected as a reflector. The reflected signal is detected by the receiver and the delay between sender and receiver is logged. This delay corresponds to a travel time and thus converts into a 'depth to reflector'.

Georadar penetration is best in materials of low electrical conductivity such as sands and gravels with fresh pore water, where penetrations of up to 50 m can be achieved. Near-surface clays, buried metal artefacts, saline soils or pore waters can result in poor penetration. Davis and Annan (1989) provide further reading on the theory of Georadar.

Portable Georadar kits typically require two operators, one to carry a ski-like sender and one to bear a similar receiver. One of the operators will also carry a processing unit, the most sophisticated of which can generate an on-screen picture of the subsurface reflectors as the pair traverse the terrain.

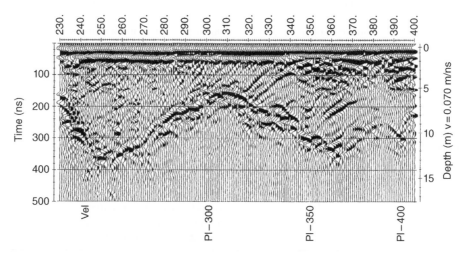

Figure B2.3(i)   *Georadar profile from Norway, showing the interface between overburden and underlying bedrock. From Mauring et al. (1994). Reproduced by permission of Norges geologiske undersøkelse, Trondheim*

investigation is 0.44 of the current electrode separation in a Wenner array, and 0.46 in a Schlumberger array, although the optimum depth of investigation is much less in both cases – at 0.17 and 0.19 of the current electrode separation, respectively (Barker, 1999). The relationship between the apparent resistivity and the electrode separation can be analysed to give the thickness and resistivity of the geological layers. This analysis can be performed manually using type

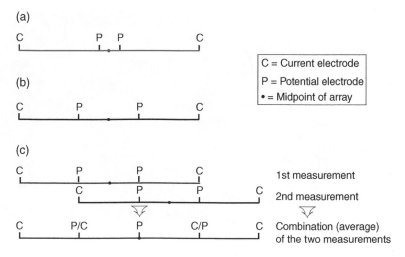

Figure 2.11 *Electrode arrays. (a) Schlumberger array; (b) Wenner array; (c) Offset Wenner array*

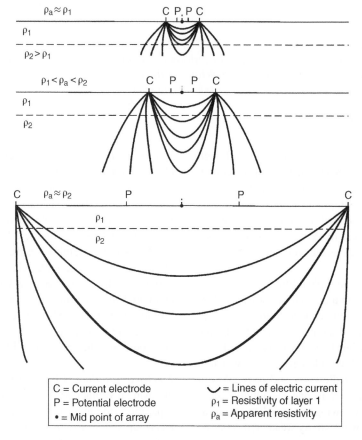

Figure 2.12 *Principle of vertical electrical sounding*

Figure 2.13   *Vertical electrical sounding in the Red Sea hills region of northern Sudan. The objective here was to identify the thickness of wadi gravel aquifer overlying low permeability volcanic bedrock. The resistivity instrument and operator are in the centre of the picture, partly shaded by the tree, whilst the person on the left is connecting a cable drum to one of the electrodes. The bucket of water (just to the left of the resistivity instrument) was used to water the contact point between each electrode and the ground surface in this very arid region. Photo by Bruce Misstear*

curves or automatically on a computer. Although commercial instruments for measuring resistivity cost several thousand dollars, it is possible to construct relatively simple instruments for as little as $250. Clark and Page (2011) describe how one such inexpensive resistivity meter was developed for use by drilling teams carrying out VES in East Africa.

Figure 2.14 shows a typical three-layer resistivity curve for a VES in a crystalline rock area. This can be interpreted qualitatively as follows: the crystalline bedrock is hard, non-porous and has a high resistivity; it is overlain by a clayey weathered zone that is saturated and has a low resistivity, which in turn is overlain by shallow dry soil or regolith of high resistivity. More than one quantitative interpretation is possible: owing to a problem known as *equivalence*, the low resistivity weathered zone can

be modelled as a layer 20 m thick of resistivity 70 ohm m, or a layer 10 m thick of resistivity 35 ohm m, the ratio of layer thickness to resistivity being equivalent in both cases (Barker, 1999). We therefore need borehole control to reduce ambiguities in situations such as this (Figure 2.10).

VES is most useful when the geology consists of horizontal layers, with relatively little lateral variation. However, surveys tend to be slow compared to electromagnetic techniques that can provide similar data. Also, there can be significant problems in achieving good contact between the electrodes and the ground in arid areas.

Electrical profiling. Electrical profiling is carried out using a Wenner array with a constant electrode separation (the method is also known as *constant electrode separation traversing*). Linear traverses

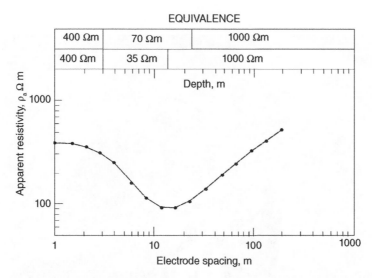

Figure 2.14 *VES curve illustrating the problem of equivalence. From Barker (1999), in 'Water resources of hard rock aquifers in arid and semi-arid zones' edited by J.W. Lloyd. Reproduced by permission of J.W. Lloyd and UNESCO*

are made across the area of interest, measuring the apparent resistivity at each station interval. The method is suitable for detecting lateral variations in geology such as the presence of a buried valley or the location of a vertical fault. The profile lines are set out at right angles to the expected orientation of the fault or other feature of interest, as determined from a geology map, air photo interpretation or field reconnaissance. An electrode separation is selected for the required depth of investigation; VES may be carried out to identify a suitable electrode separation for the profiling. This method was widely used in groundwater exploration up until the 1980s, but has now been largely replaced by the faster and cheaper ground conductivity profiling method (Section 2.4.2).

Electrical imaging. Electrical imaging, or *tomography*, combines the principles of electrical profiling and VES. Imaging is carried out by installing a large number of electrodes along a line of survey and then connecting these electrodes to the resistivity instrument using a multi-core cable (Figure 2.15). A series of profiles is made along the survey line, increasing the electrode separation

after each profile. This builds up a two-dimensional picture of apparent resistivity and electrode separation. The data are first converted into a 'pseudo-section' of apparent resistivity and depth, by using the calculated median depth of investigation for the electrode separation at each measurement station. These data are in turn converted into true formation resistivity and depth values by a technique known as inversion (Loke and Barker, 1996).

Figure 2.16 shows a resistivity section for a sand and gravel aquifer in Ireland. The position of the water table can be identified to the right of the valley, by the transition from high to low resistivity at around −5 m. The section also indicates that there are considerable variations in resistivity within the gravel aquifer, reflecting heterogeneity in these deposits.

Two-dimensional images along different profile lines can be combined to produce 3-D images. Time-lapse imaging introduces a fourth dimension, in which surveys are carried out at different time intervals to investigate changes in resistivity associated with changes in moisture content. Applications of time-lapse imaging include:

(a)

(b)

Figure 2.15 *Electrical imaging near Wagga Wagga, New South Wales, Australia. Here the objective was to investigate the presence of saline groundwater. Photo (a) shows Bruce Misstear with the resistivity instrument (ABEM Terrameter), whilst photo (b) shows the line of electrodes and multicore cables (and colleague Ian Acworth). Photos by Ian Acworth (a) and Bruce Misstear (b)*

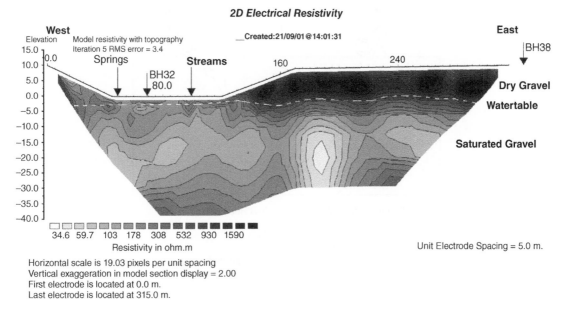

2D Electrical Resistivity

West

East

Elevation   Model resistivity with topography   __Created:21/09/01@14:01:31

Iteration 5 RMS error = 3.4

Springs   Streams   160   240

BH32
80.0

BH38

Dry Gravel

Watertable

Saturated Gravel

34.6 59.7 103 178 308 532 930 1590
Resistivity in ohm.m

Unit Electrode Spacing = 5.0 m.

Horizontal scale is 19.03 pixels per unit spacing
Vertical exaggeration in model section display = 2.00
First electrode is located at 0.0 m.
Last electrode is located at 315.0 m.

Figure 2.16 *Two-dimensional resistivity image across a gravel aquifer, County Kildare, Ireland. Reproduced by permission of Peter O'Connor, Apex Geoservices Ltd*

investigation of groundwater recharge; viewing the response of an aquifer to pumping; studying the links between groundwater and estuaries; identifying transport between fast and slow flow paths in fractured rock aquifers (Singha *et al.*, 2014). The concept of electrical imaging has been developed further to include cross-borehole electrical resistance tomography, involving arrays of electrodes set in a series of boreholes (Daily *et al.*, 2004; Nimmer *et al.*, 2007). This method can be employed to develop a full three-dimensional resistivity model of the block of aquifer material being investigated. Cross-borehole geophysics can also be carried out using seismic refraction and radar methods (Reynolds, 2011).

2.4.2 Electromagnetics

Electromagnetic (EM) methods measure the ground conductivity, which is the inverse of resistivity. EM methods, therefore, have similar applications to resistivity methods. The main difference between the survey techniques is that with EM methods it is not necessary to make a good physical

contact with the ground. Therefore EM surveys are more rapid and less expensive to carry out.

In an EM survey, an alternating electric current is applied to a wire coil or loop (the transmitter coil). This generates a primary electromagnetic field, which is modified as it passes into the ground (Figure 2.17). If there is a good conductor present, such as a saturated zone, the primary field produces eddy currents that generate a secondary field. The secondary field is detected by the alternating current it induces in a second wire coil (the receiver coil). The difference between the transmitted and received electromagnetic fields yields information on the nature and geometry of the conductor (Keary *et al.*, 2002).

Ground conductivity profiling. The main EM survey technique is ground conductivity profiling, which is similar to electrical profiling. The depth of investigation depends on the spacing between the two coils and on their orientation. Different coil orientations alter the direction of the inducing field. The maximum effective depth of

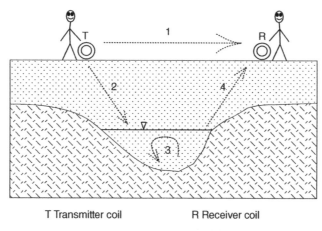

T Transmitter coil   R Receiver coil

1 Primary electromagnetic (EM) field
2 Modified primary EM field
3 Eddy currents in conductive (saturated) zone
4 Secondary EM field

Figure 2.17   *Principle of electromagnetic survey*

investigation with the coils in a horizontal position (vertical dipole mode) is approximately twice that with the coils vertical (horizontal dipole mode), being 1.5 times the coil spacing compared to 0.75 times the coil spacing, respectively (Barker, 1999). However, surveys made with the coils horizontal are much more sensitive to errors in the coil alignment.

Figure 2.18 shows the application of conductivity profiling to siting wells in a weathered and fractured crystalline aquifer in West Africa (Hazell *et al.*, 1988). The instruments used were the Geonics EM31 and EM34. The former has the two coils in horizontal mode at 3.66 m spacing on a single instrument, requiring only a single operator. It is thus a very rapid investigation tool. With a maximum effective depth of investigation of around 6 m, it could have been used in this example to choose a site for a shallow, hand-dug well that avoided near-surface bedrock. The EM34 was used to investigate deeper zones of weathering/ fracturing suitable for a higher yielding drilled well. The EM34 has two separate coils connected by a cable, and requires two operators. In the example illustrated (Figure 2.18) the surveys were

carried out with a horizontal and vertical coil orientation, both using a coil separation of 20 m. Several survey lines allowed the contouring of the ground conductivity data, leading to the location of the borehole at the intersection of two linear anomalies interpreted as fractures. For similar weathered and fractured crystalline rock terrain, Acworth (2001) recommended that the best results for borehole siting can be achieved from a combination of ground conductivity profiling with electrical imaging.

Time-domain electromagnetics. Ground conductivity profiling relies on a principle known as *frequency-domain EM*. One of the problems with this technique is that the secondary field generated from poor ground conductors is often small in comparison to the primary field and is therefore difficult to measure accurately in the presence of the larger signal. In *time-domain EM* (TDEM, also known as transient EM), the primary field is pulsed rather than continuous and the secondary field is measured after the primary field is switched off. The eddy currents produced by the primary field in conductive material tend to propagate downwards with time

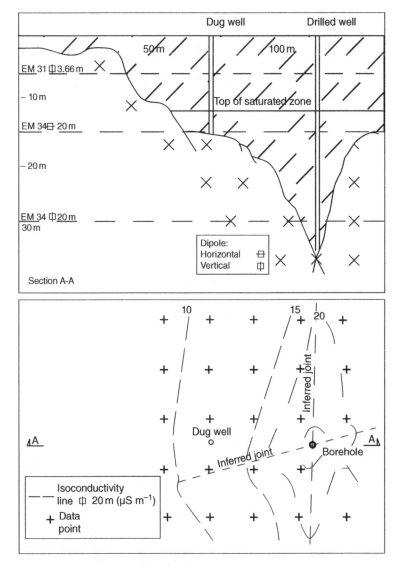

Figure 2.18 *Electromagnetic survey for well siting in weathered crystalline rock aquifers, Nigeria. Adapted from Hazell et al. (1988) and reproduced by permission of the Geological Society of London*

(Figure 2.19). Measurement of the decay rate of the secondary field after switch-off results in a type of depth sounding.

The primary field is produced from a transmitter coil laid on the ground surface. The secondary field can be measured in the same coil or in a second, receiver, coil. Depths of investigation of several hundred metres can be achieved with coils of

a few tens of metre in diameter. This gives TDEM a significant advantage over VES, which requires an electrode separation that is at least twice the depth of investigation (Section 2.4.1).

An example of a TDEM survey is illustrated in Figure 2.20 (Young *et al.*, 1998). The survey comprised several hundred TDEM soundings along a coastal plain in northern Oman. The geology of

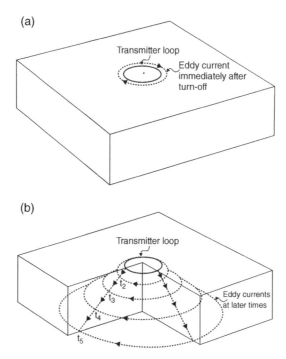

Figure 2.19   *Principle of time-domain electromagnetics showing (a) the form of the eddy current immediately after turning off the primary magnetic field and (b) the downward propagation of the eddy currents at later times. Reproduced by permission from 'An Introduction to Applied and Environmental Geophysics' by J.M. Reynolds (1997). Copyright 1997 John Wiley & Sons, Ltd*

the plain consists of a complex sequence of alluvial deposits several hundred metres thick, overlying crystalline (ophiolite) and limestone bedrock. The survey was successful in distinguishing layers within the alluvium, including the important upper gravel aquifer [Figure 2.20(a)], and in determining the depth to the saline interface [Figure 2.20(b)].

According to Barker (1999) a full sounding takes about one hour, much of this time being spent in setting up the equipment and in laying out the cables. More rapid surveys can be achieved using TDEM systems in which the coils are towed behind tractor vehicles (Thomsen *et al.*, 2004).

2.5   Drilling investigations

The drilling investigations often follow a phased approach, with initial exploration drilling followed by more detailed investigations (Box 2.4).

Exploration drilling is used to confirm the provisional interpretation of the hydrogeology derived from the desk studies, field reconnaissance, well survey and geophysics. The siting of the exploration boreholes in a situation where there are few data is difficult, but should be governed by the principle that every borehole should be drilled to provide an answer to a question: is there an aquifer? how thick is it? how thick are the cover materials? It is seldom cost-effective to drill large numbers of exploration boreholes in the first phase, for money spent on unnecessary exploration could deplete the budget for the subsequent more detailed, targeted investigations. Most exploration boreholes are vertical. However, inclined boreholes can be more effective where the aim is to investigate groundwater occurrence in large numbers of vertical or sub-vertical fractures (Box 2.5).

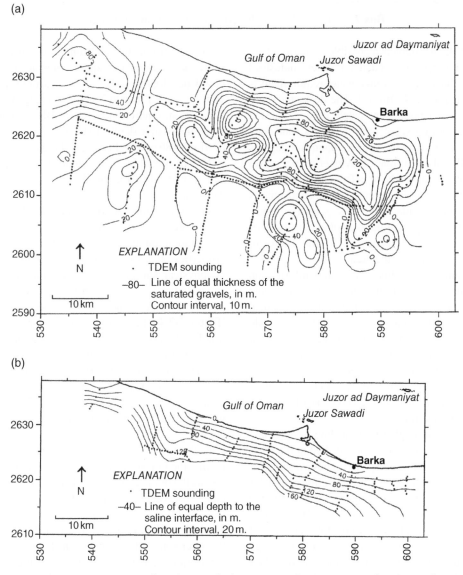

Figure 2.20 *TDEM surveys on an alluvial coastal plain, Oman (from Young et al., 1998). The upper map shows the thickness of saturated gravels whilst the lower map shows the depth to the saline interface. Reproduced with kind permission from Springer Science + Business Media: Hydrogeology Journal, 'Exploration of an alluvial aquifer in Oman by time-domain electromagnetic sounding', 6, 1998, 383–393, Young ME, de Bruijn RGM and bin Salim Al-Ismaily A*

The detailed investigation phase is intended to supplement the exploration phase by infilling the data gaps and by providing more quantitative information on the aquifer properties and resources, and hence on potential well yields. As with the initial investigations further drilling would be accompanied by other surveys including geophysics.

Box 2.4   Example of a phased groundwater investigation

Figure B2.4(i) depicts a possible scenario where a water supply is needed for a new industrial development. The average water requirement is estimated to be 450 m^3 d^{-1}. The desk study and field reconnaissance revealed little about the hydrogeology except that there is a sandstone aquifer dipping to the south at about six degrees beneath a shale formation. A river with a gravel floodplain crosses the area from the northeast to the southwest. The outcrop of the sandstone is about 400 m wide, which suggests (by trigonometry) an aquifer thickness of about 41.8 m or a drilled vertical thickness of 42.0 m. There are no existing wells in the area.

Assuming that this is the only information available then an investigation programme would be required, involving both geophysical surveys and exploratory boreholes. Suitable geophysical techniques might include VES or TDEM for determining the depth and thickness of the sandstone beneath the shale, and ground conductivity profiling or ground penetrating radar for investigating the geometry of the river alluvium (Section 2.4).

Our initial exploration drilling programme might include five boreholes drilled at the locations shown in Figure B2.4(i). These would provide the basic hydrogeological information for the area, including geological control for the geophysical surveys. Boreholes BH1 and BH2 in the river gravels would give the thickness of the gravels, provide samples of their lithology and water quality, and give an indication of their variability. This information should be sufficient to show whether the gravels are a viable shallow aquifer and provide data on which water wells could be designed. Boreholes BH3, BH4 and BH5 give information on the sandstone aquifer, particularly on its lithological variability, groundwater quality, potentiometric surface and flow directions. Borehole BH5 gives information on the unconfined zone and BH3 on the confined zone. Borehole BH3 is important to

verify the dip of the sandstone at depth and to show the nature of the aquifer close to the proposed development. The dips measured at the surface range from 5 to 7 degrees, a range which, 700 m from outcrop at borehole BH3, means a range in possible depths to the top of the aquifer of 61 m to 86 m. The lithological log of borehole BH3 will prove the actual depth to the aquifer and provide data for the design of future test wells and observation boreholes.

This initial investigation would be followed by a more detailed investigation to give further information on the aquifer geometry, aquifer properties and the groundwater quality. The geophysical surveys in the exploration phase suggest that the eastern part of the sandstone aquifer is much finer than the western part [Figure B2.4(ii)]. Drilling results from exploration borehole BH5 and a new observation borehole BH6 confirm this facies change. Observation boreholes BH7 and BH8 provide further lithological data and permanent water quality and water level monitoring points in the unconfined and confined parts of the sandstone aquifer.

In this case study, there is a single demand point – the industrial site – so, initially, only one test well would be proposed: TW9, with two satellite observation boreholes (BH11 and BH12). One additional observation borehole (BH10) would be drilled in the river gravels to measure the interaction, if any, between the deep sandstone aquifer and the alluvium during the pumping test on TW9. In a more general groundwater resources study, we would probably construct a further test well near BH6 to test the finer part of the aquifer, and also one in the alluvium to test that formation.

The pumping test on well TW9 shows that it can supply the 450 m^3 d^{-1} average demand required by the industrial development, and that it has sufficient capacity to meet this demand by pumping for only 8 hours a day.

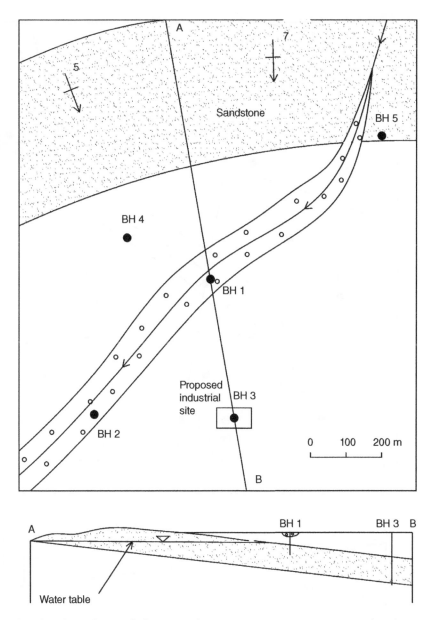

Figure B2.4(i) *The first phase of the groundwater investigation programme for the proposed industrial site*

The water sampling also confirmed that the water quality is suitable for the intended use. We therefore decide to make TW9 the production well. The next stages of the investigation would be to consider the ability of the aquifer to sustain this supply in the long-term, and to assess the pollution risk.

One approach for estimating the sustainability of the supply is to consider the maximum width of the zone of contribution (ZOC) to the well and

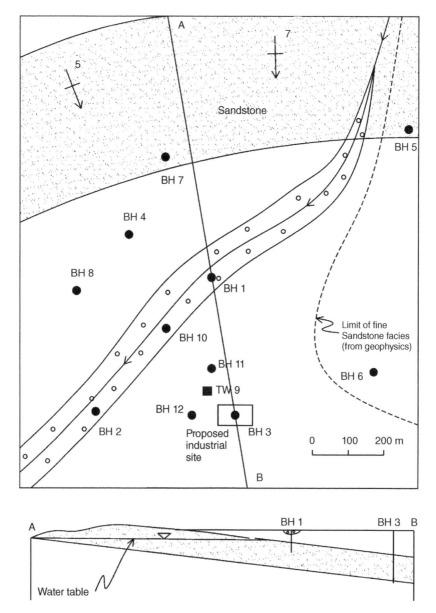

Figure B2.4(ii) *The second phase of the groundwater investigation programme for the proposed industrial site*

then calculate the available recharge within this zone at outcrop. The following data are used in the calculation of the ZOC (Section 2.8.2):

a. average water demand $Q=450\,\mathrm{m^3\,d^{-1}}$
b. aquifer thickness $b=41.8\,\mathrm{m}$

c. transmissivity $T=800\,\mathrm{m^2\,d^{-1}}$ (from the pumping test on TW9)
d. hydraulic conductivity $K=T/b=19.1\,\mathrm{m\ d^{-1}}$, but reduce to $15\,\mathrm{m\ d^{-1}}$ because of unknown, but probably lower, K in the finer lithology east of the well site

e. hydraulic gradient $i = 0.001$ (estimated from water levels in the observation boreholes)

Applying Equation (2.12), and neglecting any effects of dispersion, the maximum width (really the half-width) y_L of the ZOC where the aquifer is confined is:

$$y_L = \pm \frac{Q}{2Kbi} = \frac{450}{2 \times 15 \times 41.8 \times 0.001} = 359 \text{ m}$$

The full width of the ZOC is 718 m. Therefore, the area of outcrop that might be expected to contribute recharge to the well is:

$$718 \times 400 = 287\,200 \text{ m}^2$$

The recharge required over this area to meet an annual water demand for the industrial development of 164,250 m³ a⁻¹ is:

$$\frac{164\,250}{287\,200} = 0.572 \text{ m a}^{-1}$$

A soil moisture balance could be carried out to see if this volume of recharge is likely to occur (Section 2.6.1).

The risk of pollution to TW9 would be low, at least initially, since the sandstone aquifer is protected by a thick shale layer. Any future pollution is most likely to come from the outcrop area and we would therefore need to monitor the groundwater quality in observation borehole BH7.

Applying the Darcy equation, and assuming an effective porosity for the sandstone aquifer of 10 percent, the time of travel t from the outcrop to TW9 located at a distance d of 600 m can be calculated as:

$$t = \frac{d}{v_s} = \frac{dn_e}{Ki} = \frac{600 \times 0.1}{15 \times 0.001} = 4000 \text{ days}$$

This figure is obviously sensitive to the hydraulic gradient, which would increase towards the well. Assuming that the gradient is affected by the cone of drawdown extending to say 300 m from the well, then the travel time from outcrop for a distance of 300 m would still be 2,000 days, or 5.5 years. Monitoring of the groundwater quality in BH7 and BH4 should therefore provide plenty of advance warning of any pollution problems moving from the outcrop area towards the well. We would need to be aware, however, that the existence of any preferential flow horizons or pathways could result in 'first arrivals' of contaminants well in advance of the average predicted flow time.

In the real world the scope of the groundwater exploration programme would, of course, be subject to budgetary constraints. The severity of these constraints will vary according to the nature of the development and the financial resources of the client. In many situations, the hydrogeologist will have to make do with a far less detailed investigation than the one described here. However, money spent on carrying out a comprehensive exploration programme can result in savings in the long term, by enabling the most productive aquifer zones with the best groundwater quality to be targeted for development.

It is important in resource studies to establish the relationships between the groundwater, recharge and surface water bodies. Observation boreholes are needed to follow the response of groundwater levels to recharge events through the annual hydrological cycle. Observation boreholes adjacent to streams can be used to observe the interaction between groundwater and surface water. Frequent observations of water levels by means of an automatic recorder are desirable in at least a few of these boreholes. Nowadays, water level recorders normally comprise a pressure transducer connected to a data logger (Figure 2.21). However, traditional chart recorders connected to water level floats are still used in some situations (Figure 7.5). Different types of water level recorder are described in Section 7.2.4, with further details given in Brassington (2007).

Box 2.5   Orientation of drilled wells

Water wells do not need to be vertical. Indeed, with modern down-the-hole-hammer (DTH) drilling rigs, drilling can be commenced at an angle by setting the drilling mast at the appropriate slant. Other, more sophisticated rigs enable the angle of attack of the drilling bit itself to be controlled during drilling.

Why drill a non-vertical well? There are at least two reasons. When drilling in fractured rock we may wish to drill into a vertical fracture zone. However, if vertical drilling commences into this vertical zone [diagram (a) in Figure B2.5(i)], any slight deviation in the zone means that the drill string will leave the fracture zone. Moreover, the upper part of the well will be very vulnerable to surface contamination, being drilled in highly fractured rock. If, on the other hand, drilling commences at an angle a little away from the outcrop of the zone [diagram (b) in Figure B2.5(i)], the well can be designed such that it intersects the zone at a given depth with greater certainty, and such that the upper few tens of metres are drilled in intact rock.

The azimuth of the well can also be designated at an acute angle to the fracture zone, such that the interval where the well intersects the zone is much longer than it would have been had the azimuth been perpendicular.

Also, if the aquifer contains a number of fracture sets, each with a characteristic orientation and fracture density, the optimal well orientation can be calculated so as to intersect the maximum number of fractures, using a combination of calculus and trigonometry. Banks (1992a) describes the theory behind optimizing borehole angles, but, in general, the well will tend to be orientated perpendicularly to the most permeable (most densely spaced or most open) fracture set, in order to intersect the maximum number of fractures of this set. In a situation with three mutually orthogonal (two vertical, one horizontal) fracture sets of equal fracture spacing, it can be shown that the optimum angle of drilling is at 54.7° from the vertical, at an azimuth of 45° from the two vertical fracture sets.

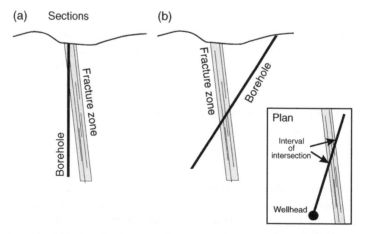

Figure B2.5(i)   *The advantages of drilling an angled borehole to intersect a fracture zone. In the plan view (inset), the more acute the angle between the borehole and the fracture zone, the longer the interval of intersection but the greater the risk of 'missing' the fracture zone*

Figure 2.21 Small water level monitoring unit (incorporating pressure transducer, power source and data logger) being installed in an observation borehole in County Monaghan, Ireland. Photo by Bruce Misstear

The observation boreholes also can be used to obtain water samples for analysis to measure the variation in regional groundwater quality (Section 2.7).

Values of the aquifer characteristics transmissivity and storativity are needed to predict the effect of a particular pumping policy on the groundwater regime, and (if not determined from reliable previous studies in the area) are obtained by means of pumping tests on test wells (Chapter 7). Test wells, with their satellite observation boreholes, are expensive installations and their number in any particular investigation is likely to be limited by budgetary constraints. In a uniform aquifer, test wells can be widely spaced – several or many kilometres apart – but in an aquifer that is variable, or in a situation where

there are several interfingering aquifers, test wells may have to be close together before the results of their pumping tests can be used to predict aquifer behaviour with any confidence. The minimum number of test wells needed is one for each significant lithology change in the aquifer being studied, and one for any specific interaction that needs to be observed.

We must also remember that a pumping test, even a long test lasting several weeks, is probably testing only a few square kilometres of aquifer and that the knowledge of the aquifer characteristics obtained from such a test is by no means perfect (Chapter 7). Also, aquifer behaviour over several years may have to be predicted from observations taken over a period of only a few weeks.

The quality of water from the test wells should be monitored on site during the tests, to detect any changes with time. At least one sample should be taken for full laboratory analysis. This sample, taken towards the end of the test, will show the quality of the water and its suitability for the intended use (see Section 2.7 and Chapter 8). In some aquifer lithologies (notably crystalline rocks), the exposure of pumped groundwater to fresh rock surfaces and drilling cuttings means that water sampled during a pumping test immediately after drilling may not be wholly representative of the long-term pumped water quality. It is recommended that a sample be taken, say, six months following completion of drilling to confirm the findings of initial water samples (Banks *et al.*, 1993b).

2.6 Groundwater resources assessment

The pumping tests of the drilling investigation phase provide data on the aquifer characteristics and well performance. These tests generally only indicate the potential well yields *in the short-term*. In designing a well or well field we must always consider the long-term sustainability of the supply. This requires an estimate of recharge and of the overall water balance of the aquifer system. In many countries information on

---

Box 2.6   Recharge concepts

Groundwater recharge can be defined as 'the downward flow of water reaching the water table, adding to groundwater storage' (Healy, 2010). There are two main types of recharge. *Direct recharge* occurs through vertical infiltration of precipitation where it falls on the ground. It is also known as *diffuse recharge* because it is distributed over a large area. This is the main form of recharge in temperate and humid tropical climates. *Indirect recharge* (also known as *focused recharge* or *point recharge*) occurs through infiltration of surface runoff and is the main recharge mechanism in arid regions. Indirect recharge can also be important in some hydrogeological environments in humid zones, such as in karst limestone aquifers where indirect recharge occurs from losing rivers and via swallow holes and other solution features (Figure 2.28). We also need to take account of 'non-natural' forms of recharge: in some rural arid regions return flows from irrigation may be the largest component of recharge (Foster *et al.*, 2000), whereas in urban areas a significant proportion of the recharge may be indirect recharge from leaking sewers and water mains, soakaways or pit latrines (Lerner, 1997).

The factors that influence the amount and type of recharge include:

- precipitation (volume, intensity, duration);
- topography;
- vegetation (cropping pattern, rooting depth) and evapotranspiration;
- soil type;
- permeability and thickness of superficial deposits;
- flow mechanisms in the unsaturated zone (fractures, sink holes, preferential flow pathways);
- aquifer characteristics;
- influent rivers;
- karst features;
- irrigation schemes;
- urban areas.

Several of these factors also influence the determination of groundwater vulnerability (Section 2.8.1). Misstear *et al.* (2009a) and Hunter Williams *et al.* (2013) describe how vulnerability maps can be used for making preliminary estimates of groundwater recharge.

Some methods of recharge estimation, such as soil moisture fluxes, assume that all water moving below the soil zone eventually contributes to recharge, but this may not be the case where there are lateral flows in superficial deposits or within fractured bedrock above the water table. It can therefore be useful to distinguish between *potential recharge* and *actual recharge* (Hulme *et al.*, 2001; Fitzsimons and Misstear, 2006).

---

recharge and on the available exploitable resource will be available from the government or regional agency that manages the water resources of the area under consideration. If the proposed abstraction is small and there is sufficient resource capacity known to be available by the relevant agency, then we may not have to undertake a separate assessment of groundwater resources. In situations where the resource is under pressure, where adverse environmental impacts are possible or where there is insufficient information

available to determine the extent of the resource and the possible impacts of the proposed well scheme, then we must carry out an evaluation of the recharge and resource availability. Box 2.6 includes a definition of recharge and outlines some basic concepts, making distinctions between *potential* and *actual* recharge, and between *direct* and *indirect* recharge.

The various approaches for estimating groundwater recharge can be grouped as follows (Misstear, 2000):

- inflow estimation;
- aquifer response analysis;
- outflow estimation;
- catchment water balance and modelling.

In broad terms, most inflow methods provide information on potential recharge whereas aquifer response, outflow estimation and catchment water balance methods produce actual recharge values, usually integrated over a relatively large area. Recharge is very difficult to estimate reliably, and more than one method should be used (Scanlon *et al.*, 2002; Misstear *et al.*, 2009b). The most suitable methods to apply in a given situation will depend on the information required. Whereas in investigating groundwater contamination it may be essential to consider recharge over small areas and short time frames, in well siting we are more often interested in quantifying groundwater resources over larger areas and longer time frames.

Approaches for estimating recharge are described in detail in Lerner *et al.* (1990), Bredenkamp *et al.* (1995), de Vries and Simmers (2002), Scanlon *et al.* (2002), Rushton (2003) and Healy (2010). A few of the approaches are considered briefly here, especially those that make use of readily-available data and hence that are easy to apply in a groundwater investigation for locating wells.

2.6.1   Inflow estimation: direct recharge

There are many different techniques for assessing the potential inflows to an aquifer. Some require the collection of detailed information on soil moisture conditions (zero flux plane approach, Darcy flux calculations in the unsaturated zone, lysimeters) and are outside the scope of a normal groundwater investigation programme for locating well sites. Two of the most widely used inflow approaches for estimating direct recharge involve soil moisture budgets and tracers.

Soil moisture budgets. Soil moisture budgets involve the calculation of soil moisture surpluses and deficits, and hence actual evapotranspiration, from precipitation and potential evapotranspiration

data. Potential evapotranspiration – the maximum evapotranspiration that can theoretically occur, dependent on energy fluxes and temperature – can be calculated from climate data using a variety of methods, but the method recommended by the FAO (1998) employs a form of the Penman-Monteith equation because this has been found to give satisfactory results in most situations.

The concept of *soil moisture deficit* (SMD) envisages the soil as a reservoir for moisture (Figure 2.22). SMD is measured in mm water (or litres m^{-2}). Water is added to the soil reservoir by precipitation (rainfall or melting snow, corrected for surface run-off) and removed from it by actual evaporation. When the reservoir reaches a level of saturation at which it can no longer hold additional water against the force of gravity (SMD = 0), the soil is said to be at field capacity, and it can then release any surplus water to form downward-draining recharge to the aquifer. When the SMD is close to zero, plants can evapotranspire water from the soil efficiently and actual evapotranspiration occurs at a rate corresponding to potential evapotranspiration. When the SMD falls below a certain threshold – the *root constant* of the Penman-Grindley model (Grindley, 1970) or the *readily available water* (RAW) threshold of the FAO (1998) – the actual evapotranspiration occurs at a reduced rate (Figure 2.23). Evapotranspiration ceases altogether when the SMD reaches a second threshold, known as the *wilting point* (Penman-Grindley) or the *total available water* (FAO). These thresholds depend on the rooting depth of the crop and the type of soil. Soil moisture budgets are best made using a daily time step, since longer time steps can lead to an underestimation of potential recharge (Howard and Lloyd, 1979; Rushton, 2003).

A soil moisture budget yields a figure for the moisture surplus. It does not indicate how much of this will give actual recharge or how much will be 'lost' to interflow or to surface runoff (if runoff has not been accounted for already in the budgeting exercise). If the calculations are made for an aquifer that is covered by thin, permeable soils, and where slopes are gentle, then it is reasonable to

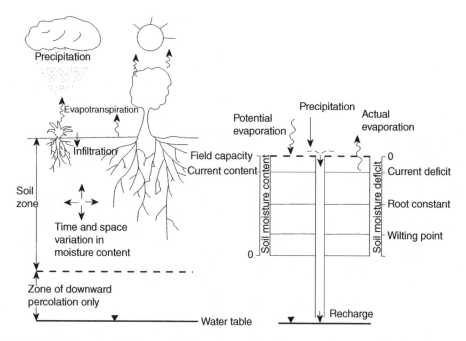

Figure 2.22 *Recharge and soil moisture deficit concepts (from Lerner et al., 1990, published by Heise Verlag). Reproduced by permission of the International Association of Hydrogeologists*

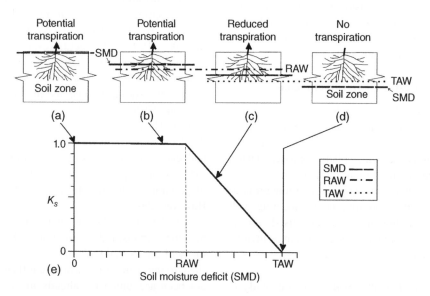

Figure 2.23 *Reducing evapotranspiration due to increasing soil moisture deficit (from Rushton, 2003). The coefficient K_s describes the actual evapotranspiration as a proportion of the potential evapotranspiration. See text for explanation of SMD, RAW and TAW. Reproduced by permission of John Wiley & Sons Ltd*

assume that all or most of the moisture surplus leads to actual recharge (provided that significant interflow is not suspected, or where the aquifer is effectively full and has insufficient storage to take additional recharge). On the other hand, if the aquifer is covered by thick, low permeability superficial deposits, then it will be necessary to multiply the moisture surplus by a factor (the *recharge coefficient*) to allow for surface runoff and other losses. A simple numerical modelling study in southern Ireland by Fitzsimons and Misstear (2006) suggested that the actual recharge occurring to aquifers covered by thick, low permeability tills may range from less than 5% to 30% of the moisture surplus estimated by a soil moisture budget. This approach has been taken further to produce a series of recharge coefficients for different hydrogeological settings in Ireland, leading to the development of a national groundwater recharge map (Misstear *et al.*, 2009a; Hunter Williams *et al.*, 2013).

Hydrogeochemical tracers. Groundwater recharge can be estimated using both environmental (natural) and applied tracers. Lerner *et al.* (1990) separate the methods into *signature* methods and *throughput* methods. What these terms actually mean requires more explanation. Applied tracers are normally only used in the signature methods, whereby a parcel of water containing the tracer is tracked and dated. Throughput methods involve a mass balance of tracer, comparing the concentration in precipitation with the concentration in the unsaturated zone (or sometimes with the concentrations below the water table).

Chloride is probably the most widely used environmental tracer for the throughput method (Edmunds and Gaye, 1994; Healy, 2010). All rainfall contains a certain amount of dissolved chloride, most likely from small amounts of marine salts incorporated from sea spray aerosols or dissolution of windborne dust (or, increasingly, industrial emissions of, for example, HCl gas). The concentration of chloride in rainfall typically decreases with distance from the coast. When rain falls on the vegetation layer or the soil, some of the water will be evaporated or transpired. As chloride is not used to a great extent by plant metabolism, and does not react with the soil or soil moisture, it can be regarded as conservative. The concentration of chloride in the recharge water increases with degree of evapotranspirative concentration. Thus, if we can neglect (or otherwise account for) surface run-off, the quantity of recharge is inversely proportional to the measured concentration of the tracer (chloride) in the unsaturated zone below the zone of active evapotranspiration. The method is particularly effective in arid zones where recharge rates are low and there is significant tracer concentration through evaporation. The simplified relationship between chloride concentration, precipitation and recharge is given (Van Tonder and Xu, 2000) by:

$$w = \frac{(P \times C_P) + D_C}{C_w} \qquad (2.1)$$

where w is the recharge rate (mm a^{-1}), P the precipitation rate (mm a^{-1}), C_P the average chloride concentration in precipitation (mg l^{-1}), C_w the chloride concentration in recharge (mg l^{-1}), and D_c is any dry fallout or deposition of chloride (mg m^{-2} a^{-1}) which is usually assumed to be zero.

To derive a value for C_P it is not sufficient to take a spot sample of rainfall. Chloride concentrations vary with rainfall intensity, season and prevailing wind direction. It is thus necessary to monitor chloride in rainfall over a significant period to derive a weighted average. Provided that rainfall recharge is the main source of chloride to the aquifer, the method *may* also be applied to concentrations of chloride in the saturated zone (Box 2.7). In this case, Van Tonder and Xu (2000) recommend using the harmonic mean concentration of chloride in groundwater from a number of boreholes as the value C_w. Equation (2.1) assumes that precipitation is the only source of chloride to groundwater, that chloride is conservative and that the observed situation reflects a long-term steady state. Errors in the mass balance may arise if there are additional chloride inputs from decaying vegetation, dissolution of palaeoevaporite minerals,

---

Box 2.7   Recharge calculation using chloride concentration in groundwater

An arid region has a mean annual rainfall of 150 mm. The mean chloride concentration of the rainfall is 2 mg l⁻¹. If the mean chloride concentration in the groundwater is 35 mg l⁻¹, what is the annual recharge rate (w)?

We can apply Equation (2.1) to solve this problem if we assume that (a) there is no surface run-off, (b) chloride is conservative (non-reactive) and (c) that the only source of chloride to the groundwater is from direct recharge of rainfall. The recharge rate is thus estimated as:

$$w = \frac{150 \times 2}{35} = 8.6 \, \text{mm a}^{-1}$$

The increase in chloride concentration from rainfall to groundwater is assumed to be solely due to evapotranspirative concentration.

---

localized run-in or from fertilizers, or if chloride is significantly taken up in vegetation.

Tritium (^{3}H), a radioactive hydrogen isotope, was widely used in the 1960s and 1970s as an environmental tracer for dating recharge. Groundwaters rich in tritium were assumed to have been recharged after fusion bomb tests commenced in 1952. However, since 1963, the amounts of tritium in the atmosphere have decreased (due to atmospheric fusion bomb testing being prohibited). Also, owing to radioactive decay (tritium has a half-life of 12.3 years), contrasts between the radioactively decayed post-1952 peaks and more recent, tritium-poor recharge are difficult to discern, and the value of tritium as an environmental tracer is diminishing with time.

More recently, chlorofluorocarbons (CFC's), sulphur hexafluoride (SF_6) and the tritium/helium (^{3}H/^{3}He) ratio have been used as environmental tracers in the saturated zone of unconfined porous-media aquifers (Scanlon *et al.*, 2002; Misstear *et al.*, 2008; Darling *et al.*, 2012). (These gases cannot be used in the unsaturated zone where they may interact with the atmosphere.) Groundwater age increases with depth, at a rate which depends on aquifer geometry, porosity, and recharge rate. CFC concentrations in the atmosphere increased from the 1930s up to the 1990s, but have been declining since then, and so CFC's, like tritium, are becoming less useful as environmental tracers. The ^{3}H/^{3}He ratio does not require measurements of atmospheric tritium and can be used to determine water ages up to 30 years (Healy, 2010).

Here, the discussion of tracers has focused on groundwater recharge assessment. Their use in well pumping tests is considered in Section 7.9.

2.6.2   Inflow estimation: indirect recharge

Indirect recharge from a river to an aquifer occurs where the water level in the river is higher than the hydraulic head or water table in the aquifer with which it is in hydraulic contact. The flux of indirect recharge depends mainly on the head gradient between river and aquifer, the permeability and thickness of the river bed deposits (separating the river from the aquifer) and the thickness and properties of the underlying aquifer. According to Lerner *et al.* (1990) indirect recharge from rivers is the most difficult type of natural recharge to estimate. They recommend the following general approach:

1. Consider how much water can be accepted by the aquifer underlying the river, taking account of the thickness of any unsaturated zone and the speed with which a groundwater mound could dissipate.
2. Estimate the transmission capacity of the unsaturated zone, paying particular attention to any low permeability layers that might lead to a perched water table.
3. Consider the river flow and river bed processes, leading to a water balance for the channel system.

Leakage through a river bed (or other surface water body) can be measured using seepage meters. Although these are relatively simple and cheap to use (they may comprise a water filled flexible bag attached to a tube or cylinder emplaced in the river bed), they provide point data and so measurements may be required at many locations to obtain representative values for the whole catchment (Scanlon *et al.*, 2002).

Alternatively, indirect recharge can be estimated by gauging the flow of the river at various points along its length. The indirect recharge is then calculated from the river flow loss per unit length, corrected for any evaporation and abstraction losses.

2.6.3 Aquifer response analysis

The response of the aquifer to recharge can be investigated by examining seasonal fluctuations of water level, and by considering the steady-state groundwater throughput under average water level conditions. Recharge can be estimated quantitatively from water level fluctuations using a relationship of the following form (Kruseman, 1997):

$$R = \left(\Delta h \times S_y \right) + Q_a + \left(Q_{out} - Q_{in} \right) \qquad (2.2)$$

In this water balance equation, R is the recharge, Δh the change in water table elevation, S_y the

specific yield, Q_a the groundwater abstraction during the period under consideration, and Q_{out} and Q_{in} are any other lateral subsurface outflows and inflows during the same period. There is an inverse relationship between Δh and S_y: for a given recharge event, a large head change indicates a low S_y. Bredenkamp *et al.* (1995) describe a number of methods for taking account of $\left(Q_{out} - Q_{in} \right)$, involving analysis of hydrograph recession when no recharge is occurring. Scanlon *et al.* (2002) and Healy (2010) point out that aquifer response analysis is best applied over short time periods in regions having shallow water tables that display sharp rises and declines in water levels. The main difficulty in applying aquifer response analysis is in determining a representative value for specific yield, especially in fractured consolidated or crystalline aquifers. The method is easiest to apply in relatively homogeneous unconsolidated aquifers. Figure 2.24 shows a well hydrograph for a sand and gravel

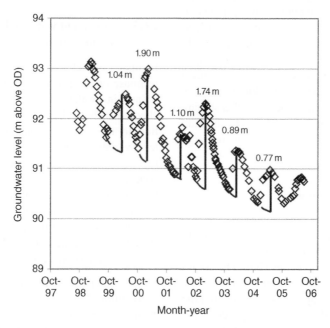

Figure 2.24   *Estimating recharge using a groundwater level hydrograph. The broken lines are the extrapolated groundwater recessions. With kind permission from Springer Science + Business Media: Hydrogeology Journal, 'Estimation of groundwater recharge in a major sand and gravel aquifer in Ireland using multiple approaches', 17, 2009b, 693–706, Misstear BDR, Brown L and Johnston PM*

aquifer in Ireland. The recharge analysis was carried out on an annual basis since there were insufficient data available to determine the water level responses to individual rainfall events. An annual approach produces values of *net recharge* (Healy and Cook, 2002), but these should be close to the gross recharge amounts provided that $(Q_{out} - Q_{in})$ is small.

A simple method for estimating recharge is to calculate the aquifer throughput under steady-state conditions using Darcy's equation (1.3). The hydraulic gradient used should be based on water levels measured in different boreholes at the same time of year. If there are significant seasonal fluctuations in water levels resulting in significant variations in the hydraulic gradient (such as might occur in a shallow, fracture-flow aquifer system), then the calculation can be repeated for different conditions to estimate the approximate seasonal variations in throughput.

This simple method assumes that we know something about the likely location of the recharge: whether all the recharge input is at the up-gradient, unconfined exposure of an otherwise confined aquifer, or whether the recharge occurs uniformly over the entire exposed area of an unconfined aquifer. The Darcy calculations apply to a confined aquifer where the saturated thickness remains constant. For an unconfined aquifer with a sloping water table, the Dupuit-Forchheimer equation (1.14) can be modified to include recharge (Box 2.8).

2.6.4   Outflow estimation

This normally involves the separation of the baseflow component from runoff at suitably located surface flow gauging stations. Over a long period, the aquifer outflow should be equivalent to the inflow, after any abstractions – or other losses such as evapotranspiration from the water table or underflow to other aquifers – are taken into account. The proportion of baseflow to total runoff is often referred to as the baseflow index (Institute of Hydrology, 1980; Shaw *et al.*, 2011).

Traditionally, baseflow separation has been done manually using a variety of techniques, relying heavily on the experience of the hydrologist or hydrogeologist concerned. However, automated techniques have become more practicable with the wide availability of desktop and laptop computers (Institute of Hydrology, 1989; Nathan and MacMahon, 1990; O'Brien *et al.*, 2014). Automated techniques have the advantage that they can use much longer periods of data than manual methods. Whatever baseflow separation method is employed, whether it is manual or automated, it is important to ensure that the results are credible by comparing the baseflows with observations of groundwater levels in the aquifer discharge area (Misstear and Fitzsimons, 2007).

The usefulness of baseflow analysis for estimating recharge is very much dependent on the frequency – and reliability – of gauging stations in relation to the size of the aquifer unit. In many countries the gauge density is insufficient to be able to assign the baseflow estimates to individual, small aquifer units.

2.6.5   Catchment water balance and modelling

Estimates of recharge derived from inflow, aquifer response or outflow approaches should be incorporated in a water balance for the aquifer system to check that the recharge figure is sensible in relation to the observed meteorology, groundwater level variations, abstractions and streamflows. The recharge estimate should be consistent with the conceptual model of the groundwater system: if they are not consistent, then either the recharge estimate or the conceptual model may need to be changed – or both of these. In conceptualizing the system, the investigator should try and answer a number of questions, for example:

- What happens to the recharge under existing conditions?
- Are there additional unaccounted outflows, such as lateral outflows to other aquifers, leakage to an adjacent aquifer via an aquitard or ungauged underflows (beneath a river) out of the system?

Box 2.8   Calculation of recharge using Dupuit-Forchheimer principles

For the situation depicted in Figure B2.8(i), the flow (q) per unit width through the full vertical section of aquifer according to the Darcy equation and the Dupuit assumptions is given by Equation (1.12):

$$q = -KH\frac{dH}{dx}$$

where K is the hydraulic conductivity, H the elevation of the water table relative to the base of the aquifer and $\dfrac{dH}{dx}$ is the hydraulic gradient. By continuity, q should also equal the recharge rate, w, over the distance x:

$$q = wx$$

Equating these two equations, and rearranging, we obtain:

$$wxdx = -KHdH$$

Integrating for the boundary conditions $H=H_o$ at $x=0$ (the groundwater divide), and $H=H_L$ at the river where $x=L$, then:

$$H_0{}^2 = H_L{}^2 + \frac{wL^2}{K} \quad \text{or} \quad w = \frac{K\left(H_0^2 - H_L^2\right)}{L^2}$$

Thus, knowing the head values, we can estimate the recharge rate. As an example, let the distance from the groundwater divide to the river bank equal 2 km, the elevation of the water table at the divide $(H_0) = 7.5\,\text{m}$, the elevation of the water table at the river bank $(H_L) = 1.6\,\text{m}$ and the hydraulic conductivity (K) of the sand and gravel aquifer $= 25\,\text{m d}^{-1}$. Then:

$$w = \frac{25 \times \left(7.5^2 - 1.6^2\right)}{2000^2} = 0.34\,\text{mm d}^{-1}$$

This is equivalent to an annual recharge of 122 mm. Such an analysis ignores the seepage face at the river, and assumes a uniform recharge rate. Nevertheless, it can be useful for making comparisons with river baseflow estimates (see Section 2.6.4).

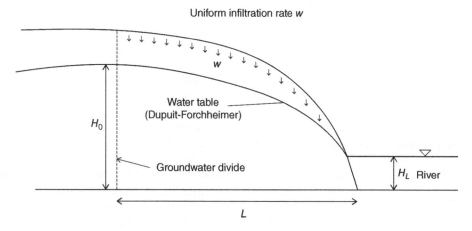

Figure B2.8(i)   *Unconfined groundwater flow, with recharge*

- Does the level of long-term groundwater abstraction suggest additional unaccounted inflows, for example: recharge from leaking water mains or sewers, or from irrigation return flows; subsurface inflows from neighbouring catchments; or contributions from an underlying aquifer via an intermediate leaky layer?

The water balance for the area of interest can be expressed as:

$$P + Q_{in} = AE + Q_{out} \pm \Delta\sigma \qquad (2.3)$$

where P is the precipitation, Q_{in} the inflows from surface water (rivers, lakes, irrigation, leaking pipes, etc) and groundwater, AE the actual evapotranspiration, Q_{out} the outflows from surface runoff, groundwater discharge, any surface-water or groundwater abstractions, and $\Delta\sigma$ the change in water stored in the soil, unsaturated zone, any surface reservoirs and the aquifer. The relationship can be simplified if the water balance is performed for a surface water catchment where $Q_{in} = 0$. It is also usual to perform the water balance for a sufficiently long period so that short-term changes in water storage can be discounted.

We can also apply a similar water balance to a given aquifer unit where:

$$w + Q_i = Q_o + Q_{dis} + Q_{abs} \pm \Delta\sigma_g \qquad (2.4)$$

where, here, w is recharge to the aquifer, Q_i is any other lateral or vertical inflows from adjacent aquifer units, Q_{abs} is groundwater abstraction, Q_{dis} is discharge of groundwater to the surface as springs or river baseflow, Q_o is any other outflows to adjacent aquifer units (or direct evaporation from the water table) and $\Delta\sigma_g$ is the change in groundwater stored in the aquifer.

The water balance enables an estimate to be made of the potential 'groundwater surplus' that may be available for new production wells. The normal aim is to achieve sustainable management of aquifer units, such that $\Delta\sigma_g = 0$ in the long term (although, occasionally, "groundwater mining" or long term depletion of storage *may* be justified by overriding economic or developmental considerations). Traditionally, the available 'surplus' was often taken to be the difference between total recharge ($w + Q_i$) and existing groundwater abstractions; in other words the available surplus was ($Q_o + Q_{dis}$). The environmental benefits of groundwater discharges (Q_{dis}) to rivers and wetlands were often ignored or, at least, were not given a high priority when planning new groundwater schemes. It is now increasingly common practice for regulators to assign minimum flows that must be maintained at springs, rivers and wetlands. In Britain, for example, it is not uncommon for the 'available groundwater resource' of an aquifer to be set at only 30 to 50% of the total long-term recharge. In Chile, Muñoz and Fernández (2001) recommended that sustainable abstraction should not exceed some 37% of the total recharge of the Santiago Valley aquifer.

It is important to appreciate that the component elements of the water balance equation are not fixed or independent quantities. For example, the recharge w may depend on the water level in the aquifer (in turn related to $\Delta\sigma_g$). As the water level drops, new recharge may be induced (w increases), or Q_{dis} or Q_o may decrease. Thus, despite its seeming simplicity, such a water balance equation may be difficult to use directly in aquifer management. Some form of numerical model will often be beneficial to conceptualise, investigate and quantify the relationships between the various component parameters (Anderson and Woessner, 1992; Rushton, 2003; Healy, 2010). In cases where the boundary conditions are well known, where there are good data on aquifer characteristics, outflows and long-term water level variations, then distributed numerical models are very useful tools for testing the sensitivity of abstraction effects to variations in recharge. Inverse modelling techniques are increasingly used in evaluating groundwater recharge. During inverse modelling exercises, recharge and hydraulic conductivity values are usually estimated together; unique estimates of recharge are then obtained using groundwater fluxes such as baseflow measurements in rivers (Sanford, 2002). In situations where data on aquifer properties are scarce, lumped rather than distributed models may be more appropriate for investigating the components of the water balance, including groundwater recharge (O'Brien *et al.*, 2013).

2.7 Groundwater quality

2.7.1 Introduction

The detailed investigation phase should include an assessment of the existing groundwater quality in order to:

- establish that the water quality is suitable for its intended use;
- identify potential corrosion or incrustation problems that need to be taken into account in the design, operation and maintenance of the well (Chapter 4 and Chapter 9);
- provide information on the groundwater flow system.

The broad aim of the investigation should be to establish the distribution of groundwater quality across the aquifer, both horizontally and with depth. Seasonal trends, longer-term natural trends or likely trends in water quality induced by pumping may also need to be investigated. The scope of the sampling exercise will be influenced by the quantity of data already available from the desk study and by the nature of the groundwater development scheme under consideration. Whereas the design of a well field intended for a town water supply will inevitably require a comprehensive assessment of groundwater quality as part of the detailed investigation phase, a well scheme for construction dewatering, for example, may involve only a limited chemical sampling programme to check on the likely corrosiveness of the water on the pumps and pipe work, and to satisfy the needs of the discharge consent.

Groundwater quality is a function of the natural characteristics of the water together with any changes that occur as a result of human or animal activities. It can be assessed in terms of its main physical, chemical and biological characteristics. Since an appreciation of groundwater chemistry is relevant to most groundwater investigations, a short introduction to the *natural chemical composition* of groundwater and its quality with respect to potable and irrigation uses is included in the following paragraphs, while Section 2.8 outlines an approach for assessing the risk of pollution at proposed new well sites. Procedures for the collection of water samples for *microbiological and chemical pollutants* (including trace organics), as well as for the chemical substances that occur naturally in groundwaters, are dealt with in Chapter 8. That microbial pollutants are described in Chapter 8, rather than here, should not imply that they are in any way less important a feature of groundwater quality than groundwater chemistry; it is just that microbial analyses tend to be part of the testing of a *completed* well, rather than part of a well siting exercise. The preferred strategy for dealing with microbiological and other contaminants is to prevent their entry into a well by choosing the well site carefully and by constructing the well properly.

2.7.2 Chemical composition of groundwater

The natural chemical composition of a water evolves during its passage from the atmosphere (rainfall), through the soil zone (recharge) to the saturated aquifer. It will be affected by a multitude of processes: physical, chemical, biological and microbiological.

Rainfall is normally mildly acidic (due to dissolution of carbon dioxide to produce a weak carbonic acid solution), but may be significantly acidic (with pH less than 2) in areas of high industrial emissions of nitrogen oxides (NO_x), hydrochloric acid (HCl) or sulphur dioxide (SO_2). It contains a weak solution of natural salts derived from marine sea-spray aerosols, windblown dust or atmospheric processes. In temperate, maritime areas, the dissolved salts in rainfall are often dominated by sodium and chloride (from marine aerosols). Anthropogenic emissions may provide rainfall with an enhanced content of sulphate, chloride or nitrate (from SO_2, HCl or NO_x), or even an enhanced content of base cations if soluble particulate oxides are emitted by power stations or cement works.

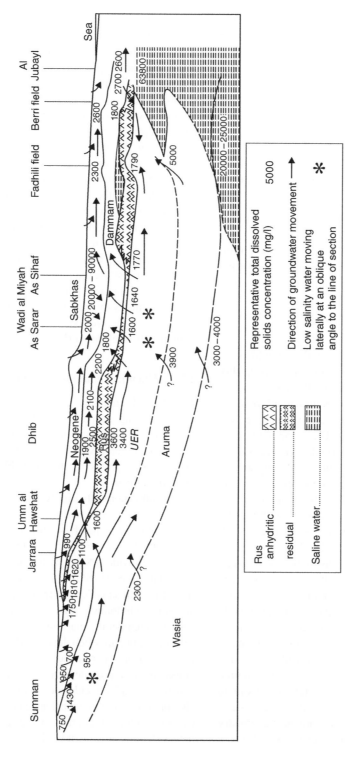

Figure 2.25 *Hydrogeological section of the Umm er Radhuma aquifer system, Saudi Arabia, showing highly mineralized groundwaters in the eastern (down dip parts) of the aquifer system, as well as in some of the shallow evaporative discharge areas (sabkhas). Based on a draft by David Ball and published in Bakiewicz et al. (1982), reproduced by permission of the Geological Society of London*

When water percolates into the soil zone, it is subject to evapotranspirative concentration of most solutes. Some solutes (potassium, nitrate) may decrease as they are consumed by plants. Microbiological respiration results in an increased dissolved carbon dioxide (CO_2) loading in potential recharge water and lowered pH. The recharge water which percolates down through the unsaturated zone to the water table is thus an acidic, typically oxygenated, dilute solution of various solutes.

This acidic, oxidizing recharge water then undergoes what may be termed 'water-rock interaction' with the minerals of the aquifer matrix (which are dominantly basic and reducing) to result in the groundwater's characteristic hydrochemical fingerprint. The most influential water-rock interaction processes are dissolution/precipitation reactions (e.g., fluorite dissolution), acid-base reactions (such as carbonate and silicate hydrolysis), ion exchange reactions and redox reactions (e.g., pyrite oxidation or sulphate reduction). Different reactions take place at different rates and water chemistry can thus be a valuable indicator of groundwater residence time and vulnerability.

Recent groundwaters are very often (but not always) dominated by calcium (and magnesium) bicarbonate, largely due to the kinetically-rapid dissolution of any calcite or dolomite mineralization present. More mature waters tend to become dominated by sodium and bicarbonate, due to a combination of ion exchange, calcite saturation and precipitation and continued weathering of other silicate minerals such as feldspars (Frengstad and Banks, 2000). Ancient, 'stagnant' groundwater is often highly mineralized and dominated by sodium and chloride (occasionally calcium chloride) and, to a lesser extent, sulphate ions.

In aquifers in humid temperate regions where there is significant recharge, groundwater throughflow rates may be relatively quick and residence times short, so that the groundwater does not become highly mineralized. In contrast, aquifers in arid areas with low recharge rates, long flow paths and long residence times can contain very old groundwaters that have significantly higher dissolved solids concentration than sea water. This is illustrated by the hydrogeological cross-section of the major Umm er Radhuma aquifer system in Saudi Arabia (Figure 2.25).

The main chemical constituents of (unpolluted) groundwater are (a) the major cations: calcium, magnesium, sodium and potassium; (b) the major anions: bicarbonate, sulphate and chloride; and (c) some more "minor" elements such as silicon (Table 2.3). Nitrate is usually included as a major

Table 2.3 *Major and trace elements and ionic species in groundwater*

Typical concentration ranges	Elements or ionic species
Major elements/ionic species	
$>100\,mg\,l^{-1}$	Bicarbonate (note: $1\,meq\,l^{-1}$ alkalinity $= 61\,mg\,l^{-1}$ bicarbonate)
$10–100\,mg\,l^{-1}$	Sodium, calcium, sulphate, chloride, nitrate
$1–10\,mg\,l^{-1}$	Magnesium, potassium, silicon
Trace elements/ionic species	
$0.1–1.0\,mg\,l^{-1}$	Strontium, fluoride
$0.01–0.1\,mg\,l^{-1}$	Phosphorus, boron, bromide, iron, zinc
$0.001–0.01\,mg\,l^{-1}$	Lithium, barium, copper, manganese, uranium, iodine
$0.0001–0.001\,mg\,l^{-1}$	Rubidium, lanthanum, vanadium, selenium, arsenic, cadmium, cobalt, nickel, chromium, lead, aluminium, yttrium

The concentration ranges given in the table are typical values for natural groundwaters, but much higher or lower concentrations can occur in certain situations.

Adapted from Edmunds and Smedley (1996). Reproduced by permission of the Geological Society of London.

anion as it can occur as a significant natural component of groundwater in some temperate agricultural areas and in some arid areas with high evapoconcentration factors (Banks, 2014c). Together, these nine constituents typically comprise some 99% of the total solute content of most groundwaters (Edmunds and Smedley, 1996). The remaining 1% consists of minor constituents, including trace metallic elements. These can be extremely important with respect to the suitability of the water for its intended use (see below).

2.7.3   Groundwater for potable supply

The World Health Organization 2011 guideline values for drinking water are included in Appendix 4, with the naturally occurring chemicals that are of health significance being listed in Table A4.2. It should be remembered, especially in the context of providing emergency water supplies in a disaster situation, that the health risks associated with the chemicals listed in Table A4.2 are not just a function of their concentrations in drinking water, but are also influenced by the water consumption, diet, age and susceptibility to disease of the people affected. Thus, prolonged exposure of children and old people with poor diets to drinking water with high contaminant levels is likely to have much more serious consequences than short-term exposure to contaminated drinking water would have for healthy adults. Indeed, the World Health Organization (2011) states that 'most chemicals arising in drinking-water are of health concern only after extended exposure of years, rather than months.'

Three of the most significant inorganic constituents of groundwater in terms of health are nitrate, arsenic and fluoride, and these parameters are discussed below. The presence of the radioactive elements uranium, radium and radon in groundwater is considered in Box 2.9, while the trace elements beryllium and thallium are described in Box 2.10.

Nitrate. High nitrate concentrations in groundwater may be the result of pollution from landfills, latrines, sewage and intensive agriculture (organic and inorganic fertilizers, deep ploughing, especially

of virgin grassland). However, nitrate also occurs naturally in recharge water, mainly as a result of biological fixation of nitrogen in the soil zone. High levels of evapoconcentration in arid climates can increase nitrate concentrations significantly. Nitrate is one of the few examples of a chemical substance that can lead to health problems as a result of relatively short-term exposure (World Health Organization, 2011). The main health concern about nitrate relates to the condition in very young children known as *infantile methaemoglobinaemia*, or blue baby syndrome. This arises where nitrate (NO_3^-) in drinking water is reduced to nitrite (NO_2^-) in the stomach. High concentrations of nitrite then bind to haemoglobin in the blood and hinder the effective uptake of oxygen. The risk of methaemoglobinaemia increases where the child is also suffering from a gastrointestinal infection, and therefore reported cases of methaemoglobinaemia are often associated with private well supplies which can be susceptible to microbial contamination (Knobeloch *et al.*, 2000; Fewtrell, 2004; World Health Organization, 2011). The World Health Organization (2011) guideline values for nitrate and nitrite in drinking water are $50\,mg\,l^{-1}$ (or $11.3\,mg\,l^{-1}$ as N) and $3\,mg\,l^{-1}$ (or $0.9\,mg\,l^{-1}$ as N), respectively.

Large scale treatment of nitrate is expensive and complex, and typically involves reverse osmosis. Anion exchange may also be employed, while methods involving reductive denitrification by zero-valent iron have also been trialled.

Arsenic. Arsenic (As) in groundwater supplies has become a major health issue in recent years, notably in respect of the large populations exposed to arsenic contamination in parts of south and east Asia. It is estimated that 60 million people are at risk from high arsenic levels in groundwater in this region, of whom about 40 million live in Bangladesh and West Bengal (World Bank, 2005). One estimate puts the annual number of arsenic-related deaths at 43 000 in Bangladesh alone (Flanagan *et al.*, 2012). Other countries in this region with large exposed populations include China (Inner Mongolia, Xinjiang and Shanxi),

Box 2.9   Uranium, radium and radon in groundwater

U, Ra and Rn are all radioactive elements that emit α-particles on decay. They all occur naturally in groundwater. Uranium occurs as a number of isotopes, the most abundant of which is ^{238}U. This decays, via intermediary nuclides, to ^{226}Ra, which in turn decays to the most common radon isotope (^{222}Rn). As all are part of the ^{238}U decay chain, one might expect some degree of co-variation in groundwater. However, the three elements have very different physical and chemical properties and their hydrogeochemical distribution is far from identical.

Uranium is poorly soluble under reducing conditions, but is highly soluble under oxidizing conditions, occurring primarily as uranyl species, such as UO_2^{2+} (Frengstad and Banks, 2014). It tends to occur at high concentrations especially in oxic, unweathered acidic crystalline rock aquifers such as gneisses and granites [Figure B2.9(i)], but can also occur in U-rich sandstones and other sediments. Concentrations

of $14\,mg\,l^{-1}$ have been recorded in groundwater in Finnish granites (Asikainen and Kahlos, 1979), while a median concentration of $3\,\mu g\,l^{-1}$ was found in a set of 476 Norwegian crystalline bedrock groundwaters (Frengstad *et al.*, 2000). ^{238}U, the dominant natural isotope, has a very long half-life of 4.5×10^9 years and is thus not highly radioactive. Many authorities believe its chemical toxicity (primarily affecting the kidneys) is of greater concern than its radiotoxicity (Milvy and Cothern, 1990). World Health Organization (2011) suggests a provisional guideline value of $30\,\mu g\,l^{-1}$ for drinking water. Uranium in drinking water may be treated by coagulation processes, reverse osmosis and by ion exchange.

Radium generally has a higher radioactivity than uranium (^{226}Ra has a half-life of 1620 years). It is chemically analogous to barium, occurring dominantly as a bivalent cation in water. Its sulphate (like barite) is highly insoluble, and

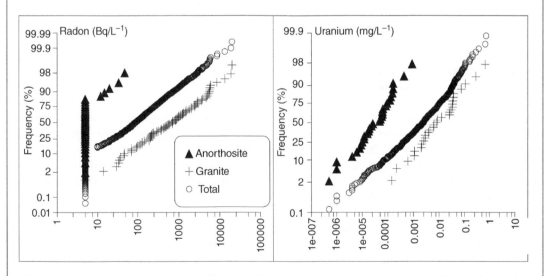

Figure B2.9(i)   *Cumulative frequency diagrams showing distribution of radon and uranium in Norwegian bedrock groundwaters. The circles show the total data set, the crosses the Precambrian granites (high risk lithology) and the triangles the Precambrian anorthosites (low risk)*

co-precipitation with barite in sulphate-rich waters is observed in some mining and oilfield situations. Hence, radium typically occurs at elevated concentrations in sulphate-poor waters; for example, in reducing waters where sulphate-reduction has depleted sulphate concentrations.

The most common *radon* isotope, ^{222}Rn, has a half-life of 3.8 days, and is formed by the decay of ^{226}Ra. It is an inert gas and its occurrence in water depends on the occurrence of uranium (or, more strictly, its immediate parent, radium) in the host rock, and on hydrodynamic factors, rather than on the water's hydrogeochemistry. Radon typically occurs at greatest concentrations in poorly weathered acidic crystalline aquifers such as gneiss and granite [Figure B2.9(i)]. Concentrations of up to 77 500 Bq l^{-1} have been measured in groundwater in Finland (Salonen, 1994). Radon is highly soluble, but this solubility decreases with increasing temperature. Radon is believed to pose a risk of gastric cancer via direct ingestion (Mose *et al.*, 1990), but also a risk of lung cancer if it degasses from household water and is inhaled. In the Scandinavian environment, ingestion is reckoned to be the most important exposure mechanism in infants (Statens strålskyddsinstitutt, 1996) and inhalation in adults. The Norwegian Radiation Protection Authority recommends an action level of 500 Bq l^{-1} for potable water, a value that is exceeded by 14% of Norwegian crystalline bedrock wells (Banks *et al.*, 1998). Radon can be most effectively removed from potable water by aeration and/or storage prior to household entry. A number of compact aeration units and meshes are available for this purpose. More generally, the World Health Organization (2011) notes that the dose from radon in drinking water is more likely to be received by inhalation than by ingestion, and hence that it is more appropriate to measure radon in air than in drinking water. The World Health Organization reference level for radon in homes is 100 Bq m^{-3}. The European Union has recently (European Commission, 2013) recommended in its directive 2013/51/EURATOM that some form of remedial action should be justified if radon in potable water exceeds 1,000 Bq l^{-1}, and that possible risks should be evaluated in it exceeds 100 Bq l^{-1}.

Vietnam (Red River delta) and Myanmar. Elsewhere in the world, high levels of arsenic have been found in aquifers in the Bolivian Altiplano, the Chaco-Pampean plain in Argentina, the Lagunera district of Mexico and western United States, to name but a few major examples (BGS and DPHE, 2001; Banks *et al.*, 2004). Arsenic is carcinogenic and affects especially the skin, lungs, kidney and bladder.

A naturally occurring metalloid, arsenic occurs geologically in sulphide and arsenide minerals, in metal arsenites or arsenates and absorbed as an accessory element on ferric oxides or oxyhydroxides. It has a complex geochemistry and can occur in groundwater in both reducing and oxidizing conditions, at low and high pH. It is mainly present in two dissolved ionic forms, as arsenite ($As^{III}O_3^{3-}$) and arsenate ($As^{V}O_4^{3-}$).

Several geological environments can be classified as having a risk of elevated arsenic in the hydrosphere:

1. Areas, such as major deltas, where sediments are accumulating and being rapidly buried. Burial eventually results in reducing conditions, and the reductive dissolution of ferric oxyhydroxides to soluble ferrous iron. This dissolution also releases any arsenic that was adsorbed to the ferric oxyhydroxides.
2. Areas rich in sulphide or arsenide mineralization (which may oxidize to release dissolved sulphide, acid and a range of chalcophile metals and metalloids).
3. Some volcanic areas and areas of thermal springs.
4. Areas of fumarolic native sulphur deposits.

Box 2.10 Beryllium and thallium in groundwater

Historically, there has been a tendency for drinking water regulations to ignore elements that occur in such small quantities that they are not readily analysable. Two such elements that occur at nanogram per litre concentrations in most groundwaters are beryllium and thallium. It is only with the advent of inductively coupled plasma mass spectrometry (ICP-MS) techniques that we have been able to analyse for such elements in water on a routine basis. In fact, both these elements are rather toxic.

Long-term exposure to beryllium can cause intestinal lesions, and damage bone and lung tissue. There is also a suspicion of carcinogenicity. It occurs geologically in basic rocks, often in connection with feldspars and pegmatites. Although the World Health Organization has not established a guideline value for beryllium because 'it is rarely found in drinking-water at concentrations of health concern', the United States now operates with a maximum permitted concentration of $4 \mu g\,l^{-1}$ in drinking water.

Interestingly, the former Soviet Union and now Russia operate a very low drinking water limit of $0.2 \mu g\,l^{-1}$ (Kirjukhin *et al.*, 1993, Minzdrav, 2007). Concentrations of up to $6.6 \mu g\,l^{-1}$ were found in a survey of Norwegian crystalline bedrock groundwater (n=476; Frengstad *et al.*, 2000), with a median of $0.012 \mu g\,l^{-1}$. The highest values were found in granitic lithologies.

Thallium is used as a rat poison and human exposure can result in a range of damage to the nervous system, blood, liver, kidneys, testes and hair. Thallium is associated with ores of copper, zinc, cadmium and gold, and also occurs in other rock types, especially in potassium and rubidium minerals. The United States has a maximum permitted concentration of $2 \mu g\,l^{-1}$, but recommends that thallium should not exceed $0.5 \mu g\,l^{-1}$ in drinking water. Again, the Soviet/Russian drinking water limit is set at a very low level of $0.1 \mu g\,l^{-1}$. Concentrations of up to $0.25 \mu g\,l^{-1}$ were found in a survey of Norwegian crystalline bedrock groundwater (n=476), with a median of $0.007 \mu g\,l^{-1}$.

Major deltas and alluvial plains composed of young sediments are particularly susceptible to groundwater arsenic problems (World Bank, 2005; Ravenscroft et al., 2009). The Ganges delta of Western Bengal and Bangladesh is a rapidly accumulating sedimentary environment, where it is believed that arsenic is largely released by reductive dissolution. Arsenic occurrence here may be vertically stratified, depending on the redox conditions in the aquifer: for example, an oxidative zone of low As may overlie a zone of reductive dissolution (high As), in turn overlying another low As zone. Some 7 to 11 million new boreholes had been sunk over the course of 30 years before the arsenic issue was finally identified in the early 1990s. Of these boreholes, nearly half exceeded the current World Health Organization provisional guideline value for As of $10 \mu g\,l^{-1}$ $(0.01\,mg\,l^{-1})$ (BGS and DPHE, 2001; Reimann and Banks, 2004). The number of shallow private wells has continued to increase into the present century, despite the arsenic hazard (World Bank, 2005).

Possible treatment methods for arsenic include (i) reverse osmosis, (ii) anion exchange (arsenate only), (iii) sorption onto activated alumina or (iv) oxidation to As^{V} followed by coagulation/precipitation/filtration using aluminium or ferric salts (Johnston and Heijnen, 2001). Small scale, low cost treatment technologies focus on the latter (oxidation/filtration) technique. One method involves filtration of water through iron-oxide coated sand, while another involves storage of water in transparent flasks in sunlight to promote oxidation of arsenite to arsenate (Ahmed, 2001). An alternative strategy to water treatment in Bangladesh is to replace shallow wells contaminated with arsenic by deeper wells. Arsenic concentrations are highest in the top 50 m of the aquifer sequence and reduce appreciably at depths greater than 100 m. Below 200 m, the chance of constructing an 'arsenic-safe' well in Bangladesh is around 99% according to Ravenscroft *et al.* (2005).

Fluoride. Fluoride (F^-) is an ion which, like many other elements, can have beneficial effects for human health at modest concentrations, but can be toxic at higher levels. It is considered by many to be beneficial to health, conveying resistance to dental caries (especially in children), when it is present in drinking water at 'optimum' concentrations of $0.5–1 mg l^{-1}$. However, fluoride can give rise to mild forms of dental fluorosis at concentrations between 0.9 and $1.2 mg l^{-1}$ (World Health Organization, 2011). Skeletal fluorosis can occur at concentrations of $3–6 mg l^{-1}$, with 'crippling skeletal fluorosis' usually only developing when fluoride concentrations in drinking water exceed $10 mg l^{-1}$ (World Health Organization, 2011). The current World Health Organization guideline is set at $1.5 mg l^{-1}$ (Table A4.2 in Appendix 4); however, the WHO (2011) notes that where the total fluoride intake from water and other sources is likely to be more than $6 mg day^{-1}$, a lower standard should be considered.

Fluoride is present in fluorite (CaF_2), and it also commonly occurs in apatite and adsorbed on anion-exchange sites in other common minerals of crystalline (including volcanic) and sedimentary rocks, such as sheet silicates (e.g., micas) and amphiboles. High concentrations occur in groundwater in some volcanic rocks (those of the East African rift being particularly notorious), and many granites and gneisses. High concentrations in groundwater can occur in other rock types, however, and are often associated with slow flow rates and hydrochemical 'maturity', with high pH and alkalinity, sodium bicarbonate chemistry and low calcium concentrations. The calcium content is important because it acts as a control on the solubility of the mineral fluorite. At high calcium concentrations fluorite dissolution does not occur; indeed fluorite may be precipitated, removing fluoride from the groundwater. High pH may also

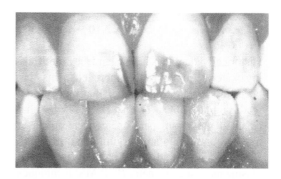

Figure 2.26 *Dental fluorosis in an adolescent boy in western Norway. The drinking water contained some 7 mg l⁻¹ fluoride. Reproduced by permission of Asgeir Bårdsen, University of Bergen*

be an important factor, as OH^- may displace F^- from anion exchange sites on some minerals. Cases of high fluoride concentrations in groundwater include those reported from southern India (Handa, 1975), western and southern Norway (Banks *et al.*, 1998), County Monaghan in Ireland (Misstear *et al.*, 2008) and the rift valley of east Africa (Gaciri and Davies, 1993). Figure 2.26 is a photo of mottled teeth in a 13-year-old boy who was exposed to high fluoride levels in drinking water in western Norway.

Fluoride can be removed from drinking water by treatment methods involving sorption to activated alumina, reverse osmosis or anion exchange. Small scale, low cost domestic treatment methods that have been trialled include sorption onto charred fish bone, onto ceramic fragments made of certain types of laterite and onto fluoride-poor apatite. Large-scale treatment for fluoride is typically accomplished by some form of coagulation and co-precipitation using aluminium salts (e.g., the so-called Nalgonda process: Nawlakhe *et al.*, 1975; Poulsen, 1996).

Other Parameters. Inorganic substances that are not harmful to health (at least in modest concentrations), but which may cause problems in terms of the appearance or taste they give to the groundwater supply, include total dissolved solids and chloride (taste), hardness (scale formation, poor soap-foaming abilities) and iron and manganese (appearance and staining). High dissolved solids content can occur in many hydrogeological environments but is especially common in arid regions (due to evaporative concentration) and in coastal aquifers (presence of fossil or modern seawater). Water hardness (specifically carbonate hardness) is a characteristic feature of most unconfined limestone and other carbonate aquifers, while non-carbonate hardness is typical of sulphate-bearing formations. Iron (Fe) problems are most often associated with weathering of pyrite under oxidizing conditions or reductive dissolution of other iron-bearing minerals (e.g., ferric oxides) under oxygen-poor conditions, and affect a wide range of aquifers. Groundwaters with high iron may often also contain high manganese (Mn) concentrations. Like iron, manganese may affect the acceptability of the drinking water supply: both dissolved iron and manganese have a tendency to oxidize on contact with air, producing discoloured orange (Fe) or black (Mn) precipitates.

2.7.4   Groundwater for irrigation

The most important characteristics of water that affect its suitability for irrigation are:

- the total dissolved solids content;
- the amount of sodium, relative to other base cations;
- the concentrations of elements that are toxic to plants (such as, for example, boron).

The salinity of soil water is usually several times greater than the salinity of the applied irrigation water. This increase is due to evapotranspiration. High soil water salinity is detrimental to crop performance because it causes a reduction in the ability of the crop to extract the water necessary for growth. The effects vary for different types of crop (cereals tend to more salt tolerant than vegetables, for example, whilst palms and certain fodder crops are relatively salt resistant), and are also influenced by the soil and drainage conditions – soil water salinity will continue to increase if there is inadequate subsoil drainage to allow sufficient

flushing of the root zone. In certain arid regions, the combination of intensive irrigation, high evapotranspiration rates and poor drainage has led to such extensive salinization of the soil that the land has had to be abandoned (Ghassemi *et al.*, 1995).

Salinity also has an impact on the infiltration properties of the soil, since low salinity water is corrosive and can lead to leaching of soluble salts from the soil, thus reducing its stability. The infiltration properties of the soil are also affected by the relative amount of sodium compared to other base cations in the applied irrigation water. High sorption of sodium by the soil may result in a breakdown of the soil structure, and hence to a reduction in infiltration capacity. The criterion most commonly used to define the relative amount of sodium is the sodium adsorption ratio (SAR), also known as the RNa (Ayers and Westcot, 1985):

$$SAR = \frac{Na}{\sqrt{\dfrac{Ca + Mg}{2}}} \qquad (2.5)$$

where the concentrations of the sodium (Na^+), calcium (Ca^{2+}) and magnesium (Mg^{2+}) ions are expressed in milliequivalents per litre (meq l^{-1}). A high SAR indicates a tendency for sodium in the irrigation water to replace the calcium and magnesium adsorbed onto the soil. The SAR criterion can be modified to take account of changes in calcium concentration *in the soil water* that may occur as a result of precipitation or dissolution reactions following an irrigation. This modification is known as the adj RNa (Suarez, 1981) but the FAO guidelines on water quality for agriculture indicate that the older SAR criterion is still acceptable (Ayers and Westcot, 1985). The combined effects of SAR and salinity (as electrical conductivity) on water infiltration are illustrated in Figure 2.27.

Three of the most common ions in irrigation waters that can be toxic to certain plants are sodium, chloride and boron. Sodium and chloride are normally only detrimental to plant growth at concentrations above a few meq l^{-1}. Boron, on the other hand, can be toxic to sensitive species such

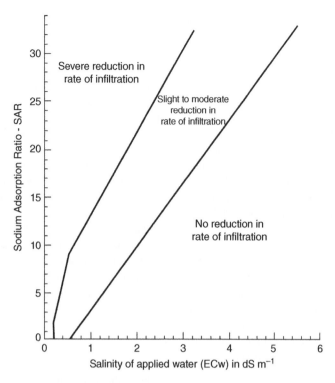

Figure 2.27   *Relative rate of water infiltration as affected by salinity and SAR of irrigation water. Note: 1 dS m⁻¹ = 1000 μS cm⁻¹. From Ayers and Westcot (1985), Water quality for agriculture, FAO Irrigation and Drainage Paper 29. Reproduced by permission of the Food and Agriculture Organization of the United Nations*

as citrus at concentrations of less than $1\,mg\,l^{-1}$. Boron occurs in most groundwaters at low concentrations. High concentrations may be found in groundwaters influenced by strong evapotranspirative concentration or by evaporite minerals (Hem, 1985) and also in polluted groundwaters – for example, boron is a constituent of detergents and so can be used as an indicator of sewage pollution (Misstear *et al.*, 1996).

The Food and Agriculture Organization (FAO) guidelines for irrigation water quality are included in Appendix 5 (Ayers and Westcot, 1985). They show the degree of water use restriction required according to the irrigation water salinity, SAR and toxic ions content. The table includes a fourth category of 'miscellaneous effects': these relate to excessive nutrients (nitrate), precipitates on leaves (bicarbonate) or potential corrosion of the

irrigation system pipe work (pH). It should be noted that these guidelines are not prescriptive, since any restrictions on irrigation water use will be influenced by local conditions including soil type, crop type, climate and irrigation management practices.

2.8   Pollution risk assessment and prevention

Having investigated the existing groundwater quality, we also need to assess the future risk of pollution at the proposed well site. Potential sources of pollutants are many and varied and include point sources such as waste disposal sites, industries, fuel storage tanks, latrines, septic tank systems and farmyards, and diffuse pollution sources such as intensive agricultural practices (inappropriate use of fertilizers, manures and deep ploughing), road salting and urbanized areas (see Table 8.3 in Chapter 8).

When locating wells for drinking water, the risks from both chemical and microbial pollutants should be considered. Particular parameters to be aware of include the inorganic chemicals arsenic, fluoride and nitrate (Section 2.7.3), synthetic organic chemicals such as pesticides and chlorinated solvents (which can be toxic or carcinogenic in low concentrations), and microbial pollutants that can have serious and immediate health consequences. Pollution by microorganisms (bacteria, viruses and protozoa) is especially common in private household wells since these are often both poorly sited and constructed (Macler and Merkle, 2000; De Simone *et al.*, 2009; Hynds *et al.*, 2013). If there is a significant risk of pollution from existing or future developments then an alternative well site should be considered, or – and this is a less satisfactory option – contingency plans prepared for water treatment.

The general pollution risk can be assessed using similar techniques to those applied in national or local groundwater protection schemes. These are normally based on a source-pathway-receptor risk assessment model, in which the source is the potentially polluting activity, the pathway is the route by which pollutants may move from the source to the receptor, and is characterized by the groundwater vulnerability, and the receptor is the aquifer or well, or the interface with a consumer or ecosystem. In applying these concepts to a proposed well site, a suitable strategy is to characterize and quantify potential pollutant sources, the subsurface transport pathway and the receptor (the well and its abstraction regime), and then:

1. Determine the groundwater vulnerability.
2. Define protection areas around the proposed well site (wellhead protection areas).
3. Estimate the pollution risk in relation to the groundwater vulnerability and the wellhead protection areas.

2.8.1 Groundwater vulnerability

Vulnerability can be defined as the 'intrinsic geological and hydrogeological characteristics that determine the ease with which groundwater may be contaminated by human activities' (Daly and Warren, 1998). The vulnerability of an aquifer to contamination may depend on many factors including the leaching characteristics of the topsoil, the permeability and thickness of the subsoil, the presence of an unsaturated zone, the type of aquifer and the amount and form of recharge. The relative importance of these factors will vary depending on the hydrogeological environment and the nature of the pollution risk. For example, the leaching characteristics of the topsoil will be the dominant factor in a temperate region where subsoils are thin or absent, the aquifer is fissured with an unsaturated zone that provides little opportunity for attenuating pollutants, and where the main land use is intensive arable agriculture. In contrast, the thickness and permeability of the subsoil will be a major influence on vulnerability in a heavily glaciated region, especially one in which the main pollution threat is from point sources such as landfills or septic tank discharges that can bypass the topsoil (Misstear and Daly, 2000). The nature of the recharge may also be an important factor: for example, in desert areas, where the main recharge is from infiltration of wadi runoff, the properties of the wadi bed deposits and the underlying unsaturated zone will be important controls on vulnerability. Figure 2.28 illustrates an extreme vulnerability situation in karstic terrain, where a sink hole in a chalk aquifer is fed by surface runoff from surrounding areas of low permeability crystalline rock.

There are many systems available for assessing and mapping groundwater vulnerability. Several rely on a qualitative categorization of whether the vulnerability is high or low, taking account of the most relevant factors that control vulnerability in the particular environment concerned. Qualitative systems have been incorporated into many national groundwater protection schemes (e.g., Department of Environment and Local Government

Figure 2.28   *A stream disappearing into a sink hole in a chalk aquifer, County Antrim, Northern Ireland. This is an example of extreme groundwater vulnerability. Photo by Bruce Misstear*

et al., 1999; Thomsen and Søndergaard, 2007; Environment Agency, 2013). There are other systems available that apply a numerical rating system in mapping vulnerability. Perhaps the best known of these index systems is DRASTIC (Aller *et al.*, 1987). The study area is mapped into units with similar characteristics and in each mapped unit a score out of 10 is applied to the seven factors that make up the acronym DRASTIC: <u>D</u>epth to water; net <u>R</u>echarge; <u>A</u>quifer media; <u>S</u>oil media; <u>T</u>opography; <u>I</u>mpact to vadose zone; hydraulic <u>C</u>onductivity. The scores are multiplied by a weighting to give the DRASTIC index for that mapped unit. The index gives a relative vulnerability rating. Since its introduction, DRASTIC has been modified to include additional parameters such as land use (Panagopolous *et al.*, 2005) and

fractured media (Denny *et al.*, 2007), whilst other index systems such as EPIK (Doerfliger *et al.*, 1999) and COP (Andreo *et al.*, 2009) have been developed for karstic environments. Quantitative approaches can also include the application of numerical models to simulate flow rates and travel times in the unsaturated zone (Schwartz, 2006). Whatever the system employed for vulnerability mapping, it is important to appreciate that vulnerability maps are intended as screening tools, and that site-specific investigations will be required to identify the detailed groundwater pollution risks. As Foster *et al.* (2013) point out, vulnerability maps "should be considered as the 'first step' and not the 'last word'".

It is worth noting, finally, that some hydrogeologists have argued against the common definition of vulnerability, pointing out that those aquifers which are most readily polluted (e.g., karstic limestones) are also those that are most quickly flushed of groundwater pollution. They would argue that a better definition of vulnerability would also take account of the persistence of a pollution problem once it has entered that aquifer environment.

2.8.2 Wellhead protection areas

The basic principles behind wellhead protection areas are illustrated in Figure 2.29. The *zone of contribution* (ZOC) to a water well is the groundwater catchment area that contributes water to the well. It is sometimes referred to as the *source catchment* or *capture zone*. In unconfined aquifers,

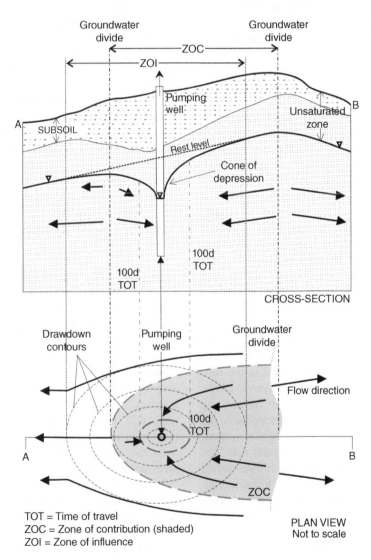

Figure 2.29 *Principles of zone of contribution and zone of influence. Adapted from US Environmental Protection Agency, 1987*

its area is proportional to the magnitude of the abstraction and inversely proportional to the recharge rate to the aquifer. The ZOC can thus vary with time – it may increase during dry periods and decrease during the main recharge season. Unless the water table is horizontal prior to pumping the ZOC is *not* the same as the *zone of influence* of the pumping well (the zone contained by the radius of influence of the well). Within the ZOC, different *zones of transport* (ZOTs) can be defined according to the average time it takes for the water in the aquifer to flow from the boundary of the ZOT to the well (the time of travel).

In many groundwater protection schemes, a series of concentric *source protection zones* will be defined. The range of permitted activities within each zone becomes more restricted with increasing proximity to the well.

Some countries specify an innermost protection zone of radius a few metres around the wellhead. This is designed specifically to protect the immediate wellhead environment against physical damage or pollution. It will often be physically demarcated by a fence and permitted activities within it will be highly restricted. Poor wellhead construction increases the susceptibility of a well to pollution, even in areas where the aquifer vulnerability is low (Hynds *et al.*, 2012).

Beyond this, a ZOT corresponding to a 50-day transport time is typically delineated to provide protection against microbiological pollution. A 50-day zone is chosen on the basis that over 99% of faecal bacteria are eliminated in the subsurface after 50 days (Adams and Foster, 1992). In some schemes, including that in Ireland, the microbiological protection zone is extended to the 100-day ZOT owing to concerns that some microorganisms (viruses or protozoan cysts) may survive for longer in the subsurface than faecal bacteria, and also because the bedrock aquifers are fissured and offer less scope for attenuating pollutants than primary porosity aquifers (Misstear and Daly, 2000).

Additional zones are delineated to protect the well against chemical pollutants. These include the ZOC itself and sometimes a zone intermediate between the inner 'microbiological protection zone' and the entire ZOC. This intermediate zone may be based on a ZOT of 300 or 400 days (Adams and Foster, 1992; Environment Agency, 2013), corresponding only very broadly with typical degradation times for some organic pollutants in the subsurface (these will vary hugely depending on contaminant type and prevailing subsurface conditions). The risk of chemical pollutants affecting a well is greater inside than outside this intermediate zone, leading to more restrictions on land use within the zone.

When calculating ZOTs, it is important to remember that most common methods result in a ZOT based on *average* transport time. However, the existence of open fractures or highly permeable preferential aquifer horizons will result in large macro-dispersion effects, causing unexpectedly quick *first arrivals* of contamination ahead of the average transport time (see Boxes 1.6 and 1.8).

A detailed review of the methods available for delineating wellhead protection areas is given in US Environmental Protection Agency (1994). A few of the commonly applied methods are summarized here.

Calculated radii. The simplest method of calculating a ZOT is to assume that groundwater flows radially to the well from all directions; that is, that the ZOT is circular. This is a dangerous assumption and is *only* valid if the initial water table or potentiometric surface is horizontal. A small deviation from horizontality will invalidate this assumption, and it is thus applicable only in a minority of cases. These simple calculations of radii will therefore typically overestimate the protection areas required down hydraulic gradient from the well and underestimate the protection areas up gradient.

If groundwater flow is, however, perfectly radial, the radius of a particular ZOT can be estimated from a simple cylindrical flow model in which it is assumed that the time of travel from the circumference of the cylinder to the well at its centre is such that the volume pumped equals the pore volume of the cylinder:

$$r_{ZOT} = \sqrt{\frac{Qt}{\pi b n_e}} \qquad (2.6)$$

where r_{ZOT} is the radius of the wellhead protection area for a particular ZOT, Q the pumping rate, t the time of pumping (or time of travel), b the effective aquifer thickness, and n_e the effective porosity. The effective aquifer thickness b is sometimes taken as the length of well screen.

An approximate estimate of the ZOC to a well in an unconfined aquifer can be made by calculating the recharge area (A_r) necessary to sustain the annual pumping volume (Q):

$$A_r = \frac{Q}{w} \qquad (2.7)$$

where w is the annual recharge in $m^3\ m^{-2}$. If we then assume that the ZOC is circular, we can translate this area into a figure for the radius of a ZOC around the well.

$$r_{ZOC} = \sqrt{\frac{A_r}{\pi}} \qquad (2.8)$$

In reality, of course, the ZOC and ZOTs are rarely circular (see above). Paradis *et al.* (2007) describe a HYBRID method in which the calculated circular recharge area is translated into an ellipse aligned with the direction of groundwater flow.

Analytical methods. The presence of an initial hydraulic gradient can be allowed for in the application of simple analytical methods for calculating wellhead protection areas, including the Darcy equation and the uniform flow equation. As indicated in Section 1.3.1, the Darcy equation expressed in terms of seepage velocity is:

$$v_s = \frac{Ki}{n_e} \qquad (1.5)$$

where v_s is the linear seepage velocity, K the hydraulic conductivity, i the hydraulic gradient and n_e the effective porosity. The distance d to the ZOT boundary up hydraulic gradient of the well for a particular time of travel t can therefore be estimated (neglecting the effect of increased

hydraulic gradient in the immediate vicinity of the pumping well) from:

$$d = v_s t = \frac{Kit}{n_e} \qquad (2.9)$$

The uniform flow equation can be used to delineate the ZOC to a pumping well in a confined aquifer with a sloping potentiometric surface. For coordinates x and y in Figure 2.30, the equation describing the boundary of the ZOC to a pumping well is given by Todd and Mays (2005) as:

$$-\frac{y}{x} = \tan\left(\frac{2\pi Kbi}{Q}y\right) \qquad (2.10)$$

where K, b, i and Q are as defined earlier. It should be noted that i is the regional hydraulic gradient before pumping the well and that the tangent in Equation (2.10) is in radians. The maximum distance of the ZOC down gradient of the well is termed the stagnation point, x_L, and is given by:

$$x_L = -\frac{Q}{2\pi Kbi} \qquad (2.11)$$

The maximum half-width y_L of the ZOC can be calculated from:

$$y_L = \pm\frac{Q}{2Kbi} \qquad (2.12)$$

The uniform flow equation can also be applied to an unconfined aquifer provided that recharge is neglected and that the drawdown is small compared to the aquifer thickness. The effect of recharge is to 'close' the ZOC at its upstream end, giving it a finite area, as opposed to the theoretically open-ended, infinite ZOC in the case of no recharge. The area of the ZOC in the case of recharge is given by Equation (2.7) above. It should be noted that these equations do not consider the effects of dispersion or preferential flow pathways (see Box 1.6).

Box 2.11 gives an example of where the uniform flow equation is used to minimize the risk from a specific pollution threat when selecting a well site.

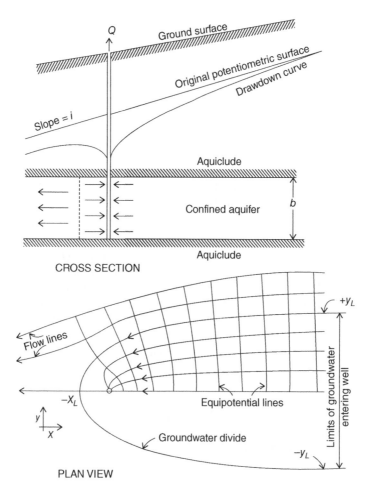

Figure 2.30 *Flow to a well in a confined aquifer with a sloping potentiometric surface. From Todd (1980); reproduced by permission of John Wiley & Sons Inc*

Numerical models. For large groundwater schemes for which there are good data, two- or three-dimensional numerical models can be developed to simulate flow paths to pumping wells and hence to delineate the ZOTs and ZOCs. Information on suitable models can be found in hydrogeology and groundwater modelling texts such as Freeze and Cherry (1979), Anderson and Woessner (1992), Fetter (2001), Rushton (2003) and in the US Environmental Protection Agency handbook on groundwater and wellhead protection (1994).

Wellhead protection in heterogeneous and fractured aquifers. There are many situations where the aquifer is so heterogeneous and anisotropic that the delineation of wellhead protection areas is highly problematic using the techniques described above. Some countries essentially recommend a 'common sense' approach (Banks and Robins, 2002), basing the shape of ZOCs and ZOTs on mapped fracture zones and lineaments, although such advice can be both nebulous and misleading.

The use of standard analytical techniques, or even numerical models with particle-tracking capabilities,

---

Box 2.11 Stagnation point calculation

A well is to be located in an unconfined sand and gravel aquifer that is 40 m thick and has an average hydraulic conductivity of 10 m d⁻¹. The slope on the water table is 0.003. There is an old gravel pit in the area which has been infilled, and which may be a source of groundwater pollution. If the well has a design yield of 15 l s⁻¹, estimate how far the well should be located from the gravel pit to minimize the risk of pollution.

The well should be located so that its stagnation point lies up gradient of the potential pollution source. For a design discharge $Q = 15\,l\,s^{-1} = 1296\,m^3\,d^{-1}$, a hydraulic conductivity $K = 10\,m\,d^{-1}$, an aquifer thickness $b = 40\,m$ and a hydraulic gradient (prior to pumping) $= 0.003$, the stagnation point x_L can be calculated from Equation (2.11):

$$x_L = \frac{Q}{2\pi Kbi} = \frac{1296}{2 \times 3.1416 \times 10 \times 40 \times 0.003}$$
$$= 172\ m$$

Therefore, the well should be sited more than 172 m up hydraulic gradient of the gravel pit.

This simple calculation assumes steady-state conditions: it would be prudent to increase the separation between the well site and the gravel pit to allow for fluctuations in the slope of the water table and possible increases in the pumping rate.

The uniform flow equation assumes a confined aquifer which is not the case here. However, it can be shown that the likely drawdown in the well would be small compared to the saturated aquifer thickness and hence that the error in assuming confined conditions (with horizontal flow through a constant aquifer thickness) should not be too large. Rearranging the Logan equilibrium approximation [Equation (1.23)] in terms of drawdown s_w, and substituting Kb for aquifer transmissivity T, the drawdown can be estimated as:

$$s_w = \frac{1.22Q}{Kb} = 3.95\ m$$

This represents only about 10% of the saturated aquifer thickness.

---

to define ZOTs in fractured rock aquifers should be approached with extreme caution. Guérin and Billaux (1994) demonstrated that, although a standard porous continuum model can be adequate for simulating bulk groundwater flow through a fractured aquifer, it might be wholly inadequate to simulate contaminant transport. This is because of the high transverse and longitudinal dispersivity of such aquifers and the existence of preferential flow conduits along a small number of fracture pathways (Boxes 1.6 and 1.8). Bradbury and Muldoon (1994) used discrete fracture network models to demonstrate that ZOTs in fractured aquifers significantly exceed the calculated ZOTs for similar abstractions in homogeneous porous aquifers, both longitudinally and transversely.

Karstic limestone aquifers pose their own problems in terms of source protection, as the

distribution of the flow conduits and their interlinkages are extremely complex and hence difficult to model. In such situations the best approach is to try and map the ZOC using tracers and other field techniques as appropriate. It is not usually useful to delineate a separate 50- or 100-day ZOT since the time of travel in a karst aquifer can be so rapid that the whole ZOC can fall within a relatively short ZOT (Deakin, 2000).

2.8.3 Estimating the pollution risk for a new well site

The vulnerability map and wellhead protection area can be overlaid one on top of the other to show the pathway-receptor risk for a specified pollution hazard. An example from the national groundwater protection scheme in Ireland is shown

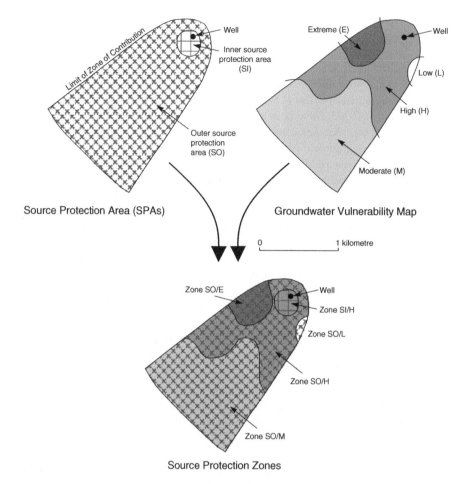

Figure 2.31 *Combining a wellhead protection area with a vulnerability map (Department of the Environment and Local Government et al., 1999). Public domain, reproduced by permission of the Department of Environment and Local Government, Environmental Protection Agency and Geological Survey of Ireland*

in Figure 2.31. Here, the groundwater vulnerability has been mapped as either extreme (E), high (H), moderate (M) or low (L). The wellhead protection area comprises an inner zone (termed the source inner, or SI) and an outer zone (SO), the latter corresponding to the ZOC. Thus a protection zone within the SI that has a moderate vulnerability is delineated as SI/M.

In siting a well, the nature of the hazard posed by the pollution source needs to be considered. The hazard would include the nature, volume, toxicity and persistence of the pollutants released into

the subsurface. It would clearly be inadvisable to locate a new well where there is a major potential pollution source such as a landfill site within the SI/M in the example mentioned above. However, it may be considered that the risk would be acceptably small if the potential pollution source within this zone was from an isolated domestic septic tank which has a drainage field constructed according to good practice. In such a situation the well should be located upgradient from the septic tank, and the hydrogeologist should take account of local guidance on minimum set back distances.

In Ireland, for example, the recommended minimum distance between a septic tank drainage field and a well varies from 15 m for an upgradient domestic well to 60 m for a public water supply well (Environmental Protection Agency, 2009). Guidelines on set-back distances are usually based on bacterial transport and attenuation, whereas Moore *et al.* (2010) provide more conservative guidance based on viruses, since viruses can be more persistent in the subsurface. In the Moore *et al.* (2010) guidelines, which are aimed at regional councils and other agencies in New Zealand, the separation distances required to remove infectious viruses between the drainage field and the well range from less than 50 m to more than 300 m, the distance depending on the nature and thickness of the unsaturated zone and the type of aquifer.

2.9   Planning the well scheme

Planning the well scheme will depend to some extent on the ratio of demand to the ability of a single well to supply that demand. It will also depend on the intended use of the groundwater: how critical would it be to have spare capacity if the main supply were temporarily unavailable owing to breakdown of the pump or some other failure of supply? The capacity of an individual well will have been established by the pumping tests of the drilling investigation phase. In the case of a small demand for potable water where one well could theoretically supply the demand, the production pumping station should ideally be designed with two wells: one for production and one for standby. Where more than one well is needed to cope with the potable water demand, the well field should include at least one standby well. In the case of water for irrigation, it may not be so critical to have a standby well – the availability of a standby pump may suffice.

In planning a well field there is usually a trade-off between the capital and running costs of the scheme. The closer the wells are spaced, the greater is their mutual interference and resulting drawdown, and therefore the greater are their running costs; but, on the other hand, the capital costs of the pipelines connecting wells in the well field are decreased. The distances between wells in the well field should be optimized by matching the capital and running costs of the wells against the pipeline costs (Sections 3.1.5 and 4.6).

The interference effects of wells pumping in a well field can be estimated using the *principle of superposition*. This principle can be applied to any number of wells in a well field. Figure 2.32 illustrates the simple case of three pumping wells. The drawdown in each well is calculated assuming that it is the sole pumping well, and then the interference drawdowns from the neighbouring pumping wells are added (superimposed) to give the total drawdown in each well. For steady-state conditions in three wells pumping at the same rate, the principle of superposition can be applied to the Thiem equation (1.21) to give:

$$s_w = \frac{Q}{2\pi T}\ln\frac{r_e}{r_w} + \frac{Q}{2\pi T}\ln\frac{r_e}{r_1} + \frac{Q}{2\pi T}\ln\frac{r_e}{r_2}$$

(2.13)

which reduces to:

$$s_w = \frac{Q}{2\pi T}\ln\frac{r_e^{3}}{r_w \times r_1 \times r_2}$$

(2.14)

In this equation, s_w is the equilibrium drawdown in a well of radius of r_w; Q is the pumping rate in this well and in two other wells located at distances r_1 and r_2 from this well; r_e is the radius of influence of each of the pumping wells; and T is the aquifer transmissivity. An example calculation of interference drawdowns for steady-state conditions is included in Box 2.12. The principle of superposition can also be applied to non-equilibrium pumping conditions, for example using the Theis or the Cooper-Jacob equations (see Section 7.4.5 and Section 9.2.1). MacMillan (2009) describes an analytical method for designing well fields which is based on a modification of the Theis equation. He demonstrates that well spacings are most sensitive to a derived parameter $r_{HA/3}$, which is the radial distance from a hypothetical well where

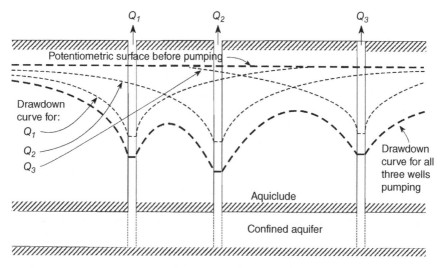

Figure 2.32   *Principle of superposition when several wells are pumping*

Box 2.12   Well interference effects

An industrial estate is supplied from a well field of three wells. The well locations form an equilateral triangle having a length of 30 m on each side. The 200 mm diameter wells fully penetrate a confined limestone aquifer having a known transmissivity of 250 m² d⁻¹. The potentiometric surface in the aquifer is 20 m above the top of the confined aquifer.

When the wells are all pumped at the same rate of 10 l s⁻¹ (860 m³ d⁻¹), the equilibrium drawdown (as defined by the Thiem equation) in each of the wells is 7.5 m.

The industrial estate needs to increase the capacity of the well field in order to meet increasing demand. One option is to drill a new well at the centre of the well group, equidistant from the other wells. Assuming the new well has the same diameter and is pumped at the same rate as the existing wells, what effect would this new well have on the specific capacities of the original wells? What would be the maximum drawdown in the well field?

1. Firstly, we calculate the radius of influence of the wells. Substituting the known values into Equation (2.14):

$$s_w = \frac{Q}{2\pi T}\ln\frac{r_e^3}{r_w \times r_1 \times r_2}$$

the only unknown r_e can be calculated:

$$7.5 = \frac{860}{2\pi \times 250}\ln\frac{r_e^3}{0.1 \times 30 \times 30}$$

$$\ln\frac{r_e^3}{90} = 13.7$$

$$r_e = 431\,\text{m}$$

2. Having determined r_e we can now calculate the new drawdown if a fourth well is added at the centre of the well field. The drawdown in each well due to pumping of the three existing wells and one new well is given by:

$$s_w = \frac{Q}{2\pi T}\ln\frac{r_e^4}{r_w \times r_1 \times r_2 \times r_3}$$

where r_3 is the distance of the new well from the existing wells. Using simple trigonometry r_3 can be calculated as 17.3 m. Therefore the new drawdown in each of the old wells is:

$$s_w = \frac{860}{2\pi \times 250} \ln \frac{431^4}{0.1 \times 30 \times 30 \times 17.3}$$

$$s_w = 9.26 \text{ m}$$

and the drawdown in the new (fourth) well is given by:

$$s_w = \frac{860}{2\pi \times 250} \ln \frac{431^4}{0.1 \times 17.3 \times 17.3 \times 17.3} = 9.86 \text{ m}$$

3. Next, we calculate the change in specific capacity. The specific capacities of the existing wells are:

$$\frac{Q}{s_w} = \frac{860}{7.5} = 115 \text{ m}^3 \text{d}^{-1} \text{m}^{-1}$$

With a fourth well added, the specific capacities of these old wells become:

$$\frac{Q}{s_w} = \frac{860}{9.26} = 93 \text{ m}^3 \text{d}^{-1} \text{m}^{-1}$$

which represents a reduction of $22 \text{ m}^3 \text{ d}^{-1}$ m^{-1} or 19%.

4. The maximum drawdown occurs in the new well in the centre of the well field. However, the above calculations have ignored the effects of well losses. It can be easily shown that the calculated drawdown of 9.86 m for well 4 in (2) above comprises 4.58 m due to the well itself pumping and 5.28 m interference drawdown from the surrounding three abstraction wells. If we now assume that there is a further 25% drawdown at this pumping well due to well losses (which would equate to a fairly good well efficiency of 80%; see Section 9.2.1), then the drawdown at well 4 can now be estimated as: $(4.58 \text{ m} \times 1.25) + 5.28 \text{ m} = 11.01 \text{ m}$. As the potentiometric surface is 20 m above the top of the aquifer a maximum drawdown of 11 m will mean that the aquifer remains confined during well field operation (this is desirable because a change to unconfined conditions could lead to the introduction of air into the well and increase the potential for clogging).

drawdown is equal to one third of the available drawdown if all of the well field output was supplied from that theoretical well.

Such analyses generally assume homogeneous and isotropic aquifer conditions, conditions that are rarely met in reality. In heterogeneous aquifers, the wells will normally be located within the most productive aquifer zones, although it should be noted that it may be necessary to space these wells further apart than wells in lower transmissivity zones because their cones of drawdown (and hence interference effects) will extend further. The well siting should also take account of hydraulic barriers present in the area. Wells near a recharge boundary such as a river should normally be located parallel to that feature, whereas wells near an impermeable boundary should be located along a line at right angles to that boundary, and as far

from the boundary as practicable (Heath, 1983). Finally, the additional well drawdowns that occur in pumping wells due to turbulent flow and pipe friction as water moves into and up the well ("well losses" – see Section 4.5) also need to be considered when making predictions of well drawdowns in the design of well fields. The hydrogeologist can simply make an allowance of, for example, 25% additional drawdown for well losses (for relatively efficient wells) or follow a more detailed procedure such as that described in Section 9.2.1 for incorporating well losses in long-term drawdown predictions.

The treatment works at a well field intended for potable water supply will often have been designed before the well field comes on-line (Kristensen *et al.*, 2014). The chemical and microbiological data on which the works are designed will have been obtained by analyses of water from the pumping tests during the drilling investigations. It should, however, be remembered that water quality can evolve with time and with increasing extent of the zone of drawdown. A pumping test of limited duration *may* not be able to adequately predict how that water quality will evolve. Once the scheme is implemented it will be necessary to monitor the quality of both the raw water and the treated supply to ensure that the quality criteria on which the treatment works were designed do not change with time.

3

An Introduction to Well and Borehole Design

The design of the well or borehole must be chosen before drilling or manual construction begins, as it will govern the choice of drilling or construction method, and the drafting of a drilling contract where applicable. The design, therefore, has to be based on existing information, and the more comprehensive this information is, the more successful will be the design. Investment in a detailed groundwater investigation (Chapter 2) will be repaid in a successful and long-lasting well. The hydrogeological information needed for the design of a production well is summarized in Box 3.1. The choice of well type will also be strongly influenced by the required well discharge rate (to meet peak, daily and annual water demands) and by socio-economic factors such as manpower and materials, fuel or electricity supply, the need for community participation, operation and maintenance logistics, and budgetary constraints. Well affordability and maintenance will be especially important considerations for communities in economically disadvantaged areas (Box 3.2).

An introduction to the design of drilled wells is given in Section 3.1, while the following sections of this chapter deal briefly with the design of hand-dug wells, infiltration galleries, radial collector wells, observation boreholes and exploration boreholes. The important subject of pumps is discussed at the end of the chapter in Section 3.7. Specific aspects of well design, including construction materials, hydraulic design and economic optimization are covered in greater detail in Chapter 4. That chapter also introduces specialist well designs for applications such as managed aquifer recharge and shallow heating and cooling systems.

3.1 Drilled wells

3.1.1 General design principles

The main components of a drilled well are described in Box 3.3. Three key components in the design of a well are the pump, the lining pipe which houses it (the pump-chamber casing) and the intake section of the well, which may be unlined where the aquifer is stable, or installed with a well screen and gravel pack where the aquifer material requires support.

Water Wells and Boreholes, Second Edition. Bruce Misstear, David Banks and Lewis Clark.
© 2017 John Wiley & Sons Ltd. Published 2017 by John Wiley & Sons Ltd.

Box 3.1   Summary of hydrogeological information needed for the design of a production well

The hydrogeological information needed for the design of a production well will include:

1. Aquifer type and lithology: crystalline, consolidated or unconsolidated.
2. Regional groundwater levels and hence aquifer condition: confined or unconfined.
3. Aquifer location: depth and thickness.
4. Aquifer characteristics: transmissivity, hydraulic conductivity and storativity.
5. Aquifer boundaries: location of impermeable boundaries and recharge boundaries (e.g., rivers).
6. Recharge and the available groundwater resource.
7. Groundwater quality: suitability for intended use; corrosion or incrustation potential; pollution risk.
8. Details of geological formations that overlie the target aquifer(s): thickness, stability, groundwater characteristics.

The actual data requirements will depend on the type and capacity of the well being designed. For example:

a. It will almost certainly be necessary to have a comprehensive hydrogeological dataset for the design of a high-yielding drilled well, whereas all that may be required for the design of a shallow hand-dug well is a general appreciation of the aquifer type, the depth to the water table, the nature of the superficial materials, the groundwater quality, and the proximity and nature of any pollution threats.

b. The design of an infiltration gallery near a river will rely on good information on the properties of the shallow alluvial aquifer, and especially on the degree of interconnection between the groundwater and the river, but there may be no need to collect information on the deeper geology and aquifers.

c. The design of an exploratory borehole or observation borehole will obviously require fewer data than the design of a production well. Nevertheless, the design of these boreholes will still require some knowledge of the aquifer type and the nature of the overlying strata, since these will affect the borehole depth and the drilling and formation sampling methods to be used.

The potential owner or operator of a water supply well requires that:

- the well (or group of wells) should have sufficient yield to meet the demand;
- the water quality should be suitable for the intended use of the well;
- the well should be reliable, requiring little maintenance;
- the construction and operating costs should not be excessive.

It is also important for the well owner, and the regulatory authority, that the well does not impact significantly on neighbouring wells or on the environment. These user requirements underpin the following basic design and construction principles for a well:

1. The location for the well should be selected after carrying out a systematic groundwater investigation as described in Chapter 2. There is little point in spending a lot of time and money in applying the design principles below if the well site has not been chosen carefully.
2. The well should be of sufficient diameter, depth and straightness for the pump, and for monitoring and maintenance equipment.
3. It should be stable, and not collapse.

Box 3.2 Well affordability and maintenance in economically disadvantaged communities

In the industrialized world, a high quality of drinking water is regarded as a fundamental of existence, and it is one that we are increasingly willing to pay relatively large sums for. In many parts of the world, however, a reliable supply of pure drinking water is a luxury. Aid workers from the developed nations often express surprise when rural communities in poorer nations fail to maintain the new water supply schemes that have been installed by the international humanitarian aid industry. There may be lots of reasons why the scheme has failed:

1. The recipient community may never have been consulted about the design of the scheme, or were not involved in choosing its location (see Box 2.1 in Chapter 2), or, in the worst cases, perhaps were never even asked if they wanted a new scheme at all.

2. There may be a societal prejudice against communal schemes. Some cultures have a strong tradition of a water source for each family or household.

3. All water supply schemes bear a cost. Even hand pumps need to be maintained and require spare parts. By constructing a new water supply scheme, we may be placing an economic burden as well as a social good on a community. Can the community afford to maintain the scheme, pay the running costs and capital amortization? If they can afford this, do they want to afford it and prioritise this expenditure over other needs?

4. Insufficient consideration may have been given to the adequate management of the scheme – its maintenance, its financing. Was there a clear plan for regular collection of subscriptions?

5. There may have been inadequate technical provision for maintenance. Are there trained mechanics in the community and can the community afford to employ them? Is there an efficient spare parts network, such that hand-pump seals and valves can be purchased locally at low cost?

6. The technology may not have been appropriate to the skills and budgets of the community. A simple dug well with bucket and rope may be the most appropriate technology in some communities. In others, a motorized pump scheme may be best, provided it is well managed.

7. There may have been a lack of knowledge or appreciation of the health benefits of clean water amongst the rural community. It is also important to recognize that household hygiene and sanitation are important factors for health as well as the purity of the drinking water. Moreover, household air quality is often seriously compromised by inefficient and polluting wood or dung-fuelled stoves – the World Health Organization estimates that about 3 billion people use solid fuels for heating and cooking, leading to 4 million premature deaths a year, mainly from respiratory and cardiovascular diseases (World Health Organization, 2014).

8. The recipient community may not place the same value on clean drinking water as industrialized nations (who may also underestimate the costs involved in acquiring it). In Ethiopia, the World Bank estimated the equivalent monetary health benefit of clean water as only some 34.6 Birr (about. 4 USD) per person per year (DHV 2002), while the estimated equivalent cost of carrying water in rural Ethiopia was approximately 1 Birr per hour (or 42 Birr m^{-3} if 24 litres are carried per trip from a source lying 1500 m distant). In the light of this, it is quite understandable that rural inhabitants may choose to continue to use nearby impure sources in preference to distant clean ones. DHV (2002) offered the following analysis: "There are three possible reasons for the present low rural (water supply) coverage (in Ethiopia). The engineering reason is that not enough resources have been allocated to construction and maintenance. The sociological reason is that users were not consulted and so the wrong facilities were built in the first place. Both the engineer and the sociologist also imply that demand for clean water would be 100% if only people were sufficiently educated. The economist, on the other hand, believes that centuries of trial and error have led rural people to inhabit places that provide the optimum combination of costs and benefits."

4. It should prevent excessive amounts of aquifer material (sand or clay particles) from entering the well during pumping.
5. It should abstract from the aquifer zone of highest yield potential.
6. The well should be efficient hydraulically, ensuring that the energy losses as water moves into and up the well are not excessive.
7. The construction materials should resist corrosion and incrustation, thereby reducing maintenance or rehabilitation liabilities, and should not adversely affect water quality.
8. The well and aquifer should be protected from contamination – mainly from the surface, but cross-contamination between aquifers should also be avoided.

9. The depth, diameter and construction materials for the well should be selected such that the cost of the well is reasonable. In extensive uniform aquifers, the design can be optimized on economic grounds.

The dominant influences on the design of a well are the discharge rate required and the type of aquifer system being exploited. The maximum discharge rate must be decided before a water well can be designed because it will dictate the size of pump required which, in turn, will govern the minimum internal diameter of the pump-chamber casing. A fundamental rule in well design is that the pump chamber must be large enough to accommodate the pump and, in the case of submersible pumps, the pump motor, cabling and pump shroud.

Box 3.3   Components of a drilled well

Key well components are illustrated in Figure B3.3(i) and include:

- the *pump-chamber casing*, which provides stability to the well and protects the pump against debris falling into the well from the sides of the borehole;
- the intake section, which may be lined with slotted casing (*well screen*) as in the example shown, or left open-hole if the borehole walls are free-standing and there is no risk of future collapse;
- the *centralizers*, which are installed around the casing and screen to hold them in the middle of the borehole, and thus provide a regular annular space for the gravel pack;
- the *gravel pack*, which is installed in some wells in unconsolidated aquifers to prevent fine particles from entering the well, while allowing easy passage of the water;
- a *bottom plug* (also known as bail plug or tailpipe), which is a short length of casing, capped at the bottom, installed at the base of the well screen to act as a sediment trap (but

this bottom plug may also collect bacteria that contribute to well biofouling – see Section 9.1.3);
- the *conductor casing* in the upper few metres of the well, which provides stability during drilling and, if permanent, may also support the weight of the pump;
- the *grout seal* around the conductor casing, and between the conductor casing and the main pump-chamber casing, which prevents movement of contaminants downwards from the surface through the annular space to the well and aquifer;
- *dip tubes* for monitoring water levels inside the well and in the gravel pack between the screen and borehole wall;
- the pumping equipment and wellhead distribution works.

The lengths of pump-chamber casing and well screen when joined together in a well are usually known as the *casing string*. The pump and rising main are sometimes referred to together as the pump string.

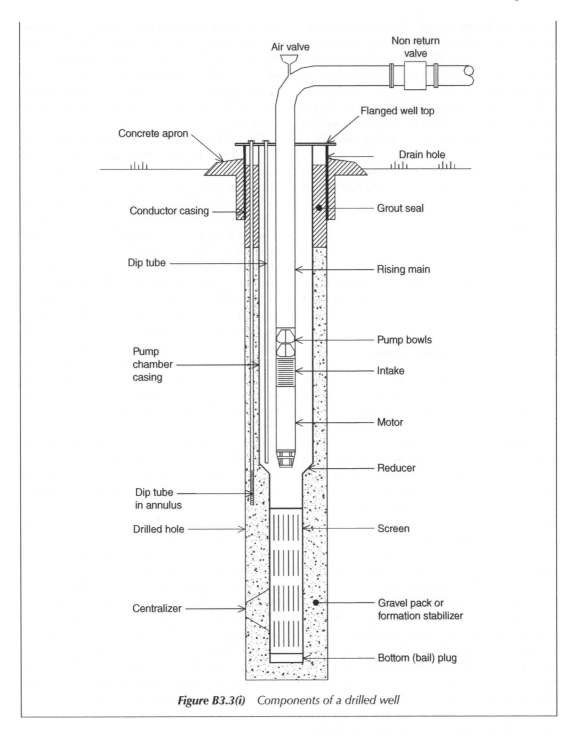

Figure B3.3(i) *Components of a drilled well*

There must also be enough clearance for the dipping tube (Box 3.3) and for water flow sufficient to cool the pump motor. The choice of pump is thus critical to the design of the well, and is considered separately in Section 3.7.

The sequence of steps to be followed in designing a well is summarized in Box 3.4. This checklist should be consulted before a drilling contract is prepared.

As we saw in Section 1.2.1, aquifers can be divided into three broad classes for the purposes of well design:

a. crystalline aquifers;
b. consolidated aquifers;
c. unconsolidated aquifers.

A simple flow-chart to aid non-geologists in the classification of aquifer types in the field is given in Figure 3.1.

3.1.2 Wells in crystalline aquifers

Groundwater occurs in crystalline aquifers in the secondary porosity that is created by weathering and fracturing. Since both weathering and fracturing decrease with depth there will be a depth beyond which the cost of drilling outweighs the prospect of significantly increasing the yield of a borehole. This maximum depth will vary from place to place depending on the geological and geomorphological history of the site. However, unless a specific, deep structural feature can be identified, the maximum drilling depth is unlikely to exceed 100 m (Clark, 1985) and is often considerably less than this depth. In drilling crystalline aquifers in Sweden, boreholes are normally abandoned after around 100 m if the required yield has not been achieved (Gustafson, 2002), while in India the effective base of the weathered crystalline aquifer is typically at a depth of only 15 to 25 m (Foster, 2012). In the (mainly) crystalline rock aquifers in Uganda, the average depth of drilled wells ranges between 45 and 90 m, although it is reported that many of the wells are unnecessarily deep owing to a lack of clear drilling specifications and/or supervision (Sloot, 2010).

In sub-arctic or temperate zones, a significant weathered zone may be absent. The optimum depth at any location has to be determined from previous drilling experience in the area, or from surface geophysical surveys which can indicate the base of the main weathering and fractured zone (Section 2.4).

The yield of water wells in crystalline aquifers is low, with median values of about 10–20 m^3 d^{-1} and rarely exceeding 250 m^3 d^{-1} (Gustafson, 2002), so large diameters are not needed in such wells. A pump with 100 mm outside diameter (OD) will cope with the available discharge from almost all water wells in crystalline aquifers, and so a 150 mm or preferably 200 mm internal diameter (ID) pump-chamber should be adequate in their construction (see Section 3.1.3 below for further discussion of the desirable clearance between a pump and the pump chamber casing). For low-yielding wells to be fitted with small diameter hand pumps, well completion diameters may be as small as 100 mm.

To protect the well against surface pollution, especially microbial pollutants, the upper section of the well will need casing and grouting to prevent ingress of surface water, and the rest of the well can remain unlined. It is common to have only 2 or 3 m of such casing, but it is recommended here that the casing should extend at least 10 m below the surface into stable rock and, preferably, to below the pump setting depth so as to avoid placing the pump in the open hole section of the well [Figure 3.2(a)]. There is a potential problem that a long section of grouted casing might seal off the more prolific shallow water-bearing zones where the weathered rock is unstable, yet these zones may also be highly vulnerable to pollution. Consequently, there is often a trade-off between well yield and protection of groundwater quality when designing wells in crystalline aquifers. This trade-off has been analysed by Misstear (2012) using data for crystalline rock aquifers in Ireland. Comte *et al.* (2012) found that the reduction in hydraulic conductivity with depth in these crystalline rock aquifers followed an inverse power relationship:

$$K = Ad^{-B} \qquad (3.1)$$

Box 3.4   Sequence of steps in designing a well or borehole

Main steps in well and borehole design	Checklist of information needed for each of the main steps	Relevant sections of book
Establish basic design parameters	1. Purpose of the borehole or well – exploration, observation or production well? 2. Water demand – peak, daily, annual? Select design discharge for a production well. 3. Type and diameter of pump; power source? 4. Suitable drilling site – access, power, flooding, contamination risk, drainage? 5. Geology and aquifer type – aquifer characteristics, boundaries, regional groundwater levels, recharge? 6. Groundwater quality, pollution risk, potential corrosion/incrustation problems? 7. Legal/regulatory requirements affecting the well or borehole design?	2.1–2.9, 3.1, 3.7
Design the structure	1. Depth and diameter of well or borehole? 2. Casing – for part or entire section above the aquifer? 3. Screen or open hole completion for aquifer? 4. Natural gravel pack, artificial gravel pack or formation stabilizer?	3.1
Select construction materials	1. Casing material – steel, plastic, fibreglass or other? 2. Screen material – steel, stainless steel, plastic, fibreglass or other? 3. Type of screen – wire-wound, bridge slot, louvre, machine slotted? 4. Gravel pack? Identify grading based on particle-size distribution of aquifer samples. 5. Screen slot size? 6. Hydraulic suitability of the well design? Are screen entrance and well upflow losses acceptable? 7. Grouting materials? Are special cements or other materials necessary?	4.1–4.5
Establish sampling requirements	1. Depth intervals for collection of disturbed formation samples? 2. Method for collecting disturbed formation samples? 3. Depth intervals for collection of 'undisturbed' formation samples (cores)? 4. Method for collecting cores? 5. Collection of water samples during drilling? Intervals? Methods?	6.2, 6.3
Choose drilling method	1. Drilling method – percussion, rotary, sonic, auger, etc? 2. Drilling fluids? 3. Type of drilling rig? Depth and diameter limitations?	5.1–5.8
Choose development method	Are well design and materials suitable for: acids, hydrofracturing, air surging, jetting?	5.9

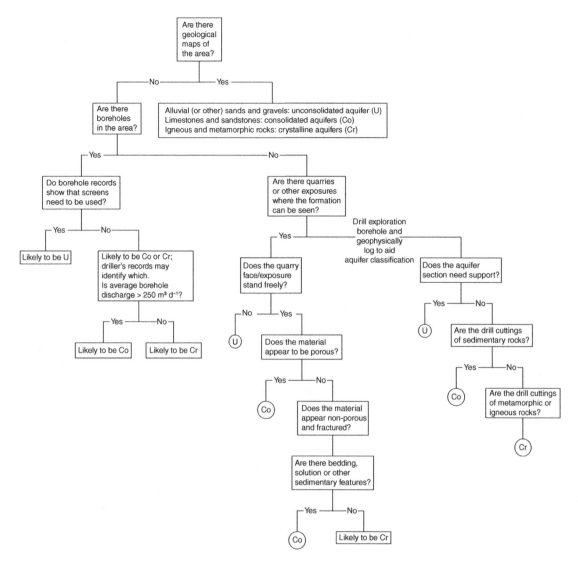

Figure 3.1   *Classification of aquifers in the field (as an aid to non-geologists)*

where *d* is depth below top of bedrock and *A* and *B* are coefficients relating to the particular rock units. The Logan equilibrium approximation equation (1.23) described in Section 1.3.3 was used by Misstear (2012) to assess the theoretical reduction in well yield for these rock units that would result from an increase in the length of the pump chamber casing. The Logan relationship can be expressed as:

$$Q = \frac{Kbs_w}{1.22} \qquad (3.2)$$

where Q is well discharge, K is hydraulic conductivity, b is aquifer thickness and s_w is drawdown. By assuming a value for the available drawdown (s_w), the potential reduction in well yield (Q) can be investigated for the situation where K declines with depth according to Equation (3.1), and where

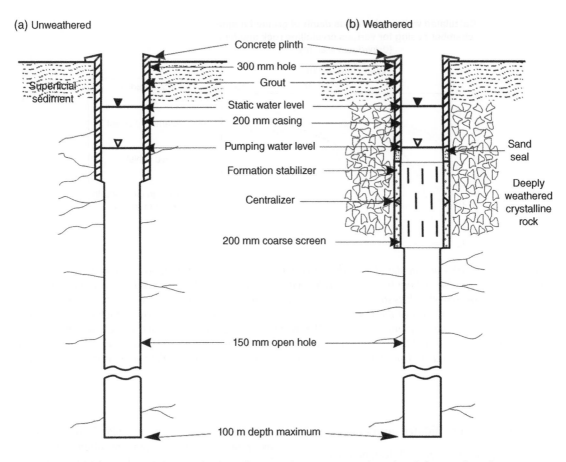

(a) Unweathered

(b) Weathered

Concrete plinth

300 mm hole

Grout

Superficial sediment

Static water level

200 mm casing

Pumping water level

Sand seal

Formation stabilizer

Deeply weathered crystalline rock

Centralizer

200 mm coarse screen

150 mm open hole

100 m depth maximum

Figure 3.2 *Well designs for crystalline aquifers. (a) unweathered and (b) weathered*

the available aquifer thickness (*b*) reduces as the length of the pump-chamber casing is increased. In Figure 3.3, the well yield versus casing length for four different bedrock types is illustrated for the case where the drawdown equals 20 m. Taking the meta-turbidite aquifer as an example, the calculated well yield reduces from approximately 13 m³ d⁻¹ with 3 m of pump-chamber casing to about 5 m³ d⁻¹ with 10 m of casing. These calculations are clearly simplistic since the Logan equation (and the Thiem equation from which it is derived) assumes idealized isotropic and homogeneous conditions which clearly do not pertain in fractured crystalline rock aquifers; moreover, no allowance has been made for the additional

contribution to well yield from induced downward leakage from the higher rock layers behind the well casing. Nevertheless, the calculations and graphs do help to illustrate the potential conflict between greater well protection and reduced well yield.

Where the weathered or unstable zone is relatively deep, and thus less vulnerable to surface pollution, then a screened section can be incorporated in the design [Figure 3.2(b)]. If the screen is merely to support shattered rock then the screen slots may be very coarse; where the weathered zone consists of loose sand and other incoherent material then the slot size should be chosen to match the appropriate grain size (Section 4.3).

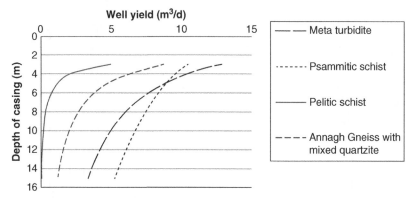

Figure 3.3 *Examples of calculated well discharge rates versus length of grouted pump-chamber casing in crystalline rock aquifers, where hydraulic conductivity decreases with depth according to an inverse power law. The reduction in well yield is shown for the situation where the length of pump-chamber casing is extended from 3 m to 15 m below top of the bedrock, where the total well depth below top of rock is 60 m and the assumed drawdown due to pumping is 20 m. The hydraulic conductivity data used in the calculations were obtained from Comte et al. (2012) (reproduced from Misstear, 2012, 'Some key issues in the design of water wells in unconsolidated and fractured rock aquifers', Acque Sotterranee, Italian Journal of Groundwater, AS02006: 9–17)*

3.1.3 Wells in consolidated aquifers

Single aquifers. A water well designed to exploit a single aquifer of limited thickness should normally be drilled to penetrate the whole aquifer. In an area where there is a large number of existing wells, the potential discharge rate of the new well can be estimated from previous experience, but in undeveloped aquifers, test wells will be needed to provide this information (Section 2.5). The potential discharge rate and the amount of lift required can be used to size the pump (Section 3.7), from which we can determine the minimum ID of the pump-chamber casing. The amount of lift, or total dynamic head developed by a pump, can be obtained from the sum of the following: the depth to static water level, the potential drawdown of the water level in the well, the elevation of the discharge point above ground level (or the equivalent head represented by a pressure tank), plus an allowance for head losses in the system.

Minimum clearances of only 25 mm between the pump and the casing are often suggested. However, a clearance of at least 50 mm (that is, a casing with an ID 100 mm greater than the maximum OD of the pump) should normally be allowed for in the design, as this will:

- provide more tolerance in the event that the borehole is not straight (the tolerance will also be affected by the diameter of the rising main and its flexibility);
- allow ample flow past the pump motor for cooling (applies to submersible pumps only);
- permit the installation of dip tubes for manual water level monitoring (permanent pressure transducers can be installed on the pump rising main, but a dip tube is needed for manual checks using an electrical contact water level dipper).

In high-yielding water supply wells, especially where incrustation problems are anticipated, it may be desirable to install a pump bypass tube. Such a tube, which may be up to 100 mm in diameter, will aid in the monitoring and diagnosis of potential incrustation problems by permitting the intake section of the well (open hole or screen) to be inspected by CCTV camera survey or be geophysically logged (Section 6.4), and for water samples to be collected at discrete depths, all while the production pump is in use (Howsam *et al.*, 1995).

In shallow aquifers, water wells are usually drilled in two stages. A borehole is drilled through the superficial deposits at a diameter of at least 50 mm greater than the pump chamber OD – a diameter difference of 100 mm is desirable to give more annular clearance for grouting, although smaller drill diameters are possible – and the casing is set and grouted into solid rock [Figure 3.4(a)]. As with wells in crystalline aquifers, this grouted surface casing should extend several metres into the solid rock to seal the well against surface pollution. The Institute of Geologists of Ireland (2007) recommend that the grouted casing should extend at least 10 m into rock, or to 20 m below ground level, whichever is the greater. Drilling then continues to total depth for an open hole completion. As with crystalline aquifers, a screen section can be incorporated in the design if it is intended to exploit shallow, unstable, zones of highly fissured rock, but it is again essential to ensure that the annular space around the casing above the screen is sealed with grout to reduce the risk of pollutants entering the well. As noted in Section 2.8.2, the absence of a grout seal around the upper casing or inadequate headworks can lead to pollution of the groundwater supply even where the geological vulnerability of the aquifer may be low (Hynds *et al.*, 2012).

Water wells in deep aquifers may be drilled in several stages, with intermediate casing strings installed and grouted in place between the pump chamber and the aquifer [Figure 3.4(b)]. High groundwater pressures may be encountered when drilling in deep aquifers and it is important that the overlying casing strings are properly grouted in place before the deep aquifer is penetrated (see Box 5.3 in Chapter 5 for information on grouts and

grouting procedures). The pump-chamber casing diameter chosen is a compromise between the minimum needed to house the pump and the diameter needed to allow the intermediate casing strings to pass through. The diameter of the final section through the aquifer has to be chosen with care, for it has direct effects not only on the potential discharge of the well but also on the diameters of all of the casing strings, and therefore on the well cost. A minimum diameter of about 150 mm is recommended as this allows passage of most work-over tools, such as jetting tools, into the well for maintenance and rehabilitation operations (Chapter 9). An increase in diameter of drilling does not give a proportional increase in the yield of the well (Box 3.5), and large well diameters are therefore rarely justified on hydraulic grounds. In the past, some very large diameter (more than 900 mm) drilled wells have been constructed to house more than one pump, but we would not normally consider such well designs appropriate nowadays, either hydraulically or economically.

The design of a well in some thick, consolidated aquifers is similar to that in a crystalline aquifer because there is no well-defined base to the aquifer. Also, the groundwater flow in most consolidated aquifers is largely through fissures or zones of enhanced permeability, so criteria for optimizing well design, based on a uniform aquifer (Sections 3.1.5 and 4.6), cannot be applied. The depth of borehole needed for a target discharge will depend on the distribution and size of water-bearing fractures – an aquifer feature that can rarely be predicted with any certainty. Well design has to rely on experience of previous drilling in the area, including geophysical logs of existing boreholes (Section 6.4). In the absence of detailed information, some broad generalizations can usually be made concerning fracture distribution: the fractures are often bedding features and tend to be more frequent and open near the surface. In limestone aquifers, those fractures near to the water table may be widened through dissolution of carbonate minerals. This karstification process can produce very large conduits with groundwater flow velocities of several hundred metres per hour.

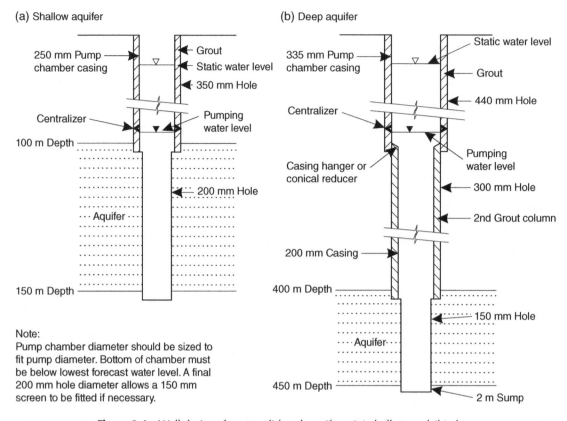

Figure 3.4   *Well designs for consolidated aquifers. (a) shallow and (b) deep*

Well construction in such aquifers can be problematical (Kresic, 2014): not only is it difficult to predict the locations of the main fissures prior to drilling, but if major cavities are encountered then drilling and sampling may experience problems of loss of circulation (Section 5.2.2). Even if a well is successfully completed in a karst aquifer, it may be vulnerable to rapid infiltration of pollutants from the surface. The performance of a low yielding well in a carbonate aquifer can sometimes be improved by applying acid treatments, which aim to widen the apertures of individual fractures and increase the interconnectivity between different fractures in the vicinity of the well (Section 5.9.4). The possible need for acidization should be considered when choosing a suitable length of grout seal behind the well casing.

Another factor to be considered when there is fracture flow within the aquifer, is erosion of sediment from the fracture walls by rapid flow along the fractures. This sediment in the water discharged from the well can damage the pump and silt-up surface works. The judgement on long-term stability is difficult to make, and can only be made on experience with other water wells, either in the area or in similar aquifers elsewhere. If there is any doubt, then the well should be designed to incorporate the necessary screens and gravel pack, just as if the aquifer was composed of an unconsolidated sediment. The specifications for screens should be finalized at the design stage, before a drilling contract goes to tender. It is both difficult and expensive to adapt a borehole not designed for screens, and to keep it

---

Box 3.5   Relationship between well discharge and well radius

The theoretical relationship between well discharge rate and well radius can be demonstrated using the Thiem equation for steady-state flow in a confined aquifer [Equation (1.21)], rearranged in terms of discharge rate (Q):

$$Q = \frac{2\pi T s_w}{\ln \dfrac{r_e}{r_w}}$$

where s_w is the drawdown in the pumping well, T the aquifer transmissivity, r_e the radius of influence of the pumping well and r_w the well radius. Let us consider the case of a well with a radius $r_w = 100\,mm$, which is being pumped under steady-state conditions in a confined aquifer which it fully penetrates. If the values for the other terms in the above equation are assumed as: $r_e = 300\,m$, $s_w = 5\,m$ and $T = 500\,m^2\,d^{-1}$, then:

$$Q = \frac{2\pi \times 500 \times 5}{\ln \dfrac{300}{0.1}} = 1960\,m^3 d^{-1}$$

If the well radius r_w is doubled from 100 mm to 200 mm, while keeping the values of s_w, r_e, and T the same, the value of Q now becomes $2150\,m^3\,d^{-1}$; that is, an increase in discharge of only about 10 percent. The reason for such a small increase is, of course, because the discharge rate in the Thiem equation is proportional to the *log of the well radius*, and so is relatively insensitive to changes in the value of the radius itself.

We must add a couple of caveats here. The calculation is based on the theoretical behaviour of drawdown in a pumping well in a homogeneous aquifer: it does not take account of well losses, which could have an appreciable impact on yield if too small a well diameter is chosen (Section 4.5). Secondly, an increase in diameter of a well in a fractured aquifer could have a significant effect on yield if that diameter increase leads to the well intersecting a fracture or fractures that would have been missed by a smaller diameter bore.

---

open while the screen is selected, ordered and delivered to site.

Multiple aquifers. A production well in a multiple aquifer system (Figure 3.5) will normally be completed as a screened well since the intervening low permeability layers (aquitards) will tend to be unstable and need casing-off. In theory, if the aquifers really are consolidated then the screen design is not very important but, in practice, to reduce the risk of sand-sized particles entering the well, the screens should be chosen on the assumption that the aquifer material is unconsolidated. The screen sections should end at least one metre from the top and bottom of each aquifer to avoid incursion of aquitard material. The annulus around the screen can be infilled with gravel pack material to act as a formation stabilizer to prevent undue collapse of material down the annulus and against the screen. The emplacement of a formation stabilizer may be difficult in deep boreholes. The accurate setting of screens in a multiple aquifer system is essential and, in most cases, will require geophysical borehole logging to provide the necessary depth control (Section 6.4).

In a multiple aquifer system where there are significant water pressure (head) and/or water quality variations between the aquifers, it is much better to design wells to exploit an individual aquifer rather than screen several aquifer layers. By following this approach, the well construction should not cause groundwater to flow between the aquifers, which in turn could lead to cross-contamination effects. If there are potential concerns about cross-contamination, then it is important to consult with the appropriate regulatory authorities that deal with groundwater abstractions and aquifer protection, as they will normally provide advice on this aspect.

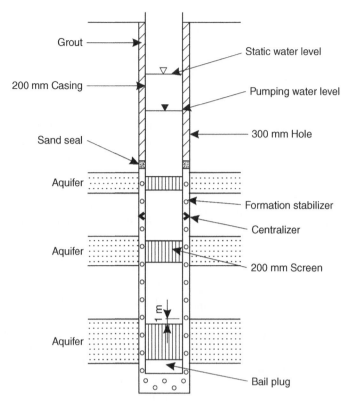

Figure 3.5   *A well design for a consolidated multiple aquifer system*

3.1.4   Wells in unconsolidated aquifers

The most common unconsolidated aquifers are alluvial deposits along river flood plains or terraces, and these range from gravel beds along small rivers, to multi-aquifer systems several hundred metres thick – along such major rivers as the Indus, for example. The design of water wells in unconsolidated aquifers has much in common with that of wells in consolidated strata, except that the former invariably require screening to prevent formation collapse. A water well design varies with the number of aquifers to be exploited and the depth of those aquifers, but each design will incorporate the following features: total depth, drilled

diameter, casing selection, screen selection, casing and screen installation and gravel pack design. These issues are addressed in general terms below; further discussion of construction materials, dimensions, hydraulic design and economic optimization are included in Chapter 4.

Total depth. In an aquifer of limited thickness the well would normally be fully penetrating so as to maximize yield. The depth of a well in a very thick aquifer is governed by the required discharge and by cost constraints. The thickness of aquifer to be drilled to give a design discharge can be estimated using a simple equilibrium formula such as the

Logan equation (1.23) described in Section 1.3.3 and Section 3.1.2:

$$T = Kb = \frac{1.22Q}{s_w} \qquad (1.23)$$

where T is the aquifer transmissivity, K the hydraulic conductivity, b the saturated thickness of aquifer to be screened, Q the design discharge rate and s_w the drawdown. The factor 1.22 is for ideal flow conditions and we usually replace this by a higher factor such as 2.0 to allow for additional well drawdown resulting from partial penetration effects and well losses (Section 4.5), and so this equation, when expressed in terms of b, becomes:

$$b = \frac{2Q}{Ks_w} \qquad (3.3)$$

As an example, if we have a well where a discharge of $2500\,\text{m}^3\,\text{d}^{-1}$ is proposed, where a drawdown of 25 m is acceptable (e.g., to maintain confined aquifer conditions during pumping), and where the mean hydraulic conductivity of the aquifer material is $10\,\text{m}\,\text{d}^{-1}$, then:

$$b = \frac{2 \times 2500}{10 \times 25} = 20\,\text{m}$$

That is, we are likely to need a minimum penetration of 20 m of saturated aquifer in this well. This approach presupposes that the aquifer material is relatively uniform; it cannot be applied in very heterogeneous formations. Also, well losses are actually proportional to the square of the discharge rate, so the allowance for well losses included in the linear multiplier of 2.0 in Equation (3.3) will not be adequate in the case of inefficient wells pumping at high discharge rates. A detailed discussion of well losses is included in Section 4.5.

Drilled diameter. The drilled diameter is dictated by the casing design chosen for the well. The drilled hole at any depth should have a minimum diameter about 50 mm greater than the OD of the casing and screen string, although larger clearances

are needed for grouting operations and installing a gravel pack. For grouting, the ANSI/NGWA Water Well Construction Standard (ANSI/NGWA-01-14, 2014), specifies that the drilled diameter should be at least 76 mm (3 inches) larger than the casing OD (when grouting with a tremie pipe), whilst Sterrett (2007) recommends that the borehole diameter should be at least 102 mm (4 inches) larger than the casing. The diameter of a borehole to be equipped with an artificial sand or gravel pack has to be at least 150 mm greater than the screen OD to accommodate the pack, which should be at least 75 mm thick (Section 4.4).

Casing/screen size and location. The casing in shallow wells is in one string, with an ID giving sufficient clearance for the pump and monitoring/ access tubes. The screen may be the same diameter as the casing or smaller [Figure 3.6(a)]. The latter can result in savings in capital costs.

In deep wells, the casing and screen may be in several sections, with reducers between each section [Figure 3.6(b)]. If the various casing/screen sections are to be installed in separate operations, as in the design for consolidated aquifers in Figure 3.4(b), then it is essential to ensure that they will nest inside each other. The intermediate casing must be large enough for the screen to pass, and be small enough to pass through the pump-chamber casing. The pump chamber again has to be large enough to accommodate the pump and monitoring/access tubes.

The bottom of the pump chamber commonly defines the depth below which the pump should not be lowered; therefore it must be at a sufficient depth below the static water level to allow for the anticipated drawdown, the length of the pump (below its intake) and a safety factor. The latter is based on a judgement on the long-term behaviour of the water table in the area being exploited plus an allowance for some deterioration in well performance and hence increase in drawdown over the lifetime of the well. This safety factor can be greater than the other two factors put together.

The decisions to be made on screen selection involve the length, diameter and type of screen to

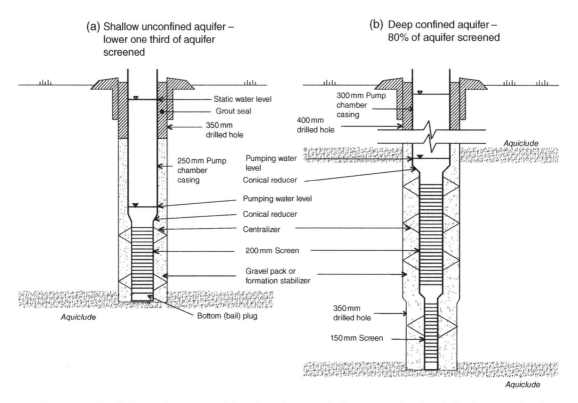

(a) Shallow unconfined aquifer – lower one third of aquifer screened

- Static water level
- Grout seal
- 350 mm drilled hole
- 250 mm Pump chamber casing
- Pumping water level
- Conical reducer
- Centralizer
- 200 mm Screen
- Gravel pack or formation stabilizer
- Aquiclude
- Bottom (bail) plug

(b) Deep confined aquifer – 80% of aquifer screened

- 300 mm Pump chamber casing
- 400 mm drilled hole
- Aquiclude
- Pumping water level
- Conical reducer
- 350 mm drilled hole
- 150 mm Screen
- Aquiclude

Figure 3.6 *Well designs for unconsolidated aquifers: (a) shallow, unconfined and (b) deep, confined*

be used. The screen is usually more expensive than the casing, so the efficient use of the screen can reduce the capital cost of a well. The screen should not extend above the lowest anticipated pumping water level. This avoids the upper section of the aquifer and screen becoming aerated as the aquifer is dewatered, and reduces the risk of incrustation of the screen and aquifer (Section 9.1.2).

In the case of a confined aquifer of limited thickness, it is generally recognized that most of the aquifer should be screened. Driscoll (1986) and Sterrett (2007), for example, recommend that 80 to 90 percent of the water-bearing zone in a homogeneous confined aquifer should be screened, while the ANSI/NGWA Water Well Construction Standard (ANSI/NGWA, 2014) specifies that at least 80 percent of a homogeneous confined aquifer shall be screened, adding that "professional judgment shall be used in determining the length of

screens in all aquifers". Keeping the total screen length at less than 100 percent of the aquifer thickness allows for casing to be set into the top of the aquifer and a bail plug (see Box 3.3) to be put at the bottom of the screen. Where the aquifer is unconfined, the choice of screen length is a trade off between a longer screen length giving a higher specific capacity but at the same time leading to a reduction in the available drawdown. Driscoll (1986) and Sterrett (2007) recommend screening the lower one third to one half of a homogeneous unconfined aquifer; similar guidelines are given by the ANSI/NGWA standard (2014).

For a thick (e.g., more than 30 m thick) homogeneous aquifer, we can apply Equation (3.3) to estimate the required screen length. If the aquifer is heterogeneous, the well design can incorporate short sections of screen placed opposite the most permeable layers, with appropriate screen slot sizes for each layer (Section 4.3), and blank casing against the intervening aquitards, similar to the design for a consolidated aquifer system in Figure 3.5. If the aquifer and aquitard layers are very thin, then a well cannot be screened in this way. Here, the whole well should be screened and gravel packed below the pump-chamber casing. The grain size of an artificial gravel pack (Section 4.4) would normally be selected to suit the finer-grained sand beds, to minimize the danger of drawing sediment into the well. This will only work if the aquitard material is cohesive, which would usually be the case if it has a high clay content. However, if it is mainly a non-cohesive silt then such a design would run the risk of drawing silt into the well. In this situation, our best approach would be to try and set blank casing opposite the silt layers, even if this meant casing-out some of the thin sand layers also.

The diameter of the screen should be at least 150 mm to allow ready access for work-over tools in future maintenance operations, although smaller diameter screens are sometimes used for low yielding wells in remote areas where the prospects of future well maintenance are slim. Screen diameters in excess of 300 mm are rarely needed on hydraulic grounds (Section 4.5, Table 4.5).

Screens are manufactured to many designs in a wide range of materials. The choice of screen type is considered in Section 4.3.

Gravel pack choice. A gravel or sand pack is introduced around the screen of a water well to produce an envelope of material with enhanced permeability and stability adjacent to the screen. The enhanced permeability reduces well losses (Section 4.5) and incrustation of the screen (Section 9.1.2), while the physical stability reduces the amount of sediment drawn into the well by pumping. Two main kinds of gravel pack are used, natural and artificial, depending on the type of aquifer being drilled.

A natural gravel pack is produced by development of the unconsolidated aquifer formation itself (Section 5.9). An aquifer is suitable for the development of a natural gravel pack if it is coarse-grained and poorly sorted. Artificial gravel packs are used in unconsolidated aquifers where the aquifer material is either very fine or well-sorted (that is, of uniform grading). A significant advantage of the artificial pack is that, because the pack material is coarser than the geological formation, screens with larger slot sizes can be used. Artificial packs are useful in allowing thin-bedded heterogeneous aquifers to be screened much more safely than with direct screening and natural development. The criteria used for selecting a natural or artificial gravel pack, and the design of an artificial pack, are described in Section 4.4.

3.1.5 Economic considerations in well design

The economics of well design in most situations depend on common-sense guidelines to avoid over-design and hence unnecessary expenditure, because the total depth of the well and the screen length are dictated by the aquifer geometry. Simple guidelines include:

1. Do not drill deeper than necessary.
2. Do not drill at larger diameter than necessary:
   - Do not design a gravel pack thicker than needed.
   - Do not design a screen or casing of greater diameter than necessary.
3. Do not use expensive materials where cheaper ones will do.
4. Do not use more screen than is necessary – although there is a trade off between reducing screen length and increasing pumping water level, as explained below.

In crystalline and some fractured consolidated aquifers drilling two shallow wells often makes better economic sense than drilling one deep well, as the chances of encountering water are greater (not least because of the decreasing fracture aperture at depth).

The water demand for an individual household is unlikely to exceed $1 \, m^3 \, d^{-1}$, so it is not normally necessary to design a high capacity well for such a house. A household can be supplied though a 75–100 mm OD, or even smaller, pump. The

---

Box 3.6   Cost-effective wells in economically-disadvantaged communities

In economically-disadvantaged communities like those often found in rural areas of sub-Saharan Africa, it is especially important to try and make the well construction process as cost-effective as possible. It has been observed that development programmes serving such communities "find it difficult to support capital costs in excess of $3,000 waterwell" (Foster, 2012), whereas examples of well prices for wells in five African countries given by Danert *et al.* (2008) are generally higher than this figure, ranging between $2,700 and $11,700 per well.

The concept of "cost-effective boreholes" has been promoted by the Swiss-based organization, the Rural Water Supply Network (RWSN), so as to improve access to sustainable groundwater supplies. The RWSN code of practice identifies nine key principles to achieve cost-effective drilled wells, which can be summarized as (RWSN, 2010):

1. Construction and supervision of boreholes should be undertaken by professional and competent organizations, which should adhere to national standards.
2. Borehole siting should include a hydrogeological desk study and field reconnaissance, and take account of the preferences of the local community (see Box 2.1).
3. The construction method chosen for the borehole should be the most economical, should involve appropriate drilling techniques, and the well depths should be neither over- or under-specified.
4. Appropriate procurement procedures should be followed to ensure that contracts are awarded to experienced and professional consultants and drilling contractors.
5. The design and construction of the borehole should be cost-effective, based on a minimum specification to achieve a target borehole life of 20 to 50 years.

6. Arrangements need to be put in place for adequate supervision, contract management and payment of the drilling contractor.
7. High-quality data from each borehole should be collected and submitted to the relevant government authority.
8. A hydrogeological database should be established and updated by the relevant Government institution, and the data made freely available to aid future drilling programmes.
9. The functionality of completed boreholes should be monitored.

Proper well siting has already been covered in Chapter 2 and issues such as drilling techniques, supervision, data collection and storage, and well monitoring will be discussed in later chapters of this book. Principles 3 and 5 above are of most relevance to this introductory chapter on well design. Key design features that contribute to the capital cost of the well include the well depth, well diameter and the lining materials. It is not uncommon practice for boreholes to be drilled beyond their optimum yield depth, incurring unnecessary costs. Doyen (2003; cited in Danert *et al.*, 2008) reported that cost savings of around 25% could be made for boreholes in Kenya if the boreholes were terminated at the optimum yield depth. With respect to well diameter, even though a minimum lining (casing and screen) diameter of 150 mm is desirable to enable access for well maintenance tools (Section 3.1.4), a smaller diameter of 100 mm can result in significant cost savings and be sufficient for cost-effective wells fitted with hand pumps (or in some cases small diameter submersible pumps). For most situations, relatively inexpensive plastic materials will be adequate for the casing and screen. In some countries, especially in Asia, locally available materials such as bamboo have been used for constructing low-cost shallow wells Shakya *et al.* (2009).

installation of a pump chamber with an ID of more than 200 mm would be unwarranted.

Similarly, for village water supplies in economically-disadvantaged areas it is important to ensure that the well design and construction are carried out in as cost-effective a manner as possible. This will generally entail constructing a small diameter borehole, lining this with 100 mm diameter grouted casing (and screen if necessary) to the optimum depth for yield, and then installing a handpump (Box 3.6). It is worth noting that, in Afghanistan, some humanitarian organizations follow a deliberate policy of installing a completion casing/screen string of only 100 mm diameter. This allows the installation of a hand pump, but not a motorized pump and thus avoids any danger of excessive abstraction and overexploitation of marginal aquifers.

The well design in a thick, uniform, unconsolidated aquifer is open to optimization techniques because the aquifer geometry dictates neither the total depth nor the screen length in the well. The primary design aim in this situation is to obtain the required discharge rate at the lowest cost. The cost of the water pumped from a well depends on both the capital costs and running costs of the well (Stoner *et al.*, 1979). These costs are interdependent because the components of the well design can affect the running (pumping costs). In a uniform aquifer, it is evident from Equation (3.3) that an increase in well depth and screen length – and hence in the capital cost – will reduce the drawdown for a given design discharge, and hence the running (pumping) cost. Similarly, a decrease in well diameter will decrease the capital costs but may increase the running costs through increased well losses (Section 4.5). There is an economic optimum – a least cost solution – for each design variable of the well. This optimization approach is described in detail in Section 4.6. Economic optimization principles can also be applied to the layout of the well field. Here the objective is to choose the well configuration that results in the minimum cost for transferring the water to a central location in the well field, while at the same time avoiding, or at least minimizing, interference effects between the wells (Swamee *et al.*, 1999).

Economic optimization of well design is especially relevant to a project in which large numbers of wells are to be constructed, such as a major irrigation or drainage project. In many potable water supply schemes, however, the well costs are small in comparison to the costs of the water treatment, storage and distribution, so the economic optimization of the well design may only have a small impact on the total cost of the scheme. Also, the optimization approach does presuppose a uniformity of aquifer conditions and a control over economic factors that may not be present in that part of the world where we are working.

3.2   Hand-dug wells

In industrialized societies, the hand-dug well has been largely superseded by the drilled well as a source of groundwater supply. The hand-dug wells that remain are often preserved because of their cultural or religious significance (Box 3.7). In developing countries the hand-dug well is still a common means for abstracting groundwater and, as a source of drinking water, it is still probably more important than the drilled well, and certainly more healthy than most surface water alternatives. In many societies, the well has an important role in community life, especially for the women and children who often are responsible for collecting the water (Figure 3.7).

The implementation of good well-design criteria therefore can have global effects on water supplies and public health. The following two sub-sections provide a short introduction to design criteria for achieving a good well yield and good water quality, respectively. Manual well construction methods are described in Section 5.8. Further information, including detailed guidance on crucial health and safety issues in construction, can be found in several publications that deal with hand-dug well construction, including those by Watt and Wood (1977), Banks and Less (1999), Collins (2000) and Davis and Lambert (2002).

Box 3.7   The Holy wells of Ireland

"No religious place in Ireland can be without a holy well" (attrib. Rev C Otway 1894; quoted in Varner, 2009).

Holy wells are an important cultural feature throughout the former Celtic lands of Western Europe, including Ireland (Robins and Misstear, 2000). Figure B3.7(i) shows a well dedicated to the seventh century Irish saint, St Saran. This well is one of perhaps 3000 holy wells in Ireland (Logan, 1980). These wells have a significant influence on Irish place names. According to Healy (2001), the Irish Townlands index gives 163 place names in Ireland that begin with the word *tobar*, or some variant of this Gaelic word for well. St Saran's well was previously known as *tobarsaran*.

Holy wells were widely believed to be a source of healing. An example is St Moling's well near the River Barrow in Co. Carlow [Figure B3.7(ii)], where several thousands of people made pilgrimage during the plague year of 1348 (Clyn, 2007). Many of the holy wells are still visited regularly, especially on the pattern or saint's day. Visitors say prayers to the saint, and leave votive offerings such as rags, statues and coins [Figure B3.7(iii)]. Great healing powers are attributed to holy wells, notably for curing eye ailments but also for rheumatism, backache, toothache and whooping cough. At St Moling's well [Figure B3.7(ii)], according to tradition, a child's head should be immersed three times in the flowing water for protection against influenza (Rackard and O'Callaghan, 2001). Although the wells are usually associated with a Christian saint, many pre-date the Christian era, and were subsequently adopted by the new religion when Christianity arrived in Ireland in the fifth century. A good example

Figure B3.7(i)   *St Saran's holy well, County Offaly, Ireland. Photo by Bruce Misstear*

Figure B3.7(ii) *St Moling's well Co Carlow, Ireland, a site of pilgrimage in the Middle Ages and where a pattern is still held each year. Photo by Bruce Misstear*

of this is given by the large number of wells dedicated to Saint Brigid, who is probably the second most important Irish saint after the country's patron saint, Patrick [Figure B3.7(iv)]. Brigid is also the name of the Celtic goddess of fertility, whose responsibilities included sacred wells (Healy, 2001).

Holy wells seem to be found in a wide range of hydrogeological environments: in unconsolidated and bedrock aquifers, in lowland bogs and on mountaintops, on islands and on shorelines. Some appear not to be wells at all, but instead are shallow hollows fed by rainwater or are ponds and lakes. With the diversity of hydrogeological settings, the hydrochemistry of well waters can be expected to vary considerably from site to site. Although the cures attributed to such waters are probably fanciful in many cases, there may be links between the presence of certain minerals and the reported health benefits in some instances. For example, elevated levels of lithium (up to $55\,\mu g\,l^{-1}$) have been detected in a holy well called *Tobar na nGealt*, in County Kerry (Buckley, *pers comm*). This name translates as "well of the madman" and the well has a long-standing reputation as a place where

Figure B3.7(iii) *Rags, cloths, photographs and other tokens wrapped around the branches of a tree next to St Brigid's well (Figure B3.7(iv)), County Kildare, Ireland. It is believed by many people that holy wells have magical powers: a cloth, which might previously have been tied around a rheumatic joint, is left beside the well in the hope that the ailment will be cured. Photo by Bruce Misstear*

people with mental health problems have sought cures. Interestingly, lithium has often been prescribed by doctors as a treatment for depression. Other researchers have expressed doubt on the hygienic benefits of numbers of possibly sick people sharing water sources of dubious quality (Kirschner *et al.*, 2012).

Figure B3.7(iv)   *St Brigid's well, County Kildare, Ireland. The well is located at the rear of the fenced area, behind the arch adorned with the white cross. The well is constructed around a small spring in a gravel aquifer and its discharge is channelled through two hollowed stones under the arch. The tree with the votive offerings shown in Figure B3.7(iii) is the second large tree from the left at the top of the picture. Photo by Bruce Misstear*

Figure 3.7   *Women and children gathered around a hand-dug well in Southern Blue Nile province, Sudan. Photo reproduced by permission of Laurence Gill*

3.2.1   Design for yield

Hand-dug wells have to possess a large enough diameter to accommodate the well diggers, but should not be wider than necessary. Excessive diameters will increase the volume of spoil to be removed, the time needed to dig the well, the cost, and the risk of surface pollution – all without significantly increasing the yield of the well.

The lining of a hand-dug well is most commonly of concrete rings or masonry, which are made permeable below the water table, but can include large-diameter metal or wooden casing and screen. In stable rocks, the well is often left unlined at depth. The method of installing linings

varies, but two examples will give an indication of the common approaches; whichever approach is followed, it is essential that the well digging team adopts good health and safety practices (Section 5.8 and Appendix 3). In soft formations where the water table is relatively shallow, the aquifer is lined with permeable concrete rings (caissons), while above the water table the well is lined with impermeable rings. The rings are typically about 0.9 to 1.5 m in diameter, between 0.5 and 1 m high and capable of being handled by a small crane or a tripod with a lifting block and tackle. The concrete rings are usually up to about 100 mm thick and can contain steel or other reinforcing material (reinforcement is especially important for rings that are not made on site, but have to be transported to the well location, often across rough terrain). The bottom ring may have a cutting edge attached.

A hole is dug from the surface to accommodate the bottom ring which is positioned carefully in a vertical position. Digging proceeds from beneath this first ring, which follows the excavation under its own weight. Extra rings are added from the surface as necessary. Excavation continues as far as is required, or is possible, below the water table, by lowering the water level in the well by pumping. The discharge of the pump being used to keep the hole dry during the excavation should give a good indication of when the desired yield has been obtained, and hence when no further excavation is required. If a dewatering pump is not available, then it is unlikely to be possible to excavate the well more than 1 m below the water table. On completion, the joints between the rings are cemented and the surface works are built. The bottom of the well is stabilized with a layer of about 200 mm of coarse gravel (Figure 3.8). Some wells may require deepening towards the end of the dry season when the water table is naturally at its lowest. In this event, the new section of the well is usually lined with permeable concrete rings of smaller diameter which can be telescoped inside the original well lining. As an alternative, the yield of the well may be increased by constructing horizontal boreholes or

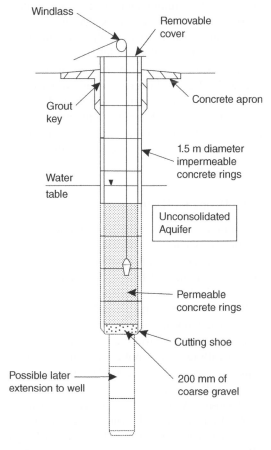

Figure 3.8 *A hand-dug well design for a shallow water table*

adits out from the well below the water table, thus converting the well into a radial collector well (Section 3.4).

Where the water table is relatively deep and the ground is stable (Figure 3.9), the shaft to the water table can be excavated in one or more sections, then lined by masonry or, if the ground stability is uncertain, the masonry can be installed as the excavation progresses. On reaching the water table, the masonry lining is pointed to give it strength and water tightness. The annulus behind the lining is filled with cement grout to prevent surface water percolating downwards and contaminating the well. As an alternative to masonry, the well above the water table can be lined with an in-situ concrete lining, the concrete being poured behind temporary metal shutters that form a ring with a diameter about 200 mm smaller than the excavation diameter. With the masonry or in-situ concreting methods, the well to the water table is usually completed in sections of not more than 5 m, to avoid having a deep unsupported and potentially dangerous excavation. Below the water table, the excavation continues for as far as is needed, or possible, using permeable concrete caissons as in the first design.

In very marginal shallow aquifers the hand-dug well has a distinct advantage over a drilled well in that it acts as a reservoir into which the aquifer can leak continuously in order to meet a peak demand which is much greater than the instantaneous aquifer yield. In rural communities in the tropics this peak demand often occurs during the main cooking and washing times in the early morning and evening. The well water level is thus able to recover during the daytime and overnight. The well can be deepened to create a reservoir volume capable of meeting demand or holding the daily yield of the aquifer. Where the aquifer below the hand-dug well is confined, a small diameter borehole may be drilled in the base of the well to increase the inflow.

3.2.2 Design for health

A hand-dug well is far more vulnerable to pollution than a drilled well. Dug wells tend to be shallow and open to infiltration of polluted surface water, their lining and headworks are commonly badly finished so that spilled water or animal wastes can flow back into the wells, and their tops are often left open, allowing rubbish to fall in. Moreover, hand-dug wells are frequently badly situated with respect to pollution sources such as latrines and septic tanks – a problem which applies to many drilled wells also.

A hand-dug well can be protected from these hazards by using the following design criteria to limit pollution of groundwater. The detailed design will need to take account of local conditions. It is especially important to liaise with local people

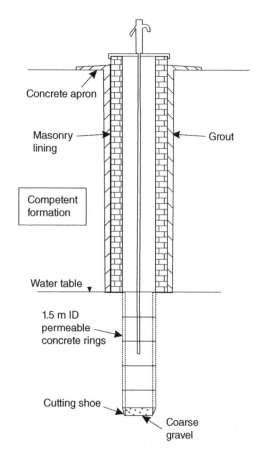

Figure 3.9 *A hand-dug well design for a deep water table*

Concrete apron

Masonry lining

Grout

Competent formation

Water table

1.5 m ID permeable concrete rings

Cutting shoe

Coarse gravel

Figure 3.10   *Concrete apron and drainage channel around a hand-dug well, Uganda. Photo by Bruce Misstear*

concerning the location of the new well (Box 2.1) and the design of the surface works to ensure that these meet their requirements. A proper consultation process will increase the likelihood that the new wells will be used and maintained properly by the community.

1. The well should be located up the hydraulic gradient (usually uphill) from any latrines or privies (Section 2.8). The minimum distance between the well and the pollution sources should be 30 m (Collins, 2000). This minimum distance refers only to a single well and latrine, whereas latrine density and pollution loading can be very high in villages, and greater separation distances may be needed. Also, recommendations on separation distances are usually based on considerations of bacteria removal in the subsurface, whereas greater separation distances are required for virus removal (Section 2.8.3).

2. The upper lining of the well must be impermeable, cemented in place, and should extend to the water table.

3. The surface works of the well must shed spilled water away from the well, for example, by a surrounding concrete apron sloping away from the well (Figure 3.10).

4. Ideally, the surface works should be such that the water is used several metres away from the well, and not at the wellhead. Most importantly, animal troughs should be located away from the well.

5. The surface works must prevent effluents from entering the well. A concrete apron should be constructed around the well, and securely keyed to the well lining.

6. The surface works must prevent rubbish from entering the well. Ideally, the well should have a cover with a removable lid.

7. The method used to lift water should discourage surface spillage and contact between

contaminated private containers and the well water. A hand pump (Section 3.7) installed through a watertight well cover offers the least risk of spillage, but hand pumps are not always available and, in any case, it may be difficult to keep them in good repair in isolated communities. A bucket and windlass arrangement is still a common option. Several designs of windlass are available – a simple roller used in West Africa was designed to squeeze the water from the hoist rope and prevent transmission of guinea worm (a common parasite in the region) to the hands of the water haulers.

The well should be disinfected on completion and then periodically thereafter. The well should be pumped to evacuate the bulk of the contaminated water and then filled with chlorinated water. Sufficient chlorine should be added (usually in the form of hypochlorite) to ensure that the water has a significant surplus of chlorine. The well should be left overnight and then pumped to waste until the taste of chlorine has disappeared.

The condition of any well will deteriorate over time, and therefore needs to be monitored. Sanitary survey forms for hand-dug well installations are included in the World Health Organization's drinking water guidelines (Volume 3, World Health Organization, 1997) and in the World Health Organization/UNICEF handbook on rapid assessment of drinking water quality (World Health Organization/UNICEF, 2012). Separate forms are available for different types of well installation. The level of contamination risk and need for remedial action is determined from the answers to a series of questions on the condition of the well lining, the concrete apron around the well, the adequacy of the drainage away from the well, and so forth.

3.3   Infiltration galleries

An infiltration gallery is a horizontal well which is often designed to abstract groundwater in a thin alluvial aquifer (an aquifer that has a high permeability but a limited saturated thickness, and hence a low transmissivity). Compared to a vertical well, an infiltration gallery can provide a relatively large yield for a small drawdown in water level. Because of this small drawdown, a gallery design can be effective in situations not only where the aquifer thickness is limited, but also where the water quality is stratified; for example, in exploiting a thin fresh water layer that overlies saline water in a coastal aquifer, where a vertical production well would not be suitable because of its larger pumping drawdown, which would lead to upconing of the saline water and deterioration of water quality.

Many infiltration galleries are designed to exploit alluvial aquifers adjacent to or underneath a river bed (Figure 3.11). The use of an infiltration gallery to abstract river water from the river bank or from beneath a river bed can reduce water treatment costs by filtering the induced recharge through a natural fine-grained aquifer. In the Spey valley in Scotland, for example, infiltration galleries and bank-side vertical wells abstract water

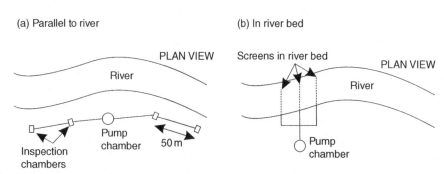

Figure 3.11   *Infiltration gallery layouts. (a) parallel to river and (b) in the river bed*

which has a much better water quality in terms of colour and turbidity than the river water that drains from this peat-rich catchment (Jones and Singleton, 2000).

There are several types of infiltration gallery. A common type is constructed from drainage pipes or well screens which are laid in a shallow trench that has been excavated below the water table by hand or mechanical digger. The area around the screens is usually backfilled with a gravel pack to act as a filter. The criteria for choosing the screen slot size and gravel pack grading are similar to those for vertical wells in unconsolidated aquifers (Section 4.4). The gallery may be tens or even hundreds of metres in length. A pump chamber is joined to one end of the gallery or, alternatively, may be located at some intermediate point along the gallery (Figure 3.11).

A second type of gallery consists of a large diameter chamber, constructed manually, which is lined with masonry, with openings between some of the bricks. The walls may also be constructed with porous concrete. In competent strata, sections of the gallery may be left unlined. In the English Chalk aquifer, there are sites where galleries have been constructed as horizontal adits to connect between a number of vertical wells. Many of these well and adit systems date from the nineteenth century, and they are often very productive (with yields of more than $10\,000\,\mathrm{m^3\,d^{-1}}$).

There is a special, and very long-established, type of gallery construction which is found widely in the Middle East, southwest Asia and north Africa, known variously as the falaj (Oman and the United Arab Emirates), qanat (Iran), kareez (Afghanistan and Azerbaijan) and khettara (Morocco). In this design, the water flows to the point of abstraction under gravity, and there is no need for a pump chamber [Figure B3.8(i) in Box 3.8]. Although very successful systems in many respects, their yields are vulnerable to the lowering of the water table in their source area, either through drought or due to adjacent abstractions from drilled wells (Banks and Soldal, 2002).

For a gallery constructed along a river bank parallel to the river, so that it receives most of its flow from the river, the length of screen required for a particular design discharge can be estimated by applying the Dupuit-Forchheimer solution for one-dimensional steady-state flow in an unconfined aquifer (Section 1.3.2). For the boundary conditions identified on Figure 3.12, Equation (1.14) becomes:

$$q = \frac{K}{2x}\left(H_r^{\,2} - H_g^{\,2}\right) \qquad (3.4)$$

where q is the flow per unit length of gallery, K the hydraulic conductivity of the river alluvium, x the perpendicular distance between the gallery and the river bank, and H_r and H_g are the water level elevations in the river and gallery, respectively. The total yield of the gallery Q is obtained by multiplying Equation (3.4) by the length of the gallery. The length of the gallery, L_g, can therefore be calculated from:

$$L_g = \frac{2xQ}{K\left(H_r^2 - H_g^2\right)} \qquad (3.5)$$

This analysis ignores any contribution to the flow in the gallery from the aquifer on the landward side (the area to the right of the gallery in Figure 3.12), and so is a rather conservative design formula.

The main disadvantages of infiltration galleries are that they are difficult to protect against pollution – being long, shallow constructions in areas often possessing a high groundwater vulnerability (Section 2.8) – their yields can be affected by a fall in the water table or river level during dry periods, and they can be problematical to maintain. Like vertical wells, infiltration galleries may become clogged through incrustation. However, maintaining a gallery in good condition can be more difficult than maintaining a vertical well: there are smaller hydraulic heads to work with when cleaning a gallery, and it is seldom practicable to generate sufficient energy to disturb blockages in the filter and formation outside the screen. Potential maintenance problems can be reduced by following good design criteria. The screen or drainage pipe should be made from

Box 3.8 The falaj system of Oman

The falaj (plural *aflaj*) system of exploiting groundwater was probably introduced into Oman from Persia about 2500 years ago (Lightfoot, 2000), and many of the 3000 or so aflaj in operation today are centuries old. The aflaj are of great cultural as well as engineering importance, and in 2006 five of Oman's aflaj were added to the UNESCO list of World Heritage Sites (Chakraborty, 2012; Walsh, 2013).

The main features of a falaj are illustrated in Figure B3.8(i). Water flows by gravity from the source area, known as the motherwell (*umm al falaj*), through a tunnel section towards the village. The tunnel generally has a gradient of between 1:500 and 1:2000, and is lined with masonry or stone blocks, which are sometimes cemented (Wilkinson, 1977). Near the village, where the falaj is relatively shallow, the tunnel may be constructed using cut-and-cover methods. At the head of the village, the place where the falaj water first appears at the surface is known as the *sharia*, and it is here that the water is used for drinking [Figure B3.8(ii)]. The falaj water is then conveyed in an open or covered channel through the village, with off-takes for bathing water and for religious ablutions. Downstream of the main housing area the falaj divides into a series of smaller channels which distribute the water to the irrigation fields. Water

is usually distributed to individual farms on a rotational, time-share basis. Historically, the time allocations during day light hours were determined using the sundial principle, whilst allocations at night-time were based on the movements of the stars (Nash and Agius, 2011). Sophisticated social structures have developed over the centuries to organize this water distribution.

The type of falaj illustrated in Figure B3.8(i) is known in Oman as the *dawudi falaj*, and is equivalent to the qanat of Iran, the kareez of Afghanistan or the khettara of Morocco. (The name dawudi falaj apparently derives from an Omani legend in which Solomon the son of David – Sulaiman ibn Daud – visited Oman on his magic carpet and instructed the aflaj to be built). The other falaj types are the *ghayli falaj*, which is a channel that carries wadi baseflow – usually from locations where the baseflow is brought to the surface by constrictions in the wadi – and the *ayni falaj* which is supplied directly from a spring; many of the aflaj in the Northern Oman mountains are of this ayni type.

The dawudi falaj is usually constructed in an alluvial gravel aquifer, near to, but not in, the main wadi channel (the occasional flood flows in the wadis are extremely erosive and could destroy a falaj). The motherwell is often located

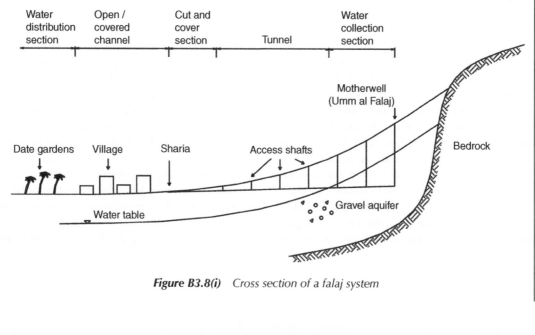

Figure B3.8(i) *Cross section of a falaj system*

Figure B3.8(ii) *Falaj Daris in Nizwa, one of the aflaj included on the UNESCO list of World Heritage Sites. Bruce Misstear is standing just downstream of the sharia section of the falaj, where the water first appears at the surface at the head of the town. Photo by Gillian Misstear*

Figure B3.8(iii) *Maintenance of a partially-collapsed section of a dawudi falaj, northern Oman. Photo by Bruce Misstear*

close to the mountain front, where the ground surface and the water table are at higher elevations than they are in the plains downstream. The motherwell is typically about 20 m deep, but can be up to 60 m deep. In some aflaj there is more than one motherwell, and so a falaj can have several branches connecting into the main tunnel section. The alignment of the tunnel sections can often be traced on the ground by following the remnants of the old spoil heaps left at the original access shafts. In Oman the tunnel sections are usually less than 5 km in length, and rarely exceed 10 km (in Iran, by contrast, there are qanats up to 30 km in length).

The flow in a dawudi falaj varies with the fluctuation of the water table in the motherwell area. There may be an interval of several years between the recharge events that occur from flash floods in the nearby wadi, during which time the falaj flow may reduce significantly, or the falaj may even dry up altogether. In modern times this situation has been exacerbated by interference drawdown effects from drilled wells. Wells, with motorized pumps, have the capacity to substantially deplete the source aquifer – unlike aflaj, which only drain by gravity the water that is available, and so operate in balance with the long-term recharge. In Oman, the construction of new wells is now prevented in the motherwell area of a falaj. In some instances, wells have been drilled downstream of the motherwell for the purpose of augmenting the falaj flow during dry periods. The flow of a falaj may also decline or cease altogether when a section of the tunnel collapses. Falaj maintenance is therefore essential. This may involve removal of caved material and relining [Figure B3.8(iii)].

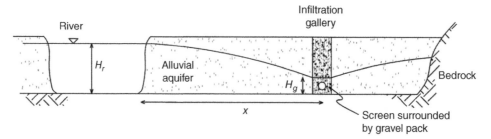

Figure 3.12 *Flow to an infiltration gallery located on a river bank*

corrosion-resistant material, and should be of sufficient diameter and open area so that head losses are not excessive for the design discharge (Section 4.5). In long galleries, inspection chambers should be installed at intervals (e.g., 50 m) to allow access for cleaning tools.

3.4   Radial collector wells

A radial collector well (sometimes known as a Ranney well, after the oil engineer Leo Ranney; Hunt, 2003) can be considered as a combination of a hand-dug well and an infiltration gallery. A large chamber or well is sunk into the aquifer, and horizontal boreholes are drilled radially outwards from this central well. Casing and screen are then jacked into the boreholes to form horizontal water wells, known as laterals. This design increases the effective diameter of the well and therefore its yield. The main advantage a radial collector well has over an infiltration gallery is that it can be constructed to greater depths below the water table; its main drawback is higher construction cost. There are many possible configurations for the number, depth, length and orientation of the laterals, and the design of collector well systems can be aided by the use of numerical models (Moore *et al.*, 2012).

Radial collector wells are most commonly installed in thin alluvial aquifers where the central well can be several metres across and the well laterals can be several tens of metres in length. The central well is completed like a hand-dug well, and can be lined with concrete rings or masonry. The design of the laterals follows the same criteria as a vertical water well in similar aquifer material. As with infiltration galleries, the radial collector well design is especially suitable for sites close to rivers, because the laterals can be directed both along the bank and beneath the river, to induce recharge from the river. In the major Káraný well field in the Czech Republic, radial collector wells are used to abstract groundwater from the alluvial gravel aquifer of the Jizera river valley. The river water, after pre-treatment, is recharged to the aquifer via a series of infiltration basins and then recovered from the radial collector wells (up to 50 m deep) after an average residence time of 30 days (Kněźek and Kubala, 1994).

Radial collector wells are also used sometimes in weathered crystalline aquifers to increase the yields of open wells. The laterals can be oriented along lines of preferential flow in the weathered and fractured rock. A project in the dry zone of northern Sri Lanka involved the conversion of 32 hand-dug wells into radial collector wells (Ball and Herbert, 1992). Horizontal boreholes up to 30 m in length were drilled from the base of the wells along the more permeable horizons of weathered crystalline rock. The average yields of the wells were more than doubled.

3.5   Observation boreholes

Observation boreholes tend to fall into four categories: those intended to provide a regional monitoring network; those drilled in a group around a test well for a pumping test; those drilled for pollution investigations; and those designed

Figure 3.13   *An observation borehole located near to an artificial recharge dam in northern Oman. Photo by Bruce Misstear*

for another specific purpose – for example, monitoring of groundwater levels close to an artificial recharge structure (Figure 3.13). In remote areas, exploration boreholes can be completed as observation boreholes and, in the case of boreholes drilled for a pumping test, the first will be drilled as an exploration borehole to provide lithological data on which the other boreholes can be designed.

The design of the observation borehole will depend on the purpose of the borehole, the site location, the geological succession, the nature of the aquifer under observation and the depth of borehole to be drilled. The design criteria in many cases will be simpler than those used for production wells in similar situations. As pump chambers are not required, observation boreholes can be completed to a much slimmer design than water wells. The completed internal diameter, however, must be large enough to accommodate any proposed sampling

or monitoring equipment. A minimum ID of 50 mm will be suitable for some depth samplers and low-capacity pumps, although an ID of 100 mm will facilitate a greater range of sampling equipment and geophysical logging tools, and will also make the borehole easier to develop (Section 5.9). An ID greater than 100 mm may make it difficult to purge the borehole adequately prior to sampling with a low-capacity pump (see Box 8.6 in Chapter 8). Small borehole completion diameters can be used for water level monitoring – pressure transducers and manual dip probes are usually less than 20 mm in diameter (Figure 2.21) – and may also be adequate for some of the specialized slim sampling devices available on the market. The drilled diameter of the borehole is usually 50–100 mm greater than the maximum OD of the casing string. The larger the annular clearance, the easier it is to install a borehole seal or gravel filter if this is required.

For groundwater level monitoring, the majority of observation boreholes are constructed using materials similar to those in water wells (Sections 4.1 to 4.3). Observation boreholes up to about 200 m deep can be constructed with relatively cheap plastic casing and screen, and more expensive materials are rarely justified. In tropical countries, plastics can deteriorate rapidly through ultraviolet degradation if stored in the sun, and can also suffer damage when transported over long distances on bad roads. All plastics should be shielded from the sun, and in some cases steel may be the preferred option because of its superior resistance to abuse (Section 4.2).

In pollution studies, it is important to choose construction materials that do not react with the pollutants being monitored. Specifically, the casing and screen materials should neither adsorb pollutants from the groundwater, nor leach substances into it. Steel casings can react with metallic elements in the well water, while certain plastics can react with organic compounds. Polyvinyl chloride (PVC) is widely used in pollution investigations and is considered suitable for most applications, but it should not be used where, for example, organic compounds are present as non-aqueous phase liquids (Fetter, 1999). It is important that the PVC used in an observation well is of a standard suitable for water wells (Section 4.2.2). If there are any doubts about the inertness of the proposed casing with respect to the pollutants to be monitored, it is advisable to consult the manufacturer.

It should be remembered in designing observation wells that both the hydraulic head and the groundwater quality may vary with depth, as well as laterally over an area (see Box 1.3 for an explanation of head as a three-dimensional concept). The hydraulic head shown by an observation borehole, screened over the full thickness of the aquifer, will represent some average of the heads distributed through the aquifer, and vertical flow in the borehole may take place. A similar average head will be given by a series of short screens distributed through the aquifer. In designing an observation borehole for monitoring groundwater

pollution, a long screen section should be avoided, since this may allow mixing between clean and polluted water. Screen lengths should normally be less than 2 m.

A sedimentary aquifer system may comprise several aquifer layers separated by aquitards or aquicludes. This presents a problem in borehole design that can be approached by either a single or multiple completion, with the latter design being strongly preferred for most applications.

Single completion. The observation borehole is completed with a single screen set adjacent to each aquifer [Figure 3.14(a)]. The head observed is then some 'average' value of head in all of the aquifers, and the water quality represents some mixture of the contributing groundwaters, perhaps dominated by a particularly poor quality inflow horizon.

If proposing a single completion design in a multiple aquifer system, for either an observation borehole or a water well, the potential risks to the aquifer system must be considered. Even a relatively short single completion design can lead to inflows and outflows between the well and the aquifer system. Figure 3.15 shows the flows (both simulated and actual) in an observation well constructed in a heterogeneous confined aquifer in South Carolina. Water enters the well at its base and leaves through the upper section of the 12 m long screen, the total upward flow being about $0.4 \, m^3 \, d^{-1}$ (Elci *et al.*, 2001). A single completion well can provide a pathway for groundwater to move between aquifers and hence potentially for cross-contamination to occur within the aquifer system. A single completion design, therefore, should be avoided if the aquifers exhibit strongly contrasting heads or water quality characteristics.

Multiple completion. In this design a nest of piezometers is installed in one borehole. Each piezometer has a single short length of screen set against a single aquifer, and separated from the other screens by an impermeable annular seal [Figure 3.14(b)]. The annular seal is usually made from bentonite or a bentonite and cement mix, and

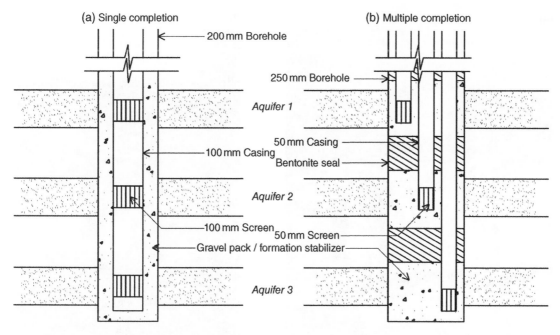

Figure 3.14   *Observation borehole designs: (a) single completion and (b) multiple completion*

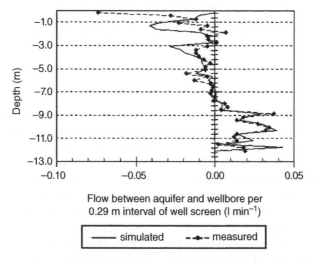

Figure 3.15   *Flow between an aquifer and screen in a single completion observation borehole. From Elci et al. (2001), Ground Water, 39(6): 853–862. Reproduced by permission of John Wiley and Sons*

is placed on top of the gravel pack or formation stabilizer that surrounds each screen. A high-solids bentonite is less likely to affect the water quality than a seal incorporating cement (ADITC, 2015).

This design has major advantages over a single completion. The definition of the head distribution, and therefore the understanding of the groundwater flow system, is much improved. Similarly, the

(a)

(b)

Figure 3.16 *Borehole clusters: (a) small diameter observation boreholes installed to different depths in dry river bed near Alice Springs, Australia, (b) cluster of larger diameter boreholes constructed in glacial overburden and crystalline rock aquifers in County Donegal, Ireland. Here, larger completion diameters were chosen to enable pumping tests to be carried out in the boreholes. Photos by Bruce Misstear*

variation in groundwater quality through the thickness of the aquifer system can be measured from this multiple completion. Importantly, multiple completion should prevent short-circuiting of waters of different hydraulic heads up or down the borehole (which can happen with a single completion).

The multiple completion design also has drawbacks. It is more complex and much more expensive than a single completion. It may be impossible to satisfactorily install the nest of piezometers in deep observation boreholes, because of the difficulty of setting the impermeable seals accurately. Even in shallow boreholes, the seals may not always work properly. Installing a nest of tubes in a borehole also means either that the tubes have to be of small diameter or the borehole has to be much larger than normal.

An alternative to multiple completion in a single borehole is to drill a cluster of boreholes, each screened in a different aquifer (Figure 3.16). This design is more reliable but the main problem is the cost of drilling multiple boreholes. Again, the technique is normally applied to shallow aquifer systems or specialized groundwater investigations.

Slim piezometers have been developed to study the head distribution in relatively shallow aquifers. Several designs are available, but they tend to follow similar basic principles, involving a short screen tip set at the end of a piezometer tube. The tube is usually made of plastic, with an ID between 12.5 or 25 mm, while the screen can be a perforated section of pipe wrapped with filter gauze, or a purpose-made porous ceramic tip.

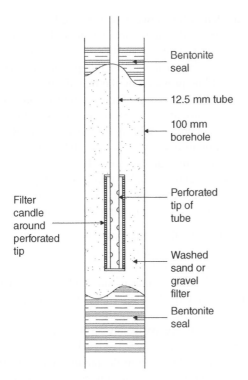

Bentonite seal

12.5 mm tube

100 mm borehole

Filter candle around perforated tip

Perforated tip of tube

Washed sand or gravel filter

Bentonite seal

Figure 3.17   *Casagrande type piezometer tip*

The well-known 'Casagrande' type of piezometer (Johnson, 1965) is illustrated in Figure 3.17.

Several proprietary designs of multilevel groundwater monitoring systems are available for collecting water samples from discrete aquifer horizons, in both consolidated and unconsolidated aquifers. Some of these systems involve permanent observation well installations using specialized multi-port tubing and packer arrangements (see Section 8.4).

3.6   Exploration boreholes

Exploration boreholes are drilled to provide information about the geology and the groundwater beneath the site. They are not permanent structures, unless completed as observation boreholes, and after drilling and sampling they should be backfilled. The correct completion of an exploratory borehole, especially in a groundwater pollution

investigation, is extremely important. A borehole left open can be a conduit for pollution from the surface to the groundwater body. Quite apart from polluting the groundwater, such conduits could give a false impression of the velocity of pollutant movement and the degree of pollutant attenuation in a pollution study. Therefore, it is recommended that, after sampling and testing have finished, all exploration boreholes are backfilled by pressure grouting through a tremie pipe set to the bottom of the hole or other suitable method (Section 9.4).

The proposed depth of an exploration borehole may be specified in terms of a geological horizon, but an estimate of the target depth must be made before drawing up a drilling contract, as it affects the choice of drilling rig and possibly the method of drilling (Chapter 5). The diameter of an exploration borehole, when it is not to be converted into a permanent well, is not usually too critical. The drilled diameter will be determined largely by the depth of the hole, the type of samples needed, and any particular testing requirements. A minimum diameter of 150 mm will allow access for most sampling equipment, but particularly for geophysical logging tools.

The recovery of formation samples and the establishment of a lithological log are extremely important parts of exploratory drilling. Sampling techniques are discussed in Section 6.2; sampling is not restricted to exploratory drilling, and similar methods are used in observation boreholes and water wells.

3.7   Pump selection

The topic of pumps is potentially enormous and cannot be covered in detail in this book. However, the hydrogeologist or water engineer *must* be aware of the crucial importance of pump selection for water well design, for at least two main reasons:

i. The type and, particularly, the diameter of the pump intake will often determine the diameter of the well casing (Section 3.1.1). This will, in turn, be a decisive factor for the entire design and construction of the borehole.

ii. The type of pump selected must be suited to the resources (both practical and financial) of the people who will use and operate the well. If not, it cannot be maintained and the well will soon fall into disuse (Box 3.9).

It is common to divide pumps into two types (Driscoll, 1986):

1. *Variable displacement pumps*, where the pumping head has a strong inverse relationship to discharge rate (Figure 3.18). These include:
   a. Centrifugal pumps (based on rotating impellers or turbines):
      • vertical turbine pumps;
      • submersible turbine pumps (including electrical submersible pumps);
      • most suction lift pumps.
   b. Jet pumps (not discussed in detail here).
   c. Air-lift pumps (not discussed in detail here, but see Section 5.9 for a description of air-lift pumping during well development).
2. *Positive displacement pumps*, where the discharge volume is relatively constant for a given pump speed and does not depend strongly on the head against which the pump operates (Figure 3.18). Positive displacement pumps include:
   a. Rotary pumps, where fixed volumes of water are displaced by snug-fitting rotating gears, vanes or screws (e.g., some suction pumps, helical rotor pumps).
   b. Peristaltic pumps (not discussed further here, but may be used for groundwater sampling rather than bulk pumping; see Section 8.4.5).
   c. Piston (reciprocating) pumps, including many hand pumps.

In order to select a pump, it is important to appreciate that:

1. For a given power input, and especially for variable displacement pumps (Figure 3.18), the quantity of water pumped decreases with increasing pumping head (height). There is thus a relationship between power (and hence cost), water yield and pumping height (Box 3.10). This is formalized in the concept of a *pump curve*. For each type of pump, such a curve should be provided by the manufacturer and it describes the pump efficiency under optimum conditions [new, unworn pump; non-turbid water; correct polarity of wiring (!)]. Real performance of pumps may fall below that indicated by manufacturers' figures.

2. The bigger the pumping head and the higher the yield, the greater the diameter of the pump and the greater its power rating and weight.

To illustrate these two principles, two examples of pump curves are shown in Figures 3.19(a) and 3.19(b). Figure 3.19(a) shows the curve for the 76 mm (3-inch) diameter, single-phase electrical submersible Grundfos SQ 1-35 pump designed for domestic supply. It is designed for pumping against heads of up to approximately 35 m, at flow rates of $<0.5 \, \text{l s}^{-1}$. It weighs only 5.5 kg and has a power consumption of $<1 \, \text{kW}$. From the pump curve, we can see that if it is pumping water up 25 m, it will achieve a flow rate of some $0.35 \, \text{l s}^{-1}$, with a power supplied by the electric motor to the pump (P2) of 0.29 kW. The electrical power consumed by the motor (P1) will be a little higher. If the pumping head is decreased to 15 m, the flow rate increases to $0.44 \, \text{l s}^{-1}$, with the power consumption staying similar.

Let us now consider Figure 3.19(b), for the 200 mm (8") diameter, 3-phase SP 215-6 multistage electrical submersible pump, weighing in at almost half a tonne. This is mainly designed for use by commercial suppliers of groundwater and its pump curve shows characteristics of a different order of magnitude. It can achieve pumping heads of up to 200 m, with flow rates of 30 to $70 \, \text{l s}^{-1}$. The price of this impressive performance is the power consumption of around 100 kW. This pump curve shows that, for a pumping head of 140 m, a flow rate of $63 \, \text{l s}^{-1}$ can be achieved.

We have chosen, in this section, to consider five of the pump types commonly used in water wells:

 i. vertical turbine pumps;
ii. electrical submersible pumps;

Box 3.9   Pump choice in impoverished rural communities

In Western industrialized nations, the electrical submersible pump is often the tool of choice for groundwater abstraction, as it offers many advantages in terms of ease of installation and removal, compactness, low noise, minimal head-works requirements, and because electricity supplies are readily available – and reliable. For impoverished rural communities in the developing world, however, the electrical submersible pump potentially has disadvantages: it requires several complex electromechanical devices (a motor, powering a generator, powering an electrical pump) to function, and to be maintained. Due to the mechanical → electrical → mechanical energy conversion, there is also a significant loss of power efficiency. Furthermore, to maintain such a system, the community requires access not only to mechanical skills – which may be readily available – but also to electrical skills, which are often harder to find.

The installation of a purely mechanical pump such as a helical rotor or vertical turbine may thus be preferable in certain situations: a mechanical pump is both potentially more efficient (no electrical energy intermediary step) and probably easier to maintain. It need not be powered by a dedicated motor but (via a belt drive – see Figure 2.9) can be powered by tractor engines or even petrol engines from domestic vehicles. The skills to maintain a basic diesel or petrol engine can often be acquired relatively readily (although maintaining regular fuel supplies can be a serious difficulty).

Alternative pumping strategies obviously include wind power and solar power. Wind power can be harnessed to drive rotary pumps or reciprocating (piston) pumps. The capital cost is, however, substantial, and the scheme depends on wind speed within a given range (i.e. the supply is likely to be discontinuous). Wind power is possibly most appropriate in agricultural applications (Figure 1.4).

Small-scale solar powered electrical submersible pumps, which are ideally suited for health centre or school use in the developing world, are becoming increasingly common. These may be of variable discharge or positive displacement (helical rotor) type and are reported to be rather robust. The main disadvantage is that, when breakdown occurs, their repair or replacement may be costly.

For the majority of impoverished rural communities the hand-pump is the most common type of pump fitted to a hand-dug or drilled well. In sub-Saharan Africa alone, there are about one million hand pumps, with about 60000 new hand pumps being installed each year (Sansom and Koestler, 2009). The three most common hand pumps in sub-Saharan Africa are the India Mk II, Afridev and Vergnet (MacArthur, 2015). There has been a trend towards standardization in the last few decades: in 2014, approximately 13 types of hand pumps were being installed in sub-Saharan Africa compared to 35 types in Burkina Faso alone in 1985 (MacArthur, 2015). Standardization can lead to benefits in terms of pumps being designed and manufactured to proper technical specifications (Baumann and Furey, 2013).

Even with greater standardization, hand pump functionality remains a major problem. One of the consequences of pump break-downs is that the community may revert to using an unimproved water source [Figure B3.9(i)]. In many African countries, more than 20% of hand pumps are non-functional at any one time, and in some countries the proportion exceeds 60% (Rural Water Supply Network, 2009). A detailed survey of waterpoint functionality in 51 districts in Tanzania found that 46% of the waterpoints were non-functional (WaterAid Tanzania, 2009); although functionality decreased with increasing age, the survey also showed that 25% of sources only two years old were

Figure B3.9(i) *A broken hand pump over a well in Uganda, with people reverting to an unprotected water source in the background. Photo by Bruce Misstear*

non-functional. In an on-line discussion 'Hand pumps: where now?' (Furey, 2014), the number one issue discussed was water quality, notably corrosion and high iron levels. Use of galvanized iron pump parts in corrosive groundwater was reported as a major failing (alternative materials discussed in the blog were PVC/ uPVC and stainless steel). In Uganda, wear on the rubber piston seals of hand pumps was the subject of recent research, with the researchers finding that the durability of the seals could be improved by the addition of surface coatings of carbon and silicon based materials (Lubwama *et al.*, 2015).

iii. motorized suction pumps;
iv. helical rotor pumps;
 v. hand pumps.

3.7.1 Vertical turbine pumps

The vertical turbine pump consists of one or several impeller units, rotated (and suspended) by a drive shaft enclosed within a pump column that also functions as a rising main to convey water to the surface [Figure 3.20(b)]. Each impeller will provide a given maximum lift. Increased lift can be

gained by mounting several impellers in series on top of each other in the downhole "pump bowl" to form a multi-stage unit. The drive shaft (which may be oil-lubricated or water lubricated, implying differing designs – see Driscoll, 1986) is rotated at the surface by a motor: this may be a directly mounted electric motor, or a fuel-driven motor (petrol, diesel or natural gas) via a right-angle gear drive (Figure 3.21) or a drive belt (Figure 2.9).

The main advantages of vertical turbine pumps are their reliability and the accessibility and

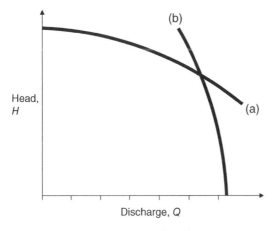

Figure 3.18 *Typical pump curves showing that the discharge rate for a variable displacement (centrifugal) pump (a) is much more sensitive to change in head than is the case for a positive displacement (helical rotor) pump (b)*

Box 3.10   Pump power ratings

The output power of a pump (i.e., the rate of work actually done by the pump) is given by (Davis and Lambert, 2002):

$$P_{out} = Q\rho_w g h$$

where P_{out} is the pump output power in watts, Q is the discharge in $\text{m}^3\ \text{s}^{-1}$, ρ_w the water density in kg m^{-3}, g the acceleration due to gravity ($9.81\,\text{m}\,\text{s}^{-2}$) and h the operating pressure head in m.

The input power to the pump P_{in} will need to be higher than P_{out} and the ratio (or overall efficiency η) will depend on the pump efficiency, power transmission and motor efficiencies.

$$P_{in}\eta = P_{out}$$

The efficiency η can often be in the range 0.4 to 0.6 (40 to 60%).

---

hence ease of maintenance of the motor assembly (above ground) – and the motor assembly can be replaced relatively easily. The main disadvantages are that they require a reasonable degree of clearance (well diameter) or a high degree of borehole verticality and straightness for installation. Also, for maintenance of the pump assembly itself, skilled craftsmen are required and the process of pump disassembly and removal is not straightforward.

3.7.2   Electrical submersible pumps

An electrical submersible pump [Figure 3.20(a)] has essentially the same pump bowl assembly as a vertical turbine pump, and this can be single stage or multi-stage. Instead of being rotated by a drive shaft from the surface, however, the turbines are driven by an electrical (one-phase or three-phase) motor mounted underwater below the turbines (Figure 3.22). The pump is thus connected to the surface by two or three elements:

- a power cable, supplying electricity to the pump;
- a rising main, which may be rigid (e.g., steel, flanged or flush couplings) or may be made of flexible plastic – the advantages of the flexible rising mains are the rapidity and ease of installation and removal of the pump;
- and, with flexible rising mains, a safety cable or rope to support the pump in case of rising main failure.

The pump motor needs to be cooled by water flowing past it. In cases where the well's flow is derived from above the pump, a shroud may be placed over the pump to artificially direct the water flow over the motor before entering the pump intake (which is usually located above the motor and below the turbines). Submersible pumps typically vary in size between 76 mm (3 inch) and 200 mm (8 inch), although small ones (<50 mm) are available for minor domestic applications or for well sampling.

The main advantages of the electrical submersible pump are: compactness, ease of installation and removal (if a flexible rising main is used),

(a)

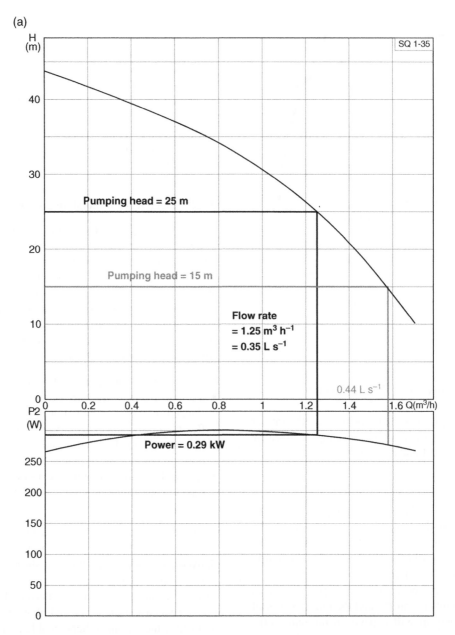

Figure 3.19(a)   *Pump head-discharge-power curve for the 76 mm (3-inch) diameter, single-phase electrical submersible Grundfos SQ 1-35 pump designed for domestic supply. Reproduced by permission of Grundfos (Ireland) Ltd*

(b)

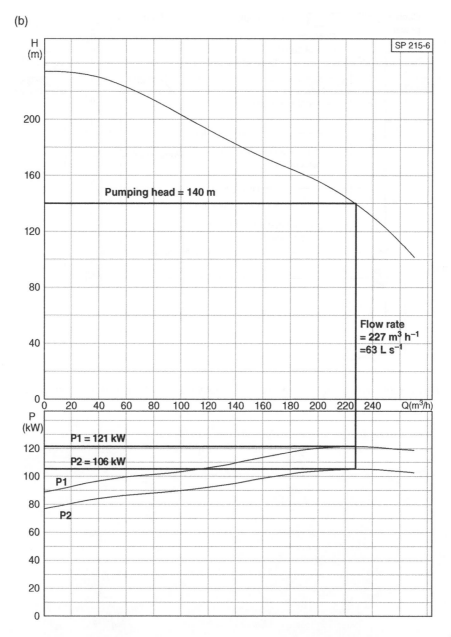

Figure 3.19(b)   *Pump head-discharge-power curve for the 200 mm diameter (8-inch), 3-phase Grundfos SP 215-6 multi-stage electrical submersible pump. Reproduced by permission of Grundfos (Ireland) Ltd*

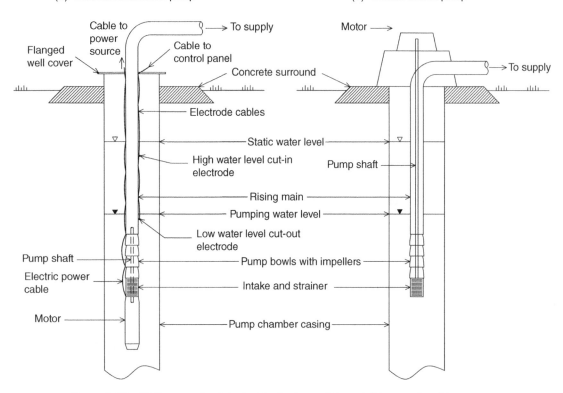

(a) Electric submersible pump

(b) Vertical turbine pump

Cable to power source

Flanged well cover

Cable to control panel

To supply

Motor

Concrete surround

To supply

Electrode cables

Static water level

High water level cut-in electrode

Pump shaft

Rising main

Pumping water level

Low water level cut-out electrode

Pump shaft

Electric power cable

Pump bowls with impellers

Intake and strainer

Motor

Pump chamber casing

Figure 3.20 *Well set-ups for (a) electrical submersible and (b) vertical turbine pumps*

Figure 3.21 *A vertical turbine pump installed in an irrigation well in Myanmar. The diesel prime mover is connected to the pump by a right-angle gear drive. This pump installation is temporary, for the purpose of a pumping test. Photo by Bruce Misstear*

Figure 3.22 *An electric submersible pump being installed in a well in Western Australia. Photo by Bruce Misstear*

lack of noise, no need for absolute verticality of borehole (especially if flexible rising main is used), minimal well-top installation. Disadvantages include: motor is downhole and needs to be removed for repair; need to ensure a stable and reliable supply of electricity, either from mains or from a generator (not a trivial consideration in many developing or wartime nations); the need to consider protection of the pump from overheating or running dry (water level cut-outs); need for both electrical and mechanical skills to maintain pump;

generally poorer overall power efficiency than vertical turbine pumps (Box 3.10).

3.7.3 Motorized suction pumps

These may be of variable displacement centrifugal type or positive displacement rotary type. All pumps operate by developing a partial suction behind the rotor/impeller; that is, in the pump intake. When we talk about a suction pump, we are merely referring to a pump where a suction hose or pipe leads from the pump intake (above water level) down to below the water level. Because the earth's atmospheric pressure at sea level is equivalent to about 10.4 m of water, this is the limit to the suction head that can be achieved. A suction pump can thus *never* pump water from a depth of more than 10 m below the pump housing, and (because of inefficiencies in design and other factors) its suction is usually not more than 5 or 6 m. Millions of people, notably in south Asia, use motorized suction pumps to abstract shallow groundwater for irrigation or drinking water supplies.

Figure 3.23(a) shows how the suction depth limitation can be overcome in wide-diameter dugs wells (for example, in Afghanistan or Oman), where the suction pump is mounted on a platform down the well. This is not necessarily a recommended solution: one needs to take great care to vent exhaust fumes from the pump's diesel motor. Also, the lowering of the suction pump may be a response to over-exploitation of the aquifer, and may thus exacerbate the problem of a diminishing resource further.

Another application of the suction pump is for the direct extraction of water from driven well points in high permeability, unconsolidated sediments with a shallow water table. This is the principle of many dewatering well arrays [Figure 3.23(b)].

In summary the main advantages of the suction pump are: ease of use and maintenance; no downhole components, and hence can be used in very narrow diameter or poorly accessible wells. The main disadvantages are: limited suction height

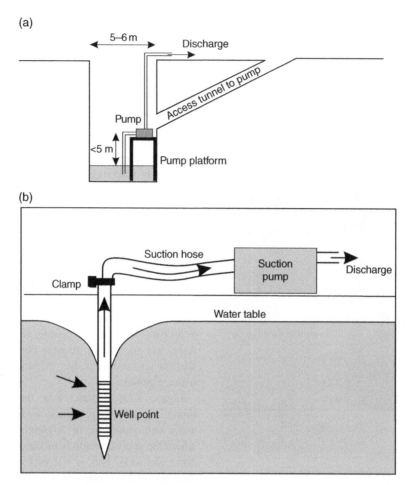

Figure 3.23 *Possible uses of motorized suction pumps. (a) placement on a platform within a wide-diameter dug well. Care should be taken to ventilate exhaust fumes thoroughly. (b) pumping of a well-point in loose, high permeability unconsolidated sediments*

(and hence pumping range) of less than about 6 m; need for priming to obtain initial suction; sensitive to entrainment of air in pump intake.

3.7.4 Helical rotor pumps

These are essentially the positive displacement version of the vertical turbine pump. Instead of a rotating turbine, the drive shaft powers a helical rotor in a rubber sheath or "stator" (Figure 3.24). As with the vertical turbine pump, the driveshaft can be powered directly by an electrical motor, or by a belt to a diesel or petrol engine – or by handle

as a hand pump. The advantages and disadvantages are similar to those for a vertical turbine pump. The main differences are that the helical rotor pump:

- is more tolerant of suspended sediment (fines);
- cannot be operated against a closed valve;
- cannot be operated in series (but can be operated in parallel);
- delivers a rate of water in relation to its speed: it is thus less sensitive to the need for a constant drive speed than a vertical turbine pump and is suited to coupling to a wide variety of motors;
- delivers a relatively constant head.

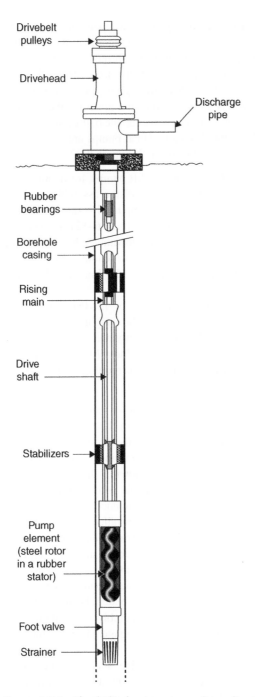

Drivebelt
pulleys

Drivehead

Discharge
pipe

Rubber
bearings

Borehole
casing

Rising
main

Drive
shaft

Stabilizers

Pump
element
(steel rotor
in a rubber
stator)

Foot valve

Strainer

Figure 3.24 *The helical rotor pump. From Davis and Lambert (2002), reproduced by permission of Practical Action Publishing*

3.7.5   Hand pumps

The simplest form of hand-pump is a bucket on a rope. This may be winched by hand (or animal) or be raised by a long cantilevered arm (the *shaduf*, Figure 3.25). This simple technology should not be ignored. It is easy and almost cost-free to maintain. Experience has shown that many rural communities lack the resources to maintain more sophisticated hand-pumps or motorized pumps (Box 3.9). This is seldom due to lack of technical skill, and is more often due to unwillingness or inability to afford maintenance, or the lack of an effective spare parts distribution network. In such situations, bucket solutions may be the preferable option: they do not have

Figure 3.25 *Shaduf used for lifting water from a temporary well in a dry river bed to an irrigated field beside the river bank, northern Nigeria. Photo by Bruce Misstear*

Figure 3.26   *Bailer tube used in a rural Kenyan drilled well. Photo by David Banks*

to be primitive or unhygienic. Dug wells where buckets are used should:

- have a dedicated bucket from which water is decanted to private domestic containers;
- be cordoned off from animals;
- have a covered well top, with a snug-fitting hole for the bucket to pass into the well (and this should also be covered when not in use).

Alternatively, it is possible to use a bailer tube, fitted with a basal flap-valve, on a rope or chain, fitting snugly down a narrow diameter borehole (Figure 3.26).

Many ingenious variants of manual pump are available, but most are reciprocating pumps based on the motion of a piston inside a cylinder. The piston is moved up and down manually via a string of pump rods in the rising main. On the upstroke, the piston sucks in water through a one-way foot valve to the cylinder. On the down-stroke, this volume of water flows through the piston via another one-way valve. On the next upstroke, the cylinder lifts the volume of water up the rising main towards the pump discharge. Hand pumps are divided into three types (Davis and Lambert 2002):

1. Shallow or suction lift pumps, for depths of up to 6–7 m, where the piston assembly is above the surface in the pump headworks (Fig. 3.27).
2. Medium lift pumps (lifts of 6–15 m), where the piston assembly is downhole, but where the weight of water is low enough to be lifted directly by hand (e.g., the Bangladeshi 'Tara' pump).
3. High lift pumps (lifts of 15–45 m, and sometimes more), where the weight of water is too great to be lifted directly and either a lever (as in the 'Afridev' and 'India Mk II' pumps), or gearing or a flywheel is required to assist in the water lift (Figure 3.28).

Typical discharge rates for hand-pumps depend on the lift, but are often in the range $0.3–0.5\,\mathrm{l\,s^{-1}}$ ($1000\,\mathrm{l\,hr^{-1}}$ is a commonly cited typical figure), and they often employ a rising main diameter of some 60 to 65 mm, allowing them to be fitted inside a 100 mm borehole (this is less than the 50 mm clearance recommended in Section 3.1.3 for a motorized pump installed inside a pump-chamber casing). The Afridev, and other models, may be licensed for production at a large number of factories in various countries. The designs may vary a little according to local conditions but, more importantly, the quality of materials and workmanship may also vary considerably, necessitating any potential purchaser to engage in some form of quality control.

Considerable knowledge and effort has often been expended in hand-pump design. Some models are designed for ease of maintenance, with piston assembly and pump rods being removable for

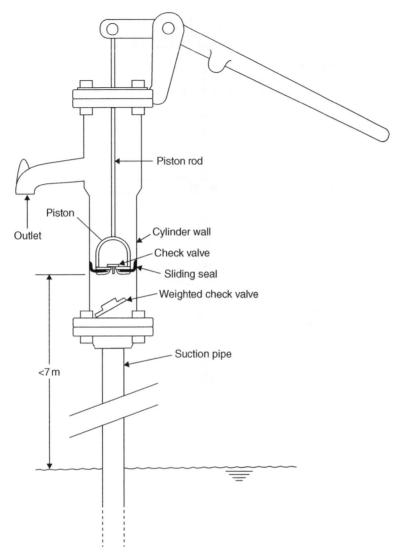

Figure 3.27 *Shallow lift (suction) hand pump. From Davis and Lambert (2002), reproduced by permission of Practical Action Publishing*

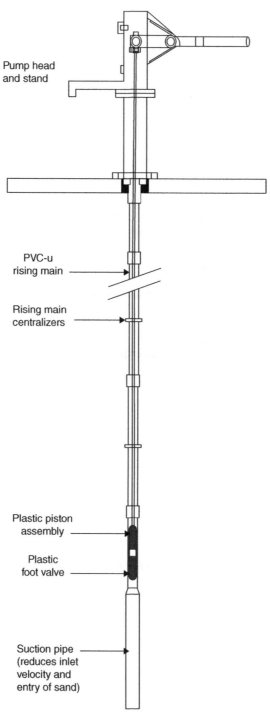

Pump head
and stand

PVC-u
rising main

Rising main
centralizers

Plastic piston
assembly

Plastic
foot valve

Suction pipe
(reduces inlet
velocity and
entry of sand)

Figure 3.28   *High-lift AfriDev type hand-pump. From Davis and Lambert (2002), reproduced by permission of Practical Action Publishing*

Figure 3.29 *Removal of pump rods during hand-pump maintenance, Southern Blue Nile province, Sudan. Photo reproduced by permission of Laurence Gill*

maintenance through the rising main (Figure 3.29). Hand-pumps are not maintenance-free options, however, and there are numerous examples throughout the developing world where the hand-pumps are broken. The capital cost of some hand-pumps is substantial, and their affordability as well as their reliability and ease of maintenance should be carefully considered when recommending them for use in impoverished communities (Box 3.9).

4

Issues in Well Design and Specialist Applications

In Chapter 3 we introduced the different types of wells and boreholes and described the general principles of their design. We will now look in more detail at several important design issues of specific relevance to drilled wells:

a. well and borehole construction materials, including the casing, screen and gravel pack (Sections 4.1 to 4.4);
b. the hydraulic suitability of the well design, by considering the various components of head loss in and around a pumping well (Section 4.5);
c. economic optimization of well design (applicable to wells located in an extensive uniform aquifer; Section 4.6).

We will also cover aspects of well design for specialist applications including wells for heating and cooling (Section 4.7), well doublets (Section 4.8), groundwater recharge (Section 4.9), and aquifer storage and recovery (Section 4.10).

4.1 Choice of construction materials

The materials most commonly used as casing and screen in drilled wells and boreholes are steel (mild or low-carbon steel, carbon steel and stainless steel), plastic [especially unplasticized polyvinyl chloride (PVC), but also acrylonitrile butadiene styrene (ABS), polyolefins and other plastics] and fibreglass (glass-reinforced plastic, or GRP). Hand-dug wells are normally constructed with concrete, stone or brick, but sometimes involve more exotic materials such as bamboo (widely used in southeast Asia).

The choice of construction materials will be influenced by many factors, including their strength, jointing system, durability, chemical inertness, ease of handling, cost, local availability and familiarity. National regulations, specifications or guidance documents on water well construction will also influence our choice. The relative importance of the various factors will differ according to the type and purpose of the well or borehole being designed. A well to be installed for a temporary construction-dewatering scheme, for example, will not require the same quality and durability of materials as a permanent water supply well. An observation borehole for sampling groundwater quality will require casing and screen materials that do not significantly absorb, or leach out, inorganic or organic contaminants being sampled, so as to minimize interference effects (Section 4.1.4).

4.1.1   Strength

The main facets of strength that we need to consider when selecting casing and screen are:

- radial compressive strength;
- longitudinal compressive strength;
- longitudinal tensile strength.

Radial compressive strength is probably the most important strength characteristic. It is needed to resist deformation (buckling) or collapse of the casing or screen due to earth pressures in unconsolidated formations, differential drilling fluid or water pressures across the casing string, pressures from emplacement of cement grouts in the annulus behind the casing, and pressures caused by well development. It should be able to withstand the maximum hydraulic load to which it is subjected, that is, about 10 kPa for each metre the casing extends below the water table (assuming a scenario in which the well is emptied of water during development while the water on the outside remains at the static level), plus additional pressures from earth forces, especially where swelling clays are to be cased out, and grouting. Well development procedures such as surging, jetting and the use of explosives (Section 5.9) may lead to internal pipe 'bursting' pressures, which should also be taken into account.

The radial compressive strength is indicated by the collapse resistance of the casing. The collapse resistance of a casing depends on its diameter, wall thickness and the elastic properties of the material

(given by its Young's modulus and Poisson's ratio). With plastic casings especially, the elastic properties and hence collapse resistances are affected by temperature (Section 4.2.2). Formulae for calculating critical collapse pressures are available in publications such as National Ground Water Association (1998), American Water Works Association (ANSI/A100-06; AWWA, 2006) and Sterrett (2007). However, the designer is always advised to check with the manufacturer as to the strength and suitability of a casing for a particular application, since the actual collapse pressures are site specific (Section 4.2.1).

Longitudinal compressive strength is required to resist any longitudinal earth forces if the ground around the well is likely to subside, including forces resulting from consolidation of an artificial gravel pack. Where subsidence problems are expected, then the design can include a compression section (slip joint) that permits vertical compression of the casing string (Roscoe Moss, 1990; NGWA, 1998). Also, if subsidence occurs the casing must be strong enough to support the full weight of the pump and wellhead fittings.

Longitudinal tensile strength is required to support the weight of the full casing string during installation (Section 5.2.7). Additional loadings may occur if it is necessary to remove the casing and screen, either during installation or after completion of the well; for example, the bottom section of the borehole may collapse before the screen and casing have reached their target setting depths, thus requiring the removal of the string before the borehole can be re-drilled and the casing string re-installed. The additional longitudinal tensile forces will depend on the nature of the geology, the type, length and diameter of the casing and screen, the presence of a gravel pack, and so on.

4.1.2   Jointing system

The type of jointing system is important when selecting the construction material because it affects both the material strength and the ease with which the casing string can be installed or removed. The joints are normally the weakest part of the

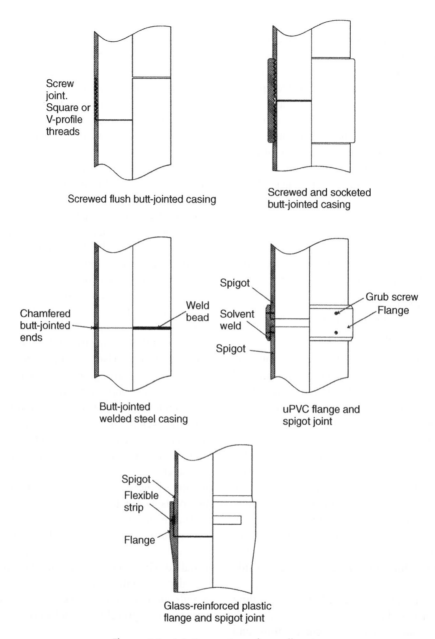

Figure 4.1 *Jointing systems for well casing*

casing string, yet must be strong enough to carry the full weight of the casing string. The jointing system should be such as to ensure a strong, straight, watertight and durable connection between the lengths of casing. There is a huge variety of jointing systems available for steel and plastic casings, the main ones of which are described in general terms below and illustrated in Figure 4.1.

Threaded couplings can be used for both steel and plastic casings, for both flush-jointed and external-upset couplings (screwed and socketed joints). Flush-jointed casings are easier to install and pull back, but they may have less strength than external upset-couplings because there is less effective thickness of material at the joint. Drawbacks of external-upset couplings are that they require a larger drilled diameter to provide the same minimum annular clearance for installing the casing string, gravel pack and grout seal, and the casing string may be more difficult to pull out if the geological formation collapses on top of the joints before the full casing string has been installed. The threads come in a variety of shapes: v-shaped, square, round and trapezoidal threads.

Welded joints are used for steel casings, especially for some of the larger casing sizes. The welding process requires expertise and care to ensure that the integrity of the pipe is not compromised, so it should only be performed by experienced personnel. Plastic pipes can be 'solvent-welded' (glued) together; it is important that the joints are allowed to set before the casing string is lowered into the borehole.

Spigot and socket with key locks, and variants of this system, are used for some plastic and fibreglass casings and can provide an effective and strong seal (Figure 4.1).

4.1.3   Durability

For most water well applications it is essential to select materials that resist corrosion. Electrochemical corrosion involves loss of metal into solution as a result of currents flowing between areas of different electrical potential in the well. Microbes that operate under anaerobic conditions also play a part in the corrosion process (Section 9.1.3). Corrosion is to be avoided since it can lead to:

- enlargement of screen openings and consequent sand pumping, leading to further abrasion of the well liner and pump;
- re-deposition of corrosion products leading to incrustation and blockage of screen openings and of the gravel pack and formation surrounding the well;

- perforation of the casing and casing joints, potentially allowing polluted water to enter the well;
- reduction in the strength of casing and screen through progressive removal of material, leading eventually to collapse of the well.

4.1.4   Chemical inertness

For production wells, it is important that the materials used for well construction do not leach chemicals into groundwater in amounts such that the potability of the water is adversely affected. However, this problem should not normally arise provided that the materials used conform to accepted international standards for water well casing and screen (Section 4.1.5). The situation with respect to choosing suitable materials for the construction of groundwater quality observation boreholes is more complex.

When installing observation boreholes for monitoring groundwater quality, the borehole construction materials should be non-reactive chemically. Using the terminology of Aller *et al.* (1991), with an observation borehole we wish to avoid obtaining 'false positive' results from contaminants that have leached into the groundwater from the borehole materials, and also to avoid 'false negative' results by failing to detect contaminants in the groundwater because these contaminants have been removed through sorption by the casing. Full chemical inertness with respect to all contaminants is probably impossible to achieve and in any case may not be necessary because, provided that the borehole is purged properly before sampling, the contact time between the construction materials and the groundwater being sampled is limited. Having said this, the cost of an observation borehole is usually small in comparison to the costs that will be incurred in water sampling and analysis during monitoring, hence it is not sensible to skimp in the choice of construction materials. Where there are doubts about the suitability of a material for a particular monitoring application, it is advisable to contact the manufacturer.

Table 4.1   *Construction materials for groundwater quality monitoring*

Material	Advantages	Disadvantages
Mild or low-carbon steel	Strong, rigid, inexpensive	Liable to corrode and leach metallic constituents. May sorb metals, especially on corrosion sites
Stainless steel	Strong, corrosion-resistant, good chemical inertness with respect to organics	Expensive. May leach chromium or other metallic elements
PVC	Rigid, light, inexpensive, good chemical inertness with respect to inorganics	Not as strong as steel. May react with organic compounds in groundwater. Not suitable where non-aqueous phase liquids are present. Temperature sensitive compared to steel
Polyolefins, including polypropylene and high-density polyethylene (HDPE)	Light, inexpensive, good chemical inertness with respect to inorganics	Less rigid than PVC. May react with organic compounds in groundwater. Temperature sensitive compared to steel
Fluoropolymers	Light, very good chemical inertness	More expensive, less rigid and more difficult to handle than most other plastics. Low tensile strength of joints. May sorb certain organic compounds

Some of the main advantages and disadvantages of the common construction materials for groundwater quality monitoring boreholes are summarized in Table 4.1. Mild or low-carbon steel is not normally used because it is vulnerable to corrosion. Stainless steel (Types 304 or 316) is suitable, albeit expensive, for most monitoring applications, but it may not be appropriate to use stainless steel where trace metals are present in the groundwater to be sampled.

Of the many plastic materials, PVC offers a good balance between practicality and performance (Fetter, 1999; BSI, 2010b). Only casing-standard PVC should be used, as this is unplasticized – and hence often described as uPVC or PVC-U casing – whereas flexible plasticized PVC tubing is much more likely to be reactive. Pipes should be jointed with leak-tight threaded couplings; solvent-welded joints should be avoided. PVC is not suitable where non-aqueous phase liquids are present (Fetter, 1999). Plastic pipes made of PVC and, especially, polyethylene may be vulnerable to permeation, whereby organic chemicals such as fuel hydrocarbons and chlorinated solvents may diffuse through the plastic.

Permeation has led to instances of drinking water contamination where plastic water distribution pipes have been exposed to heavily contaminated soil and groundwater (Goodfellow *et al.*, 2002). Therefore, when designing observation boreholes in contaminated areas, plastic materials that are susceptible to permeation should be avoided.

Fluoropolymers have high chemical inertness, but are more expensive and weaker than other plastics such as PVC and high-density polyethylene (HDPE). They are therefore mainly suited to shallow, small diameter monitoring installations. The International Standards Organization indicates that fluoropolymers are ideal for aggressive (acidic or alkaline) environments and where organic compounds are present (ISO 5667-22:2010).

4.1.5   Standards

Various standards are available for steel and plastic casings and jointing systems. Examples of American, British and Australian standards are included in Table 4.2. The U.S. standards – American Petroleum Institute (API), American Society for Testing and Materials (ASTM),

Table 4.2 *Examples of water well casing standards*

Material		United States	Australia	UK
Steel casing		API Spec. 5 L and 5 LS ASTM A53, ASTM A139, ASTM A589 ANSI/AWWA C200	AS 1396, AS 1579	BS 879 Part 1
Stainless steel casing		ASTM 312, ASTM A778		
Plastic casing		ASTM F480	AS/NZS 1477, AS/NZS 3518	BS 879 Part 2
Joints	Steel	ANSI/AWWA C206	AS 4041	BS 879 Parts 1 and 2
	Plastic	ASTM F480	AS/NZS 3879	

Note: Many of these standards are also designated by the year of publication; the reader should consult the most recent version.
Sources: AWWA (2006), National Uniform Drillers Licensing Committee (2012), NGWA (2014)

American Water Works Association (AWWA) and American National Standards Institute (ANSI) – are commonly used in countries for which there are no local standards. The standards serve three main purposes (NGWA, 1998):

- they specify factory testing standards and methods for measuring the properties of the casings;
- they eliminate the need for much of the detailed information that it would otherwise be necessary for the well designer, regulator or supplier to obtain from the manufacturer;
- they provide a type of warranty of construction quality.

4.2   Casing

4.2.1   Steel casing

Steel is the traditional material used for water well casing primarily because of its strength. It can be installed to great depths and pressure-grouted – and it can withstand fairly rough treatment on site, including driving and jacking. Even if an alternative material such as plastic is used for the main casing string, steel is almost always chosen for the surface conductor casing because of this robustness. The casing is normally produced in flat sheets and then fabricated into cylindrical tubes, although seamless steel pipes that are manufactured in cylindrical form are also used as casing. The jointing system can be threaded couplings or butt welding (Figure 4.1).

Steel casing is made in several grades and weights. The type of casing has to be chosen to suit local conditions: heavy, high-grade carbon steel is for use in deep wells whereas lower-grade steel is often used in shallow wells (although plastic materials are usually a better option for shallow wells – see Section 4.2.2 below). The definition of a shallow or deep water well will vary with the hydrogeological conditions within a particular area. As a generalization for this book, wells up to 200 m deep are regarded as shallow, while those over 200 m are deep.

Indications of the approximate relative strengths of steel compared to plastic and fibreglass casing materials are included in Tables 4.3 and 4.4. It is clear that steel casings have much greater compressive and tensile strengths than plastic casings. It should be noted that the actual strength of any casing depends on the specific product and the site conditions. Therefore the well designer is advised to obtain data from the manufacturer for each casing application, and particularly for deep wells. This is especially important when selecting screens, since their manufacturing and design details are far more complex than those for casing, and also because screen is normally weaker than casing and so particular care must be taken to select the correct material.

The main disadvantage of steel is that it can be vulnerable to corrosion. Local standards may incorporate a minimum thickness criterion, and this may include an allowance for progressive loss of metal based on local experience. Resistance to corrosion can be improved through the application of bituminous or other coating materials. These coatings can be effective unless and until they are broken, so it is extremely important when handling coated casing to avoid scratching or scraping it. Special rubber-faced tools are available for installing such casing. The use of stainless steel is the best defence against corrosion but it is expensive compared to plastic or fibreglass alternatives.

4.2.2   Plastic and fibreglass casing

Plastic is widely used in shallow aquifers because it is cheap and corrosion-free. Nowadays, most plastic casings are manufactured with unplasticized polyvinyl chloride (PVC, also referred to as uPVC or PVC-U), and other plastic casing materials such as acrylonitrile butadiene styrene (ABS), rubber-modified styrene and polypropylene are much less common. Fibreglass (glass-reinforced plastic, or GRP) casing is stronger than conventional PVC casing and is also corrosion-resistant. However, according to the National Ground Water Association, fibreglass products 'have not been received as favourably for potable water well construction due to perceived and actual problems with fibreglass sloughing' (NGWA, 1998). Smith and Comesky (2010) also caution against the use of fibreglass casings and screens because of their variable quality worldwide. The use of different materials often depends on local availability and practice; plastics are used widely in the Middle East whereas fibreglass casings are commonly found in Pakistan and Bangladesh, especially in irrigation and drainage wells.

Some of the plastic casings installed in water supply wells are flush-jointed, with square or trapezoidal threads. Solvent welding is often used for joining small diameter plastic casings in observation wells, although some PVC pipes have spigot and socket connections (Figure 4.1). Fibreglass

lengths of water well casing are joined by external upset couplings with special mechanical key locks.

In the United States, plastic casings are often specified according to their Standard Dimension Ratio (SDR), which is the ratio of the outside diameter of the pipe to the minimum wall thickness (ASTM F480; NGWA, 1998; AWWA, 2006). This can be helpful in selection, since casings of a given material with the same SDR value will possess the same theoretical radial compressive strengths, irrespective of the diameter. Similarly, the smaller the SDR number, the thicker and hence stronger will be the casing of a given diameter. Common SDR values are 13.5, 17, 21 and 26. An alternative system involves a schedule rating, in which the schedule defines a minimum wall thickness. For a particular Schedule number, because the wall thickness is constant, the strength of the casing will tend to decrease as the casing diameter increases.

The plastic and fibreglass casings are significantly weaker and more fragile than steel casings (Tables 4.3 and 4.4). Threads can be destroyed by abrasion and the casing cracked by shocks; the casings therefore must be handled carefully during transport and installation. These casings also deform more easily than steel and it has been known for casing to be irreversibly crushed by external hydrostatic pressures during well

Table 4.3   *Typical collapse strengths of selected casing materials[a]*

Material	Casing wall thickness (mm)	Collapse strength (kPa)
Steel casing	7.9	5245
PVC[b]	13[c]	790
ABS[b]	13[c]	690
Fibreglass	6.0	690

Notes:
[a] Collapse strengths are for 250 mm nominal diameter casings
[b] Collapse strengths are for room temperature, and reduce at higher temperatures (see text)
[c] These are approximate wall thicknesses for SDR 21 casing (see text for explanation of SDR)

Sources include Clark (1988), NGWA (1998), AWWA (2006), Sterrett (2007)

Table 4.4   *Tensile strength of selected casing material*

Material	Specific gravity	Tensile strength (MPa)
Low-carbon steel	7.85	414
Stainless Steel (Type 304)	8.0	552
PVC	1.4	55
ABS	1.04	31
Fibreglass	1.89	115

Note: the strength of plastic casings reduces at high temperatures (see text)

Source: Based on data from Driscoll (1986), after Purdin (1980)

development or maintenance operations, when excessive drawdown may occur.

The strength of plastic casing reduces at higher temperatures; the amount of reduction depends on the material but, based on research by Johnson *et al.* (1980), an average figure of 6.2 kPa per degree centigrade above 21 degrees centigrade is often quoted (Roscoe Moss Company, 1990; NGWA, 1998). This reduction in collapse resistance at higher temperatures may be an important design factor in certain situations – where heat is generated in deep water wells by exothermic grouting reactions (Molz and Kurt, 1979; NGWA, 1998), or in geothermal wells, for example. Plastic materials can also degrade when exposed to sunlight, and so pipes should always be covered during storage and transport. Because plastic casing bends easily, the casing string should be installed centrally in the well using centralizers. This will help the verticality and alignment of the well (Section 5.2.7), and also ensure a consistent annular clearance for the gravel pack.

The collapse pressures in Table 4.3 are for unsupported casing. The collapse resistance of a casing can increase appreciably when the radial stiffness of the surrounding material is taken into account (Kurt, 1979). However, for the purposes of design, it is safer to use the values for unsupported casing.

Plastic casings tend to be mainly used in shallow wells – wells less than 200 m deep, and most

commonly in wells less than 100 m deep – although there are examples where thick-walled plastics and fibreglass casings have been installed to depths of more than 250 m in consolidated aquifers. As shown by the specific gravity values for different casing materials in Table 4.4, plastic and fibreglass casings are much lighter than steel, which has considerable advantages in transport costs and ease of installation.

4.3   Screen

The purpose of the screen is to allow water to flow into and up the well efficiently (that is, without incurring large additional head losses in the well), while at the same time preventing – in combination with a gravel pack (Section 4.4) – sand and other fine material from clogging the well or damaging the pump. As we have seen in Section 3.1, well screens are required in all wells in unconsolidated aquifers, and are also installed in many wells in consolidated and crystalline aquifers when part of the borehole geology is unstable.

The choice of a particular screen type for a well will depend on a combination of factors: the strength and corrosion resistance, the slot design and open area (that is, the proportion of a screen face made up of open slots), and cost. The first two factors are a function of the materials used for screen manufacture – which include steels, plastics and fibreglass. Mild steel screens are susceptible to corrosion and incrustation, while stainless steel screens are corrosion-resistant, strong and expensive.

4.3.1   Slot design and open area

The perforations in screens (the slots) are very varied in design and methods of production (Figure 4.2). They include slots cut by an oxyacetylene torch or machine-saw in blank casing; bridge and louvre slots, pressed from steel plate and later rolled and welded into cylindrical tubing; wire-wound (or spiral-wound) screens where a wedge-shaped wire is wrapped in a continuous

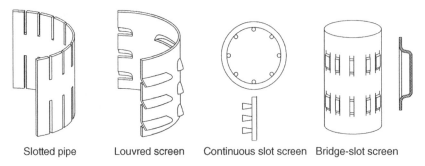

| Slotted pipe | Louvred screen | Continuous slot screen | Bridge-slot screen |

Figure 4.2   *Slot designs for screens*

Figure 4.3   *Plastic screens, with 2 mm slot size in the screen on the left of the picture and 1 mm slot size in the screen on the right. Photo by Bruce Misstear*

spiral around a cage of vertical rods and welded to each rod in turn; and perforated pipe base, wrapped with a sleeve of plastic mesh to act as a filter.

The torch-cut slots in mild steel casing are not recommended for water wells because the width of the slots cannot be accurately controlled and the open area of the screen is usually only 2 or 3%. Machine saw-cut slots are common in plastic and fibreglass screens, and typically have open areas of between 7 and 12%, although open areas of over 15% can be achieved with the larger slot sizes. They are commonly used with slot sizes of between

1 mm and 3 mm (Figure 4.3). PVC well casing of 200 mm or greater nominal diameter can be provided with slots as fine as 0.2 mm but for any plastic screen, even at a slot width of 1 mm, the ratio of the slot width to the wall thickness is sufficiently low that it raises concerns both about the slots becoming blocked with fine material, and also about the effectiveness of well development and maintenance techniques (Section 5.9).

The louvre slot screens are of very strong construction and have open areas typically in the range 3 to 10%. The bridge slot screens are not as strong in terms of collapse strength (because of the vertically-aligned slots), but have higher open areas than louvre screens (up to 15% or more for the larger slot sizes). Both types of pressed screens are available in mild and stainless steel materials.

The wire-wound screens are designed to give very large open areas (30 to 40%, typically, but can be more than 50% in small diameter screens with large slot widths) and to minimize slot blockage. They are available in stainless steel and plastic, the former being significantly stronger. The v-profile of the slots, and the slot widths, can be reproduced evenly and accurately to smaller widths than other screen types. The slot shape, in which the size of the aperture increases towards the inside of the screen, helps the movement of sand particles into the well. Together with its large open area, this makes this type of screen design particularly good for well development.

Where it is necessary to maximize the yield from an aquifer of limited thickness, stainless steel wire-wound screen is probably the best because of its high open area. The bridge slot and louvre slot screens may be particularly suitable where strength is the main design criterion – in deep aquifers or in situations where shock loads may be expected. The main advantages of plastic or fibreglass screens are their lightness and relative cheapness, but their inferior strength compared to steel means that they are mainly used in shallow wells. Perforated pipe wrapped in filter fabric material is suitable for many observation borehole applications, but it is not recommended for high-yielding water supply wells since it does not facilitate aquifer development or well rehabilitation.

The open area of a screen governs the rate at which water can enter the well. Intuitively one would expect that the higher the open area, the freer the flow into the well. However, this may be a simplification because it does not take account of the effective open area of the aquifer (the effective open area of a cross-section of aquifer is not the same as its porosity). Because the open cross-sectional area of closely packed (rhombic packed) spheres is only about 10% (Nold, 1980), the effective open area of an aquifer is unlikely to

exceed this value, especially given that even a relatively homogeneous gravel aquifer is not comprised entirely of spheres of constant diameter, and that small particles will infill the voids between the larger grains. In describing the results of trials on the hydraulic efficiency of well screens, Clark and Turner (1983) concluded that this effective limit of about 10% aquifer open area was the basic reason why the hydraulic performances of the screens tested were similar once the open area exceeded that value.

The open area, together with the diameter and length of the screen, determine the screen entrance velocity for a particular well discharge rate. The screen entrance velocity is one of the design considerations when examining potential well losses and hence the hydraulic suitability of the well design. Although there is much dispute in the literature as to the importance of maintaining a low screen entrance velocity by selecting a screen with a high open area (Section 4.5.4), such a screen does have the advantages that it is easier to develop and is less likely to become blocked than a screen with a small open area (and small slot width – Section 4.3.2). The ANSI/NGWA water well construction standard (NGWA, 2014) recommends that the open area should be as large as possible 'to facilitate development, to reduce biofouling, to maximize production of water, and to maintain the entrance velocity at a rate which provides laminar flow'. Here, we recommend that the open area of a screen ideally should be as great, or greater than the aquifer in which it is set, so a minimum screen open area of 10% is desirable. In practice, because a proportion of the slot open area will become blocked over time by pack or aquifer material, the concept of an 'effective' screen open area is usually adopted in design. This is difficult to quantify, but it is often assumed that 50% of the initial open area may become blocked.

4.3.2 Slot width

The width of slots in a screen should be selected with regard to the grain size of the aquifer or gravel pack (see Section 4.4 for grain-size distribution curves and pack design and Section 6.3.2 for a description of how to perform grain size analyses on drilling samples). The screen slot size and pack grading are determined from a standard sieve analysis of lithological samples from the aquifer. In the development of a natural gravel pack the screen should have wide enough slots to allow a certain proportion of fine material through for development of the pack. The criteria used vary between authorities. Driscoll (1986) and Sterrett (2007) suggest a slot size that will allow 50% to 60% of the aquifer material to pass through (referred to as the D50 and D60 sizes, respectively), the lower figure being recommended where the groundwater is very corrosive or if there is uncertainty over the lithological sampling. The ANSI/NGWA water well construction standard (NGWA, 2014) follows these guidelines, with the more conservative D50 figure being recommended where the aquifer material is uniform. In coarse sand and gravel aquifers, Driscoll and Sterrett indicate that a less conservative range of D50 to D70 may be used for selecting the slot size. The Australian Drilling Industry Training Committee (ADITC, 2015) bases its recommendations around the D50 size.

We suggest here that the choice of slot size in most cases should be based on the rather conservative average D40 of the grain-size analyses of at least six aquifer samples. In homogeneous and poorly-sorted aquifers a more relaxed criterion could be adopted, such as the D50. The slot choice for a screen in a very heterogeneous aquifer should be based on the analyses of the finer parts of the aquifer, not on 'average' aquifer material. In a layered aquifer the screen can be of multiple construction with a slot size to match each layer. In such a construction the fine screens should extend about a metre into the coarse sand to avoid sand pumping, and only in the thicker layered aquifers can sampling be accurate enough for this design to be practicable.

The size of slots in artificially gravel-packed wells is generally recommended to be around the D10 of the pack material (Driscoll, 1986; ASCE, 1996; Sterrett, 2007; NGWA, 2014). The AWWA (2006) specifies a slot width between the D20

and D5 of the gravel pack; that is, a slot size that will retain between 80 and 95% of the pack. The Australian NUDLC standard (2012) specifies a similar slot size, between the D20 and D0. The ADITC recommends an even finer slot size, which should be '20% smaller than the smallest pack material size so that all of the pack is held outside the screen' (ADITC, 2015). Bakiewicz *et al.* (1985), on the other hand, opted for a slot size greater than the D10 in their design of gravel-packed wells in Pakistan. They proposed that the slot width should be less than half the D85 of the gravel. This was based on the criterion that the D85 of the grading, if retained by the slots of the screen, will retain the remainder of the grading; a safety factor of two was added. This would indicate a slot size around the D40 of the gravel pack. It is suggested here that D40 of the gravel be considered as a maximum for the slot size, but that the more conservative D10 rule be followed for the slot width when a uniform pack is being installed.

4.4 Gravel pack design

4.4.1 Natural gravel pack

A natural gravel pack is produced by the development of the aquifer formation itself. Development techniques (Section 5.9) are used to draw the finer fraction of the unconsolidated aquifer through the screen, leaving behind a stable envelope of the coarser, and therefore more permeable, material of the aquifer. The development process, if successful, should produce a filter in which the coarsest particles are adjacent to the screen, with the grading reducing in size from the well face to the undisturbed formation at some distance from the borehole.

An aquifer is suitable for a natural pack if it is coarse-grained (usually taken to mean that it has a D10 greater than 0.25 mm) and poorly sorted. The grain size and sorting of sediments are illustrated by grain-size distribution curves derived from sieve analyses of formation samples, as explained in Chapter 6. Figure 4.4 shows examples of

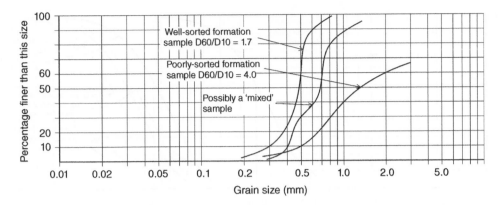

Figure 4.4   *Grain-size distribution curves*

grain-size distribution curves for a well sorted and a poorly sorted aquifer. Both curves have a characteristic 's' shape, with that of the well-sorted sample being much steeper. Definitions as to what exactly constitutes a well-sorted or a poorly-sorted sample vary, but the distinction adopted here is that a poorly-sorted aquifer is one with a uniformity coefficient greater than 2.5, where the uniformity coefficient is given by the D60 divided by the D10 size (that is, the 60% 'finer-than' size divided by the 10% 'finer-than' size). Figure 4.4 shows a third grain-size distribution curve that appears to have a 'double s' shape. Such a curve may represent a sample derived from two different horizons, and should not be used in the design of the gravel pack.

Following the discussion of screens in Section 4.3, the slot size recommended for the screen to develop a natural gravel pack for the poorly-sorted aquifer in Figure 4.4 would be the D40 size, that is, 1 mm. However, this D40 criterion would not be suitable in cases where it would result in a very narrow slot size. Screens with very narrow slots are available, but may be susceptible to blockage. It is therefore suggested that a natural pack design is not used for an aquifer when the D40 criterion would dictate a slot width of less than 0.5 mm (here, an artificial pack would be appropriate). It is also recommended that a natural pack should not be used when the uniformity coefficient is much less than 3. In a well-sorted sand aquifer, there is so little difference

between the D40 and the D80 sizes (Figure 4.4) that sampling errors or a slight mismatch of the slot size could lead to a sand-pumping well; an artificial gravel pack would avoid this danger.

4.4.2   Artificial gravel pack

Artificial gravel packs are used where the aquifer material is fine, well-sorted or layered and heterogeneous. Their main advantages are:

- they allow screens with larger slot sizes to be used, because the pack material is coarser than the formation;
- they reduce the risk of sand pumping and screen blockage;
- they increase the effective diameter of the well, since the gravel pack has a much higher permeability than the surrounding aquifer;
- they may reduce the required well development time.

Their disadvantages include:

- they require a larger drilling diameter to provide the required annular clearance for installing the pack;
- suitable pack material is difficult to obtain in many situations.

Many designs of gravel pack have been proposed, relating the grain-size distributions in the aquifer and pack (e.g., Hunter-Blair, 1970;

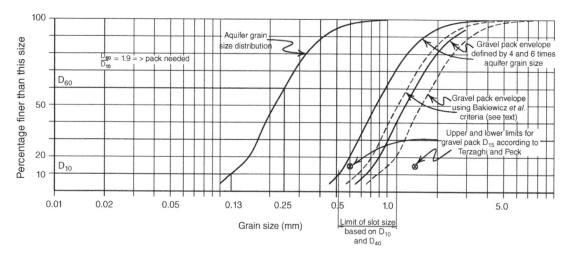

Figure 4.5 *Gravel pack design*

Bakiewicz *et al.* 1985; Driscoll, 1986; Roscoe Moss, 1990; NGWA, 1998; Sterrett, 2007; ADITC, 2015). The basis of these designs is to select a gravel-size distribution that will allow only the finer aquifer material to enter the well during development. Many of the designs relate back to the well-known work by Terzaghi and Peck (1948) on filter design in soil mechanics. Terzaghi and Peck's suggestions for a filter design can be expressed as:

$$\frac{Gravel\ pack\ D15}{Aquifer\ D85} < 4 < \frac{Gravel\ pack\ D15}{Aquifer\ D15} \quad (4.1)$$

where *Aquifer D15* is the D15 size of the coarsest layer of the formation and *Aquifer D85* is the D85 of the finest layer of the formation. These criteria are illustrated in Figure 4.5, along with the recommendations on gravel pack design discussed below.

There is a general consensus that the grading of the gravel pack should be:

1. uniform (and therefore typically have a similar uniformity coefficient to the aquifer);
2. coarser than the aquifer grading by a specified factor or range of factors;
3. based on the finest aquifer layer to be screened.

Recommendations generally differ only on points of detail concerning the gravel-pack grading. Sterrett (2007), for example, recommends filter gradings determined by multiplying the aquifer D30 size by factors of between three and eight, with factors of four to six being generally suitable where the aquifer is an unconsolidated sand (and factors of three to six being recommended when the aquifer is fine-grained and uniform). A curve with a uniformity coefficient of 2.5 or less should then be drawn through the gravel pack D30 to give the required grading.

Bakiewicz *et al.* (1985) produced their filter design for wells in unconsolidated aquifers in Pakistan using the following criteria: i) the pore size of a uniform material is approximately equal to 40% of its D10 size; and ii) the D85 of a grading, if retained by the screen, will retain the rest of the grading. Thus the maximum D10 of the gravel pack should be equal to the aquifer D85 size divided by 0.4. By assuming tolerances of plus or minus 8% around the target grading to determine the upper and lower envelopes of the gravel pack, this maximum D10 size can be used to determine the target D18 size and the minimum D26 of the gravel pack envelope. The grading curve for the pack is then drawn through these points using the same uniformity coefficient as the aquifer (Figure 4.5).

Roscoe Moss Company (1990) bases its recommended gravel pack design on multiplying the D50 of the aquifer by between four and six to obtain the D50 of the pack; the grading is then obtained by drawing two lines through these points parallel to the aquifer grading. The ADITC (2015) recommends a uniform gravel pack based on five times the modal size of the formation, that is, the size that there is most of in a sieve analysis; this size is often found between the aquifer D20 and D30 size.

A useful review of the filtration theory for gravel packs is provided by Houben and Treskatis (2007). When the ratio of pack size to aquifer formation size is less than four, the permeability of the pack is reduced because the gravel is finer than necessary to hold back the aquifer particles. When the ratio exceeds six, pack permeability is also reduced because aquifer particles can enter the pack and clog the pore spaces; if the ratio exceeds twelve, the aquifer particles will pass through the pack and so it will not perform its function as a filter. Thus a pack size four to six times the aquifer size should avoid both sand pumping and unnecessary reductions in permeability.

Because errors in formation sampling are common, and also because it can be extremely difficult to obtain an accurately-graded gravel pack in many situations, the best approach is to specify a design envelope for a gravel pack, rather than a particular grading. It is therefore recommended here that the artificial gravel pack grain-size distribution should be similar to that of the aquifer being screened, and should lie within an envelope defined by four and six times the aquifer grain size (Figure 4.5). The slot width for the screen should be approximately the D10 size of the artificial gravel pack, and not larger than the D40 size (Section 4.3).

A gravel pack only a few mm thick will retain the formation in undisturbed conditions, but intentional disturbance of the pack and formation during well development means that a thicker pack is needed to avoid sand-pumping. A thicker pack will also be able to cope with unintentional thinning of the pack in cases where the screen is not adequately centred. However, gravel packs much greater than 150 mm in thickness may create difficulties in development, particularly if mud-flush rotary drilling has been used and a mud cake has to be removed (Section 5.2). Also, the thicker the pack, the greater will be the drilled diameter of the well and therefore the cost. A pack thickness of approximately 75 mm (requiring a borehole diameter 150 mm greater than the maximum external diameter of the well screen) is considered optimal (Sterrett, 2007).

Laboratory experiments have shown that dual filter packs, constructed with coarser material installed next to the well screen and finer material placed next to the aquifer, can offer advantages in terms of their hydraulic efficiency (Kim, 2014), but such designs are considered impractical for most field installations. In a conventional graded pack, installation by tremie pipe from the bottom of the hole upwards (rather than by simply pouring the gravel in from the top) is recommended to avoid the different grain sizes in the pack settling at different rates – leading to lamination in the pack – and to ensure the gravel completely fills the annulus. Where a thick series of aquifers of different grain sizes is being packed, the pack against each aquifer can be tailored to suit that aquifer. However, the accurate emplacement of such a pack will be extremely difficult because of pack settlement during subsequent development. The screen and pack for a fine aquifer layer in such a situation must extend at least one metre into the adjacent coarser aquifers, to avoid sand pumping. Where a thin sequence of aquifers of different grain sizes is being packed, a single pack design based on the finest aquifer layer should be used.

The material used for a gravel pack should be natural, sub-rounded siliceous sand or gravel. Iron-rich sand or limestone gravel should be avoided, because solution and precipitation of iron or calcium salts are likely to cause problems.

Some screens are available with resin-bonded artificial gravel pack already installed. Although this screen and pack will act as a filter, it cannot be developed and may be susceptible to blockage.

Similarly, small-diameter screens suitable for observation boreholes can be purchased with a geotextile filter mesh wrapped around the slotted pipe. These can be convenient in situations where it is impractical to install an artificial gravel pack.

Box 4.1 summarizes the steps in gravel pack and screen selection.

4.5   Hydraulic design

Groundwater moves very slowly in porous (as opposed to fractured) aquifers under most natural conditions and the flow usually conforms to the laminar condition required by the Darcy equation (Section 1.3.1). Near to a discharging well, however, the velocity can be very high and the flow is turbulent, or non-Darcian (see Appendix 2). The situation may become worse if the aquifer is damaged and has not been developed properly. At the well, there will be further friction and momentum changes as water enters the screen and flows up towards the pump.

The head losses from turbulent flow or pipe friction increase the drawdown in the pumped well, and together are generally referred to as well losses. Jacob (1946) suggested that the drawdown in a well consists of two parts: aquifer loss and well loss. The aquifer loss is that drawdown to be expected from the flow in the aquifer and is directly proportional to the discharge rate. For steady-state conditions in a confined aquifer the aquifer loss can be described by the Thiem equation (1.21). The well loss is the drawdown caused by the turbulence or friction effects in or adjacent to the well, and Jacob suggested that it is proportional to the square of the discharge rate. The validity of the well loss proportionality has been the subject of much debate – Rorabaugh (1953), for example, suggested that the exponent can be between 2.4 and 2.8, with an average of 2.5 – but the squared correlation does appear to hold in many cases, and the Jacob equation is the basis of most analyses of well performance from step-drawdown pumping tests (Section 7.4.3). The Jacob equation is often given as:

$$s_w = BQ + CQ^2 \qquad (4.2)$$

where s_w is the drawdown in the pumping well, Q the discharge rate, and B and C are the coefficients of aquifer and well loss, respectively. The term BQ thus describes the drawdown attributable to aquifer loss and CQ^2 the well losses. This equation represents something of a simplification of what is undoubtedly a complex flow situation at a well, since the BQ term may include some component of laminar well loss (as recognized in Jacob's original paper) – for example in the gravel filter – and the CQ^2 term may include turbulence effects in the aquifer, particularly when the flow is restricted to fissures or to thin, highly permeable zones. Nevertheless, the well efficiency is usually expressed as the ratio of BQ to the total drawdown in the well.

Roscoe Moss Company (1990) gives an expanded version of the Jacob equation that includes a term for the turbulent head losses in the gravel pack:

$$s_w = BQ + B'Q^n + CQ^2 \qquad (4.3)$$

where B' is the turbulent filter zone loss coefficient and n an exponent varying between 1 and 2. As with the Jacob equation, this equation applies to a well that penetrates the entire thickness of the aquifer. It also assumes that the well is fully developed, and that there is no additional head loss in the formation near the well from wall-cake, or other damage caused by the drilling (see Section 4.5.2).

For steady state flow to a pumping well, the total head loss (ΔH) between the aquifer and the head in the pump chamber casing above the screen, where the pump is situated, can be summarized as:

$$\Delta H = \Delta h_a + \Delta h_p + \Delta_{dz} + \Delta h_g + \Delta h_e + \Delta h_{su} + \Delta h_{cu} \qquad (4.4)$$

where Δh_a is the aquifer loss, Δh_p the additional head loss due to the well only partially penetrating the full thickness of aquifer, Δh_{dz} the head loss in the damage zone, Δh_g the gravel pack loss, Δh_e the slot (screen entrance) loss, Δh_{su} the well screen upflow loss and Δh_{cu} the head loss as water moves above the screen to the pump. The different

Box 4.1   Summary of screen and gravel pack selection

1. The need for a screen and pack will depend on the aquifer type:
   - *Crystalline aquifer.* No screen or gravel pack is normally needed (Section 3.1.2). However, we may require a screen if the main water-bearing zones are in unstable fractured or weathered rock, in which case we should include a formation stabilizer.
   - *Consolidated aquifer.* Screen and pack are often not required (Section 3.1.3). When needed for an unstable aquifer, such as a friable sandstone, or for a multiple aquifer system, then we should use a formation stabilizer, or follow the design guidelines for a gravel pack (Section 4.4).
   - *Unconsolidated aquifer.* Screen and gravel pack are needed. The choice of a natural or artificial gravel pack will depend on the aquifer grain size and its uniformity (see 4 below).

2. We should choose a screen diameter so that:
   - head losses are small for the design discharge - the well upflow velocity should be less than $1.5\,\mathrm{m\,s^{-1}}$ (Section 4.5);
   - the screen is large enough to accommodate work-over tools – a 150 mm minimum diameter is recommended for most water supply wells;
   - cost is optimized (Section 4.6).

3. Our choice of screen material, screen length and slot design will depend largely on site conditions:
   - If the aquifer is thick, then a long screen with limited (but greater than 10%) open area may be chosen. The screen length may be up to 90% of the full thickness of a confined aquifer, or the lower one-third of an unconfined aquifer (Section 3.1.4). For a very extensive, uniform aquifer, we can use Equation (3.3) to calculate screen

   length, or the screen length can be optimized economically (Section 4.6).
   - If the aquifer is thin, then a screen with a large open area should be selected.
   - If the aquifer is over 200 m deep, then a strong screen must be chosen – usually carbon steel or stainless steel, but thick-walled plastic or fibreglass may be suitable in some situations (Section 4.3).
   - If the groundwater is corrosive (Section 9.1.2), then corrosion-resistant screens should be used such as plastic, fibreglass and stainless steel.
   - If the groundwater is encrusting, then a screen with a large open area should be used to offset the effects of screen blockage.

4. Our choice of gravel pack depends on the uniformity of the aquifer formation, as determined from the particle-size distribution curves (Section 4.4 and Section 6.3):
   - If the aquifer is poorly sorted (uniformity coefficient >2.5), then a natural pack can be developed. The screen slot width should be the average D40 of the aquifer samples (based on the finest layer in a heterogeneous aquifer system). If the D40 size gives a slot width <0.5 mm, then we can use an artificial gravel pack to allow a larger slot width (Section 4.4.2).
   - If the aquifer is well sorted (uniformity coefficient <2.5), then an artificial pack is needed. Design the pack by plotting two curves parallel to the particle-size distribution curves of the finest aquifer sample, but four to six times coarser than that curve (Section 4.4.2, Figure 4.5). The grading of the gravel pack should lie between these two new curves. The screen slot width should be approximately the D10 size of the gravel pack, and not larger than the D40 size.

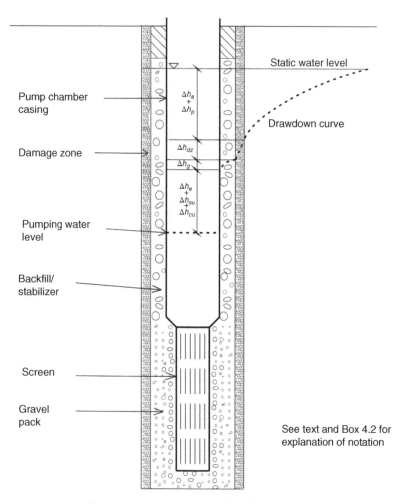

Figure 4.6 *Components of head loss at a well*

components of head loss are shown schematically in Figure 4.6.

As noted above, aquifer loss can be described by the Thiem equation for steady-state flow in a confined aquifer. The other terms in Equation 4.4 make up head losses associated with flow to and inside the well, although the penetration loss Δh_p is not included in Jacob's well loss term CQ^2 since it does not result from turbulent flow.

The main elements of head loss are considered in the following sections. For convenience, the notation used in the discussion is summarized in Box 4.2.

4.5.1 Partial penetration effects

In a well that only partially penetrates an aquifer, some of the flow lines are longer and the water is entering a smaller area of screen than with horizontal radial flow to a fully penetrating well. This leads to additional head loss. Huisman (1972) gives an expression for this additional head loss, for a well in an isotropic confined aquifer pumping under steady-state conditions:

$$\Delta h_p = \frac{Q}{2\pi Kb}\frac{(1-p)}{p}\ln\left(\frac{\varepsilon L_s}{r_w}\right) \qquad (4.5)$$

> **Box 4.2**   Notation used in discussion of hydraulic design in Section 4.5
>
> A_{eo} effective open area of well screen
> α parameter relating to screen type and screen diameter
> β parameter relating to screen type and screen diameter
> b aquifer thickness
> B aquifer loss coefficient
> B' gravel pack loss coefficient
> C well loss coefficient
> C_c contraction coefficient for an orifice
> C_v velocity coefficient for an orifice
> D screen diameter
> D_c casing diameter
> D_r diameter of a casing reducer
> ε parameter related to the penetration ratio and the eccentricity of the screen
> F_{skin} skin factor
> f pipe friction factor
> f_c friction factor for an individual section of the casing string
> g acceleration due to gravity
> Δh_{cu} casing upflow head loss
> Δh_{dz} damage zone head loss
> Δh_e screen entrance head loss
> Δh_g gravel pack head loss
> Δh_{gg} gravel pack head loss (non-Darcian)
> Δh_p partial penetration head loss
> Δh_{skin} skin zone head loss
> Δh_{su} screen upflow head loss
>
> k_r loss coefficient for a reducer
> K hydraulic conductivity
> K_{dz} horizontal hydraulic conductivity of the damage zone
> K_h horizontal aquifer hydraulic conductivity
> K_{hp} horizontal hydraulic conductivity of the gravel pack
> K_{skin} horizontal hydraulic conductivity of the skin zone
> K_v vertical aquifer hydraulic conductivity
> L_c casing length
> L_s screen length
> μ momentum factor
> n Manning roughness coefficient
> η parameter relating to the eccentricity of the well screen
> p penetration ratio, equal to the screen length divided by the aquifer thickness
> Q well discharge rate
> q flow rate into the screen per unit length of screen
> r_{dz} radius of the damage zone
> r_p gravel pack radius
> r_w well radius
> s_w drawdown in the pumping well
> v_e screen entrance velocity
> v_u well upflow velocity
> z_c vertical distance from the centre of the screen to the centre of the aquifer

where Q is the discharge rate, K the aquifer hydraulic conductivity, L_s the screen length, r_w the well radius; and p is the penetration ratio given by:

$$p = \frac{L_s}{b} \qquad (4.6)$$

where b is the aquifer thickness. The parameter ε is a function of the penetration ratio and the eccentricity of the screen with respect to the vertical centre of the aquifer.

Barker and Herbert (1992b) give the following equation for the case of a partially penetrating well in an anisotropic aquifer:

$$\Delta h_p = \frac{Q}{2\pi K_h b} \frac{(1-p)}{p} \ln \left[\frac{p(1-p)b}{(2-\eta^2)r_p} \sqrt{\frac{K_h}{K_v}} \right]$$
$$(4.7)$$

The terms K_h and K_v represent the horizontal and vertical aquifer hydraulic conductivities,

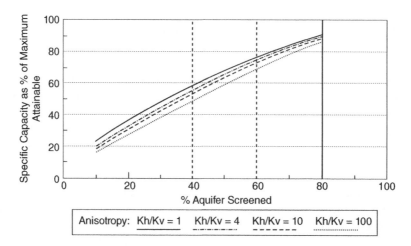

Figure 4.7 *Variation of well specific capacity with proportion of open area screened (Parsons, 1994)*

respectively, r_p is the radius of the gravel pack, and η describes the eccentricity of the well screen:

$$\eta = \frac{2z_c}{b(1-p)} \qquad (4.8)$$

where z_c is the vertical distance from the centre of the screen to the centre of the aquifer, and goes from zero (centred screen) to ±1 (screen at top or bottom of aquifer). Equation (4.7) is considered valid only for penetrations greater than 20% of the aquifer thickness.

Parsons (1994) applied Equation (4.7) to develop a relationship between the ratio of the specific capacity of a partially to a fully penetrating well, and the percentage of the aquifer screened. This relationship is shown in Figure 4.7 for the case of a well of effective radius 250 mm located in a 150 m thick unconsolidated aquifer, with the aquifer anisotropy ratio – the ratio of horizontal to vertical hydraulic conductivity – ranging from 1 (isotropic) to 100. It can be seen that the relationship is approximately linear for the typical situation where 40 to 80% of the aquifer is screened, and that the specific capacity varies from 50 to 90% of the maximum available across this range. This suggests that it is not unreasonable to use a version of the Logan equilibrium approximation equation [Equation (1.23)], with its linear

relationship between screened aquifer thickness and specific capacity, in choosing a well depth in a thick aquifer without a clearly-defined base [see Section 3.1.4 and Equation (3.3)].

4.5.2   The damage zone and well bore skin

The damage zone is the zone in the immediate vicinity of the borehole where drilling mud, filter cake or other drilling activities have altered the hydraulic properties of the aquifer. The head loss across the damage zone Δh_{dz} (between the gravel pack and the undisturbed geological formation) can be expressed using the Thiem equation as follows (ASCE, 2014):

$$\Delta h_{dz} = \frac{Q}{2\pi K_{dz} L_s} \ln \frac{r_{dz}}{r_p} \qquad (4.9)$$

where K_{dz} is the horizontal hydraulic conductivity of the damage zone, r_{dz} is the radius of the damage zone and r_p is the radius of the gravel pack (the radial distances are measured from the well to the outer edges of the damage zone and filter pack, respectively). Where present, a low permeability damage zone can be a major contributor to drawdown in a pumping well (Houben, 2015b).

The damage zone is often considered as part of the so-called skin zone around a well, which may also include the effects of partial clogging of the screen slots and the presence of an artificial gravel pack. A positive skin is where the hydraulic conductivity is reduced compared to that of the aquifer – for example due to remnant mud cake on the borehole wall and fine mud particles penetrating the aquifer – whereas a negative skin occurs when the hydraulic conductivity is enhanced, as may happen in a properly graded gravel pack (Section 4.5.3).

The head loss in the skin zone Δh_{skin} under steady-state pumping can be expressed by a relationship of the form (Kroening *et al.*, 1996):

$$\Delta h_{skin} = \frac{Q}{2\pi T} F_{skin} \qquad (4.10)$$

where F_{skin} is a dimensionless skin factor. This skin factor is a function of the hydraulic conductivity of the aquifer compared to that of the skin zone, and of the radial thickness of the skin zone compared to the radius of the well (Moench, 1997). Quantification of the skin effect, including the contribution to head loss from the damage zone, is impractical for most production wells, since the hydraulic conductivity and thickness of the skin zone are difficult to measure without a fine-scale network of observation boreholes. However, the inclusion of a water level monitoring dip tube within the gravel pack can be helpful for identifying whether clogging is occurring at the screen or within the gravel pack and/or in a damage zone close to the well (Section 9.2.1). Barrash *et al.* (2006) carried out very detailed laboratory, field and modelling investigations of the skin effect on pumping and observation wells at a research site in the United States; the wells were constructed in a coarse-grained unconsolidated aquifer. Although residual fine particles from the drilling process did contribute to the additional head loss (which we may regard as being the damage zone head loss Δh_{dz}), the main skin effects were attributed to partial clogging of the screen slots by sand grains, and also to increased head losses at the pumping wells resulting from strongly convergent radial flows within a heterogeneous, layered aquifer system.

4.5.3   Gravel pack loss

The head loss across the gravel pack Δh_g is given by (Barker and Herbert, 1992b):

$$\Delta h_g = \frac{Q}{2\pi K_{hp} L_s} \ln \frac{r_p}{r_w} + \Delta h_{gg} \qquad (4.11)$$

where K_{hp} is the horizontal hydraulic conductivity of the gravel pack and Δh_{gg} is a correction factor for non-Darcian head losses in the gravel pack (related to the square of the discharge rate). This equation assumes that all flow through the pack is from a permeable layer opposite the pack, which is in line with simple calculations by Driscoll (1986) that show that the contribution of flow from an overlying aquifer via the gravel pack is very small in relation to the total well yield. Barker and Herbert also provide a relationship similar to Equation (4.11) for non-Darcian flow in the aquifer. In most situations it would be expected that the head losses across a properly specified and installed gravel filter would be relatively small compared to other components of head loss. An exception might occur when the artificial "filter" comprises angular stone chippings rather than a proper filter composed of inert graded material as recommended in Section 4.4, a situation that may arise if the well construction is not supervised by experienced and competent personnel.

4.5.4   Screen entrance loss

The average velocity of water entering a screen v_e can be calculated from:

$$v_e = \frac{Q}{\pi D L_s A_{eo}} \qquad (4.12)$$

where Q is the discharge rate, D and L_s are the diameter and length of the screen, respectively, and A_{eo} is the effective open area of the screen, usually taken to be the manufacturer's specified open area divided by a factor of two to allow for long-term blockages of the slots (Section 4.3). Equation (4.12) shows that entrance velocity can be reduced by increasing the screen diameter, screen length or the open area.

There is much debate in the literature about the effect of screen entrance velocity on head losses. Various maximum entrance velocities have been recommended to keep head losses in the screen to a minimum. Low entrance velocities are also suggested to have benefits in terms of minimizing screen corrosion and incrustation problems. A brief discussion of the main issues in the entrance velocity debate follows, concluding with some suggestions for the well designer.

Published recommendations on maximum screen entrance velocity. One of the commonest recommendations is that entrance velocity should not exceed $0.03\,\mathrm{m\,s^{-1}}$ (e.g., Johnson, 1966; Hunter Blair, 1970; Driscoll, 1986; Detay, 1997; Sterrett, 2007). Other authors have proposed that the screen entrance velocity should be kept below or within a range of values, this range being consistent with the $0.03\,\mathrm{m\,s^{-1}}$ criterion. According to Walton (1970), for example, screens should be selected on the basis of an optimum entrance velocity, which is between 0.01 to $0.06\,\mathrm{m\,s^{-1}}$. Walton gives a table relating the optimum screen entrance velocity to the aquifer hydraulic conductivity value; thus, a low entrance velocity of $0.01\,\mathrm{m\,s^{-1}}$ is recommended for aquifers having a low hydraulic conductivity, since these are assumed to contain fine-grained material and so the screens will be more prone to clogging. Walton's approach is followed by Campbell and Lehr (1973) and by the UK's (formerly-named) Institution of Water Engineers and Scientists (IWES, 1986). IWES adds that the optimum values should be reduced by one third where the groundwater in encrusting. Using a physical model for well development experiments on wirewound screens, Wendling *et al.* (1997) concluded that the maximum screen entrance velocity should be between $0.03\,\mathrm{m\,s^{-1}}$ and $0.06\,\mathrm{m\,s^{-1}}$ in order to maintain laminar flow in the well. This research is cited by Sterrett (2007) in support of the recommendation that the entrance velocity should not exceed $0.03\,\mathrm{m\,s^{-1}}$. The Australian NUDLC standard (2012) also includes the recommended design entrance velocity criterion of $0.03\,\mathrm{m\,s^{-1}}$ for well screens (note that in the NUDLC document the term screen is used for wirewound screens, and slotted pipes are described separately).

Williams (1985), on the other hand, indicates that well efficiencies do not increase significantly if the open areas of the screen are increased above 3 to 5%. At these open areas, the upper limit of entrance velocity is suggested to be between 0.6 and $1.2\,\mathrm{m\,s^{-1}}$. Williams goes on to conclude that 'entrance velocity and screen open area are not critical design factors for most field situations'. These (and earlier) research findings by Williams are also referred to by Helweg *et al.* (1983), Roscoe Moss Company (1990) and ASCE (2014).

The ANSI/AWWA water well standard has moved away from specifying a maximum screen entrance velocity. The foreword to the 1997 edition of the ANSI/AWWA standard commented on the setting of its screen entrance velocity limit as follows: 'The upper limit of entrance velocity, 1.5 ft/second (0.46 m/second) included in this standard is a compromise upper limit for entrance velocity based on the judgment and consensus of the committee.' (AWWA, 1998). In contrast, the foreword to the current (2006) ANSI/AWWA standard notes that 'there is no singular, uniquely defined criterion for permissible velocity through the screen slot openings that is solely suitable for designing a screen without consideration of the aquifer characteristics and manner of well construction' (AWWA, 2006). An appendix to the standard includes sample calculations that show that different combinations of entrance velocity and screen open area can give similar values for the design screen length.

Similarly, the recent ANSI/NGWA water well construction standard (NGWA, 2014) does not stipulate a maximum screen entrance velocity. Rather, the document recommends that the screen should be selected with the largest slot sizes and open area possible so as 'to maintain the entrance velocity at a rate which provides laminar flow'. The standard also highlights the importance of large aperture sizes and open area for aiding well development and reducing biofouling; the non-hydraulic benefits of maintaining a low screen entrance velocity are discussed later in this section.

Houben (2015a) also notes the disagreements in the literature about recommended screen entrance velocities. He goes on to suggest that calculating the Reynolds number [Equation (1.7) and Appendix 2] may be a better way of assessing the flow regime in the vicinity of the well and hence the suitability of the well design. However, this approach faces a similar problem in that there is also disagreement in the literature about the maximum acceptable Reynolds number for flow around a well.

Calculating the head loss across a screen. The head loss across the screen slots can be examined by applying standard formulae for flow through an orifice. Using this approach Barker and Herbert (1992b) give the following expression for screen entrance loss (slot loss) Δh_e:

$$\Delta h_e = \frac{1}{2g}\left(\frac{Q}{\pi L_s DC_v C_c A_{eo}}\right)^2 \qquad (4.13)$$

The orifice terms C_v and C_c represent the velocity coefficient and the contraction coefficient, respectively, and the other terms are as identified previously.

Combining Equations (4.12) and (4.13) for the condition where the screen entrance velocity is the oft-recommended maximum value of $0.03\,\text{m s}^{-1}$, gives (Parsons, 1994):

$$\Delta h_e = \frac{1}{2g}\left(\frac{0.03}{C_v C_c}\right)^2 \qquad (4.14)$$

Using values of 0.97 and 0.63 for C_v and C_c, respectively, Equation (4.14) yields a value for Δh_e of only 1.23×10^{-4} m. This is clearly a very small head loss in comparison to other head losses that make up the well loss component of drawdown (see Section 4.5.5). If the maximum screen entrance velocity is increased to the limit of $0.46\,\text{m s}^{-1}$ recommended in the previous ANSI/AWWA A100-97 standard (AWWA, 1998), then the value of Δh_e increases to only 0.03 m. At the

upper limit of $1.2\,\text{m s}^{-1}$ mentioned by Williams (1985), the value of Δh_e is 0.2 m, a head loss that is likely to be acceptable in many situations.

The non-hydraulic benefits of low screen entrance velocity. The non-hydraulic advantages of low screen entrance velocities – reduced corrosion, incrustation, abrasion and hence less well maintenance – are much more difficult to quantify, and are also the subject of much debate. Driscoll (1986) provides a lengthy discussion of the merits of low entrance velocity, citing several earlier publications, and he presents the results of some research into corrosion of metals in flow cells, which showed higher corrosion rates for carbon steel and stainless steel screen materials at increased screen entrance velocities. Howsam *et al.* (1995) reviewed the issue of design entrance velocity as part of their guidance document on well monitoring and rehabilitation. They report that reducing entrance velocity (or design discharge) can have the following beneficial effects:

- reduction in viscous drag forces on sand particles in the vicinity of the well, and hence reduction in sand pumping;
- reduction in erosion of screen slots by gas-liquid-solid mixtures;
- reduction in supply of nutrients or reaction compounds to the screen and gravel pack;
- reduction in kinetic energy, leading to a reduction in physical and chemical changes, such as degassing;
- reduction in the loss of kinetic energy by jet dissipation inside the well.

Howsam *et al.* (1995) support earlier guidelines of a maximum entrance velocity of $0.03\,\text{m s}^{-1}$, and go on to recommend that this limit should be reduced to $0.02\,\text{m s}^{-1}$ in situations where biofouling is suspected. As pointed out earlier, the ANSI/NGWA well construction standard (NGWA, 2014), whilst not specifying a maximum screen entrance velocity, does recommend that the slot size and open area should be chosen so as to ensure laminar flow and to minimize problems with biofouling.

Parsons (1994) argues that the non-hydraulic benefits of low screen entrance velocities are often overstated, and notes that incrustation and corrosion are complex biological and chemical processes that are strongly influenced by the groundwater chemistry, bacteriological activity, the casing and screen materials, as well as by the well hydraulics. Like Williams (1981), he points out that the issue of sand-pumping is related to the approach velocity at the interface between the gravel pack and the aquifer, rather than the screen entrance velocity (however, these velocities are related: a reduction in screen entrance velocity through an increase in the diameter or length of the well will result in a reduction in the approach velocity). Helweg *et al.* (1983) also indicate that entrance velocity is not a factor controlling sand-pumping where wells have been correctly designed and developed. Williams (1985) points out that low entrance velocities may even be detrimental in that 'Well photo and TV logs show that corrosion and encrustation are greatest in screens set opposite aquifers in which production and entrance velocities are low'.

Conclusions and suggested approach. Clearly, there are many opinions on this subject, and much more research is needed to resolve these differences, especially research into the performance of different well designs under a range of field conditions. Nevertheless, we draw the following tentative conclusions from this debate:

1. it is not appropriate to adopt a single design screen entrance velocity for all situations;
2. the screen entrance head losses will generally be small if the entrance velocities are kept below 0.5 m s⁻¹;
3. there are likely to be some benefits in maintaining lower entrance velocities in situations where the well is susceptible to corrosion and incrustation, including biofouling.

Further, we would concur with the view of the NGWA manual (1998): it is more important to base the design on the professional advice and experience of the local hydrogeologist or engineer than adhere to a rigid standard. Moreover, we would emphasise here that entrance velocity is not the only criterion in the hydraulic design of the well screen and that other considerations may be more important (see Section 4.5.5).

Although there has been a trend away from specifying a maximum screen entrance velocity in some of the current water well standards, it can be helpful for the designer to have guidance on this issue, since there may not be sufficient local experience available in the particular study area to inform the design process. Therefore, in the absence of relevant local experience, we suggest that the well designer should:

a. use a screen made from corrosion-resistant materials;
b. use a screen with an open area of at least 10% (Section 4.3);
c. use a screen with a length, diameter and slot geometry such that the entrance velocity for the design discharge:
   • is less than 0.5 m s⁻¹; or
   • is less than 0.03 m s⁻¹ if serious corrosion and/or incrustation problems are anticipated and/or this is the recommended maximum entrance velocity specified by the particular screen manufacturer.

Further information on screen selection is given in Section 4.3 and Box 4.1; an example of hydraulic design, including a calculation of screen entrance velocity, is presented in Box 4.3.

4.5.5 Well upflow losses

The average velocity of water flowing up through the screen and casing to the pump, v_u can be calculated from:

$$v_u = \frac{Q}{\pi \left(\dfrac{D}{2} \right)^2} \qquad (4.15)$$

There has been far less discussion in the literature about minimizing upflow velocity in the screen and casing than about restricting screen

Box 4.3   Example showing the application of hydraulic design criteria to well design

A production well with a design discharge of $25 \, ls^{-1}$ is to be constructed in an extensive and relatively uniform sand and gravel aquifer. The aquifer is over 200 m thick, has an average hydraulic conductivity of $12 \, m \, d^{-1}$, and is confined by an overlying clay layer that extends from the ground surface to the top of the aquifer at 35 m. The aquifer's potentiometric surface lies at 17 m below ground level, and nearby observation boreholes indicate that seasonal variations in water level are small. Our task is to choose a suitable design for the well assuming that the aquifer is to remain confined during pumping.

The main hydraulic design criteria are therefore:

- $Q = 25 \, ls^{-1} = 0.025 \, m^3 \, s^{-1} = 2160 \, m^3 \, d^{-1}$
- $K = 12 \, m \, d^{-1}$
- The maximum s_w is dictated by the depth interval between the potentiometric surface and the base of the clay, which equals $35-17 \, m = 18 \, m$; we should reduce this to say 15 m, to allow for a 3 m factor of safety

The hydraulic design should proceed in a series of steps:

1. *Choose an initial diameter for the screen.* From Table 4.5, a diameter (D) of 150 mm should be sufficient for the design discharge of $25 \, ls^{-1}$ (note that 150 mm is also the *minimum* recommended screen diameter to allow sufficient room for development and maintenance operations).
2. *Select the type of screen and the desired open area* (see Section 4.3). For the purposes of this example, a slotted plastic screen with an open area of 12 percent will be assumed.
3. *Estimate the required length of screen.* An estimate can be made using a modified Equation (3.3), in which saturated aquifer thickness b is replaced by the screen length L_s:

$$L_s = \frac{2Q}{Ks_w} = \frac{2 \times 2160}{12 \times 15} = 24 \, m, \text{ say } 25 \, m$$

4. *Check the screen entrance and upflow velocities.* Allowing a factor of 50% for long-term blockage of the screen slots, Equation (4.12) indicates a screen entrance velocity of:

$$v_e = \frac{Q}{\pi D L_s A_{eo}}$$
$$= \frac{0.025}{\pi \times 0.15 \times 25 \times (0.12 \times 0.5)} = 0.035 \, m \, s^{-1}$$

This screen entrance velocity is only slightly greater than the conservative limit of $0.03 \, m \, s^{-1}$ quoted in some of the literature and is unlikely to lead to significant head loss (Section 4.5.4). The diameter and open area combination in this example therefore appears to be satisfactory (especially in view of the rather pessimistic assumption that 50% of the open area will become blocked), and we would only need to consider increasing the diameter or open area if local experience indicated that groundwater in the area is very encrusting (corrosion should not be a problem with the plastic screen selected).

The upflow velocity from Equation (4.15) is:

$$v_u = \frac{Q}{\pi \left(\frac{D}{2}\right)^2} = \frac{0.025}{\pi \left(\frac{0.15}{2}\right)^2} = 1.4 \, m \, s^{-1}$$

This is within the usual guideline of $1.5 \, m \, s^{-1}$ (and only slightly greater than the ANSI/AWWA A100-06 limit of $1.22 \, m \, s^{-1}$; Section 4.5.5). Equation (4.16) can be used to confirm that upflow head losses in the screen would be small using this combination of design discharge and diameter:

$$\Delta h_{su} = 3.428 q^2 n^2 L_s^3 D^{-\frac{16}{3}} = 3.428 \times \left(\frac{0.025}{25}\right)^2$$
$$\times 0.013^2 \times 25^3 \times 0.15^{-\frac{16}{3}} = 0.22 \, m$$

This is likely to be an acceptable upflow head loss for the screen, and the upflow loss above the screen should be comparatively small if the pump is set near the base of the pump-chamber casing.

5. _Select the diameter and length of the pump-chamber casing._ If the $25\,\mathrm{l\,s^{-1}}$ pump has a maximum diameter of say 200 mm, then a pump-chamber casing with an internal diameter of 300 mm would provide the recommended clearance of 50 mm (Section 3.1). The length of the pump-chamber is determined by adding:

- the length of casing to be left above ground (usually a metre or less);
- the depth below ground surface to the static water level in the well;
- the maximum drawdown anticipated over the life of the well;
- the length of the pump;
- an allowance for the clearance between the bottom of the pump and the base of the pump-chamber casing.

In this example, the pumping water level is fixed by the need to keep the aquifer confined, and so the pump-chamber casing would only need to extend one or two metres into the aquifer, giving a total casing length of around 37 m. The pump-chamber casing would be joined to the top of the screen by a 300 mm to 150 mm reducer section.

---

entrance velocity. Where guidance is given, this usually indicates that v_u should not exceed $1.5\,\mathrm{m\,s^{-1}}$ ($5\,\mathrm{ft\,s^{-1}}$) in order to avoid excessive upflow head losses (Driscoll, 1986; Sterrett, 2007; NGWA, 2014); the AWWA (2006) specifies a slightly lower limit of $1.22\,\mathrm{m\,s^{-1}}$ ($4\,\mathrm{ft\,s^{-1}}$) for vertical velocity in the well screen.

Bakiewicz _et al._ (1985) used the following expression, based on the Manning formula for flow in pipes, to calculate the upflow head loss in a well screen:

$$\Delta h_{su} = 3.428 q^2 n^2 L_s^3 D^{-\frac{16}{3}} \qquad (4.16)$$

where Δh_{su} is the upflow head loss (in m), q the flow rate into the screen per unit length of screen ($\mathrm{m^3\,s^{-1}\,m^{-1}}$), n the Manning roughness coefficient, and L_s and D are the length and diameter of the screen (m), respectively. The roughness coefficient n is given as 0.013 for slotted pipe and 0.018 for wire-wound screens (Bakiewicz _et al._, 1985; Parsons, 1994). Equation (4.16) shows that the upflow head losses in the screen are inversely proportional to about the fifth power of the screen diameter. They will therefore only be significant if the diameter of the screen is small in relation to the discharge. This is illustrated in Figure 4.8, which

gives the screen upflow head losses calculated using Equation (4.16) for a range of screen diameters and design discharge rates. A slotted screen with a Manning n of 0.013 and length 10 m is assumed.

Using the results of hydraulic experiments on 27 well screens of various types and diameters, Barker and Herbert (1992a, 1992b) produced the following formula for calculating the screen upflow loss:

$$\Delta h_{su} = Q^2 \left(\frac{\alpha L_s}{4} + \frac{\beta}{3} \right) \qquad (4.17)$$

The parameters α and β relate to the screen type and diameter:

$$\alpha = \frac{32 f}{4\pi^2 g D^5} \qquad (4.18)$$

and

$$\beta = \frac{32 \mu}{\pi^2 g D^4} \qquad (4.19)$$

where f is a pipe friction factor and μ a momentum factor, the latter being usually close to unity. The tests indicated an average f value of approximately 0.016 for slotted plastic and fibreglass pipes and 0.027 for stainless steel wire-wound screen.

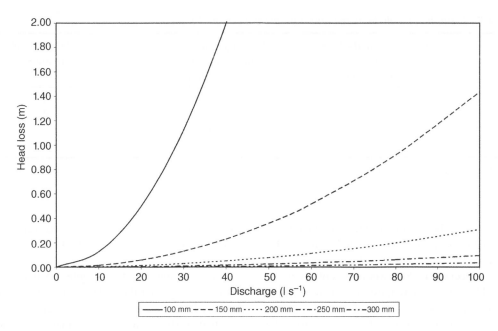

Figure 4.8 *Upflow head losses for different discharge rates and screen diameters (screen length = 10 m)*

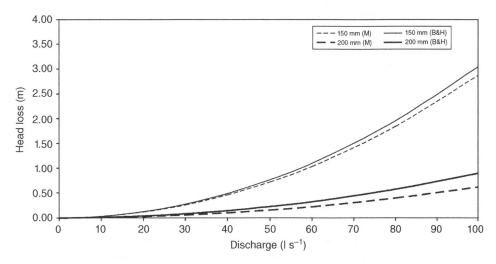

Figure 4.9 *Comparison of upflow head losses in slotted screens calculated using Manning-type and Barker and Herbert formulae (for screen length = 20 m)*

Although the stainless steel screens have higher internal roughness than the slotted pipes, Barker and Herbert (1992a) point out that they are often the most efficient hydraulically for short screen lengths because they have larger internal diameters for the same nominal diameter (and hence have correspondingly smaller values of the momentum loss parameter β). Figure 4.9 compares upflow head losses in slotted screens calculated using Equations (4.16) and (4.17). The comparison

Table 4.5 *Examples of well discharge rates for different screen diameters*

Screen diameter (mm)	Discharge rate (l s⁻¹) for maximum upflow velocity of 1.5 m s⁻¹	Discharge rate[a] (l s⁻¹) for maximum upflow head loss of 0.3 m	Equivalent upflow velocity (m s⁻¹) for upflow head loss of 0.3 m
100	12	11	1.4
150	27	32	1.8
200	47	70	2.2
300	106	205	2.9

[a] This is the discharge rate for a 20 m length of slotted screen that would give an upflow head loss of 0.3 m, calculated using Equation (4.16)

shown is for the common screen diameters of 150 mm and 200 mm, with a screen length of 20 m in both cases. It can be seen that the results are similar, especially in the lower ranges of discharge rate and head loss where these screens would normally be used. It should be noted that these calculations assume a uniform inflow rate across the length of the screen, whereas the largest inflow rate normally occurs at the top of the screen when the pump is set in the pump-chamber casing above it (Houben, 2015a). The same author points out that the head losses could actually be reduced by placing the pump within the screen (Houben, 2015b), although this practice is not commonly recommended owing to concerns about sand pumping and increased risk of air entering the screen.

Relationships such as Equations (4.16) and (4.17) can be used to propose suitable screen diameters for different discharge rates. One such example, calculated using Equation (4.16), is included in Table 4.5: this is for a 20 m length of slotted screen and a maximum allowable upflow head loss chosen somewhat arbitrarily as 0.3 m (or 1 ft). The upflow velocity is shown for comparison and, from this and similar calculations for other screen lengths, it can be seen that the 1.5 m s⁻¹ limit is a reasonable, and generally conservative, guide for minimizing head losses. Indeed, huge numbers of wells – in the Indo-Gangetic alluvial aquifers, for example – have been constructed with design upflow velocities of between 1.7 and 1.9 m s⁻¹. As suggested by Parsons (1994), however, it is worthwhile calculating

well upflow losses for each proposed well design so that the design can be modified if necessary to reduce costs.

It is also apparent from Table 4.5 that screen diameters of greater than 300 mm are only justified for very large capacity wells. Although a diameter less than 150 mm might be acceptable for a low-yielding well on hydraulic grounds, a screen diameter of at least 150 mm is desirable to facilitate well development and maintenance activities (Section 3.1.3). An exception to this guideline applies to the construction of wells in low income countries where smaller diameters may be warranted on the basis of cost (see Box 3.6).

The additional upflow head loss above the screen (Δh_{cu}) should be comparatively small when the pump intake is set close to the bottom of the pump-chamber casing, giving a short section of flow. For flows in longer lengths of pump-chamber casing, the upflow head loss can be calculated using standard formulae for flow in water pipes (see Appendix 2). Barker and Herbert (1992b) provide a more sophisticated approach for calculating Δh_{cu}, one that allows for the presence of reducer sections in the casing string:

$$\Delta h_{cu} = \frac{8Q^2}{\pi^2 g} \left\{ \sum_c \frac{f_c L_c}{D_c^5} + \sum_r \left[k_r \left(\frac{1}{D_{r1}^2} - \frac{1}{D_{r2}^2} \right)^2 - \left(\frac{1}{D_{r1}^4} - \frac{1}{D_{r2}^4} \right) \right] \right\}$$

(4.20)

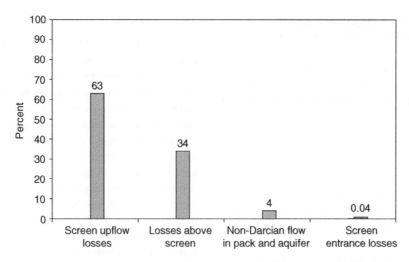

Figure 4.10   *Average theoretical contributions to the well loss coefficient C, for 17 test wells in Bangladesh (based on data in Table 2 in Barker and Herbert, 1992b)*

where $\sum_c$ and $\sum_r$ represent the summations for all casing and reducer sections, respectively; f_c is the friction factor for a section of casing; k_r is the loss coefficient for a reducer; L_c and D_c are the casing lengths and diameters, respectively; and D_{r1} and D_{r2} are the diameters of the bottom and top of each reducer section, respectively.

Using equations such as (4.11), (4.13), (4.17) and (4.20), Barker and Herbert (1992a) calculated the theoretical contribution to the well loss coefficient C for 17 wells in Bangladesh, and compared the total C value obtained for each well with the C value from a step-drawdown test. The wells tested represented a variety of screen types: stainless steel wire-wound (7 wells), mesh-wrapped plastic slotted pipe (4 wells) and slotted fibreglass pipe (6 wells). The screen lengths ranged from 18 m to 36 m and the screen diameters from 100 mm to 250 mm, approximately. The theoretical total well losses were in good agreement with the actual well losses derived from step-drawdown tests. The breakdown of the theoretical well losses is illustrated by the histogram in Figure 4.10. In this example, it is clear that the largest contribution by far is from the upflow head loss in the screen (63%) and that the screen entrance losses are negligible (0.04%).

An example of the application of hydraulic criteria to well design is given in Box 4.3.

Finally, we would add that the application of good hydraulic design principles as outlined here will be of less value if the well is not properly constructed and developed (Chapter 5). For example, the presence of a residual low-permeability damage/skin zone may be a relatively large contributor to total drawdown in a pumping well compared to the head losses associated with the screen design (Houben, 2015b).

4.6   Economic optimization of well design

4.6.1   General principles

Economic considerations in well design were introduced in Section 3.1.5. In most crystalline and consolidated aquifers, the well design, and therefore the well cost, are largely dictated by the aquifer geometry. However, in a thick aquifer of uniform permeability distribution (usually an unconsolidated aquifer), it is possible to apply optimization techniques to obtain the required discharge rate at the least cost. The optimization approach is essentially based on the premise that, for a specified pumping rate, a greater depth and screen length for a well in a uniform aquifer will result in a lower drawdown. Combining Equation (1.23) and Equation (4.2), and substituting screen length for saturated thickness of aquifer, the steady-state drawdown in a well can be expressed as:

$$s_w = \frac{1.22Q}{KL_s} + CQ^2 \qquad (4.21)$$

where s_w is the drawdown in the pumping well, Q the pumping rate, K the aquifer hydraulic conductivity, L_s the screen length and C the well loss coefficient. When well losses are small, the relationship can be simplified by eliminating the second term in the equation, whilst increasing the multiplier in the first term from 1.22 to 2.0 to allow for some well losses plus additional drawdown due to partial penetration effects in a thick aquifer:

$$s_w = \frac{2Q}{KL_s} \qquad (4.22)$$

Thus, drawdown is inversely proportional to screen length for constant values of Q and K.

In terms of the cost of a well, an increase in screen length will lead to an increase in the well depth, and therefore an increase in the capital cost of the well. However, the well operating costs will decrease, since a greater screen length will result in a decrease in drawdown and therefore a decrease in the pumping costs. Economic optimization seeks to determine the screen length – or other design variable – where the capital and total recurring costs over the lifetime of well are at a minimum. This is achieved by expressing all costs as a present value, and then partially differentiating this present value with respect to any of the design variables to determine the minimum. The methodology was developed by Stoner *et al.* (1979) and has been applied to a large number of irrigation and drainage wells, notably wells constructed in the extensive alluvial aquifers of Pakistan and Bangladesh.

4.6.2 Example

The methodology will be described with reference to the well and turbine pump set-up illustrated in Figure 4.11, using the sample well construction costs in Table 4.6. The basic units are metres (for length) and cubic metres per second (pumping rate), but the currency is not quoted – unit cost relationships would have to be determined for the particular area where the hydrogeologist or engineer is working. In this example we will look at optimizing screen length and well depth. The capital cost of the well will be considered first, then the recurring costs, and then the present value of all the costs. The optimum screen length and well depth will be determined from the lowest present value cost.

Capital cost. From Table 4.6, the capital cost C_1 is:

$$C_1 = 6500 + 130L_c + 165L_s + 500QH \qquad (4.23)$$

Now, from Figure 4.11:

$$L_c = w + s_w + 5 = w + \frac{2Q}{KL_s} + 5 \qquad (4.24)$$

and thus:

$$130L_c = 130w + \frac{260Q}{KL_s} + 650 \qquad (4.25)$$

Again, with reference to Figure 4.11:

$$H = 2 + w + s_w + 2 = 4 + w + \frac{2Q}{KL_s} \qquad (4.26)$$

and therefore:

$$500QH = 500Q(4 + w) + \frac{1000Q^2}{KL_s} \qquad (4.27)$$

Substituting Equations (4.25) and (4.27) into Equation (4.23), the capital cost of the well is:

$$C_1 = 7150 + 130w + 500Q(4 + w)$$
$$+ 165L_s + \frac{260Q}{KL_s} + \frac{1000Q^2}{KL_s} \qquad (4.28)$$

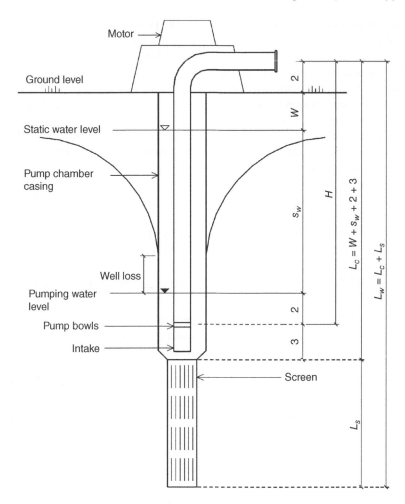

Figure 4.11 *Schematic well layout showing the notation used in the example of economic well design in Section 4.6*

Table 4.6 *Cost rates used in the well design optimization example in Section 4.6.2*

Well construction item	Unit	Rate	Quantity	Amount
Setting up rig on site	Sum	–	1	1000
Drilling borehole	m	60	$L_c + L_s$	$60(L_c + L_s)$
Supplying and installing well screen	m	90	L_s	$90L_s$
Supplying and installing casing	m	70	L_c	$70L_c$
Artificial gravel pack	m	15	L_s	$15L_s$
Developing well and test pumping	Sum	–	1	2000
Surface works	Sum	–	1	1500
Supplying and installing pump	Formula	–	–	$2000 + 500QH$

This represents the cost of well construction at present value (PV_1).

Recurring costs. The main recurring costs are the power and spares costs for the pump (the staff costs are ignored in this example, both for simplicity and also because they will be largely independent of the design variations being investigated). The annual power costs (C_2) relate to the hourly fuel consumption, unit fuel cost, pump efficiency and the annual operating hours, and can be expressed in terms of the discharge rate (Q), pump lift (H) and annual pumping hours (R). The following relationship is assumed here:

$$C_2 = 0.5QHR \quad (4.29)$$

If there are say 1200 operating hours per year, then the annual power costs are:

$$C_2 = 600QH \quad (4.30)$$

The cost of spares (C_3) can be assumed to be covered by a replacement pump (Table 4.6) during the lifetime of the well:

$$C_3 = 2000 + 500QH$$
$$= 2000 + 500Q(4+w) + \frac{1000Q^2}{KL_s} \quad (4.31)$$

The future recurring costs must be discounted back to a present value using standard formulae for interest calculations. For the annual stream of power costs (C_2), the present value (PV_2) can be calculated from:

$$PV_2 = C_2 \left[\frac{\left((1+i)^n - 1\right)}{i\left((1+i)^n\right)} \right] \quad (4.32)$$

where i is the annual interest rate (expressed as a decimal) and n is the period of years for which the costs are incurred. If we assume an annual interest rate of 5% and a well life of 20 years, Equation (4.32) becomes:

$$PV_2 = C_2 \times 12.46 = 600QH \times 12.46$$
$$= 7476Q(4+w) + \frac{14\,952Q^2}{KL_s} \quad (4.33)$$

The present value (PV_3) of a replacement pump (C_3) purchased after a period of n years can be found from:

$$PV_3 = C_3 \left[\frac{1}{(1+i)^n} \right] \quad (4.34)$$

If the pump is replaced after 10 years, and the annual interest rate is again 5%, then:

$$PV_3 = C_3 \times 0.614$$
$$= 1228 + 307Q(4+w) + \frac{614Q^2}{KL_s} \quad (4.35)$$

Adding Equation (4.33) to Equation (4.35) gives the full recurring costs (PV_{2+3}):

$$PV_{2+3} = 1228 + 7783Q(4+w) + \frac{15566Q^2}{KL_s} \quad (4.36)$$

and when these are added to the capital cost of the well (C_1) in Equation (4.28), the total present value cost of the well (PV_t) is obtained:

$$PV_t = 8378 + 130w + 8283Q(4+w)$$
$$+ 165L_s + \frac{260Q}{KL_s} + \frac{16566Q^2}{KL_s} \quad (4.37)$$

Optimization procedure. To find the length of screen corresponding to the lowest present value cost, PV_t is differentiated with respect to L_s and equated to zero:

$$\frac{dPV_t}{dL_s} = 165 - \frac{260Q}{KL_s^2} - \frac{16566Q^2}{KL_s^2} = 0 \quad (4.38)$$

which gives the following expression for the optimum value of L_s:

$$L_s = \sqrt{\frac{260Q + 16566Q^2}{165K}} \quad (4.39)$$

For a design discharge of $50\,l\,s^{-1}$ $(0.05\,m^3\,s^{-1})$ and a hydraulic conductivity K of $5 \times 10^{-4}\,m\,s^{-1}$, the optimum screen length is $26\,m$. This gives an optimum well depth (L_w) of:

$$L_w = L_c + L_s = w + s_w + 5 + L_s = w + \frac{2Q}{KL_s} + 5 + L_s \quad (4.40)$$

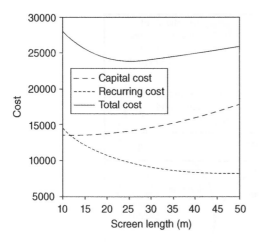

Figure 4.12 *Optimum screen length based on minimum cost*

If the depth to the rest water level w is 10 m then, with the other assumptions stated, L_w is 49 m.

For a well of this design discharge and the other conditions stated, the capital and recurring costs can be plotted for a range of screen lengths, as shown in Figure 4.12. This illustrates the obvious fact that capital costs increase with screen length whereas recurring costs reduce, and that there is a minimum total cost around a screen length of 26 m. It also shows that the total cost curve is relatively flat either side of this minimum value, whereas there are significant variations in the capital and recurring costs, thus giving the designer some scope for altering the design to increase or decrease the relative proportions of capital versus recurring costs. As noted by Stoner *et al.* (1979) in their (different) optimization example, a transfer of some of the total cost from operating to capital costs might be of considerable benefit to irrigation well owners in developing countries, where the Government or donor agency often pay for the well construction while the farmers pay the operating costs.

The optimization procedures can be applied to other design variables such as the well discharge rate and the well diameter. The optimization of well diameter takes into account the important influence diameter has on well losses (Section 4.5). A small screen diameter will produce savings in drilling and screen costs, but will lead to higher recurring costs because of higher well losses.

Each optimization exercise should include a sensitivity analysis, especially with respect to future variables such as fuel costs, pumping hours and annual interest rates, which will have a major influence on the recurring costs. This will help in selecting the right balance between capital and recurring costs, and in budgeting for well operation and maintenance.

4.7 Groundwater and wells for heating and cooling

We conventionally think of aquifers as simply being the source of one of humankind's most basic resources: water for potable, hygienic and industrial needs. However, aquifers contain another fundamental resource: heat and "coolth" (the latter term objectifying a lack of heat or heat "sink"). It is only within the last four decades that we have begun to make serious use of this resource, as fossil fuel reserves dwindle and we realise that squandering high-exergy fuels on space heating is both wasteful (Wall, 1986) and often CO_2-emitting. Now, an entire discipline of *thermogeology* has arisen to study the movement, storage and exploitation of subsurface heat (Banks, 2009a, 2012a). In fact, it turns out that the equations for the behaviour of heat in the earth are directly analogous to those for groundwater [e.g., Darcy's Law; Equation (1.3)]. Indeed, Theis's equation for radial groundwater flow to wells [Equation (7.22)] had its origin in heat transfer theory (Freeze, 1985; Banks, 2012b, 2014b).

It has long been known that the temperature of the ground is affected by complex heat exchange processes at the surface and that seasonal temperature changes occur in the shallow subsurface (Figure 4.13). Below a depth of several metres, however, the ground temperature is very constant at, or slightly above, the annual average air temperature at the location in question (Forbes, 1846; Banks, 2014a). Thus, in the UK, for example, the ground temperature at modest depth varies from around 8 to

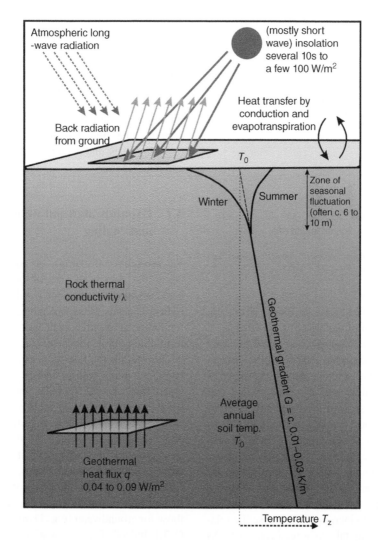

Figure 4.13   *A schematic diagram to illustrate the heat balance of the shallow subsurface. Ground temperature near the surface is largely determined by, and is slightly higher than, the annual average air temperature, although down to depths of a few metres it shows seasonal variation. At greater depths, the temperature increases, due to the geothermal gradient. Modified after Banks (2012a) and reproduced by permission of © John Wiley & Sons, Chichester, 2012*

12 °C, depending on altitude, latitude and proximity to the sea. In other words, the constant temperature of the ground and the groundwater it contains is warmer than the air temperature in winter (and hence is a good source of heat) and cooler than the air temperature in summer (and thus a good source of cooling).

4.7.1 Groundwater for cooling

It does not take much imagination to understand that, if cool groundwater can be abstracted from a spring or well in summer, and circulated either (a) directly through a network of heat exchangers in a building (radiators, fan coil units, structural cooling elements) or (b) through a heat exchanger coupled to a modern air conditioning system, heat will be removed from the building's air to the groundwater, with a space-cooling effect. The result will be a modest temperature increase in the groundwater.

The earliest documented use of groundwater for cooling is believed to be in Shanghai in the latter part of the nineteenth century, where workshops were cooled by just such a method (Volker and Henry, 1988). In the 1920s and 1930s, groundwater was used for space cooling in Long Island and Brooklyn, United States (Kazmann and Whitehead, 1980). In the UK, modern case studies of successful groundwater cooling of a pickle factory and its potential for cooling a cyclotron are published by Todd and Banks (2009) and Gandy *et al.* (2010), respectively.

4.7.2 Heating with groundwater: geothermal fluids

The use of groundwater for cooling in summer is immediately understandable, but it is perhaps less obvious how groundwater can be used for heating, when the required room temperature may be around 20 °C, which typically necessitates that heating fluids of at least 35 °C (and usually more) be circulated through radiators or underfloor heating systems.

Although shallow groundwater is typically very slightly warmer than the annual average air temperature, the groundwater temperature increases with depth. This increase is known as the *geothermal gradient* (*G*) and it is typically between 1 and 3 °C per 100 m (0.01 to 0.03 K m^{-1}) in tectonically stable areas (Figure 4.13). Thus, the approximate temperature (*T$_z$*) at a given depth (*z*) can be predicted from:

$$T_z = T_0 + G.z = T_0 + \frac{q.z}{\lambda} \qquad (4.41)$$

where T_0 is the temperature at shallow depth, *q* is the geothermal heat flux in W m^{-2} and λ is the vertical thermal conductivity of the rocks in W m^{-1} K^{-1} (and where there is assumed to be no internal source of heat in the rocks). Therefore, if the rock thermal conductivity is 2 W m^{-1} K^{-1} and the heat flux is 40 mW m^{-2}, the predicted

geothermal gradient is 0.02 K m^{-1} (2 °C per 100 m). If the shallow temperature is 10 °C, the temperature at 1 km depth would be 30 °C, while that at 2 km depth would be 50 °C. Hence, it would normally be necessary to drill to depths of at least 1½ to 2 km to obtain water that could be used for space heating and to over 4 km for fluid that could be used for electrical power generation. In some locations, where the Earth's crust is thin (Iceland, East African Rift), where the rocks contain high contents of heat-generating radioactive isotopes or where bodies of magma exist in the shallow subsurface, geothermal heat fluxes and gradients may be much higher, and usable resources may be found at much shallower depths.

Of course, it is not just a matter of drilling to great depths or in the right locations in order to obtain geothermal fluids; one must also ensure that an aquifer exists at the relevant depth (and permeabilities generally tend to reduce with depth as overburden pressures increase). Spent geothermal fluid (after heat has been extracted) is typically re-injected to the aquifer via one or more injection boreholes, to ensure that reservoir pressures are not excessively depleted. If a natural aquifer does not exist at the target depth, it may be possible to create a permeable zone in hard rocks at the target depth, by hydrofraccing, through which a fluid can be circulated (so called 'Hot Dry Rock' or 'Enhanced Geothermal' systems).

In principle, the boreholes drilled to abstract geothermal fluids are not hugely different from conventional deep water supply wells, except that (a) they are a lot more expensive; (b) they will likely need to withstand higher stresses; and (c) they may be subject to intense corrosion or incrustation due to elevated temperatures, dissolved gases and high mineral contents. The design of such specialist wells will not, however, be considered further in this book. Those interested in exploring this topic further can examine the review of geothermal resources in the UK by Barker *et al.* (2000), the drilling manual of Finger and Blankenship (2010) and the thesis by Rutagarama (2012).

4.7.3   Heating with groundwater: heat pumps

Fortunately, it is not necessary to drill to huge depths to use groundwater for heating buildings, thanks to a device called the 'heat pump', first constructed by Jacob Perkins around 1835 (Hodson, 1837) and first proposed for space heating by William Thomson, Lord Kelvin in 1852 (Thomson, 1852; Banks, 2014a).

A submersible water pump in a well transfers water from a region of low hydraulic head (below ground) to a high hydraulic head (water tower or pressure tank), from which it can be distributed and used. To do so, the pump needs to be supplied with an input of energy – usually electrical – which typically powers an impeller or turbine. A heat pump is directly analogous: it transfers heat from a region of low temperature (e.g., groundwater) to one of high temperature (a central heating system), from which it can be distributed and enjoyed as space heating. To do so, the heat pump must be supplied with an input of energy – usually electrical – which typically powers a compressor forming the core of a compression-expansion refrigerant cycle. Put in simpler terms, a heat pump

'sucks' heat out of groundwater, boosts its temperature and delivers it to the house. Note that energy is not being created or destroyed (refer to Figure 4.14).

Heat from groundwater (in thermal kilowatts, kW_{th}) plus electrical energy (in electrical kilowatts, kW_e) equals heat delivered to space heating, for example:

$$18\ kW_{th} + 6\ kW_e = 24\ kW_{th} \qquad (4.42)$$

In this example, the coefficient of performance (or COP_H, the ratio of heat output to electrical input) is $24\,kW/6\,kW = 4$, not an unreasonable figure for a groundwater-sourced heat pump scheme. One of the earliest documented examples of such a heat pump was that constructed by Dr TGN 'Graeme' Haldane, using water from the springs at Foswell, Perthshire, Scotland, in around 1927–28 (Haldane, 1930; Banks, 2015). The quantity of groundwater that is required to release 18 kW_{th} of heat can be estimated from:

$$Power\left(kW_{th}\right) = Flow\ rate\left(1s^{-1}\right)$$
$$\times Temperature\ change\left(K\right)$$
$$\times 4.2\ kJ\,l^{-1}K^{-1} \qquad (4.43)$$

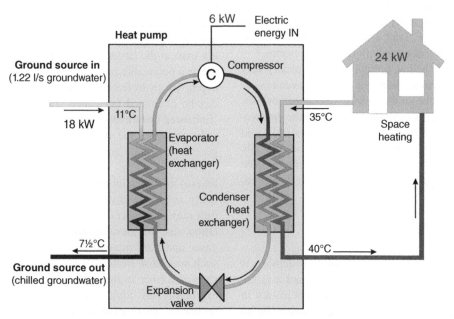

Figure 4.14   *A schematic diagram of a groundwater-based heat pump system, providing space heating to a house via a waterborne heating system. Modified after Banks (2012a) and reproduced by permission of © John Wiley & Sons, Chichester, 2012*

where $4.2\,\mathrm{kJ\,l^{-1}\,K^{-1}}$ is the volumetric heat capacity of water.

Assuming, for example, that the water has a natural temperature of 11 °C and that the anticipated temperature change is 3.5 °C within a heat pump or heat exchanger, then the required groundwater flow rate is:

$$\text{Flow rate} = 18\,\mathrm{kW} / 3.5\,\mathrm{K} / 4.2\,\mathrm{kJ\,l^{-1}\,K^{-1}} = 1.22\,\mathrm{l\,s^{-1}} \tag{4.44}$$

In other words, around $1\,\mathrm{l\,s^{-1}}$ of groundwater is ample to supply heat to a substantial house via a heat pump scheme. On emerging from the heat pump, the groundwater has been cooled to a temperature of $11\,^\circ\mathrm{C} - 3.5\,^\circ\mathrm{C} = 7.5\,^\circ\mathrm{C}$. We must now do something sensible with this flow of cool groundwater.

4.7.4 Well configurations

The $1.2\,\mathrm{l\,s^{-1}}$ of cool groundwater in the last example could simply be 'thrown away' to a nearby stream or to a sewer (Figure 4.15). In many circumstances, this may be problematic: (a) sewers have a finite capacity and a utilities company may require a payment to accept such a water flow; (b) there may not be a surface watercourse nearby; (c) the extra water may exacerbate a flooding risk; (d) the groundwater chemistry may not be compatible with the stream water chemistry. The most likely objection, however, may be from a groundwater regulator who would argue that good quality groundwater should not be 'thrown away', and who may thus refuse to grant an abstraction licence (McCorry and Jones, 2011). The water could be considered for a household potable water supply, but $1.2\,\mathrm{l\,s^{-1}}$ is a lot more water than a single household could normally use.

Consequently, many environmental regulators may insist that the cool, thermally 'spent' groundwater is returned to the aquifer from whence it came, to prevent resource depletion. In some cases, where the aquifer is unconfined and permeable and where the quantities are modest, it may be possible to return the water via a soakaway to the unsaturated zone (Figure 4.15). In other cases, it may be necessary to construct a reinjection well for this purpose (USEPA, 1999b). The reinjection well will typically be similar in design to the abstraction well and will usually be located down the groundwater hydraulic gradient from the abstraction well, to minimize the risk of the cold

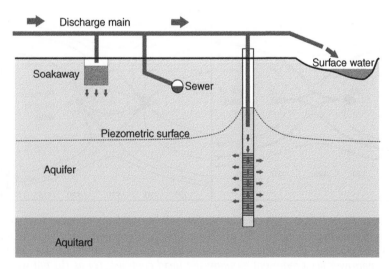

Figure 4.15 *Possible alternative modes of disposal of thermally 'spent' water from a heat pump. From left to right – soakaway, sewer/drain, reinjection borehole, surface watercourse (e.g. river)*

reinjected groundwater percolating back into the abstraction well. This arrangement is often referred to as a 'well doublet' (Figure 4.16; Section 4.8).

Finally, because groundwater can contain corrosive components (salinity, hydrogen sulphide) or can cause incrustation (calcite, iron or manganese oxyhydroxides) or biofouling, an engineer may not be keen on circulating groundwater directly through the innards of an expensive heat pump. Hence it is common practice to pass the groundwater through a heat exchanger (e.g., shell and tube, or parallel plate), which is coupled to the heat pump via a secondary circuit of controlled thermal transfer fluid. In this way, heat can be transferred from the

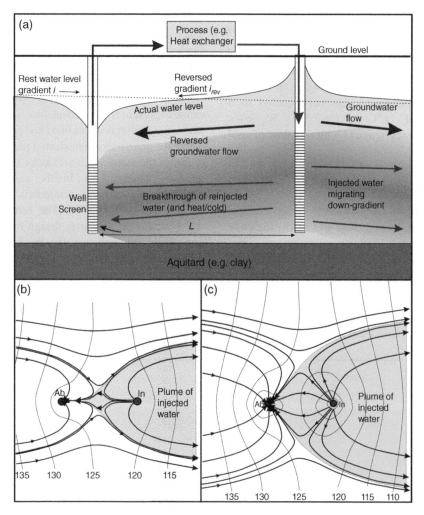

Figure 4.16　*(a) A cross-section of a typical well-doublet scheme, with the injection well located a distance L down-gradient from the abstraction well; (b) two-dimensional plan representing the thermal plume emitted by an open loop well doublet. The shaded area shows the thermal plume, the arrows show groundwater flow lines and the narrower lines represent groundwater head contours; (c) as (b) but with greater degree of feedback. Modified after Banks (2012a) and reproduced by permission of © John Wiley & Sons, Chichester, 2012*

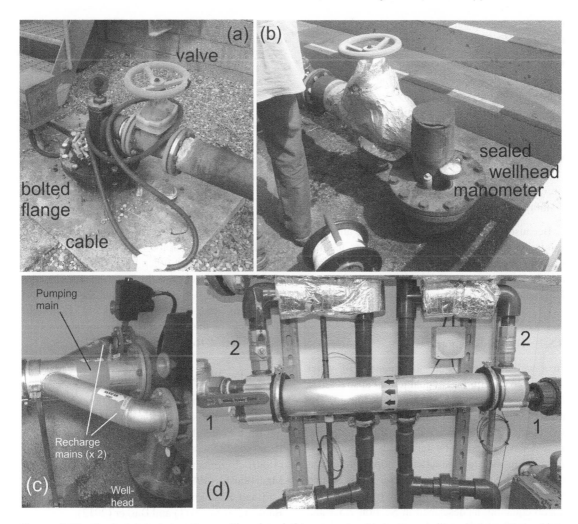

Figure 4.17 *Typical (a) abstraction wellhead and (b) pressurized injection wellhead of a well doublet. (c) Reversible wellhead installed at Arlanda airport ATES scheme, showing single abstraction main and two narrower recharge mains (photos (a) to (c) by David Banks, modified after Banks (2012a) and reproduced by permission of © John Wiley & Sons, Chichester, 2012). (d) Prophylactic shell and tube heat exchanger installed on coal mine water-based heat pump scheme near Bolsover, UK. The pipes marked '1' carry the mine water flow, those marked '2' carry the secondary heat transfer fluid*

groundwater to the heat pump, but if corrosion or incrustation occurs, it will be the heat exchanger, rather than the heat pump, that is affected. A clogged heat exchanger can either be sacrificed or cleaned. Because of its protective effect, this is often termed a 'prophylactic' heat exchanger [Figure 4.17(d)].

4.8 Well doublets

As we have seen above, a well doublet is a pair of wells comprising an abstraction and an injection well. Often, the water abstracted from first well may be subject to some process and then reinjected (in the same quantity) to the injection well (Figure 4.16). The process in question may be:

- the extraction or rejection of heat (via a heat pump or heat exchanger for the purposes of heating or cooling), or
- some kind of water treatment (as in the case of 'pump and treat' schemes for removing contaminants from polluted groundwater), or
- the addition of some unwanted substance to be disposed of to the aquifer (although most environmental regulators would look on this unfavourably).

Because, in all these applications, the aim is to minimize the amount of injected water feeding back to the abstraction well, the injection well should ideally be located directly down the hydraulic gradient from the abstraction well.

4.8.1   Hydraulic equations

In a well doublet, where the quantity of water injected to the aquifer is equal to that abstracted, a hydraulic equilibrium is eventually achieved. The closer the spacing (L) of the doublet (Figure 4.16), the quicker this is achieved and the smaller the drawdown of groundwater levels around the abstraction well (and smaller the opposite effect; i.e., the 'up-coning' of groundwater levels around the injection well). By application of the Cooper-Jacob equation, as described in Section 7.5.3 and Equation (7.40), the drawdown (s) at any point in an ideal aquifer containing two pumping wells is approximated by:

$$s = \frac{Q_A}{4\pi T}\ln\left(\frac{2.25T}{r_A^2 S}\right) + \frac{Q_A}{4\pi T}\ln t + \frac{Q_B}{4\pi T}\ln\left(\frac{2.25T}{r_B^2 S}\right)$$
$$+ \frac{Q_B}{4\pi T}\ln t$$

(4.45)

where Q_A and Q_B are the pumping rates of wells A and B, t is the time since both started pumping, r_A and r_B are the distances from the two wells, and T and S are the transmissivity and dimensionless

storage coefficient of the aquifer. In the special case of the well doublet, $Q_B = -Q_A$ and thus the equilibrium drawdown is given by:

$$s = \frac{Q}{2\pi T}\ln\left(\frac{r_B}{r_A}\right)$$

(4.46)

where Q is the well doublet operation rate. If L is the well doublet separation, then the equilibrium drawdown (s_w) in an idealized (100% efficient) abstraction well of radius r_w is given by:

$$s_w = \frac{Q}{2\pi T}\ln\left(\frac{L}{r_w}\right)$$

(4.47)

Therefore, the smaller the value of L, the closer the two wells are to each other, the less the changes in groundwater level will be and the smaller the pumping costs. On the other hand, the closer the wells are, the greater the risk of hydraulic 'feedback' of the injected water to the abstraction well (Figure 4.16).

4.8.2   Feedback and breakthrough

The mathematical treatment of flow in a well doublet is a classic problem of hydrogeology and has been treated by Hoopes and Harleman (1967), Bear (1972), Shan (1999), Kazmann and Whitehead (1980), Lippman and Tsang (1980), Clyde and Madabhushi (1983), Javandel and Tsang (1986), Luo and Kitanidis (2004) and, most recently, by Banks (2009b, 2011) and Barker (2012).

In fact, for an idealized well doublet of spacing L, with the injection well immediately down-gradient of the abstraction well, it can be shown that there is no hydraulic feedback from the injection to the abstraction well, if

$$L > \frac{2Q}{\pi b v_D} = -\frac{2Q}{\pi b K i} = -\frac{2Q}{\pi T i}$$

(4.48)

where b is the aquifer thickness, K is its hydraulic conductivity, T its transmissivity, i is the natural hydraulic gradient (a negative number, as natural groundwater flow takes place along a decreasing hydraulic gradient) and v_D the natural, regional Darcy flux. If L is less than this critical figure, there is a finite risk of hydraulic feedback. Ignoring dispersion effects, the time (t_{hyd}) until this occurs along the shortest flow-path (i.e., that directly linking the two wells) is given by:

$$t_{hyd} = \frac{Ln_e}{Ki}\left[1 - \frac{\beta}{\sqrt{\beta-1}}\tan^{-1}\left(\frac{1}{\sqrt{\beta-1}}\right)\right] \quad (4.49)$$

where n_e is effective aquifer porosity, $\beta = -2Q/(\pi KibL)$, and i is deemed to be a negative number (negative gradient), such that β is positive. Note that the $\tan^{-1}$ term is in radians. If i is zero, then:

$$t_{hyd} = \frac{\pi L^2 n_e b}{3Q} \quad (4.50)$$

If there is a finite natural hydraulic gradient, only a fraction of the injected water will be recirculated to the abstraction well (Figure 4.16); the remainder will 'disappear' down-gradient. The flow that is recirculated (Q_{recirc}) is given by:

$$Q_{recirc} = Q\left(1 - \frac{2}{\pi}\left[\tan^{-1}\left(\frac{1}{\sqrt{\beta-1}}\right) + \frac{\sqrt{\beta-1}}{\beta}\right]\right)$$

$$(4.51)$$

As a small example, consider a fully-penetrating idealized well doublet, spaced at $L = 80$ m, with well diameter of 200 mm ($r_w = 0.1$ m), operating at 10 l s^{-1} (864 m^3 d^{-1}), in an aquifer of hydraulic conductivity 2 m d^{-1}, effective porosity 18% and thickness 50 m, with a natural hydraulic gradient $i = -0.01$. Then:

- equilibrium drawdown in abstraction well [Equation (4.47)] = 9.2 m;
- critical separation [Equation (4.48)] = 550 m (so some hydraulic feedback would be expected);
- $\beta = 6.875$;
- time to hydraulic feedback [Equation (4.49)] = 79 days;
- proportion of water eventually recirculated [Equation (4.51)] = 53% (or 5.3 l s^{-1}).

When considering the injection of hot or cold water, it is important to remember that a heat signal travels more slowly than the water molecules themselves, because the heat is absorbed into the aquifer matrix and is thus retarded by a factor R_{th}. Assuming that the heat exchange between water and matrix is almost instantaneous, and that heat and water flow are two-dimensional (i.e., no vertical conduction of heat – which is a significant assumption, Banks 2011), then the thermal breakthrough time t_{th} can be calculated from Equations (4.49) or (4.50):

$$t_{th} = R_{th}t_{hyd} = \frac{\rho_{aq}c_{aq}}{n_e\rho_{wat}c_{wat}}t_{hyd} \quad (4.52)$$

where ρ_{aq} and c_{aq} are the bulk density and bulk specific heat capacity (kJ kg^{-1} K^{-1}) of the saturated aquifer material, n_e is the effective porosity of the aquifer, and ρ_{wat} and c_{wat} are the density and specific heat capacity of water (approximately 1000 kg m^{-3} and 4.2 kJ kg^{-1} K^{-1}, respectively).

4.8.3 Water chemistry

The general topic of water quality and injection wells will be addressed in Section 4.9. However, in the specific case of a well doublet being used for ground source heating or cooling, there are some very specific risks related to groundwater quality:

- *Particulate matter*. Particles in pumped groundwater are generally undesirable and can clog heat exchangers and recharge wells. The abstracted groundwater *should*, if the borehole is well-designed, contain minimal particulate matter.
- *Incrustation*. Often, deep groundwater can contain dissolved iron and manganese. If these are exposed to oxygen, they can oxidize to form iron or manganese oxyhydroxides, which can encrust or clog heat exchangers, pipework or recharge wells. The key to preventing this is to ensure a sealed abstraction-heat exchange-recharge main, such that the water is not exposed to oxygen and reducing conditions are maintained. If this can be achieved, evidence suggests that iron and manganese will tend to remain in solution. Attempts to treat the water (e.g., to remove iron) prior to heat exchange or reinjection are often counter-productive. Mine waters are particularly prone to elevated iron

concentrations in the water: Banks *et al.* (2009) document two case studies of mine water heat pump schemes, one operating relatively successfully and another where exposure to the atmosphere caused iron hydroxide precipitation and clogging of a recharge well.

- *Degassing of CO_2.* Groundwaters and, especially, mine waters can contain excess partial pressures of carbon dioxide, which tend to degas if exposed to atmospheric pressure or under-pressure. If CO_2 degassing occurs, it tends to raise the pH of the water and increase the tendency for incrustation by, for example, calcium carbonate and iron oxyhydroxide (Section 9.1.2). Pressures downstream of the heat exchanger and in the recharge main should be managed to ensure they remain positive.
- *Temperature change* can affect the tendency for mineral scaling to occur. The rate of iron oxidation typically increases with increasing temperature (Faraldo Sánchez, 2007), as does the thermodynamic tendency of carbonate minerals to precipitate (Banks *et al.*, 2009). Thus, mineral scaling is likely to be more of a problem in situations where heat is being reinjected to groundwater, rather than extracted from it.
- *Degassing of dissolved nitrogen.* Groundwater may also contain partial pressures of dissolved nitrogen slightly in excess of atmospheric (and also excess carbon dioxide – see above). If underpressure develops in a recharge pipe line, bubbles of gas can exsolve. These can then be rein-jected into the aquifer and lodge in pore throats, effectively reducing hydraulic conductivity.

In summary, the abstraction-heat exchange-reinjection systems used in well doublets should ideally be sealed from contact with atmospheric oxygen. Injection should take place below the water level in the recharge well. Pressures, espe-cially in the recharge main (downstream of the hydraulic restriction that the heat exchanger repre-sents) should be managed to maintain positive pressures, if necessary by use of a throttle on the recharge main. The potential effects of degassing of CO_2, contact with O_2 and temperature change

can all be simulated by hydrochemical modelling software such as PHREEQC (Parkhurst and Appelo, 1999; Banks *et al.*, 2009)

4.9   Recharge wells

4.9.1   Purpose

Recharge or injection wells may be constructed for different purposes:

1. For return to the aquifer of groundwater abstracted for purposes of heating or cooling. They may often form part of a well doublet in such cases – although sometimes more than one recharge well may be constructed for each abstraction well.
2. For the return of water from a groundwater pump-and-treat scheme to decontaminate polluted groundwater.
3. For disposal of liquid wastes. This practice is generally discouraged by regulators, but has been used in the past for chemical wastes and used in the Siberian city of Tomsk to dispose of high level radioactive wastes (Lgotin and Makushin, 1998; Solodov, 1998). Reinjection is a common means of disposing of waste fluids from hydrocarbon extraction and hydrofraccing operations.
4. To create a zone of high groundwater head that effectively forms a hydraulic barrier against, for example, seawater intrusion or migration of contaminated groundwater. A scheme involving the injection of reclaimed waste water to pre-vent sea water intrusion in Hollywood, Florida, United States, is documented by Bloetscher *et al.* (2005).
5. For artificial recharge of groundwater with treated surface water during times of excess surface water flow. This may be simply to enhance groundwater reserves or may also be to limit land subsidence (Volker and Henry, 1988).
6. The deliberate storage of excess treated surface water in a limited volume of aquifer, which can be recovered by abstraction during times of

high demand (Figure 4.18). In fact, fresh water can be injected into a brackish or saline aquifer to create a zone or "bubble" of fresh water around the injection well(s), and subsequently recovered. If the water is recovered from the same well, this is termed *aquifer storage and recovery (ASR)*; where a nearby well is used for abstracting the recharge water, it is referred to as *aquifer storage, transfer and recovery (ASTR*; National Research Council, 2008; Dillon *et al.*, 2009). For ASR and ASTR schemes, well-bounded or confined aquifers are preferred, to prevent vertical migration of the freshwater body (as in the case of Figure 4.19).

7. The deliberate storage of excess heat or 'coolth', as hot or cold water, in a limited volume of aquifer, which can be recovered by abstraction during times of high demand. Excess warmth (e.g., from summer cooling operations) can be injected with a flow of warm water to the ground, forming a warm 'bubble' or zone around the borehole, which can be recovered for space heating in the winter. Similarly 'coolth' (e.g., from winter heating operations) can be injected to the ground with a flow of cold water, which can be recovered for space cooling in the summer. These operations are arguably best achieved by a reversible well doublet. The 'warm' well(s) of such a doublet are pumped during winter, the groundwater is used to provide heating (often via a heat pump) and the resulting cold water is injected to the

Figure 4.18   *A dual abstraction/injection well used for aquifer storage and recovery in Las Vegas. The elevated section of pipework on the right hand side of the picture allows water to bypass the non-return valve during the injection phase. The inclined cylinder is a sand separator. During injection, the impellers of the turbine pump (to the left in the photo) are adjusted so that they do not rotate. Photo by Bruce Misstear*

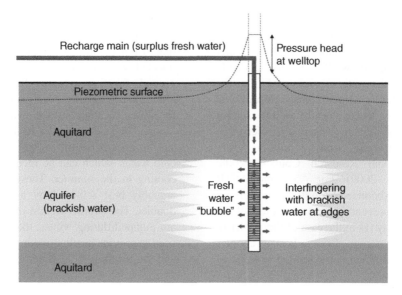

Figure 4.19   *The development of an 'injection bubble' of fresh injected water in an otherwise brackish, confined aquifer. Note that 'upconing' around the injection borehole means that the wellhead will be pressurized*

other "cold" well(s) of the doublet, thus accumulating a 'bubble' of cool water. During summer, the 'direction' of the doublet is reversed. The cold wells are pumped to provide cool groundwater for air conditioning. The cool groundwater absorbs waste heat from a heat exchanger and is reinjected to the warm wells, thus accumulating a 'bubble' of warm water (Figure 4.20). Such schemes are very efficiently operated on a large scale at Stockholm's Arlanda international airport and at Oslo's Gardermoen airport (Andersson, 2009; Banks, 2012a). This is referred to as *aquifer thermal energy storage (ATES)*.

4.9.2   Construction of injection wells

The construction of a well for injection of water to an aquifer will often be very similar to that of an abstraction well, as regards casing, filter pack and well screen (Pyne, 2005). Bloetscher *et al.* (2005) suggest that the installation of a gravel pack may increase the risk of clogging, while cable tool drilling may reduce this risk. In some circumstances,

an injection well may need to withstand high pressures, which will affect the materials selected and the strength of the cement bond between the casing and the borehole wall (Bloetscher *et al.*, 2005). Recharge wells typically have diameters of 200 to 400 mm. They may be constructed of steel but, because they are often subject to large fluctuations in water level (during injection and backwashing), they may be prone to corrosion and rusting. As rust particles can cause screen or aquifer clogging during reinjection, it is not unusual for alternative materials to be used for casing (Section 4.2), screen (Section 4.3) and even wellhead pipework, including:

- PVC (which lacks strength compared to steel; but thick-walled varieties of PVC have been installed in wells to depths of over 250 m);
- epoxy-coated steel;
- stainless steel;
- fibreglass.

The screened section of the injection well will usually be maximized to enhance hydraulic efficiency and to reduce clogging rates. The slot size will be approximately the same as that recommended for abstraction wells (Section 4.3.2) although some authors argue that slightly larger slot sizes can be accommodated.

Around an abstraction well, pumping causes a 'cone of depression' of groundwater levels. Around injection wells, the opposite occurs, namely 'upconing' of groundwater levels, described by exactly the same equations as those for drawdown, but in reverse. If the static water level in an injection well is deep, injection may be able to be carried out without the upconing reaching the surface. In such a case, the recharge well will not be pressurized. Where the groundwater level is shallow and/or the upconing is large enough, the groundwater head in the injection well might rise *above* ground level (Figure 4.19). If the aquifer is unconfined, this may lead to groundwater flooding at the surface. If the aquifer is confined, then the well head is usually sealed and a pressure head is allowed to build up within it [Figure 4.17(b)]. In this case, the sealed wellhead will typically incorporate air/vacuum release valves. The grouting (cement bond) between the casing and the borehole wall must be strong enough to withstand this excess pressure. A weak, bentonite-based grout may not suffice. Cement grouts typically release heat of hydration during curing, potentially resulting in temperatures high enough to distort or damage plastic casings, which must be borne in mind if these materials are utilized.

4.9.3 Installations

An injection well will contain one or more injection pipes: these will usually extend below the water level for two reasons (a) to prevent cascading of water down the casing, which can promote corrosion and rust; and (b) to prevent the water coming into contact with oxygen: this is important in well doublets where the groundwater may be reducing in nature and contain dissolved iron and manganese, which should not be oxidized.

If the injection rate is likely to be variable, it may be beneficial to have two narrow diameter recharge pipes rather than one large one (Figure 4.17c). At times of low injection rate, the injected water is directed down a single pipe and experiences a hydraulic resistance that maintains positive pressures in the recharge system (a 50 mm ID steel pipe flowing at $18 \, l \, s^{-1}$ typically results in a head loss of 1 m per metre length; Pyne, 2005). At higher rates, both recharge pipes can be utilized. The recharge main may additionally be fitted with throttles (e.g., constrictions of pipe diameter) or even valves to regulate and maintain positive pressure within the recharge system. A deflector may be fitted at the base of the injection main to dissipate the flow jet as it emerges into well. Consideration should be given to surge/water hammer protection.

If the well is designed both for abstraction and injection (this is often the case), a pump (usually an electrical submersible or vertical turbine pump) will be fitted. It is even possible to use a single rising main for both pumping and injection operations. This is usually easier to achieve with a vertical turbine pump: a ratchet can be fitted to the turbine to prevent backspin during injection, and this also provides a 'throttle' effect helping to maintain positive pressures. If a submersible pump is used, flow rates must be low enough not to cause excessive backspin; one should also remember that a non-return valve cannot be fitted and some kind of motor restart delay mechanism may be required. Several convenient devices can be fitted to facilitate combined pumping and injection: for example, special valves that allow the operator to switch between an electrical submersible pump and a recharge vent mounted above the pump (Cla-val, 2012). It is even possible to use the 'fall' through the recharge main to generate electricity by installing a turbine or even by using the reverse spin on a reversible pump turbine. Pyne (2005) reports that such electrical power recovery is being practised in California, United States.

Finally, it is common practice, especially where surface water is injected, to trickle chlorinate the injection well when not in operation. A modest flow of chlorinated water to the well will percolate through the screen and into the aquifer and prevent bacterial/biofilm colonization of the well (see Chapter 9). Pyne (2005) states that a trickle flow of 8-19 lmin⁻¹ of chlorinated water via a narrow diameter tube should suffice to maintain an adequate chlorine residual in the well.

4.9.4   Testing and operation

Injection wells will normally be tested in 'abstraction mode' (as with normal water wells, see Chapter 7), following clearance and development pumping, by a series of step tests and a constant rate test. This can be followed by a series of *injection tests*, comprising:

1. A series of four-five step injection tests at increasing rates, with upconing measured over time.
2. A constant rate injection test of, say, 72 hours duration.
3. For pressurized injection wells, a mechanical integrity test at 150% of the maximum operational injection pressure (or 10 bar, whichever is greater, according to Bloetscher *et al.*, 2005).

For the operation of injection wells in a hydraulically efficient manner, many of the same guidelines apply as to normal abstraction wells (see Section 4.5). Bloetscher *et al.* (2005) recommend a maximum downhole injection velocity of $3\,m\,s^{-1}$ (i.e., discharge divided by cross-section of well casing), but this should probably be less to account for bubble rise velocity (see below). To calculate head losses in the injection pipe, the Hazen Williams formula can be applied (see Appendix 2).

4.9.5   Clogging of recharge wells

Like any water well, injection wells are subject to a variety of physical, chemical and microbiological corrosion, incrustation and clogging processes, which are comprehensively described in Chapter 9. There are, however, some particular clogging issues, to which injection wells are especially susceptible, including clogging by particulates, chemical precipitates, microbes and gas bubbles.

Particulate clogging. Because we are injecting water into an aquifer, any particles within that water

are pushed into the wall of the borehole and may lodge in the well screen, the filter pack or the pore spaces of the formation. These particles include mineral particles and other physical debris: grains of clay, silt or sand, or particles of rust from pipe/casing corrosion. Surface waters may be especially problematic in this respect. Groundwaters from properly designed abstraction wells should be particle-free, unless the well screen and filter pack are poorly designed. Particulate clogging may also be caused by chemical precipitates and flocs – including residual floc of aluminium hydroxide from pre-treatment of surface waters used for injection – and by organic fragments from bacteria, algae and other organic matter (see below).

Turbidity (i.e., the transmittance of light through the water) can be a very approximate indicator of particulate matter content, but it is not a very good one. It is far preferable to measure total suspended solids (TSS) in a recharge water by one of methods described in Box 4.4.

Particulate matter can be removed by treatment (flocculation, settlement, filtration) prior to injection, although these processes may expose the water to air or chemicals (flocculation agents) which may cause their own problems. In-line cartridge or mesh filters can remove coarser particulate fragments (these filters will require replacing or cleaning on a regular basis). Alternatively, a length of large-diameter pipe near the wellhead, on the recharge line, will reduce flow velocities and may allow larger particles to settle out (Pyne, 2005). Centrifugal sand separators may also be applied. If clogging is observed, the temptation to increase injection pressures should be avoided, as this will simply push clogging materials deeper into the filter pack or aquifer.

Particulate clogging of the well screen, filter pack and near aquifer can be removed by back-washing (i.e., pumping of the injection borehole to waste) according to a regular schedule (which can range from daily to seasonal) or when clogging reaches a trigger threshold. In many cases, the back-pumping is performed immediately prior to an injection episode, for a duration of up to 2 hrs and at a rate slightly above the operational recharge

Box 4.4   Measuring total suspended solids

The Total Suspended Solids (TSS) content of a water sample encompasses the content of that water sample that can be removed by filtration. It can be divided into:

- settleable solids, that is, those that will settle out from a water sample under gravity, within a given time frame.
- colloidal solids, that is, those that will remain in suspension.

Potable water generally contains $<5\,mg\,l^{-1}$ TSS. However, as little as $2\,mg\,l^{-1}$ of TSS can cause clogging in and around injection wells and one should aim for considerably less than $1\,mg\,l^{-1}$ (Pyne, 2005; Bloetscher *et al.*, 2005). Indeed, particles only one twentieth of the pore size can cause clogging (Bloetscher *et al.*, 2005). There are various laboratory methods for determining TSS, often involving vacuum filtration of the water sample (e.g., via a Buchner funnel), followed by drying (e.g., at $110\,°C$) and weighing the filter residue. A TSS of $0.05\,mg\,l^{-1}$ or less would typically be considered low and might lead to normalized clogging rates (see Box 4.5) of less than $0.3\,m\,month^{-1}$. A TSS of $0.1\,mg\,l^{-1}$ or more would typically be considered high and might lead to normalized clogging rates (see Box 4.5) of more than $3\,m\,month^{-1}$ (Pyne, 2005).

TSS can also be measured using:

1. A Rossum sand tester (Rossum, 1954; Hix, 1995). This is a passive device whereby pumped

water enters a centrifuge vessel at a given rate before exiting. Sand particles (above a certain size) settle in the centrifuge and are collected in a vessel at the base. This equipment thus measures the larger particle fraction in the water and may not strictly correspond to the TSS.

2. An Imhoff cone (Abbott, 2013). This is a clear, inverted cone-shaped sample vessel, with a graduated measuring cylinder at its base. As settling rate is related to particle size, one can monitor the rate of sediment accumulation over time from a water sample, and calculate the particle size distribution in the water sample. This, of course, only measures settleable solids.

TSS or sand-testing should be carried out at the step-testing stage of well testing to examine the influence of discharge rate on TSS.

Clogging rates, rather than TSS, can be measured on a small scale using:

1. A membrane filter of standard diameter (47 mm) and pore size (0.45 μm), through which water is passed at a standard pressure. The rate at which the filter becomes clogged is termed the Membrane Filter Index (MFI).

2. A by-pass filter, through which a small portion of the water flow is passed. This is a cylindrical cartridge filled with a filter medium. It will trap particles at a given rate over time. Changes in pressure differential across the cartridge can also indicate the hydraulic effects of clogging.

rate (Pyne, 2005). Following back-pumping and prior to reinjection, the recharge pipeline should be purged to waste. Thus, injection wells that are to be 'back-pumped' should be fitted with a permanent pumping installation.

In a study of clogging of recharge wells constructed in a weakly-consolidated sandstone aquifer near Perth, Australia, Johnston *et al.* (2013) noted that whereas backwashing was effective in

reducing the severity of clogging by aquifer fines, it did not prevent clogging entirely, and hence that regular redevelopment of the wells was also required. Redevelopment techniques are described in Chapter 9, and include swabbing, jetting or airlifting. Backwashing and redevelopment treatments are unlikely to return the well to a 'pristine efficiency', and there will typically be a 'residual clogging' level that cannot be shifted.

Chemical clogging. This is most commonly caused by iron/manganese oxyhydroxides or by calcium carbonate, and is a particular problem for groundwater-based well doublets. Experience suggests that this is best avoided by:

- Monitoring the water quality by taking regular samples (and all reinjection installations should have unthreaded, non-ferrous sample taps at the abstraction wellhead, and on the injection line at a location of positive pressure, upstream of the last control valve).
- Maintaining positive pressures and preventing degassing throughout the abstraction-(heat exchange)-recharge system. Avoid large pressure/elevation drops on the reinjection main.
- Maintaining a sealed abstraction-(heat exchange)-reinjection line, such that contact with atmospheric oxygen is prevented, and iron and manganese remain in their soluble reduced Fe^{2+} and Mn^{2+} forms (Banks *et al.*, 2009).
- Preventing iron and manganese from oxidizing by adding reducing agents to the water. These include sodium bisulphite ($NaHSO_3$; used at Swimming River, New Jersey, United States; Pyne, 2005) and sodium dithionite ($Na_2S_2O_4$; trialled in connection with the Bullhouse minewater, Yorkshire, UK; Dudeney *et al.*, 2003). These react with oxygen to ultimately form sodium and sulphate. Environmental regulators may have objections to additives in recharge water.

Microbiological clogging. Chapter 9 describes how some micro-organisms (e.g., 'iron bacteria') can colonise well and pipe surfaces to form clogging biofilms. If inadequately treated surface water is injected to a borehole, it can contain microbes that can 'infect' a borehole and initiate biofilm development. High levels of nutrients (e.g., organic carbon, phosphorus, nitrogen, some minor elements and terminal electron acceptors such as O_2, NO_3^-, $SO_4^=$, depending on the bacterium) in the water can also promote biofilm development. Up to a certain point, modest increases in temperature (e.g., 20–40°C; Pyne, 2005) will promote bacterial growth, although the availability of nutrients is likely to be a more important limiting factor. To

hinder microbial colonization, surface waters will usually be treated/disinfected prior to injection.

Microbiological clogging can be hindered by sterilizing the water prior to injection. This is especially relevant to injected surface water (abstracted groundwater *should* be no more microbiologically active than that already in the aquifer around the injection well). Treatment of the injected water can be by chlorination to achieve a residual chlorine concentration of 1 to $5\,mg\,l^{-1}$ (Pyne, 2005). Chlorination can be problematic as (a) it can oxidize dissolved iron or manganese in the water, which can form clogging precipitates; (b) it can react with organic materials or natural bromide to form unwanted *chlorine disinfection by-products* such as trihalomethanes and haloacetic acids (some European nations practice dechlorination and organic carbon removal from waters prior to injection; Pyne, 2005). Ozonation is an alternative disinfection technique (although it still causes oxidation of dissolved Fe and Mn, and can react with bromide to form unwanted bromates), as is ultra-violet disinfection (although this requires low turbidity water). As noted above, trickle chlorination of an injection well may be practized when it is not being actively used.

Gas bubbles. As noted in Section 4.8.3, all waters will contain dissolved gases – and some groundwaters, especially, will contain high partial pressures of, for example, N_2 and CO_2 (White and Mathes, 2006). Sharp pressure changes and under-pressures in the recharge pipeline can allow bubbles of gas to exsolve. If these bubbles are entrained in the injected water flow, they can enter the well's filter pack or the aquifer, lodge in pore throats and cause significant permeability reductions. This type of clogging is sometimes referred to as a mechanical clogging process (Martin, 2013).

Gas bubble clogging can be hindered by:

- Analysing and monitoring dissolved gas concentrations (see Chapter 8; Jahangir *et al.*, 2010, 2012; USGS 2014a,b for techniques).
- Modelling and managing pressures in the abstraction-(heat exchange)-recharge system.
- Ensuring that the injection line is airtight, to avoid air being sucked in and entrained.

Box 4.5 Measuring clogging rates

Clogging rates ($\Delta\varphi$) in an operational injection borehole are often measured as the rate of increase in injection head (m month^{-1}) to achieve a given water flux (Q) into the formation per exposed area of "hydraulically active" aquifer in the borehole wall (A).

A normalized clogging rate ($\Delta\varphi_{norm}$) is obtained by normalization to a standard flux (Q/A) of 3 ft^3 ft^{-2} hr^{-1} (3 ft hr^{-1} = 0.914 m hr^{-1}):

$$\frac{\Delta\varphi_{norm}}{\Delta\varphi} = \left(\frac{0.914A}{Q}\right)^2 \left(\frac{\mu_{20°C}}{\mu}\right) \text{ for metric units}$$
$$\left(Q \text{ in m}^3\text{hr}^{-1}, A \text{ in m}^2\right)$$

$$(4.53a)$$

$$\frac{\Delta\varphi_{norm}}{\Delta\varphi} = \left(\frac{3A}{Q}\right)^2 \left(\frac{\mu_{20°C}}{\mu}\right) \text{ for imperial units}$$
$$\left(Q \text{ in ft}^3\text{hr}^{-1}, A \text{ in ft}^2\right)$$

$$(4.53b)$$

where μ and $\mu_{20°C}$ are fluid dynamic viscosities at the actual injection temperature and at a standard temperature of 20 °C, respectively ($\mu_{20°C}$ for pure water at 20 °C is 1.002 cP). A low normalized clogging rate might be <0.3 m month^{-1}, while a high one might be >3 m month^{-1} (Pyne, 2005). A normalized clogging rate of 1 m month^{-1} implies that, at a Q/A of 0.914 m hr^{-1}, the injection head would increase 12 m over the course of a year.

- Avoiding cascading water (i.e., ensuring injection below water level in recharge borehole) and maintaining positive pressures, if necessary by using throttles or control valves.
- Avoiding large pressure/elevation drops on the reinjection main.
- Ensuring that the downward flow velocity in the injection borehole casing is less than the bubble rise velocity. This means that bubbles can escape upward, rather than being entrained in the injected water. A typical air bubble rise rate is 0.3 to 0.4 m s^{-1}.

Some further comments on clogging of recharge wells. According to Pyne (2005), the various clogging processes develop at different rates. Gas bubble clogging typically develops rapidly following commencement of an injection operation, but that reaches a "plateau" with no further deterioration. Bacterial clogging develops at a more modest pace but relentlessly: if sufficient nutrients are available, clogging rates can even develop quasi-exponentially. Particulate clogging tends to be more gradual and linear. The rate of clogging is related to the particle content in the water.

As we have seen, various strategies and treatments can be adopted to minimize clogging risk and to rehabilitate clogged wells. The risk cannot be eliminated, however, and this is why some degree of clogging must probably be accepted, necessitating typically greater screen lengths for injection than for abstraction. In some well doublet schemes, two injection wells may even be provided for a single abstraction well. Pyne (2005) asserts that, even in unclogged injection schemes, the specific capacity (discharge/water level change) in injection mode is almost always less than that in pumping (abstraction) mode. The performance ratio (injection:abstraction) varies from 20% to 100%, but is often 50–80% in unconsolidated aquifers. It is still unclear why this is the case: it may be related to unknown skin effects, particle rearrangement or some form of hysteresis during aquifer matrix settlement and decompression.

To monitor performance of injection wells and the progress of clogging, flows and pressures should be continuously monitored, and consideration should be given to performing initial and regularly repeated step tests [a step injection test is simply the same as step abstraction tests (Chapter 7) but with water being injected at systematically increasing rates, rather than abstracted].

4.9.6   Seismic risk from water injection

It should be remembered that injection of water at high pressure can, under exceptional circumstances, be sufficient to open up existing fractures or even create new fractures (if the applied pressure exceeds the sum of the tensile strength of the rock and the in-situ stresses). This latter process is termed *hydraulic fracturing* or *hydrofraccing*. This can be performed deliberately to increase injection/abstraction capacity in some low-permeability aquifer environments (Section 5.9), but it can also be an unforeseen consequence of injection. The seismic risk from hydrofraccing is normally regarded as low (e.g., Westaway and Younger, 2014) but may need to be considered. Re-injection of large quantities of waste water at high pressure has also been blamed for activation of critically-stressed faults, resulting in modest seismic events (Ellsworth, 2013; Keranen *et al.*, 2013; Van der Elst *et al.*, 2013; Zang *et al.*, 2014a,b; Darold *et al.* 2015). This risk is sometimes associated with large geothermal operations (including so called 'hot dry rock' or 'enhanced geothermal system (EGS)' projects; Banks, 2012a) and the disposal of waste water from hydrocarbon production operations.

4.10   Aquifer storage and recovery

The deliberate injection of water into an aquifer for subsequent later recovery is termed *aquifer storage and recovery* (ASR). It may simply involve the injection of surplus water (usually, treated surface water, during a rainy/winter season) into an existing aquifer containing fresh groundwater. In this case, the injected water simply increases the quantity of stored water, locally increasing groundwater levels or heads. There is such a scheme in Las Vegas, where water sourced from Lake Mead behind the Hoover dam is recharged into the aquifer below Las Vegas during the winter months, creating a 'groundwater bank',

which is then available for use during the peak-demand summer period (Figure 4.18 shows one of the dual purpose abstraction/injection wells). The aquifer system in Las Vegas comprises a compressible sequence of sedimentary deposits, and the ASR scheme was partly designed to try and arrest the ground subsidence (up to about 1.5 m locally) that resulted from major groundwater withdrawals during the twentieth century (US Geological Survey, 1999). A second example of an ASR Scheme is the North London Artificial Recharge scheme (NLARS) where surplus treated mains water (in large part derived from surface water) is injected into the Chalk aquifer, for subsequent abstraction during drought episodes (Boniface, 1959; Flavin and Hawnt, 1979; Connorton, 1988; O'Shea *et al.*, 1995; O'Shea and Sage 1999; Harris *et al.*, 2005). Similar schemes have subsequently been developed in South London (SLARS; Anderson *et al.*, 2005). There is evidence that undesirable chemical compounds, such as nitrate and disinfection by-products like trihalomethanes and haloacetic acid, are removed during storage in the aquifer and are found in significantly reduced concentrations when the water is reabstracted (Pyne, 2005).

In another variant, surplus fresh water can be injected into an aquifer containing poor quality, brackish or saline groundwater (e.g., in coastal areas). The injected water forms a 'bubble' of fresh water around the injection wells which, if properly managed, mixes minimally with surrounding saline water, and can be re-abstracted (Figure 4.19). Recovery efficiencies of over 80% can be achieved in well-managed schemes, and the recovery typically improves with use, as the utilized volume is 'flushed' of saline water (Pyne, 2005).

Heat energy can also be stored in this way. Warm water (carrying a load of waste heat from industrial processes, or summer cooling operations) can be injected and re-abstracted at times of heat demand (winter). Conversely cool water (chilled by snow melting, dry coolers, and so on) can be injected to the aquifer during winter, to be re-abstracted for space cooling or dehumidification operations in

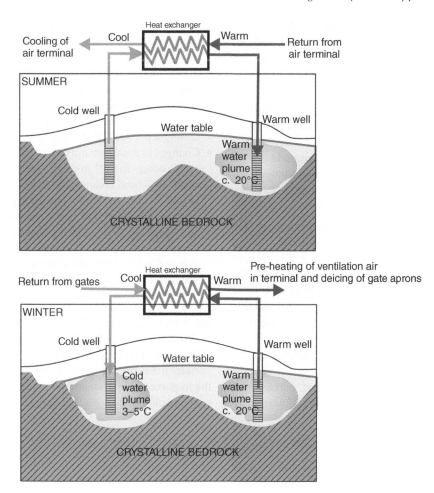

Figure 4.20 *Seasonally reversible aquifer thermal energy storage at Arlanda Airport, Stockholm - see text for explanation. Modified after Banks (2012a) and reproduced by permission of © John Wiley & Sons, Chichester, 2012*

summer. Such schemes are referred to as aquifer thermal energy storage (ATES) and are particularly attractive in climates with seasonal extremes. Examples of such schemes at Arlanda Airport, Sweden and Gardermoen Airport, Oslo, have already been mentioned in Section 4.9 (Andersson, 2009; Banks, 2012a; Figure 4.20).

Where multiple recharge wells are required, it has been found to be advisable to arrange them in a cluster, preferably in a concentric pattern. The central well(s) are recharged first and, as the fresh water 'bubble' expands around the outer ring(s) of boreholes, these are brought into service as recharge wells. This avoids pockets of 'external' (e.g., brackish/cool) water being trapped in the zones between the wells. During recovery, the outer wells are discharged first, followed by the central well(s).

Management challenges can be minimized by:

- Careful monitoring of flows and injection pressures. Methods for flow measurement can include ultrasonic, magnetic, venturi or impeller/turbine meters.
- Careful monitoring of quality and hydrochemistry of injected and abstracted water.

- Use of numerical groundwater models (which may need to account for mechanisms such as dispersion, molecular diffusion and density contrasts between fresh and saline, or between warm and cool water).

One should also be aware of common hydrochemical and clogging issues that can occur, especially when a 'bubble' of injected water comes into contact with either (i) 'native' groundwater of a very different quality, (ii) aquifer sediments or minerals that have previously been close to equilibrium with 'native' groundwater:

- Precipitation of dissolved ferrous iron in the native groundwater as ferric hydroxide, potentially resulting in aquifer clogging. This is seldom an areally extensive problem and normally occurs very close to the well (Pyne, 2005); it typically occurs where oxygen or other oxidizing species (including chlorine) are present in the injected water.
- Conversely, dissolution of iron minerals, such as ferric oxides or hydroxides or siderite (ferrous carbonate), may result in elevated ferrous iron concentrations in the recovered water. Mobilization

of iron will be controlled largely by redox (reducing) conditions and to some extent by pH (see also de Zwart, 2007).

- Similar issues with manganese. However, manganese redox reactions tend to be kinetically slower than those involving iron, and mobilization of manganese can be suppressed by maintaining a $pH > 8.5$ (Pyne, 2005).
- Changes in dominant ion chemistry (from Na^+ in saline waters to Ca^{++} in many 'fresh' waters) can cause structural changes, dispersion and coagulation of clay minerals, potentially leading to their mobilization and problems with turbidity and clogging. Pyne (2005) suggests that pretreatment of aquifer materials with solutions rich in Ca^{++} or Al^{+++} can to some extent prevent clay dispersion.

The mixing of different water chemistries and its effect on mineral solubility can be simulated using hydrochemical modelling programs such as PHREEQC (Parkhurst and Appelo, 1999), although such hydrochemical models do not take account of the important role of microbial processes in corrosion and clogging issues (Section 9.1).

5

Well and Borehole Construction

The aim of this chapter is to provide the engineer or hydrogeologist with an overview of the main methods of constructing wells and boreholes, so that he or she can prepare and supervise a drilling contract, and can also recognize equipment in depots or when brought onto the site. This chapter is not a guidance manual for drillers: for such guidance, the reader is referred to the excellent manual published by the Australian Drilling Industry Training Committee (2015).

The person letting a drilling contract will have to assess the suitability of various drilling organizations to undertake the work. Important points to consider in such an assessment are summarized in Box 5.1. The contractor's ability to prepare and implement a proper health and safety plan will be one key requirement (Appendix 3). The contractor should be required to keep a daily record of work on site, a 'driller's log' (Box 5.2). An accurate driller's log is essential for certification of the work and approval of payments. The daily driller's log also forms an important input into the final interpretative log and completion report for the borehole or well, normally prepared by the supervising hydrogeologist or engineer (Section 6.7 and Chapter 10). In many countries, there is a legal obligation on the driller or employer to provide details of the completed borehole or well to a central authority such as a geological survey or water resources ministry.

There are many different drilling techniques, but most of them are traditionally classified as either *percussion* or *rotary* techniques depending on the predominant drill action (although some methods involve a combination of the two actions). Rotary techniques can be further classified according to the method used to circulate the drilling fluid: *direct circulation* or *reverse circulation*. Top-hole and down-hole-hammer drilling are essentially a combination of direct circulation rotary drilling and percussion, sonic drilling is a percussive method involving high-frequency mechanical vibrations along with some rotary action, while auger drilling is a rotary method that does not use a circulating fluid. Jetting involves both a percussive and hydraulic drill action.

There are a number of considerations that will influence our choice of drilling method, including:

1. What is the purpose of the well – for exploration, monitoring or abstraction?
2. In what kind of geology is it to be constructed – unconsolidated, consolidated or crystalline?

Water Wells and Boreholes, Second Edition. Bruce Misstear, David Banks and Lewis Clark.
© 2017 John Wiley & Sons Ltd. Published 2017 by John Wiley & Sons Ltd.

Box 5.1   Points to consider in selecting a drilling contractor

The price quoted for the work is obviously an important factor in selecting a drilling organization, but it should not be the only factor. The following is a checklist of some of the other factors to consider when selecting the drilling contractor.

Drilling organization's experience and history

1. Has the company a local or national reputation for good work?
2. Can they provide details of similar drilling works that they have completed successfully, and on-time, recently?
3. Can they provide examples of the drilling records they prepare? Are these records comprehensive?
4. Do they belong to a professional organization of drilling companies or have some recognized quality assurance certification?
5. Are they on a list of approved contractors for an organization employing drilling contractors?
6. What routines do the company have for quality control of materials used in well construction?

7. What health and safety procedures have they in place for their drilling operations, and do they have a good health and safety record? Can they provide examples of health and safety plans?
8. Does the company carry appropriate public and professional liability insurances?

Visit to the drilling company's depot

1. Have they adequate equipment to do the work? (This includes not only a drill rig of the right capacity, but also the ancillary equipment for sampling, grouting, well development and test pumping).
2. Is the equipment in good repair and well-maintained?
3. Has the contractor adequate maintenance and back-up facilities in case of plant breakdown?
4. Is the depot close to the proposed drilling site so that back-up can be efficient?
5. Is the depot kept in a clean and workman-like state?

Box 5.2   Daily drilling records

The daily driller's record should contain the following technical information with respect to each borehole or well:

1. site location;
2. identification or reference number of borehole or well within the contract;
3. date;
4. driller's name;
5. method of boring and rig used;
6. drilling penetration rate and weight on bit;
7. depth of hole at beginning and end of day or work shift;
8. type, length and diameter of casing and screen;
9. length and diameter of open hole;

10. water strikes, and water level at beginning and end of day or work shift, with details of any fluctuations;
11. description of each lithology encountered;
12. depth below ground of each change in lithology;
13. sample depth, type and characteristics;
14. any other useful information, such as drill bit behaviour, drilling fluid characteristics (viscosity, density, electrical conductivity), loss of drilling fluid circulation, or borehole collapse problems.

The driller should also keep records of items such as fuel consumption, drilling cost and any health and safety issues, including site accidents.

3. What are the intended depth and diameter of the well?
4. How is the well to be lined?
5. What drilling equipment and expertise are available?
6. What technique(s) have been successful in this area before?
7. What topside constraints are there? Is access limited? Will there be concerns over noise, dust, vibration, and so on.

For a drilling method to be suitable, it must be able to fulfil certain basic requirements for the given geological conditions, including:

- it must permit the efficient breaking up and removal of the soil or rock from the borehole;
- the hole must remain stable and vertical to the full depth;
- the drilling fluid should not cause excessive damage to the aquifer formation;
- the method should not have an adverse impact on groundwater quality (and special care needs to be taken when selecting a drilling method for constructing boreholes for investigating certain groundwater contaminants such as trace organics);
- the hole must be large enough to permit the installation of the casing and screen, including sufficient annular clearance for the grout seal and artificial gravel pack (if a pack is required);
- the method should allow the well to be completed in the required timeframe and at reasonable cost.

Percussion drilling, sonic drilling, augering, jetting, drive sampling and manual drilling methods are used most often at shallow depths and in unconsolidated sediments, or relatively soft rocks. Rotary methods (especially direct circulation rotary) predominate in the construction of deep boreholes or wells, and top-hole or down-hole-hammer drilling is the method of choice in hard crystalline rocks. The main drilling techniques are described in Sections 5.1 to 5.7 below. Manual construction of wells is discussed separately in Section 5.8. The advantages and disadvantages of the different well construction methods are summarized in Table 5.1. The development of water wells is described in Section 5.9, while Section 5.10 deals with well-head completion works.

5.1 Percussion (cable-tool) drilling

Percussion drilling is a very old drilling method, having been used to construct wells in China as long ago as the first millennium B.C. The percussion method has been superseded to some extent by faster rotary techniques but it is still widely used for drilling small diameter geotechnical investigation boreholes and for constructing water wells in some countries. The main features of a percussion rig are shown in Figure 5.1. A string of heavy cutting tools is suspended on a cable which passes over a sheave (pulley) mounted on a mast, beneath a sheave on the free end of a spudding arm (which imparts the reciprocating motion to the tool string), over the sheave at the base of a spudding arm, and is then wound on a heavy-duty winch. This use of the steel cable gives the technique its alternative name of 'cable-tool drilling'. The cable is a non-preformed, left-hand lay, steel-wire rope. The left-hand lay of the cable tends to impart a slight rotation to the tool string and to tighten the right-hand threaded joints of the string. Power is normally supplied by a diesel engine. The whole rig can be mounted on the back of a truck or trailer, and is quite mobile.

The tool string generally includes the following units: rope socket or swivel, jars, sinker bar and drill bit (Figure 5.2). The drill bit is usually shaped like a very large heavy chisel (Figure 5.3). The rope socket has an internal mandrel which allows the tool string to swivel and so prevent over-twisting. During drilling, the cable rotates the tool string slightly, but when the string has rotated sufficiently for torque in the cable to have built up, the mandrel on the rope socket allows the tool string to swivel back. Jars are interlocked sliding bars, which allow a free stroke of about 150 mm. They are not used for drilling, merely for releasing tools by upward

Table 5.1 *Advantages and disadvantages of common drilling methods*[1]

Construction method	Typical drill diameter and depth ranges for water wells[1]	Advantages	Disadvantages
Percussion (cable-tool) drilling	Diameter: 100–750 mm Depth: >1000 m possible in consolidated formations; but mostly used nowadays for shallow wells	Low technology rigs, and therefore relatively cheap mobilization, operation and maintenance Suitable for a wide variety of lithologies Needs small work area Uses little water Good samples Water strikes easy to identify	Drilling depth is limited in unconsolidated formations because of need for temporary casing Slow method, especially in hard formations
Direct circulation rotary	Diameter: 100–750 mm Depth: > 1000 m possible	Very great drilling depths are possible Suitable for a wide variety of lithologies Fast drilling Does not need temporary casing	High-technology rig, so relatively expensive to mobilize and operate May need a large working area for rig and mud pits Can use a lot of water Filter-cake build-up can make development difficult Samples from conventional DC rotary can be poor (but good samples from coring) Loss of circulation of drilling fluid in fissures Sometimes difficult to identify aquifer
DTH, top-hammer and air-flush rotary	Diameter: 75–450 mm (100–150 mm most common) Depth: > 500 m possible with DTH in dry formations	Hammer methods are very fast in hard formations Do not need water No invasion of aquifer by drilling mud Water strikes easy to identify	Hammer methods are not normally suitable for soft, clayey formations Drilling depth below the water table is limited by hydraulic pressure Use of lubricating oil for the DTH hammer could be a problem in a contamination investigation Samples can be poor from standard hammer methods (but better samples can be obtained from reverse circulation, dual tube, DTH hammer)
Reverse circulation	Diameter: 150–800 mm Depth: up to 300 m	Rapid drilling in coarse unconsolidated aquifers at large diameters	May use large volumes of water

Method	Dimensions	Advantages	Limitations
Dual rotary	Diameter: 150–600 mm Depth: up to 400 m	Leaves no filter cake Quite good samples, especially with dual tube reverse rotary	Not suitable for hard formations (unless used with a dual tube system and DTH hammer) Large boulders can be a problem More expensive than conventional rotary rigs
Sonic drilling	Diameter: 75–300 mm Depth: up to 200 m	Good for drilling and casing through unconsolidated formations, before using conventional rotary or air-hammer for drilling in stable rock below Good samples Rapid method of sampling and drilling, especially in unconsolidated formations Good samples Little or no drilling fluid required Relatively quiet	Samples may be distorted by liquefaction and (in rock) affected by heat Borehole wall materials may be altered (densified) by liquefaction process
Auger drilling	Diameter: <50–>1000 mm, with the smaller sizes for solid stem varieties and largest for bucket augers Depth: <10–50 m	Inexpensive method for drilling exploration boreholes in unconsolidated aquifers Good samples using a hollow-stem auger and core barrel	Not suitable for very coarse formations or hard rock Drilling difficult below the water table
Jetting	Diameter: 50–100 mm, but can be up to 200 mm Depth: <30 m	Rapid, inexpensive method for constructing small-diameter wells in shallow unconsolidated aquifers	May use large volumes of water in permeable formations Poor samples
Direct push, drive sampling	Diameter: 50–100 mm Depth: <10 m	Inexpensive method for drilling exploration holes, collecting samples and installing observation wells in shallow unconsolidated formations	Not suitable for rock formations Restricted to shallow depths
Driving of well-points	Diameter: 40–100 mm Depth: <15 m	Rapid method of installing shallow wells in unconsolidated aquifers	Not suitable for rock formations
Manual construction	Diameter: > 1000 mm (hand-dug); <200 mm (hand-drilled) Depth: <30 m normally	Uses low technology, and therefore is cheap where labour is cheap	Hand-digging wells is slow Restricted to shallow depths

Note: [1] Drilling diameters and depths depend very much on the geology and drilling conditions, so the ranges given are indicative only

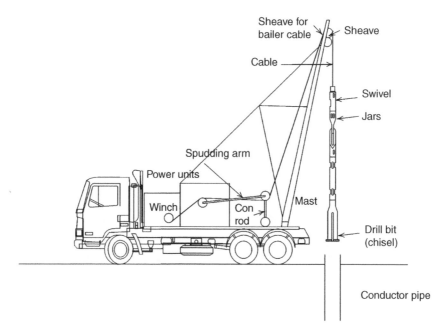

Figure 5.1   *Percussion (cable-tool) drilling rig for water wells*

jarring if the drill bit becomes trapped. The sinker bar, or drill stem, is a solid steel rod used to give extra weight above the bit, and to improve the verticality and straightness of the hole. The units of a tool string are connected by standard taper thread cable-tool joints.

The drilling of a borehole is started by installing a short, large diameter conductor or starter pipe into the ground, either by drilling or by digging a pit 1–2 m deep. The function of the conductor pipe is to prevent the surface material beneath the rig from collapsing into the hole and to guide the tool in the initial drilling. The tool string is assembled and lowered into the conductor pipe, and then drilling begins. The driller can vary the number of strokes per minute and the length of each stroke by adjusting the engine speed and the crank connection on the spudding arm. The tool string is held in such a position that the drill bit will strike the bottom of the hole sharply, and the cable is fed out at such a rate that this position is maintained.

The actual procedure of drilling will depend on the formation to be drilled. We will illustrate this by describing the procedures for hard-rock (crystalline and consolidated formations) and unstable formations.

5.1.1   Drilling in hard-rock formations

The tool string shown in Figure 5.2 is the standard string for drilling hard-rock formations such as granite (crystalline aquifer), limestone and indurated sandstone (consolidated aquifers). The drill bit will be a heavy solid-steel chisel (Figure 5.3). Drilling proceeds from the bottom of the conductor casing to below the base of the weathered zone, or to a pre-determined depth, and then a length of permanent casing is set and grouted into position. This casing length may or may not include a lower section of slotted casing to allow groundwater inflow from the highly fissured horizons often found near the top of limestone and chalk aquifers, or from the saprock found at the transition from regolith to unweathered bedrock of a crystalline rock aquifer in tropical climates (Section 3.1.2). In both cases, the formation opposite the slotted casing possesses significant transmissivity but may be unstable. It is important that the casing length is set vertically, so as to ensure that the remainder of the borehole is drilled vertically. The borehole is then continued open-hole to its full depth. As drilling proceeds the bit

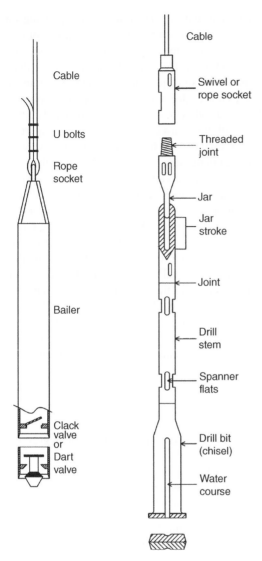

Figure 5.2 *Tool string for percussion drilling in hard rock*

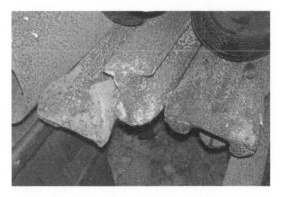

Figure 5.3 *A rack of rather worn chisel bits used for percussion drilling. Photo by Bruce Misstear*

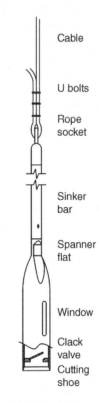

Figure 5.4 *Shell for drilling soft formations*

must be 'dressed' (reconditioned) to retain its cutting edge.

The action of the heavy chisel is to fracture and pound the rock into sand-sized fragments. The slight rotation of the bit imparted by the lay of the cable ensures that the hole is drilled with a circular section. The bottom of the hole becomes filled with rock debris which has to be removed periodically to allow drilling to proceed freely. In a dry borehole, a few litres of water are added to allow the debris to be mixed into a slurry and removed by a bailer (Figure 5.2). The slurry is removed from the hole and the operations repeated until the hole is clean; drilling then continues. Samples of slurry are taken from the bailer for description and analysis (Section 6.3).

The addition of water is not necessary for drilling or bailing once groundwater has been encountered. Below the water table the bailer contents can be analysed (after filtration) to give a first indication of the groundwater quality. The driller should note when water is struck, and measure the water level at the start and end of each shift. These data can give valuable information on the groundwater regime and the borehole productivity. A simple test involving repeated withdrawal of the bailer full of water can provide an initial estimate of the borehole yield.

5.1.2 Drilling in soft, unstable formations

The tool string used for drilling in soft, unstable formations is commonly simpler than for hard formations. The most common tool is a steel tube or shell with a cutting shoe on the bottom (Figure 5.4). The shell may be used alone or, if extra weight is needed, a sinker bar may be incorporated in the body of the shell. The top end of the shell is open and it is very similar in design and appearance to a normal bailer.

Different cutting shoes are used depending on whether the formation material is cohesive or not. With sand, the shoe may have a serrated or smooth edge to chop the formation, but a flap-valve will be set inside the shell, just inside the shoe, to retain the sand as it is drilled. Clay tends to be more cohesive than sand and, with stiff clay, a sharp-edged cutting shoe with thin chisel blades set across its aperture may be used (Figure 6.4). The shell may have windows cut in the side to help sample removal.

A major difference between unstable formations and hard rocks is that the former need support during drilling and this means that temporary casing must be used. The temporary casing has to be of sufficient diameter to allow the permanent casing and screen to pass inside it on completion of drilling, and to allow a gravel pack or grout seal to be installed in the annulus (Box 5.3). Drilling begins with a large-diameter shell, to drill a hole deep enough to set the first length of temporary casing. This casing will have a sharp-edged drive shoe screwed on the bottom, to help the driving of

---

Box 5.3   Grout seals

The purpose of grouting is to provide a low permeability seal in the annular space between the casing and the borehole wall. There are many situations where a grout seal is required, for example:

1. To cement the borehole conductor casing in place, to provide a stable platform for further drilling.
2. To cement the upper well casing in order to prevent the ingress of contaminated surface water or shallow groundwater into the well.
3. To seal off the annular space around the casing between different aquifers in a deep borehole, to prevent the flow of water between the aquifers and thus prevent cross-contamination.
4. To protect a well against ingress of groundwater from an aquifer where the water quality is poor.

5. To provide a seal on top of an artificial gravel pack.
6. To protect a casing against corrosion.
7. To seal an intermediate casing string prior to drilling into an artesian aquifer.

Grouting is also important when decommissioning or abandoning a well (Section 9.4).

Grout materials. Grouting is normally performed using cement, bentonite clay, or a combination of the two. Cement is used in situations where the strength of the seal is important, such as in setting a surface casing, or in installing a casing above an artesian zone. Portland cement is generally used, with specific gravities for the cement mix in the range 1.6 to 1.8 (Australian Drilling Industry Training Committee, 2015). Shrinkage of a cement-only grout can be a problem, however, and so a small amount of bentonite (2 to 6% usually) is sometimes added to reduce this shrinkage, and to increase the fluidity of the mix. Where loss of grout into coarse granular formations or fissured aquifers occurs, sand or other coarse materials can be added to help the grout form 'bridges' over these loss zones.

Bentonite is composed mainly of montmorillonite clay and, when hydrated, can expand to 10 to 15 times its dry volume. It can therefore provide a very effective low permeability annular seal, and is often used for grouting between screen sections in monitoring wells. Bentonite is available in powder, granular or pellet form. It is usually mixed to give a grout specific gravity of around 1.1 or 1.2. The density can be checked using a mud balance.

In pollution investigations it is important that the monitoring well construction materials do not affect the groundwater samples (Section 4.1.4). Ideally, therefore, a chemically inert material should be used for the grout seal. Neither cement nor bentonite is inert, however. Cement is highly alkaline, and may therefore affect the groundwater pH. Also, laboratory tests have suggested that glycols and some certain organic compounds can leach from the cements used as seals in monitoring wells, which could lead to false positives in groundwater samples (Smith *et al.*, 2014). On the other hand, bentonite has a high cation exchange capacity, which could potentially affect concentrations of trace metals, for example. The best way to reduce such problems is to try and prevent groundwater coming into contact with the seal materials in the intake section of the well. It is therefore necessary to place a fine sand layer above the gravel pack, so as to prevent the seal material from entering the pack.

Grouting methods. One of the prerequisites for successful grouting is to have a sufficient annular clearance around the casing. The ANSI/NGWA Water Well Construction Standard (ANSI/NGWA-01-14, 2014) specifies that the drilled diameter should be at least 76 mm (3 inches) larger than the casing OD when grouting with a tremie pipe, and at least 102 mm (4 inches) larger when using bentonite pellets as the grouting material.

The length of grout seal varies according to the particular application. For example, when installing a well in a crystalline aquifer, it is recommended here that the casing should extend at least 10 m below the surface into stable rock (Section 3.1.2) in order to prevent ingress of contaminants from the surface. The Institute of Geologists of Ireland (2007) provides a more conservative recommendation, whereby the grouted casing should extend at least 10 m into rock, or to 20 m below ground level, whichever is the greater. As noted in Section 3.1.2, a long section of grouted casing, whilst giving good protection against surface pollution, will reduce the potential yield of the well by blocking off the shallow water-bearing fractures. Thus the well designer is faced with a compromise choice between achieving greater well protection at the expense of a reduced well yield (Misstear, 2012).

Grouts can be emplaced in several ways, a few of which are described below.

1. The cement grout for a surface conductor casing is usually poured into the annular space from the ground surface. However, such a procedure is only suitable for shallow casings, where the quality of the seal can be easily established. For other applications, a 'bottom-up' grouting procedure, using a tremie pipe, is required.
2. The cement grout for a pump chamber casing, for example, can be emplaced through a tremie pipe. The tremie pipe is set initially near the bottom of the annulus and is then raised upwards as grouting proceeds. The

cement should be pumped (rather than let flow by gravity) into the tremie pipe, as this ensures that the grout can be delivered quickly, and reduces the risk of voids forming in the grout seal. Grouting continues until grout (of the correct density) appears at the surface.

3. A similar procedure to (2) above can be used for placing a bentonite grout seal. Again it is preferable to pump the bentonite mix into the tremie pipe, although bentonite grouts are often installed by placing bentonite pellets down the tremie pipe and then hydrating.

4. In deep wells, the intermediate casing can be installed with a drillable grout shoe and non-return valve at its base. The drill string with a suitable sub at its base is connected to this grout shoe and the grout is then pumped under pressure into the bottom of the borehole below the grout shoe, and pumping continues until such time as the grout appears at the surface. The drill string and sub are removed and, after the grout has set, the bottom plug is drilled out and drilling below this continues.

5. A displacement method can be used, in which the quantity of cement required for the seal is pumped inside the casing, and then water or mud is used to displace this cement out of the bottom of the casing and up the annulus to the surface. Two spacer plugs are generally used, one to separate the grout mix from the drilling fluid inside the casing, and the second to separate the grout from the displacement fluid above.

Further information on grouting methods is given in Sterrett (2007) and Australian Drilling Industry Training Committee (2015). Whatever approach is adopted, there are several important things to consider when planning grouting operations:

• The volume of grout should be estimated before the seal is installed; if, for example, the volume injected is less than the estimated volume, then this suggests either the presence of some voids in the grout or partial collapse of part of the borehole wall.
• Samples of grout should be taken to test its properties, including the setting time.
• The casing to be grouted must be located centrally in the borehole to ensure an even surround of grout – casing centralizers should be used for this purpose (Box 3.3).
• The casing must be strong enough to withstand the heat generated during the setting of a cement grout: this can be an issue with respect to grouting certain plastic casings, and it important to check the effect of this heat of hydration on the collapse strength of the material (Section 4.1).

Manufacturers' recommendations should be followed when preparing grouts. It is important that further drilling operations should not commence until the grout has set properly. Setting times for cement grouts vary, and can be more than 48 hours. Various additives are available to accelerate the curing process but note that calcium chloride, which is sometimes used as an accelerator, is corrosive to steel and so is not suitable where steel casings are used (Australian Drilling Industry Training Committee, 2015). Upon setting, the integrity of the grout seal should be tested. This can be done by applying an air pressure test to the grouted casing, or by using a sonic (cement-bond) geophysical log (National Ground Water Association, 1998).

the casing into the soft formation. The casing is lowered into the hole, and driven in firmly for a short distance with its verticality being checked carefully. While driving, the top of the casing is protected by a heavy steel ring – the drive head – screwed to the casing. The driving is usually achieved by blows of the drilling tool on the drive head.

Drilling progresses by removing the formation from inside and ahead of the casing for 1–2 m, and then driving the casing to the bottom of the hole. Extra casing lengths are added as the top of the previous casing is driven down to ground level. In very soft formations, driving may not be necessary as the casing will follow the shell under its own weight.

Unstable running sand can be a great problem with percussion drilling, because the reciprocating action of the shell inside the casing can build up a suction which pulls the sand up inside the casing. Remedies for running sand include building up a hydraulic head in the casing by filling it with water, adding drilling mud to stabilize the sand, driving casing past the unstable zone, or grouting off the running sand.

Sometimes casing is needed in formations that are hard but unstable – karstified limestones, for example. In this situation, there can be problems in ensuring that the casing is vertical: it can be driven offline by near vertical fractures. These fractures may also steer the drill bit off course.

Drilling continues until the total design depth is reached, until the casing cannot be driven further or until a grout seal needs to be installed (for example, before penetrating an artesian horizon). In the latter two cases, a second string of temporary casing has to be telescoped inside the first and drilling continues at a smaller diameter. The initial well design on which a drilling contract is based should allow for this contingency, because the permanent casing has to be able to pass inside the smallest temporary casing with an adequate annulus for the installation of any gravel pack or grout/bentonite seal (Box 5.3).

On reaching the total depth, we need to decide upon the final design of the permanent casing and screen string. The screen must be located accurately within the aquifer, and for this, the formation samples must be collected carefully and described properly (Section 6.3). The depth of samples can be measured fairly accurately with percussion drilling, but geophysical logging is worthwhile to obtain precise depths of aquifer horizons. The most useful logs will be the natural gamma ray log or radioactive logs (the temporary casing in the hole will invalidate the use of electric logs; see Section 6.4). After logging, the casing and screen are installed and the temporary casing is withdrawn. Removing the temporary casing is often quite difficult, and may require the use of jacks or a hydraulic vibrator.

In a borehole where an artificial gravel pack is required, we must analyse the formation samples in order to design the gravel pack and decide on the screen design (Sections 4.3 and 4.4). Grain-size determinations using sieves (Section 6.3.2) can be made on site or in a laboratory, the former being preferable to save time. The permanent casing and screen string is installed inside the temporary casing and the gravel pack poured into the annulus through a tremie pipe. The gravel pack may have to be flushed down the tremie pipe with water, and the casing should be vibrated to encourage settlement of the pack and avoid gravel bridging. Great care has to be taken to withdraw the temporary casing just ahead of the pack, otherwise a sand lock between the screen and temporary casing can occur, which binds the two together, and causes the screen to be removed along with the temporary casing. The depth to the top of the gravel must be checked frequently with a plumb-bob.

The greatest strain on a percussion rig is during the withdrawal of the temporary casing. The size and robustness of a rig and its fittings must be sufficient to cope with the pull it has to exert. This is particularly important in deep, large-diameter boreholes that are drilled in formations that may collapse around and grip the temporary casing.

5.1.3 Light-percussion drilling

A great many boreholes, particularly those drilled for site investigations or for shallow explorations in soft formations, are small structures. A light-percussion rig is used for this kind of borehole, with emphasis on mobility and lightness. The rig is based on a tripod instead of a single mast

Figure 5.5 *Light-percussion rig, northern Nigeria. Photo by Bruce Misstear*

(Figure 5.5) and can be collapsed to be towed behind a small truck or site vehicle. The drilling operations are identical to those with a traditional percussion rig but there is no spudding arm, and the reciprocating action of the tool string is achieved by direct operation of the cable winch. Drilling with a light-percussion rig is often referred to as the 'shell and auger' method, the term originating from a time when these rigs used to include an auger (Section 5.4).

5.2　Rotary drilling

5.2.1　Direct circulation rotary

Rotary drilling techniques were developed in the nineteenth and early twentieth centuries partly to overcome the problem with percussion drilling of having to employ temporary casing in unstable formations. Instead, rotary methods use the hydrostatic pressure of circulating drilling fluids to

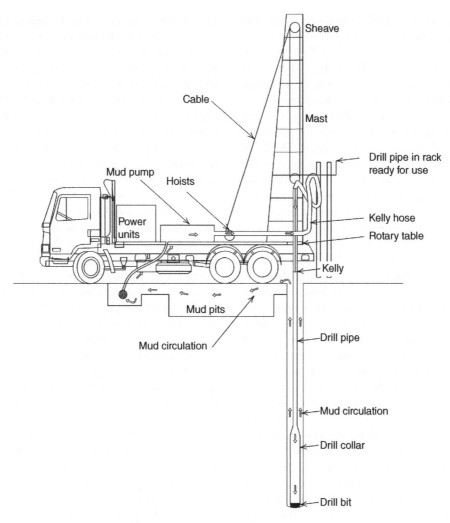

Figure 5.6 *Direct circulation rotary rig*

support the borehole walls. This use of drilling fluids enables the boreholes to be drilled to much greater depths than can be achieved by percussion rigs. In direct circulation, the drilling fluid is pumped down the drill pipe and up through the annulus around the drill string to the surface. The drilling fluid may be a bentonite mud, clean water, air, a foam-based fluid or a synthetic polymer (or some combination of these), depending on the properties required to stabilize the hole and bring the cuttings to the surface (Section 5.2.2).

Rig set-up and drilling procedure. Direct-circulation rotary drilling is carried out using one of two main designs of rotary drill rig: the *kelly-drive* rig and the *top-drive* rig. A typical set-up of a kelly-drive rig for drilling water wells, using mud as the circulating fluid, is shown on Figure 5.6. The body of the rig comprises a floor on which is mounted a diesel-engined power unit (some smaller rigs have a power transfer from the vehicle engine), a mud pump for circulating the fluid, a winch for raising or lowering the drill string and a mast from which the drill string is suspended. The drill string is made up of lengths of heavy-duty steel tubing or drill pipe, with the drill bit assembly attached to the bottom. All members of the drill string are connected by standard American Petroleum Industry (API) taper thread joints.

The drill pipe is normally of circular section, but the kelly (which is the top length) usually has a square or hexagonal external section. The kelly passes through a similar-shaped opening in the rotary table at the base of the drill mast. This enables rotary power drive to be transmitted from the main power unit, through the rotary table, to the kelly and so to the entire drill string.

The drilling mud is mixed in a mud pit or mud tank and is pumped by the mud-pump through the kelly hose to a water swivel at the top of the kelly. This swivel is the unit from which the entire drill string is suspended, and allows the mud to pass while the drill string rotates. The mud passes down the drill string to the bit which it leaves by ports in the bit faces, and then returns up the annulus between the borehole wall and drill pipe to the mud pits. A mud pit often has at least two chambers, the first and largest allows cuttings to settle from the mud, before it passes to a second chamber which acts as a sump for the mud-pump. The mud pit should normally have a volume of about three times the volume of the hole to be drilled.

The drill bit assembly commonly consists of the bit itself with, immediately above, a length of large-diameter, very heavy drill pipe called a drill collar. The collar is designed to give weight to the drill string, improve its stability and hence the hole verticality, as well as decreasing the annulus around the drill string, and so increasing the velocity of mud flow away from the bit. The weight added by the drill collar is very important, especially when the hole is shallow and there is not much weight in the drill rods.

Drilling begins by installing a length of conductor pipe to prevent erosion of the ground surface by the mud flow. The water swivel is then hoisted up the mast and the drill bit assembly screwed on to the kelly. The bit is lowered into the hole, the kelly clamped into the rotary table, and mud circulation

and rotation of the drill stem are begun. The kelly can pass vertically through the rotary table quite freely, and drilling progresses under the weight of the drill string. When the swivel reaches the rotary table, the drill string is held suspended on the hoist, and mud circulation is continued until all the cuttings are removed from the hole. Circulation and rotation are then stopped, the drill pipe suspended by friction slips in the rotary table and the kelly is unscrewed. A new length of drill pipe is fastened to the kelly and drill string, the slips removed, and circulation and rotation restored to start drilling again. The momentum of drilling needs to stop only when the final depth is reached or the drill bit needs replacing. The withdrawal of the drill string is a reversal of the drilling procedure.

The use of a rotary table to transmit the rotary drive to the drill string allows a very robust design, suitable for rigs varying from small truck-mounted rigs to large platform rigs similar to those used in the oil industry. An alternative design, which is now the main design used in rigs for water wells, is the top-drive rig (Figure 5.7). The principles of drilling are identical in the two designs, but in the top-drive there is no kelly, the drill pipe being attached directly to the rotary head. Rotary drive is transmitted to the head by a hydraulic motor mounted alongside the head. The top-drive unit is held to the rig mast by two slides, which allow the unit to move down the mast as drilling proceeds.

The size of the rig chosen, irrespective of design, increases with the depth and diameter of the borehole to be drilled. The hoist and mast must be strong enough to cope not only with the weight and vibration of the drill string, but also with the weight of the casing strings to be set when the hole is completed.

The height of the mast governs the length of drill pipe which can be attached in a single operation. Drill pipe is usually supplied in 3–6 m (10–20 ft) lengths for water well drilling, but may be in shorter lengths for smaller rigs and slim holes. The mast on the tallest rigs can cope with several lengths of pipe fastened together – so speeding up operations. The internal diameter of the drill pipe must be large enough to allow the mud to pass

Figure 5.7   *Air-foam drilling using a top-drive rotary rig, northern Oman. Photo by Bruce Misstear*

freely down the drill string, and the outside diameter should be large enough to ensure sufficient velocity in the mud passing up the annulus to carry cuttings from the bit to the surface.

Drill bits. The choice of drill bit depends on the formation to be drilled. In soft formations a simple drag bit equipped with hardened blades can be used. The drag bit action is one of scraping or planing (Figure 5.8). The commonest rotary bit is the tricone or rock-roller bit, which has three conical cutters which rotate on bearings (Figure 5.9). The drilling fluid passes through ports which are placed to clean and cool the teeth as well as carry away the cuttings. The teeth on the cutters vary in size, shape and number to suit the

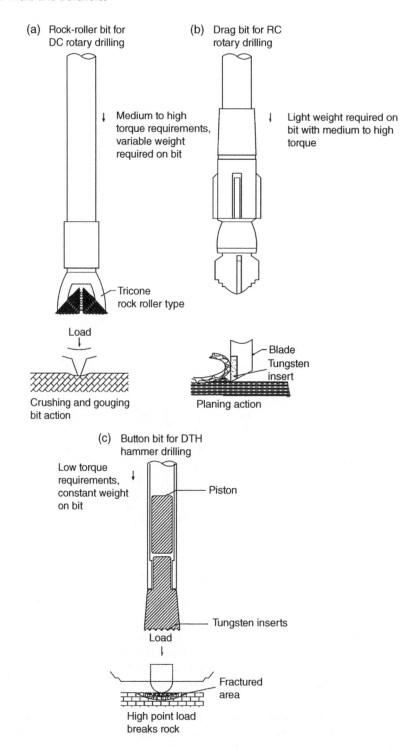

Figure 5.8   *Actions of drill bits. (a) rock-roller bit; (b) drag bit; (c) button bit*

Figure 5.9 *Rock-roller drill bit. Photo by Bruce Misstear*

Figure 5.10 *Retrieving core using the wireline system, western Ireland. Photo by Bruce Misstear*

formation being drilled – small, numerous teeth for hard formations and larger teeth for soft formations. The bit operates by crushing the rock: by overloading it at the points of the teeth and by tearing or gouging as the cones rotate.

Sometimes a pilot hole is drilled with a small-diameter bit and then, if water is successfully encountered, the hole is enlarged using a larger-diameter reaming bit. Or reaming may be necessary because the rig does not have sufficient power to drill the production hole at the full size in one go. If hole enlargement is required below an already installed casing, then an expanding under-reamer can be used. This has cutters on arms which are expanded using fluid circulation pressure.

Drag and tricone bits break the rock into fragments or cuttings, which are returned to the surface as 'disturbed formation samples'. Samples of undisturbed (or at least relatively undisturbed) formations can be obtained as cores of strata, by using special coring assemblies which comprise a tubular diamond or carbide-studded bit attached to a core barrel which, in turn, is attached to the drill string. The bit cuts a solid rod of formation which is captured in the core barrel (Section 6.2.2; Figure 6.10). When the barrel, which is often 1.5 or 3 m long, is full, the entire drill string is removed from the borehole to retrieve the core from the core barrel.

An alternative system, wire-line core drilling, avoids the need to remove the drill string after taking each core. Special wire-line drill pipe, with an internal diameter large enough for the core barrel to pass, is used for the drill string, and when the core barrel is filled, the barrel and core are retrieved by a special bayonet tool lowered on a cable down the inside of the drill string (Figure 5.10). The bayonet locks into a female head in the core barrel.

5.2.2 Fluids used in direct circulation rotary drilling

An important factor in direct circulation drilling is the choice of drilling fluid. The fluid can be air or clean water if the formations are hard and stable, as is the case in many crystalline and consolidated aquifers, or a mud or foam-based fluid where the borehole wall needs to be supported. Whatever drilling fluid is used, it is essential that its quality and condition be controlled in order to avoid introducing pollutants into the aquifer, and also to ensure that the formation damage (see Section 5.9.1) is minimized and hence that well development will be effective. For instance, water used for mixing fluids should not contain harmful bacteria, and therefore untreated surface water is not suitable.

Drilling muds. The most common general-purpose drilling fluid for unconsolidated aquifers has traditionally been a mud based on natural bentonite clay. The mud fulfils several purposes; it:

1. Keeps the hole clean by removing cuttings from the bit, carries them to the surface, and allows them to settle out in mud pits.
2. Cleans, cools and lubricates the drill bit and drill string.
3. Forms a supportive mud cake (filter cake) on the borehole wall.
4. Exerts a hydrostatic pressure through the filter cake to prevent caving of the formation.
5. Retains cuttings in suspension while the drilling stops to add extra lengths of drill pipe.
6. Supports the weight of the casing string in deep boreholes.

The properties of the mud which allow it to fulfil these functions are its velocity, viscosity, density and gel strength (thixotropy) (Box 5.4).

The mud is a suspension, partially colloidal, of clay in water. Under the hydrostatic pressure of the mud in the borehole, water is forced from the suspension into the adjacent formations. The water leaves the clay behind as a layer or cake of clay platelets attached to the borehole wall – a filter cake or mud cake. The filtration will be greatest adjacent to the more permeable formations (Figure 5.11). The hydraulic pressure exerted by the mud column depends on the mud density and, in severely caving formations, the mud density can be increased by the addition of heavy minerals such as barytes (barite, $BaSO_4$). A heavy drilling mud may be used to counteract strong pore pressures in artesian aquifers. Great care must be taken, however, to avoid excessive filter cake build up, because if it becomes too thick it can reduce the diameter of the borehole sufficiently to prevent withdrawal of the drill string.

The filtration properties of the mud, including its ability to build a filter cake, can be measured with a filter press. In this instrument, a sample of mud is forced under pressure through a piece of filter paper, enabling the thickness of filter cake left on the paper and the volume of filtrate passing through it to be measured. The American Water Works Association A100–06 standard specifies a maximum filter cake thickness of 2.38 mm with a maximum 20 cm^3 water loss in 30 min (AWWA, 2006), whilst the Australian Drilling Industry Training Committee (2015) notes that if the filter cake is thicker than 2 mm then this represents a poor quality filter cake.

The filter cake can be difficult to remove after drilling, especially when an artificial gravel pack has been installed in front of the filter cake (Section 5.9). Also, in the build-up of the filter cake, the filtrate of drilling fluid that invades the formation and the fine sediment carried by this filtrate can severely reduce the formation permeability (this is known as 'formation damage'). The removal of this filtrate and sediment by development is difficult and may not be successful. In order to avoid these problems, alternatives to bentonite are widely used, the most important being organic polymers, foam and air.

Box 5.4   Drilling mud properties

Up-hole velocity. Large cuttings tend to sink through the mud, so its upward velocity needs to be greater than the velocity at which the cuttings are sinking. The sinking velocity of a coarse sand fraction in water is about $0.15\,\text{m s}^{-1}$ but will be much less in viscous mud. The velocity of the mud moving up the borehole will depend on the speed and capacity of the mud-pump and the cross-sectional area of the annulus between the drill string and the borehole wall. Recommended up-hole velocities for muds lie in the range $0.3\,\text{m s}^{-1}$ (high viscosity muds) to $0.6\,\text{m s}^{-1}$ (low viscosity muds) (Australian Drilling Industry Training Committee, 2015).

Viscosity. The viscosity of the mud controls the rate at which the cuttings sink; if it is too high then the cuttings may not settle in the mud pit. This leads to recirculation of the cuttings, which causes excessive wear on the mud-pump and drill bit, and mixing of the formation samples. The viscosity of the mud may increase naturally through the addition of formation clay, or decrease due to the influx of water. The viscosity can be increased by the addition of lime or other flocculating agent, and decreased by the addition of dispersants. The mud viscosity can be monitored regularly by means of a Marsh funnel. The funnel is filled with over one litre of mud, and then allowed to empty into a measuring container. The time is measured in seconds for the container to fill to the 0.946 l mark; the viscosity is then expressed in seconds. The Marsh viscosity required for a drilling mud will increase with the size of cuttings to be lifted,

ranging from 35 to 45 seconds for fine sand to 75 to 85 seconds for coarse gravel (Sterrett, 2007). The AWWA A100–06 standard for water wells recommends that, during 'normal drilling operations', the Marsh viscosity of the drilling fluid should be kept between 32 and 40 seconds (AWWA, 2006).

Density. The density of a mud influences:

- the hydrostatic pressure the mud exerts on the aquifer;
- the thickness and quality of the filter cake formed on the borehole wall;
- the ability of the mud to carry the cuttings.

The required drilling fluid density will depend on the hydrostatic pressure (the product of fluid density and height of fluid column) needed to overcome the pore pressures in the aquifer. Mud densities of between 1.0 and $1.2\,\text{kg l}^{-1}$, equivalent to specific gravities of 1.0 to 1.2, are fairly typical, but heavier muds may be needed in strongly-confined aquifers.

Gel strength (thixotropy). Bentonite clay is thixotropic and this ability to gel when not disturbed is important in holding cuttings in suspension when mud circulation is interrupted for the addition of extra drill pipes. In a thin mud with low gel strength, there is a danger that cuttings will settle behind the bit assembly and trap it down the borehole. Thixotropy also influences the stabilization of the borehole, since the mud filtrate forms a stabilizing gel in the formation behind the wall cake.

Organic polymers. Polymer-based drilling fluids can be made from natural or synthetic polymers. The polymers can also be added to bentonite-based muds to reduce the amount of clay required. When mixed with water, organic polymers form a viscous fluid with many of the characteristics of bentonite-

based mud. They too lead to the formation of a filter cake, but in a different manner to bentonite; the long-chain polymers invade the pores of the formation at the borehole wall where they form a mesh, which traps the mud solids and builds up the filter cake. After a certain time, the polymer mud

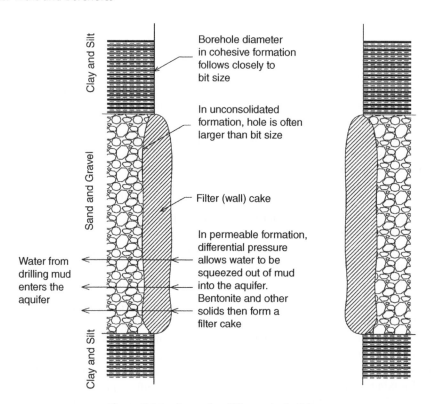

Figure 5.11   *Example of filter cake build-up*

breaks down to a low-viscosity fluid which can be removed far more easily by well development. The natural life of the polymer mud varies with the type of polymer and local conditions, but is often a few days. Breakdown can occur more rapidly if the drilling water supply is bacteriologically contaminated, or if the pH is low. The life of the polymer mud can be extended by various proprietary additives based on food-grade inhibitors, and by maintaining a high pH. The density of the polymer can be increased through the addition of salt to give a low-solids weighted drilling fluid. On completion of a borehole the breakdown of the mud can be accelerated to within an hour using an additive such as chlorine.

Organic polymers have obvious advantages over bentonite as a mud base:

1. The controlled breakdown of the drilling mud to a water-like liquid on completion of drilling assists greatly in the removal of the mud cake and in the well development.
2. Less polymer than bentonite is needed to make an equivalent mud.
3. The drill cuttings settle out more quickly in the mud pit.
4. The formation samples recovered are much cleaner. There is no bentonite to remove or distinguish from formation clays, so that lithological logs and grain-size distribution analyses tend to be more accurate.

There are also some drawbacks to organic polymer muds:

1. The mud condition has to be monitored very closely to avoid unexpected breakdown. The manufacturer's guidelines for use have to be followed closely to ensure optimum performance.

2. A biodegradable organic polymer will act as a food source for bacteria in a well, and has to be removed completely to avoid subsequent bacterial contamination of the well and increased susceptibility to biofilm formation. Again, the manufacturer's guidelines must be followed closely.
3. Although the polymers solve the problem of filter-cake removal to a large extent, they do not totally eliminate the problem of formation invasion by natural fine particles from the mud.

A major problem with drilling with either bentonite or polymer-based muds in extremely porous formations or in fissured (especially karstic) limestones, is loss of drilling fluid circulation to the aquifer. In karstic limestone, the entire mud circulation can disappear into a major fissure. This results in a loss of cutting samples but, more importantly, a loss of lubrication between the drill string and the borehole wall, loss of support for the borehole wall and a need to constantly replenish the mud supply (which may not be practicable in an area short of water). The problem can sometimes be overcome by plugging the zone of lost circulation with a commercial bridging agent or locally available bulky medium (such as wood chips, bran or straw), or emplacing grout to seal off the loss zone. Unfortunately, grouting may also grout-off productive aquifers and, in the case of large fissures, may merely result in the loss of large volumes of grout. Similarly, casing off the zone of lost circulation with permanent casing can block off a potential aquifer from the final production well. In such karstic environments, the dual rotary method – in which temporary casing fitted with a cutting shoe is advanced at the same time as the drill bit – may be a suitable option for maintaining fluid circulation and hole stability (Section 5.2.5).

Foam-based drilling fluids. As an alternative to dense drilling fluids such as bentonite and polymer-based muds, a light foam-based fluid can be used where circulation loss is excessive, or where the borehole walls do not require support during drilling. Direct circulation rotary drilling with a foam-based fluid is a common method for constructing water wells in limestone aquifers.

Proprietary chemicals, akin to domestic detergents, are added to the drilling water, together with compressed air, to give a very thixotropic foam (Figure 5.7). The mix of foaming agent, air and water used will depend on the agent and the site conditions. The mixture is passed down the drill stem under pressure but, on release of pressure in any cavities, a stiff foam will form (of the consistency of shaving foam), block the cavities and may allow circulation to be restored. The foam has sufficient viscosity and gel strength to carry cuttings to the surface, where the foam should break down. If the foam is too stiff, or the circulation rate too high, then the foam can build up to be troublesome, especially on windy days. The area of settlement pits should be large enough to cope with the foam produced.

Compressed air. Compressed air as the circulation medium avoids the use of liquids altogether (especially useful in arid areas where water is in short supply) and can be very effective for small diameter boreholes in stable rock. When drilling observation boreholes for groundwater pollution studies, where the use of drilling fluids may be prohibited because of impacts on their groundwater quality, then air drilling may be the best technique, especially where the ground is too hard for percussion drilling. The technique, however, does present several challenges:

1. The low density of the air means that the return velocities must be high to carry cuttings to the surface – recommended up-hole velocities for air drilling are around $25\,\mathrm{m\,s^{-1}}$, compared with less than 0.3 to $0.6\,\mathrm{m\,s^{-1}}$ for bentonite-based muds and $0.2\,\mathrm{m\,s^{-1}}$ for foam-based fluids (Australian Drilling Industry Training Committee, 2015). The need for a high up-hole velocity means that for large-diameter boreholes, very large capacity compressors may be needed to supply the necessary volume of air.
2. Air drilling presents few problems in dry holes, but below the water table the air pressure must

overcome the hydrostatic pressure of the water column as well as be sufficient to air-lift the water and cuttings to the surface.

3. In pollution studies, particularly studies of organic pollution, small quantities of lubricating oil from the compressor can be present in drilling air and can contaminate the formation. An oil trap on the air line is essential for drilling in these conditions. Alternatively, some form of vegetable oil can be used for lubrication, one that can be distinguished from mineral oils during analysis.

4. Air drilling may also lead to a stripping off of volatile organic compounds in the groundwater, thus potentially affecting the sampling results for these compounds in pollution studies.

5.2.3   Reverse circulation

The reverse circulation system was developed in the early twentieth century primarily for drilling large-diameter boreholes in loose formations, although the method is sometimes also used nowadays in consolidated aquifers. As the name suggests, the drilling fluid is circulated in the opposite direction to direct circulation, that is, *down* the borehole annulus and *up* the drill string. The general layout of a rig is shown in Figure 5.12. The borehole walls are supported by the hydrostatic pressure exerted by the water column in the borehole annulus and the positive flow of water from the borehole to the formation, rather than by a dense mud and filter cake (Figure 5.13).

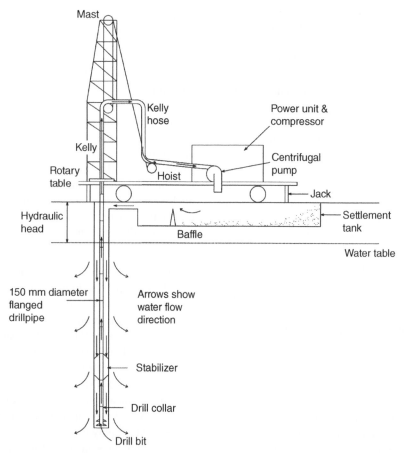

Figure 5.12   *Reverse circulation rotary rig*

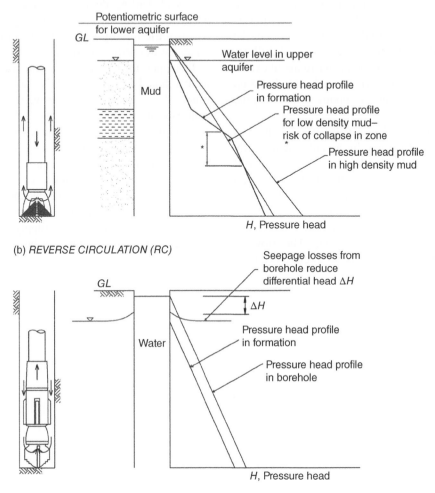

(a) *DIRECT CIRCULATION (DC)*

Potentiometric surface
for lower aquifer

GL

Mud

Water level in upper
aquifer

Pressure head profile
in formation

Pressure head profile
for low density mud–
risk of collapse in zone
*

Pressure head profile
in high density mud

H, Pressure head

(b) *REVERSE CIRCULATION (RC)*

Seepage losses from
borehole reduce
differential head Δ*H*

GL

ΔH

Water

Pressure head profile
in formation

Pressure head profile
in borehole

H, Pressure head

Figure 5.13   *Formation support requirements in rotary drilling. (a) direct circulation; (b) reverse circulation*

The reverse circulation rig is similar to a direct circulation rig, except that many of the rig components are larger. The drill pipe has a much larger internal diameter, usually 150 mm minimum, to provide a large waterway, and tends to be in short (about 3 m) lengths of heavy-duty steel tube with flanged connections. The drill bits used at large diameters are commonly composite bits of variable design, but all have an open end to allow cuttings to enter. One design has rings of conical cutters set on the side of the bit body, with the diameter of the rings increasing away from the tip to ream out the hole progressively to its full diameter. At large diameters, extra weight is needed in the drill string above that of the drill pipes, to effect penetration and maintain stability. For this reason a heavy drill collar, together with large-diameter stabilizers, is commonly fitted above the drill bit. The stabilizer, above the collar, has an ID equal to the central waterway, but an OD close to that of the borehole. The stabilizer is designed with a large diameter to allow free passage of the drilling fluid.

Rotary drive can be transmitted to the drill stem via a rotary table (as in Figure 5.12), or by a top-drive mechanism. The rate of rotation is much slower than with direct circulation, but the weight of the drill string leads to rapid drilling. Care must be taken by the driller to avoid excessive weight on the bit, which can lead to build-up of torque in the string and rupture by twisting the drill pipe.

The drilling fluid is usually water, which is pulled up the drill pipe by a centrifugal pump, commonly aided by airlift. The kelly and drill pipe lengths are short to avoid the need for high suction heads in the system. The water passes through the swivel, is discharged into a large settling pit, and then flows by gravity down the annulus around the drill string. The flow of water up the drill string is at a relatively high velocity ($3–4\,\mathrm{m\,s^{-1}}$), and is capable of lifting pieces of drilling debris of a size close to the waterway diameter. These cuttings, when in water, settle rapidly in the settlement pit before 'clean' water returns to the borehole.

To support the borehole walls, the water level in the borehole has to be kept at the surface or, if the water table is shallow, kept above the surface to maintain sufficient hydrostatic head (Figure 5.14). The minimum excess head required is about 3 m. The positive head means that there is a constant heavy loss of water to the formation and this has to be replenished, usually by tanker. The top-up water required will increase with the thickness and permeability of the formation drilled, but commonly exceeds $50\,\mathrm{m^3\,h^{-1}}$. The water supply needed for reverse circulation drilling may be a problem in arid areas.

Figure 5.14 *Tractor-mounted rotary rig raised above the ground surface in order to provide the necessary hydrostatic pressure for reverse circulation drilling. The target aquifer in this area of central Myanmar had a potentiometric surface only about 1–2 m below ground level. Photo by Bruce Misstear*

The reverse circulation system provides fairly good formation samples with little time delay between their cutting and their arrival at the surface, although sample collection from the high velocity return flow can be difficult (Section 6.2.1). The system is very suitable for drilling coarse sediments such as sands or gravels, because the gravel pebbles can be removed without any grinding. Drilling in such situations is faster than with direct circulation, averaging around $10\,m\,h^{-1}$. A further, major, advantage of reverse circulation is the use of water as the drilling fluid, which means there is no filter cake to remove. The lack of filter cake, however, means that there can be deep penetration of formations by fine material in the water, an action positively encouraged by the hydrostatic head and water loss.

A modification of this basic system, known as the reverse circulation dual tube method, uses drill pipes with an inner and outer tube. Air (or other fluid) is pumped down the outer tube to the drill bit, and the cuttings, air and water are returned to the surface up through the inner tube. The drill bit in this case is similar to a coring bit, and the method can be used to drill relatively small-diameter investigation boreholes for obtaining good formation samples. The dual tube system can also be employed for reverse circulation drilling with a down-hole-hammer.

5.2.4 Top-hole and down-the-hole hammer drilling

In the hammer methods, a rapid percussive action is transferred to a highly resistant bit at a typical rate of 1000–2000 strokes per minute (Rosén *et al.*, 2001). This is combined with a relatively slow rotation (12–40 rpm) to ensure an even drilling face. Drilling is usually carried out using compressed air as a fluid, directed down the drill stem and emerging from apertures on the bit to carry the cuttings to the surface. These cuttings are fine grained and 'dusty' and can be difficult to interpret, but some guidelines are included in Chapter 6. Water (or foam) may also be added to the air flow to cool the bit and encourage some degree of clumping of the

cuttings. For slim investigation holes, sampling can be improved through the use of a reverse circulation hammer with a dual drill pipe system.

The power applied to the rock is not directly related to the size of the rig or the rotation, but rather related to the power of the hammer. As this is typically driven by compressed air, the power of the compressor is often the decisive factor for drilling rate. The compressor may be mounted on the rig or, more usually, on a separate trailer. Where they are separate, the rig itself need only be large and tall enough to handle the desired length and weight of drill string and casing. In fact, impressively small trailer-mounted DTH rigs are available which, provided a suitable compressor is available, can be airlifted onto site to drill 100–120 mm diameter wells in crystalline rock to 80 m or so.

The hammer bits are specially designed to withstand the shocks involved in the technique. Sharp teeth would be broken, so the most common design for crystalline rocks is a button bit in which the teeth are roughly hemispherical tungsten carbide buttons set in the steel head (Figure 5.15). These teeth produce high point-loading on the rock, which induces fracturing (Figure 5.8).

Figure 5.15 *Button bit used for down-hole-hammer drilling. Photo by Bruce Misstear*

Top-hammer drilling. In top-hammer drilling, the percussive action is applied to the top of drill string. It can only be used to shallow depths of some 70–80 m, due to energy losses in transferring energy along the flexible drill string to the bit (Gehlin, 2002). Verticality and alignment are also more difficult to ensure with the top-hammer method.

Small caterpillar-mounted rigs are available, however, where eccentric bits and a top hammer mechanism are able to provide (a) shallow, small-diameter hammer drilling in hard rocks, (b) rotation jetting in unconsolidated sediments (eccentric bit) and (c) driving of well-points in unconsolidated sediments. A combination of air and water (and if necessary, foaming agents) can be utilized as fluids.

Down-the-hole hammer (DTH) drilling. The down-the-hole (DTH) hammer technique was developed in the quarry and mining industries. The rig and drill string assembly are similar to that used for direct circulation rotary drilling (Figure 5.16). The DTH drilling bit assembly is, however, a pneumatic hammer, similar in action to a common road drill, in which the compressed air supply operates a slide action to give rapid percussive blows to the bit face as it is rotated by the drill string (Figure 5.8). Small quantities of lubrication oil may be added to the air supply to lubricate the down-hole hammer (which should be borne in mind in contamination investigations).

The DTH hammer was originally designed for drilling slim holes. Nowadays, bits of 100 to 150 mm are commonly used for exploration boreholes and small diameter water wells, with bits of 200 mm or more also being available for larger diameter wells. In hard crystalline or consolidated rocks – granites, basalts and limestones – the DTH

Figure 5.16 *An inclined borehole being drilled by the down-hole-hammer method, Norway. Photo by David Banks*

hammer is in its element because it can drill several times as fast as conventional rotary drills with tricone bits in such rocks. In unconsolidated sediments, the bit's rotation and the jetting action of the high pressure fluid are probably as important as the hammer motion itself. Special systems are available where the hammer bit fits inside, and is connected to, a casing advancer, which itself is connected to a drive shoe. Casing is thus advanced as drilling progresses through unconsolidated deposits.

The main restrictions on the use of hammer rigs are soft, plastic clay-rich rocks and sediments, as the clay can clog the bit, jam the slide action and can absorb the percussive blows. Also, below water, the air has to overcome the hydrostatic pressure before it can operate the hammer, so that the depth of penetration below the water table is limited. Nevertheless, in hard crystalline rocks, water inflow rates below the water table may be so low that the well has little chance to fill with water once drilling has started. According to Gehlin (2002), commonly applied air pressures are in the range 2–2.4 MPa, corresponding to a depth of 200–240 m water (although the practical limit is significantly less than this, around 150 m).

In Scandinavia, when drilling by DTH in hard-rock terrain, drilling often progresses at 200 mm diameter through any superficial sediments and at least 2 m into unweathered bedrock (often using eccentric drilling techniques – see below). The casing is grouted in place and drilling typically continues at a diameter of 130–150 mm to the full depth (often 70–80 m for water wells). Swedish regulations state that at least 6 m of casing must be grouted into place below the surface. If loose superficial sediments occur, the casing must penetrate at least 2 m into solid bedrock (Rosén *et al.*, 2001).

Innovative variants of DTH drilling are available. Hydraulically powered DTH bits are now available, driven by water under pressure, rather than air (Rosén *et al.*, 2001). These can achieve much deeper drilling depth, penetrate at least twice as fast and use less energy than air-driven DTH bits. Hydraulic drilling requires a large supply of particle-free, low salinity water (200–300 litres per minute), however, and the cost in bit abrasion is much higher (Rosén *et al.*, 2001).

Eccentric drilling bits. Eccentric drilling techniques are usually marketed under names such as 'Odex' or 'Tubex'. These utilize button bits, combined with DTH drilling or top-hammer/rotation jetting drilling. Above the main cutting face of the bit ('pilot bit'), is a second 'eccentric' drilling element (the 'reamer'), also set with tungsten carbide buttons, which can swing out on an axis that is offset from the drilling string's main central axis. This reams the borehole to a larger diameter. The casing sits on top of the reamer and sinks as the drilling bit progresses. When the casing has achieved the desired depth, the drill string is counter-rotated and the reamer is swung back into the main drilling shoe, allowing the drill string to be withdrawn from inside the casing. Drilling can then progress (once the casing is grouted in) at a narrower diameter through the casing using a conventional bit.

5.2.5 Dual rotary

A dual rotary drill rig carries two independent rotary drives. A conventional top drive operates the main drill string, whilst a lower drive rotates an outer casing (Figure 5.17). This lower drive uses power-operated jaws to hold and advance the casing, the bottom section of which is fitted with a cutting shoe comprising carbide studs. The inner drill string and the outer casing can be rotated in opposite directions. The drill string may be fitted with a rock roller bit, drag bit or air hammer. The drilling fluid is normally air or water.

The main advantage of dual rotary is that collapsing formations can be stabilized by the use of the outer casing, avoiding the need for a drilling mud as with conventional direct circulation rotary methods. The dual rotary method is particularly suited to drilling through unstable overburden above bedrock. The outer casing is drilled until stable rock is reached, and then drilling continues through the rock using a rock roller or air hammer

Figure 5.17 *Dual rotary rig, Ireland. A drill rod is being added to the top drive whilst jaws from the lower drive clasp the outer casing. Photo by Bruce Misstear*

bit. An advantage compared to eccentric, under-reaming drilling bits is that the drill bit in DR can be raised or lowered in relation to the outer drill casing; thus, if heaving sands are encountered, the bit can be withdrawn inside the casing to provide hole stability. Advancing the casing slightly ahead of the drill bit also helps to prevent cross-contamination of drill samples from overlying formations (see Section 6.2.1).

5.2.6   Borehole testing during drilling

During rotary drilling, there are several test methods that can be used to give an initial indication of water quality and/or well yield. With direct circulation rotary mudflush, two commonly used methods are drive pointing and 'ratholing'. Drive pointing can be carried out when drilling soft formations [Figure 5.18(a)]:

1. The borehole is drilled to the top of the potential aquifer horizon to be tested. The drill tools are withdrawn and a drive point with a screen section is fitted to the base of the drill string.
2. The drill string is then lowered to the bottom of the borehole, and the drive point with the screen is pushed into the formation below the base of the hole.
3. An air eductor pipe is installed inside the drill string and airlift pumping is carried out until all the drilling mud has been removed from inside the drill string.
4. Airlifting is continued until the water being pumped at the surface is clear, sand-free, and has a stable water quality (as shown, for example, by constant temperature, pH and electrical conductivity). Water samples can then be collected for further field and laboratory analysis. These will give an idea of the water quality to be expected from this aquifer horizon (at least initially). The ease or difficulty in achieving a flow of water by airlifting will also give an impression of aquifer productivity, although this will be very approximate.
5. On completion of airlifting, the air pipe is removed and the water level is measured inside the drill string after recovery. The water level should be checked against the mud depth and density outside the drill string; additional mud can be added to counter any artesian pressures prior to further drilling.

If the formation is too consolidated for drive pointing, then the potential aquifer horizon at the base of the borehole can be investigated by ratholing. With this method, the bottom section of the borehole is drilled at a smaller diameter, and is then sampled with a screen and packer assembly fitted to the drill string, as shown in Figure 5.18(b). Water samples are recovered by airlifting, as for drive pointing described above.

Indicative yield of wells in hard rock aquifers during drilling. When drilling in hard rock aquifers (fractured/fissured limestones, metamorphic and igneous rocks) with down-the-hole-hammer

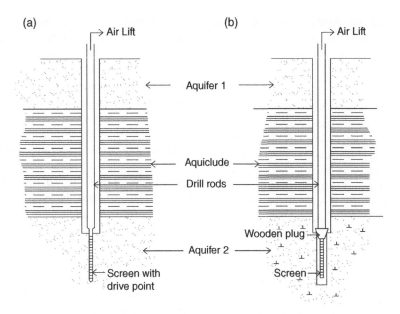

Figure 5.18 *(a) drive pointing and (b) ratholing*

(DTH) techniques, it will often be possible to gain some impression of water yield during the drilling process, as any groundwater encountered will be blown out of the hole with the drilling cuttings in a stream of compressed air (see Section 6.1.1). Some water-bearing fractures may contain a volume of groundwater but not be connected with a transmissive fracture network. In such cases, water or damp cuttings will initially be encountered on striking the fracture, but the wetness or quantity of water will decline with further drilling.

High-yielding, transmissive fracture networks will continue to yield a stream of water up-hole as drilling progresses. An experienced driller will be able to estimate this yield, and in some cases it can be confirmed by measurement with a bucket and stopwatch. Furthermore, when drilling pauses (e.g., during a lunch break or overnight break of duration Δt) the driller should measure the water level (h_1) at the end of drilling and the water level (h_2) at the end of the break. The approximate water yield (Q) can be estimated by:

$$Q = \frac{\pi\left(h_2 - h_1\right)r_w^2}{\Delta t} \tag{5.1}$$

where r_w is the well radius. This method works best for relatively short intervals (Δt), as the rate of inflow decreases as the water level approaches its static level. Also, we need to be aware of the fact that water draining down the walls of the borehole following cessation of drilling may lead to overestimates of Q. Finally, any value derived by such observations will represent a short term yield (governed largely by the transmissivity of fractures local to the borehole) and may not represent a long-term sustainable yield (which will be governed by the transmissivity, fracture connectivity, storage and recharge properties of the fracture network in the wider aquifer).

It is always important to appreciate the limitations of samples of water taken for analysis during and shortly after the drilling of boreholes in hard rock aquifers. Experience clearly suggests that the presence of freshly exposed drilling cuttings and rock surfaces, with high surface areas, can lead to overestimation of a wide range of chemical parameters. Studies of boreholes on Hvaler, Southern Norway (Banks *et al.*, 1992b; 1993b) suggest that this effect may persist for at least several weeks or months, unless vigorous clearance pumping is carried out. Analyses of water from freshly drilled boreholes in hard rock lithologies should only be regarded as indicative of long-term water quality.

Lugeon testing. A specialized form of testing, used in engineering geology and carried out during drilling, is known as Lugeon testing (Moye, 1967). Here, during drilling, a packer is placed around 20 ft (6 m) from the base of the hole, or two packers are placed at the top and bottom of the interval to be investigated, and water is injected until the flow has stabilized after a short time. Typically three different pressures are used for each section, in the range 10–75 lbs per square inch (psi). A hydraulic conductivity is calculated in Lugeon units (gallons/minute of flow per foot of test section per psi of pressure). Drilling then continues until the next 20 ft section can be tested. The conversion factor to metric units of hydraulic conductivity can be given as: 1 Lugeon $\approx 1.1 \times 10^{-7}$ m s^{-1} (Banks *et al.*, 1992a). When testing aquifer horizons, leakage around the packer can be a problem. Further information on borehole packer tests is included in Section 7.8.2.

5.2.7 Methods of casing and screen installation

The methods of casing and screen installation in rotary-drilled boreholes vary with the depth of the borehole and the ability to install the casing and screen in one operation.

A borehole to be completed with a single string of casing and screen is drilled to the design depth and then, ideally, is geophysically logged. The aquifer boundaries, upper and lower, are identified and the casing and screen string is assembled on the surface into lengths suitable for handling by the available rig. As the string is lowered down the hole, each length is joined to the top of the previous one, the joining being achieved by solvent welding, electric welding or screwed (or socket) joints, depending on the material being used (Section 4.1.2). The casing and screen string is

lowered until the screen is opposite the target aquifer. The annulus is then filled either with an artificial gravel pack or formation stabilizer, or the formation is allowed to collapse against the screen during development. This method of installation can be used for single or multiple aquifers; in both cases, close control of the depth(s) of the screen(s) must be kept to avoid misplacing them.

The casing and screen string is held in the centre of the borehole by means of centralizers fixed around the string at 10 to 20 m intervals. Centralizers are barrel-shaped cages of spring steel ribs (Figure 5.19) and, apart from keeping the string central in the hole, they serve to guide it down the borehole. Other useful attachments to the string, in a borehole drilled in a stable formation, are wall-scratchers which are strapped to the casing at intervals. They are collars of radiating spring-steel wire spokes which help to scrape off the mud filter cake to help in development, and to improve the keying of grout seals (Section 5.9.4).

A casing and screen string where the screen is smaller diameter than the pump chamber (Figure 3.6) requires special care in assembly. The two lengths of different diameter are often connected to a conical reducer and it is essential, particularly if this is made on site, to ensure that the upper and lower sections are coaxial.

The screen in a 'telescopic' multiple string completion is part of the last string to be installed. The borehole is drilled to the required depth and at the

Figure 5.19 *Casing centralizer (lying on top of two lengths of wirewound screen). Photo by Bruce Misstear*

appropriate diameter for the pump chamber, the pump chamber is set and then grouted. Drilling then proceeds through the grout seal at a reduced diameter, and any intermediate casing is set and grouted into position. Drilling then continues through the aquifer(s) at the final diameter to the full depth. The aquifer section is geophysically logged, and the position and length of the final casing and screen are confirmed. The string is assembled on the surface and lowered into the well, attached to a special sub on the drill stem of the drilling rig. When it is adjacent to the aquifer(s) the screen can be set in position or attached to the intermediate casing in several ways. Whatever method is used, it is important to keep the screen under tension to ensure that it is set vertically in the well. The simplest method is to suspend the screen so that the tail plug is just above the bottom of the borehole and the top of the screen string extends a few metres up inside the casing; the annulus between the casing and screen is left open. With this kind of completion the gravel pack or formation stabilizer can be placed down the annulus by tremie pipe, but it does potentially allow formation material to be washed into the well before the annulus is sealed. Once the screen is supported by the gravel pack or formation stabilizer, the sub and drill rods used for suspending the screen can be removed. The top of the annulus can be sealed by a flexible sheath which is hammered into a conical shape.

In deep water wells, screen strings may be suspended from the intermediate casing by oil-well hangers. The screen is lowered to the desired depth and then the hanger mechanism is operated. The hanger presses against the casing with an action that is made more secure by the weight of the suspended screen. The advantage of the sealed annulus screens is that aquifer material cannot enter the well, but their disadvantage is that – without special ports – gravel pack or formation stabilizer cannot be installed.

Once the casing string is installed, the well should be checked for its verticality and alignment, as these factors can have a significant influence on the successful operation of the well,

especially for a deep well to be equipped with a lineshaft turbine pump. Alignment, or straightness, is usually the most important factor, since many pumps will operate in a non-vertical well whereas it may not be possible to install the pump in a crooked well. Even if the pump can be installed, there may be excessive wear on the shaft if this is out of alignment. Verticality and straightness are also important for submersible pumps: the pump should be located centrally in the well to maintain a cooling flow of water around the motor.

Verticality can be measured with a plumb-bob. The American Water Works Association Standard A100–06 specifies a maximum allowable deviation from the vertical equal to 0.0067 times the smallest inside diameter of the well per 0.305 m (1 foot) being tested (AWWA, 2006). This applies to wells with lineshaft turbine pumps. Greater deviations are acceptable in wells to be equipped with other types of pump. Indeed, in Norway, where inclined boreholes in crystalline aquifers are not uncommon, submersible pumps can be installed in 'cradles', with three runners spaced around the perimeter. Alternatively, it is not unknown for small capacity submersible pumps to simply be left resting against the inclined borehole wall on flexible rising mains, where they often appear to operate satisfactorily despite their far from ideal position.

Alignment is usually checked by running a pipe or dummy inside the casing string (a dummy consists of rings set along a rigid frame). The pipe or dummy should be of slightly smaller diameter than the pump-chamber casing, and long enough so that if it can move freely up and down the well, then the pump will be able to fit also. The AWWA standard specifies a 12.2 m (40 ft) long pipe or dummy, with an OD no smaller than 12.7 mm (i.e., 0.5 in) less than the ID of the casing being tested (AWWA, 2006).

5.3 Sonic drilling

Sonic drilling was pioneered by the Romanian George Constantinesco in around 1913 and further developed in the 1940s (GeoDrilling

International, 2002). It is mainly used in geotechnical, mineralogical and contamination investigations, but there are increasing applications of this method for constructing closed-loop ground source heat boreholes (Banks, 2012a) and water wells, both production wells and observation wells. Drill diameters vary from less than 100 mm up to 300 mm, with hole depths of up to 200 m (Australian Drilling Industry Training Committee, 2015).

This method incorporates a hydraulically-powered sonic drill head (oscillator) that produces high frequency vibrations (typically around 150 Hz). These sonic vibrations are transferred via the drill rods to the drill bit, which effectively fluidises the surrounding formation particles (producing a dry powder when drilling in the unsaturated zone, or a slurry bellow the water table). This fluidization process reduces friction between the bit and the formation, thereby enabling rapid progress through unconsolidated formations. The sonic vibrations are also applied when advancing a temporary casing outside the drill bit or core sampler. When completing a well, the casing and screen (and artificial gravel pack if necessary) are installed before the temporary casing is withdrawn. The main advantages of sonic drilling are rapidity, lack of need for large quantities of drilling fluid, lack of noise and heavy vibration.

The method can also be used to drill boreholes in consolidated formations. However, in rock, fluidization of the material around the drill bit does not occur, frictional resistance is higher and hence bit rotation is also required, with circulation of a fluid such as water or air-foam to cool the bit and (unless a core barrel is being used) to lift the cuttings to the surface.

5.4 Auger drilling

Auger rigs vary from small-diameter manual augers for soil sampling to large truck or crane-mounted augers used for drilling shafts for piles or piers that are more than a metre in diameter.

The commonest auger design is the screw auger, with a blade welded in a spiral to a central solid drill stem. In small soil augers, the spiral may extend for the first few centimetres of the tip, but with continuous flight augers the auger is supplied in sections, usually 1 m long, with the spiral blade extending the full length of each section.

An alternative design of auger is the 'bucket' auger, which cuts the soil with two blades on the base and then passes the soil up into the bucket. When the bucket is full, it has to be withdrawn for emptying and, as drilling proceeds, sections have to be added to the auger stem. Large bucket augers are sometimes used for drilling water wells in the United States.

The continuous flight auger is a common auger method for drilling exploration boreholes. The

Figure 5.20 Hollow-stem auger drilling, Saudi Arabia. Photo by Bruce Misstear

formation drilled is recovered as disturbed samples on the blades of the auger. 'Undisturbed' core samples can be taken by hollow-stem augering. The mechanics of drilling are similar, but with the hollow stem auger the spiral blade is welded to a hollow tube, typically of 100 or 150 mm internal diameter (Figure 5.20). As augering progresses, a core of sediment is forced into the tube or hollow stem, to be recovered when the augers are withdrawn, or by wire-line if a wire-line core barrel is incorporated in the design. A 'split spoon' sampler can also be used, which is a type of core barrel that is driven into the undisturbed formation below the base of the auger and can then be split apart to reveal the sample.

Auger drilling is a valuable, inexpensive method for rapid formation sampling at shallow depths, or for constructing small observation boreholes (the casing string is installed through the hollow auger stem) provided that the formation is soft and cohesive. Augering is impractical in hard rocks, dry sand, coarse gravel or in stiff boulder clay; below the water table, penetration can be difficult because of formation collapse, although bucket augers have been used in mud-filled holes.

5.5 Jetting

Jetting is an effective means of constructing wells in shallow, unconsolidated aquifers. The method is rapid, can be carried out with relatively simple equipment, and is inexpensive. Well-jetting relies on a combination of percussion and fluid circulation. The operating principle is illustrated in Figure 5.21. The drill pipe, which may be open

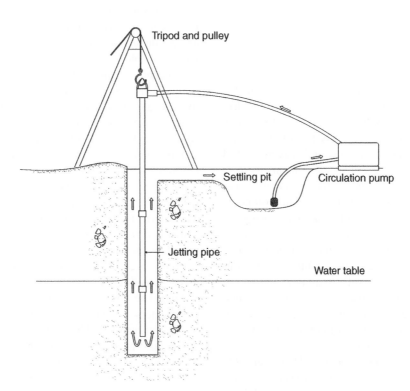

Figure 5.21 *Principle of jetting*

Figure 5.22 *A well being drilled by jetting, Indonesia. Photo by David Banks*

ended or have a drill bit at its base, is driven into the shallow soil. Water is then pumped down the drill pipe, which fluidises the unconsolidated formation, and helps the drill pipe to sink further. The water carries the cuttings up the annulus between the drill pipe and the borehole wall to a settlement pit on the surface, where the cuttings settle out and the water is recirculated through the pump. If the formation being drilled is very unstable, temporary casing can be installed (as in percussion drilling) or some drilling mud can be added to the circulating water.

Well-jetting has been successfully used to construct small-diameter wells in alluvial aquifers in Africa and Asia (Figure 5.22). The method becomes impractical if large boulders are present, and it is not suitable for drilling in hard consolidated or crystalline aquifers.

Many modern top-hammer or DTH rigs are able to operate without (or with minimal) hammer operation, and to sink boreholes in loose unconsolidated sediments by a combination of air or water jetting and a low rate of rotation. Use of eccentric bits (see

Section 5.2.4) allows for rapid construction of shallow wells in sandy or gravelly fluvial sediments.

5.6   Direct push and drive sampling

There are a number of percussive methods that are used to construct small diameter boreholes for geotechnical or environmental investigations. These are variously referred to as direct push, direct drive, drive point, drive sampling or window sampling methods. They are mentioned here briefly as they can also be employed to construct observation boreholes. Further details on the methods and their applications are provided by U.S. Environmental Protection Agency (2005) and Australian Drilling Industry Training Committee (2015).

Steel tubes are either pushed into soft soil using hydraulic cylinders, or, where the ground is firmer, are driven using a percussive hammer. Various types of samplers can be employed, including U100 core barrels, split tubes, window samplers and piston samplers (see Section 6.2). The maximum borehole diameter is usually around 100 mm, and borehole depths seldom exceed 10 m. Drive sampling rigs are often track-mounted for mobility, and are small enough to fit inside a trailer which can be towed from site to site by a four-wheel drive vehicle. For the construction of observation boreholes in unstable ground, an outer casing may be advanced simultaneously with the inner sampling device. When the target depth is reached, the inner sampler is pulled out, the casing and screen are installed inside the outer casing, which is then withdrawn, allowing the formation to collapse around the well casing and screen.

Figure 5.23(a) shows a small track-mounted drive sampling rig in operation at a site in Ireland. The rig was used to install a transect of piezometers in glacial till and the underlying weathered bedrock. The average borehole depth was about 4 m. Figure 5.23(b) shows a sample of clay being extruded from a sample tube (the cutting shoe having been removed already). The extrusion process and sample integrity are aided by fitting the sampling tube with a removable plastic liner.

(a)

(b)

Figure 5.23 *(a) Track-mounted drive sampling rig at drill site in County Louth, Ireland; (b) extruding a soil sample from the sample tube. Photos by Bruce Misstear*

5.7   Driving of well-points

A well-point is a narrow diameter (about 40–75 mm) string of steel tubing. The bottom end is a pointed spike which can be driven into loose unconsolidated sediments. The lowermost section(s) will typically consist of slotted pipe to allow the abstraction of water.

Well-points can be very rapid means of exploring the hydraulic behaviour of loose sedimentary aquifers with a shallow water table – for example in sand dunes or alluvial sands – as the well-points can be test-pumped for short periods using surface mounted suction pumps, and samples can be taken. Well-points may also be used as observation boreholes for monitoring of hydraulic responses during a large scale pumping test. Once the testing, sampling or period of observation is complete, the well-point can be jacked out of the ground and re-used elsewhere. Occasionally, where the aquifer properties are suitable, well-points may also be installed permanently for groundwater abstraction.

Well points can be readily installed by appropriately modified top-hammer rigs (and may even be able to be installed by a hand-held pneumatic hammer), especially if a pilot hole has been drilled by rotation-jetting.

5.8   Manual construction

While some boreholes can be drilled by hand, using percussion, auger, sludging or jetting techniques (Box 5.5), manual methods are used mainly in the construction of open wells and shafts. The major factors to be considered in hand-digging wells are methods to maximize the penetration rate and to minimize the danger to the excavators.

The excavation in soft formations is usually by shovel, pick and hoe, and the debris is removed from the hole by bucket and hoist. In hard rock, the rock has to be broken up before it can be extracted and this may require a chisel, jack-hammer or even explosives.

Box 5.5   Drilling by hand

In some hydrogeological environments in developing countries, especially shallow alluvial aquifers, inexpensive manual drilling techniques may be suitable as an alternative to more expensive motorized drill rigs. Danert *et al.* (2008) have reported that a manually-drilled well fitted with a hand pump serving 150 people can be provided at a cost of $1,000 (i.e. $6.67 per capita) compared to a conventional machine-drilled well serving 300 people, which costs $9,000 ($30 per capita). Most hand-drilled wells are less than 40 m deep, but in the alluvial aquifers of Bangladesh drilling depths of 300 m have been achieved (Danert, 2015). Hand-drilling techniques fall into four main categories: hand auger, sludging, manual percussion and jetting. The first two methods will be described briefly here, whilst the principles of manual percussion and jetting are similar to those described in

Sections 5.1.3 and 5.5, respectively. More detailed accounts of all the manual drilling techniques can be found in Carter (2005), van der Wal (2010) and Danert (2015).

The Vonder Rig (Banks, 1989; Elson, 1994), which was developed in Zimbabwe, is essentially a hand-powered auger that is best suited to relatively cohesive but unconsolidated sedimentary environments, or to weathered crystalline rocks (regolith) where large unweathered rock fragments are absent. The bit typically comprises either a corkscrew-like flight auger or a cylindrical bucket auger mounted on a drill string and rotating on a swivel-mounted support suspended from a tripod. This is rotated by hand, usually requiring at least four persons on the crossbar [Figure B5.5(i)]. A bailer is also used to remove drilling cuttings and slurry from the bore. The temporary casing is twisted down behind the progressing auger, using a combination of brute force and personal experience. Below the water table, if the sediment starts running into the base of the borehole, progress can be very slow and in some circumstances penetration into the saturated aquifer may be no more than a few metres. The final casing and well screen is installed inside the temporary casing when full depth has been achieved, and the temporary casing is withdrawn with a winch simultaneously with the installation of any artificial gravel pack in the annulus between the screen and temporary casing. Such boreholes are typically equipped with hand pumps such as the Afridev (see Section 3.7). The method sounds crude, but under the right ground conditions it can be surprisingly effective.

Another method of manual drilling, much used in unconsolidated silty and sandy waterlogged sediments in Nepal, is that of 'sludging' (Whiteside and Trace, 1993). This involves sinking a simple string of narrow diameter (typically 25–50 mm but occasionally larger) galvanized steel or bamboo pipe by repeatedly jerking it up and down into the sediment (sometimes aided by a bamboo lever). Until the water table is reached, water must be added (sometimes mixed with cow dung to create a simple mud) to create a slurry downhole. On each down-stroke, a slurry of water and sediment enters the casing string. On the upstroke, the operator places his (gloved) hand over the top of the string, creating a suction which lifts the slug of slurry towards the surface. On the next down-stroke the operator's hand is removed and a new load of slurry enters the base of the casing. This continues until the slurry column reaches the top of the casing string and is permitted to discharge as the operator removes his hand during the downstroke. As this method works essentially by the operator's hand to create a suction on the upstroke, the water table must be within around 6 m of the surface, although total depths of around 60 m are possible under favourable geological conditions. Rates of 20 m per hour can be achieved. Typically, a PVC string of casing and screen is installed in the completed hole and the galvanized drilling pipe withdrawn.

(a)   (b)

Figure B5.5(i) *Drilling at (left) the Chiaquelane camp for displaced persons, and (right) a village near Chokwe, Mozambique, after the devastating floods in 2000, using the Vonder rig. The left hand photo shows drilling by manual rotation of the drill-string using a crossbar. The right hand photo shows the flight-auger bit and the temporary casing, and the rig's winch. Photos by David Banks*

In unconsolidated sediments, a 'casing' string comprising reinforced concrete rings is often used to support the well. Excavation takes place from underneath the lowermost ring (which may be equipped with a special shoe), such that the string of rings sinks as excavation progresses and new rings (or masonry layers) are added at the surface (Section 3.2.1).

The diameter of the hole must be large enough for one, or possibly two, people to work down the hole, but should be as small as possible to minimize the amount of debris to be excavated. A 33% increase in diameter from 1.5 m to 2.0 m will result in a 78% increase in the volume of spoil to be removed. Nevertheless, in some countries it is still common practice to excavate shafts of several metres in diameter, especially where the aim is to provide substantial water storage in the well.

To complete the well below the water table the excavation must be dewatered. This is normally achieved using a surface-mounted centrifugal pump with its suction hose in a sump in the bottom of the well.

The safety aspects of well digging are sometimes overlooked because of the low technology used in the construction. However, such an approach is unacceptable, and it is imperative that every precaution is taken in the design and construction of the well so that risks to the well-diggers are minimized. Detailed guidance on safety in hand-dug well construction is provided in texts such as Watt and Wood (1977), Collins (2000) and Davis and Lambert (2002), but factors which should always be considered include:

1. No persons should dig a well on their own. The well digging team should have an experienced and competent supervisor.
2. All equipment for accessing the well, for digging and removing the spoil, etc should be in good condition and be checked regularly.
3. The well-diggers should be fitted with proper safety equipment, including hard hats and protective boots. When using pneumatic drills, masks and eye protectors should be worn against the dust and chippings.

4. The sides of the excavation must be shored up with strong timbers, jacks or concrete/masonry rings to prevent collapse. This is especially important when excavating the well in long lifts (Section 3.2.1).
5. Combustion engines for powering pumps should not be lowered in the well, since dangerous exhaust gases can accumulate in the well.
6. When working below the water table a back-up drainage pump should be available. Escape ladders should be provided.
7. When using explosives all precautions demanded for their use in quarries or mines should be followed. The well should be purged of all fumes before digging proceeds.

It is also essential, as with all well and borehole construction operations, to ensure that the public are not put at risk, so the area around the well excavation must be secured to prevent access. See Appendix 3 for general guidance on preparing a health and safety plan.

5.9 Well development

The action of drilling a well will inevitably lead to some damage to the aquifer immediately adjacent to the well, and result in a reduction in the well's potential performance. The primary purpose of well development is to repair the damage done to the aquifer and to restore the well's performance. The secondary aim is to develop the aquifer itself by increasing the transmissive properties of the aquifer adjacent to the well to values actually greater than they were before drilling. Not only is well development an essential process in the construction of a successful production well, it is also important when installing observation boreholes, to ensure good hydraulic connection between the borehole and the aquifer and to remove remnant drilling fluid that might impact on the collection of representative groundwater samples.

5.9.1   Well and aquifer damage

The deleterious effects of drilling fall into two categories: damage to the well face and damage to the aquifer matrix. The relative importance of the two categories will depend on the drilling technique being used.

The reciprocating piston action of a shell or drill bit in percussion drilling can seal a well face by smearing clay over the borehole wall, particularly if there are clay layers in the geological succession. In a fissured aquifer, wall-smear or the squeezing of clay into the fissures can cut off the flow of groundwater into the well. The surging action of the tool string may also force dirty water into the aquifer matrix but, because the level of water in a well during percussion drilling is usually lower than in the ground, the formation invasion is not likely to be excessive.

The build-up of a filter cake on the well face is a necessary part of direct-circulation mud-flush rotary drilling (Section 5.2.1), in order to support the borehole and to prevent loss of drilling fluid circulation. This means that the filter cake prevents the flow of water into the borehole, that is, it effectively seals off the aquifer. Although the recommended thickness of filter cake is 1–2 mm, it is often much thicker in reality and can be strongly keyed to the porous aquifer. The removal of such a cohesive layer can be difficult and require the use of violent methods. As we discussed in Section 5.2.1, the high hydraulic heads maintained in a rotary-drilled hole force some fluid from the drilling mud through the filter cake into the aquifer formation. Whilst this filtrate may be cleaner than the drilling mud, it still contains fine material and can block the pores of the aquifer matrix in a zone of invasion around the borehole (Figure 5.11).

Reverse-circulation drilling usually uses water instead of mud-based drilling fluids but, though this water may be clean at the start of drilling, it will become contaminated by the finer material from the formations being drilled. These fines will be actively flushed into the aquifer under the high positive head imposed by the drilling technique.

A borehole in an unconsolidated aquifer has to be supported, while drilling, by temporary casing,

filter cake or by the hydraulic head of the fluid column. This means that the permanent screen and pack have to be installed before well development can take place, and this results in the filter cake and damaged aquifer being separated from the well bore by a screen and sometimes also by a thick artificial gravel pack. The blinding of a well face by filter cake is serious, though usually curable by development, when the well face is open, but when the face is masked by an artificial gravel pack, the problem is much worse and full development is not always possible.

5.9.2   Developing the well

The main problem in developing a well is the removal of wall-smear, or filter cake, and mud filtrate derived from bentonite-based muds. The methods of development are dealt with below, but clearly it would be best if the need for development could be avoided altogether. The use of biodegradable organic polymer-based muds does overcome many of the problems associated with bentonite-based muds. However, it does not eliminate the need for development altogether. A filter cake is still produced, and formation invasion by fines from the formation being drilled also occurs. The breakdown of organic polymers can be accelerated by the use of additives such as chlorine. The organic polymer mud must be removed totally from the well or it may act as food for bacteria and encourage infection and biofouling in the well. The driller must ensure that polymer is not left behind the blind casing, where development cannot be effective.

Care should be taken in the disposal of all fluids and washings from development activities. With a polymer mud, its oxygen demand can be high and careless disposal to a river could result in de-oxygenation of the water and damage to the ecosystem.

5.9.3   Developing the aquifer around the well

'Aquifer development' is merely the extension of well development to the stage where the fine fraction of the aquifer itself is being removed. There is no clear demarcation between the two stages of development. The objective is to create a zone of enhanced porosity and hydraulic conductivity around the borehole, which will improve the well performance. This increase in well efficiency is a result of the decrease in the velocity of groundwater flow in the zone of enhanced porosity adjacent to the well screen. The natural flow velocity of groundwater in a porous aquifer is low, usually less than a metre a day, and the flow is laminar (Section 1.3). As groundwater approaches a pumping well it has to flow through progressively smaller cross-sections of aquifer, and therefore, for a given well discharge, will flow at increasing velocities. The groundwater velocity 1 m from a well will be one hundred times greater than at a radius of 100 m. The groundwater velocity can be high enough for the flow to become turbulent, which leads to increased energy loss, which has the effect of increasing the drawdown in the well for a specific discharge rate. This additional drawdown is termed the well loss (Section 4.5). The relationship between the groundwater velocity, the aquifer porosity and the hydraulic conductivity is shown by Equation (1.5), a modified form of the Darcy equation:

$$v_s = \frac{Ki}{n_e} \qquad (1.5)$$

where v_s is the groundwater velocity (referred to more formally as the linear seepage velocity), K the hydraulic conductivity, i the hydraulic gradient and n_e the effective porosity.

An increase in porosity owing to development will lead to a proportional decrease in v_e in this critical zone where turbulent flow is most likely to occur, and so will reduce the potential well-loss factor in the drawdown (Section 4.5). 'Aquifer development' can therefore improve the well performance but, because its effect is local to the well, it cannot produce additional groundwater resources from the aquifer as a whole.

Successful development of a poorly sorted, unconsolidated, aquifer will lead to the formation of a natural gravel pack (Section 4.4.1). Development could remove up to 40% of the aquifer close to the screen but, because the effects of development will decrease away from the well face, the gravel pack will grade from a clean gravel against the screen, to the natural material (Figure 5.24). Development will affect only the aquifer close to the well, usually within a metre of the screen.

A well in a fine, well-sorted aquifer should be equipped with a screen and an artificial gravel pack (Section 4.4.2). The development of such a well should be undertaken with care, especially if the pack is thin, because too vigorous a development could mobilize the aquifer material to break through

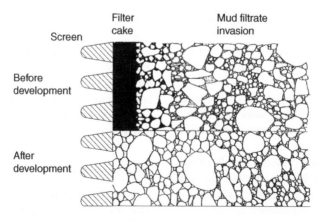

Figure 5.24   *Development of a natural gravel pack*

the pack and lead to sand pumping. The opportunity for development in such an aquifer, apart from repairing aquifer damage, is in any case limited, because there is not a distinct finer aquifer fraction.

The development of a fissured aquifer may involve cleaning out fine debris from fissures; for example, in removing fine material from fissures in a sandstone aquifer by disturbing the cohesion of the material using physical methods. The most common fissured aquifer development, however, is the widening and cleaning of fissure systems in carbonate aquifers by acidization. This development is concentrated on very restricted sections of the aquifer – the fissures – and its effect can extend a considerable distance, commonly tens of metres, from the well. In crystalline aquifers, fissures can be opened up by the physical method of hydrofracturing.

5.9.4 Methods of development

Development relies on either physical or chemical methods to clean the well face and mobilize the material to be removed from the well. The physical methods include scratching, surging and jetting, while the chemical methods include polyphosphates or other dispersing agents, and acidization. Box 5.6 summarizes the factors to consider when choosing a well development programme, while some of the common development techniques are described below. Where a well is constructed with plastic materials, great care needs to be taken to ensure that the more vigorous forms of well development do not damage the casing or screen. It should also be noted that domestic wells and observation boreholes are generally constructed with less robust

---

Box 5.6   Choosing a well development programme

There are four main factors which will influence our choice of development method:

- *Aquifer type*: unconsolidated, consolidated non-carbonate, consolidated carbonate, crystalline.
- *Well design*: open hole completion or screened; plastic or metallic screen.
- *Drilling method and rig type*: percussion, direct circulation rotary, reverse circulation rotary or other method.
- *Drilling fluid*: bentonite mud, organic polymer, air, water.

The best results are usually achieved by a combination of methods. The following steps will help in selecting the appropriate methods:

1. *Unconsolidated or non-carbonate consolidated aquifer, and the borehole is to be drilled by mud-flush rotary methods.* A mix of development methods might be used: scratching, jetting, surging and possibly dispersants. Check that:

- the contractor has the necessary equipment and chemicals, and knows how to use them; if dispersants are proposed, then polyacrylamide or other non-phosphate alternatives are preferable to polyphosphates;
- the contractor has either surge blocks or a surge pump, and then agree with the contractor the method to be used (take special care when planning surging in wells lined with plastic casing and screen).

2. *Unconsolidated or non-carbonate consolidated aquifer, and the borehole is to be drilled by percussion methods.* Then development by surging and clearance pumping may be sufficient. Again, we should ensure that the contractor has the equipment (and chemicals if needed) and knows how to use them.

3. *Limestone aquifer.* Then acidization is an option (although hydrofracturing may also be useful in crystalline marbles – see step 4 below). For acidization, we should check the following:

- Have other wells in the area been acidized? If so, check the results to see if the method is effective.
- Does the contractor have acidization experience and equipment, including safety equipment such as protective clothing, breathing apparatus, etc?
- Is suitable acid available locally? Note that sulphamic acid is often preferred to hydrochloric acid nowadays for reasons of safety.
- Are there risks to other abstractions, land uses or environmental features during acidization?
- Will the pollution control authority permit disposal of the spent acid? Check that the contractor will comply with the regulations.

4. *Crystalline aquifer.* Then hydrofracturing (or even explosives) could be considered, especially for obtaining larger yields from wells in crystalline aquifers (Box 5.7). Check:

- Does the contractor have hydrofraccing experience and equipment, including safety equipment?
- Will 'proppants' or chemical additives be required?

Clearance pumping will be needed after all development work, and so it is important to check that:

- the contractor has air-lift equipment or has, and is willing to use, a suitable pump;
- the contractor has equipment to remove wastes and spent chemicals and has a place to dispose of them legally. If permissible concentrations of spent chemicals are defined by the pollution control agency, then ensure that the contractor has the necessary permits and the equipment to measure those concentrations.

It is essential that the contractor keeps good records on the development programme, including the methods, equipment, the length of time each method was applied, the type and volume of chemicals used, the quantity of sand and turbidity in the water discharge, etc.

Finally, after applying the appropriate development methods, we need to check that development is complete – the discharge water should be clear, with little or no sand, and have stable electrical conductivity (EC).

materials than those used in municipal, industrial and irrigation wells, so special care needs to be taken when applying development methods to these wells.

Wall scratchers and wire brushes. Wall scratchers are circles of radiating steel spokes which are strapped or welded at intervals along sections of the screen and casing as these are installed in the well. The rings of the spokes act like a flue brush and scrape the well walls as the casing string is lowered down the well. They are not recommended for boreholes in uncemented aquifers, because of the danger that they may cause caving. Wire brushes (Figure 5.25) are used to clean the inside of the casing and screen string, often as part of a well maintenance operation.

Surge blocks and bailers. Surging a well is a process which attempts to set up the standard washing action of forcing water backwards and forwards through the material to be cleaned – in this case, the screen, gravel pack and aquifer matrix. The simplest method of surging is to use a bailer, which acts like a piston in the well casing. Such a bailer, because of its loose fit in the casing and its flap-valve, cannot push water back through the screen with any force. A solid surge block will provide a much more forceful surge.

The surge block (Figure 5.25) has several flexible washers, sized to fit tightly in the well casing, and is fastened either to a rigid drill stem on a rotary rig or in a heavy tool assembly on a percussion rig. The block is put below the water level inside the well casing, but not in the screen, and is

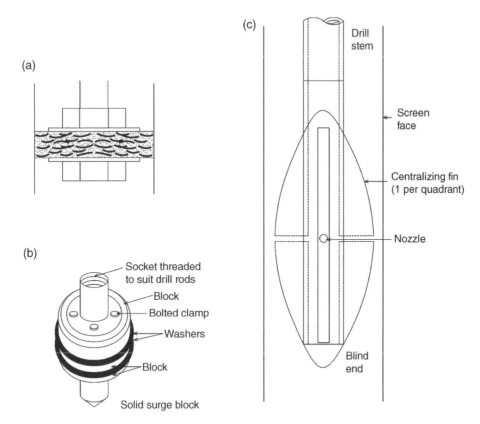

Figure 5.25   *Well development tools. (a) wire brush; (b) surge block; (c) jetting head*

then rhythmically moved up and down on a stroke length of about one metre. Water is forced out through the screen on the down-stroke and pulled back through the screen on the up-stroke. Surging will bring material into the well which will have to be periodically cleaned out by bailing.

The development should start gently to ensure that water can move through the screen, before becoming more vigorous to extend the effects into the aquifer. The suction caused by vigorous development of a blocked screen could damage the screen by making it collapse or deform, so care needs to be taken, especially in wells installed with plastic screen.

Surge pumping (interrupted overpumping). Pumping systems can be used for surging a borehole. A submersible pump is not advised because, during development, debris is pumped which could damage the pump impellers. In addition, the pump discharge is not easily controlled and, if the screen is blocked, the pump could dewater the borehole. In such a case, the hydraulic pressure behind the screen could be great enough to collapse the screen.

Overpumping, or pumping at a discharge greater than the design capacity, may not be effective where the pumping is held at a constant rate. Pumping at a constant rate, however high, will remove loose material but, because there is no surging, could stabilize a situation of partial development. Development by pumping should be in a manner designed to induce surging; this can be done using a turbine pump without a non-return valve so that water can run back down the rising main when the pump is switched off. The pump is switched on for a few minutes and then switched

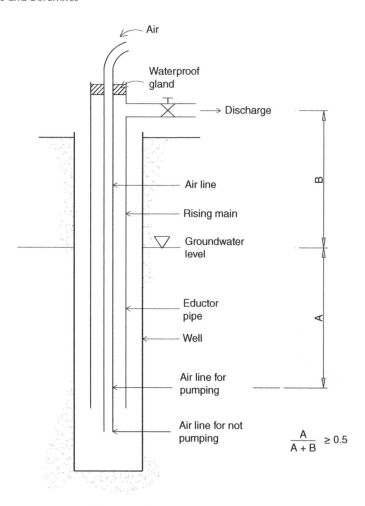

Figure 5.26   *Air-lift pumping arrangement*

off again to backwash the well for a couple of hours. This sequence is repeated at different discharge rates, for example at half, one, one and a half and two times the design discharge rate. This alternation of pumping and non-pumping phases is known as 'rawhiding' in the United States.

Air-lift pumping is very suitable for well development because there are no moving parts to be damaged by the debris drawn into the well. In normal airlift pumping, the air line from a compressor is inside the rising main, with the end of the air line at such a depth that at least 50% of the length of the air line is submerged (Figure 5.26). The rate of

pumping can be controlled by the volume of air passed down the air line. When pumping has started satisfactorily, it can be stopped and started at short intervals by shutting off air from the compressor; this will induce surging. A gate valve on the discharge pipeline can also be used for surging, by allowing pressure in the pipe system to build up, until either the air forces water out of the rising main and starts bubbling up the well, or is held in the rising main by the hydrostatic pressure in the well. In either case, when the gate valve is suddenly opened, there is a violent release of pressure and water is pulled through the screen as pumping

Figure 5.27 *Well development using air-lift pumping, Ireland. Photo by Bruce Misstear*

begins. When the valve is closed, water is forced back through the screen as pressure builds up again.

A further alternative is to set the air line just a couple of metres from the bottom of the rising main, then the pumping/non-pumping cycle can be set up by raising the air line inside the rising main (pumping) and lowering below the rising main (non-pumping). This rapid alternation of pumping and non-pumping phases is a very vigorous method of well development (Figure 5.27). However, there are instances where a more measured, gentler approach to air-lift pumping is required; for example, in cleaning out clays from fissures in limestone.

When applying the air-lift method it is important not to pump air directly into the aquifer through the screen; rather, the aim is to use the air to force water through the screen. Air injected directly into the aquifer can lead to air-locking (trapping of air bubbles in pore spaces) which can reduce the aquifer permeability (Smith and Comeskey, 2010).

Jetting. Jetting is the washing of the well face with high-pressure jets of water. On a rotary rig, water is pumped, using the mud-pump, down the drill string to a jetting tool fixed to the end of the string (Figure 5.25). The jetting tool has three or four nozzles at right angles to the drill string, so that the water jets are directed at the screen slots. The method works best with continuous slot screens having high open areas; it is less suited to louvre slot screens, for example, because the shape of the louvre slots causes the water jets to deflect, thus dissipating some of their energy. The water used for jetting should be sand-free to avoid abrasion of the screen.

The pressure of the jets will depend on the depth of the well and the nature of the screen material. The National Ground Water Association recommends a maximum working pressure of 1380 kPa (200 psi) for metallic screens and 690 kPa (100 psi) for PVC screens (NGWA, 1998). The nozzle diameter is usually between 4 and 10 mm, and water is pumped at such a rate to produce a jet velocity of between 50 and $70\,m\,s^{-1}$ (Australian Drilling Industry Training Committee, 2015). The jetting head should have centralizers attached, and the nozzles can be on the ends of stems which can be adjusted for screens of different diameter. The nozzle orifice should be as close to the well face as possible.

A jetting head on the drill string of a rotary rig can be rotated as it is raised and lowered past the section to be cleaned. The jetting should be done in short sections, and should start at the bottom of the screen (or open hole) and progress to the top. This method of well development is very effective in cleaning the screen or well face and removing filter cake, but it is less effective in restoring damaged aquifer matrix. The energy of the jets is quickly dissipated in the aquifer matrix and, unless some pumping system is incorporated with the jetting, the flow of water is into the aquifer, so driving fines further into the matrix. The pumping rate must be greater than the jetting rate, so that there is a net flow of water from the aquifer into the well.

Jetting a percussion-drilled hole is done by a jetting rig which is separate from the drilling rig. A jetting tool can be lowered down the borehole on a flexible pressure hose, down which water is fed from a surface pump. The principle is similar to that used on a rotary rig, except there are no facilities for rotating the head, so that washing may be incomplete.

Mud dispersants. The cohesiveness and plasticity of clay can be broken down by chemical dispersants. These cause the individual flakes of the clay minerals to repel each other and so break up any clay flocs. The clay in the filter cake and in the mud filtrate becomes less cohesive, more dispersed and is more easily removed by washing. However, the dispersed clay can slough over and block the well face, so mud dispersants are generally only recommended when there is a thick filter cake to be removed and where physical development methods have not been effective.

Traditionally, the dispersant most commonly used has been some form of polyphosphate. This is often supplied as a granular hygroscopic material which has to be dissolved before it is added to the well. Dosage is calculated at about 10 to $20\,kg$ per cubic metre of water in the well (Australian Drilling Industry Training Committee, 2015). The polyphosphate is left in the hole for sufficient time to react, usually around 12 hours. After this reaction time, the well is pumped to try and remove all the spent phosphate, clay and any other debris from the well.

Phosphate is a nutrient for bacteria in groundwater, and any remaining in the well can promote bacterial growth and lead to biofouling problems. For this reason, if a polyphosphate dispersant is used, then a biocide such as chlorine should be added to the well after polyphosphate treatment. However, there are alternative dispersants, such as polyacrylamide, that do not contain phosphate, and hence do not promote biofouling, and their use is recommended (Smith and Comeskey, 2010). Wastewaters from the development process should be treated and disposed of in accordance with local regulations; water containing phosphate should not be discharged to surface rivers or lakes as it may lead to eutrophication.

For observation boreholes intended for groundwater sampling, the use of chemicals such as polyphosphates and biocide agents affects the aquifer environment in the vicinity of the well and so can adversely impact on the quality of groundwater samples. It is therefore best to try and construct such wells without using muds and organic polymers, thus eliminating the need for dispersing chemicals.

Acidization. Acidization involving the use of hydrochloric or sulphamic acid is a common method for developing water wells in carbonate aquifers. Hydrochloric acid, HCl, also called muriatic acid, operates mainly by dissolving the calcium carbonate from wall-smear or from drilling debris forced into fissures while carbonate aquifers – limestone and chalk – are being drilled (Banks *et al.* 1993a). Acidization can improve the performance of a well considerably, as shown by the examples of pre-acidization and post-acidization yield-drawdown curves for a chalk well in Figure 5.28.

The acid is supplied in solution (typically at a concentration around 15% – Australian Drilling

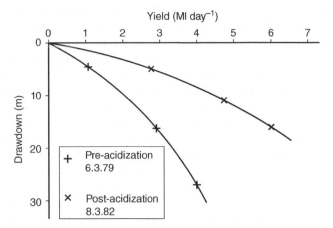

Figure 5.28 *Example of impact of acidization on performance of a chalk well. From Harker in Howsam (ed) (1990), reproduced by permission of Taylor & Francis Ltd*

Industry Training Committee, 2015), and contains an inhibitor to prevent the acid causing undue corrosion of steel casings. Industrial grade acids sometimes contain impurities that could act as groundwater contaminants, so it is important that only high grades of acid are used (such as food grade or water treatment grade acid). The use of strong acid can be a hazardous procedure and must only be undertaken by professional, competent personnel. The practitioners should possess all necessary equipment for personal safety, including masks, breathing apparatus, protective clothing and washdown facilities. Prior to acidization a health and safety risk assessment should be performed (Appendix 3). Additionally, environmental risks (such as migration of acid to wetlands, streams, lakes; and emergence of acid onto agricultural fields) and risks to other groundwater users (nearby wells and springs) should be identified, considered and addressed.

The acid solution is injected into the well, and then the wellhead is sealed with a cap equipped with a safety valve. The dissolution of the limestone by the acid generates carbon dioxide, which builds up a pressure against the wellhead and forces the acid into the aquifer formation. The amount used will depend on the volume of water in the well and the amount of carbonate material to be removed.

In a fissured karstic aquifer, the acid can develop the aquifer by cleaning and widening the fissures for a considerable distance away from the well face. This ability of the acid to travel along fissures also means that care must be taken during acidization to avoid contamination of nearby pumping wells. The pumps in all neighbouring wells, to a radius of at least 100 m, should be switched off during acidization.

A further problem with acidization is the disposal of the spent acid. In theory, hydrochloric acid should react to completion with limestone to produce a relatively harmless solution of the salt, calcium chloride. If the reaction has not gone to completion, the residue may still be acidic. Many pollution control authorities will set limits on the pH and chloride content of the effluent to be discharged. These limits must be ascertained before acidization begins. The spent acid is withdrawn from the well, slowly at first, and put in a storage tank where its pH can be neutralized and its chloride content reduced. The pH and chloride levels should be monitored by on-site meters. If necessary, this 'spent' acid may have to be sent to a hazardous waste treatment facility. Later, as the spent acid becomes more dilute and within the discharge limits allowed, it may be permissible to put it into a storm water or foul sewer. Acidization of carbonates produces carbon dioxide, and measures should be taken to avoid this heavy gas filling any enclosed spaces, where it could asphyxiate people.

Sulphamic acid is supplied as granules or pellets to be dissolved in warm water on-site, to give a strong acid solution. The reaction between the acid and carbonates is less vigorous than with hydrochloric acid, and consequently, residence times for treatment are longer. Corrosion of metal materials is less of a problem than with HCl. The main advantage of sulphamic acid is its convenience and ease of use – it is fairly safe to handle in its granular or pelletized form. Despite being somewhat more expensive than HCl, it is often preferred for well acidization nowadays because of the safety considerations (Smith and Comeskey, 2010). Phosphoric acid is another option that is also relatively safe to handle and has the advantage of being less corrosive to metallic well components than HCl or sulphamic acid. However, it can leave behind a phosphate residue that provides food for bacteria, so it should only be used in combination with a (non-phosphate) polymeric dispersant (American Society of Civil Engineers, 2014).

Hydraulic fracturing. Hydraulic fracturing (hydrofracturing or 'hydrofraccing') is a development technique for opening up fractures in consolidated and (especially) crystalline aquifers to increase well yield. The method has been successfully applied to wells in hard rock aquifers in North America, Scandinavia and Africa. Hydrofracturing involves the injection of (usually, chlorinated) water under high pressure into a short section of open borehole isolated by packers. Pressures range from around 5000 kPa to more than 20 000 kPa, the pressure used depending on the geology and on local experience as to what works. Sometimes sand is added to the injected water to act as a 'proppant' to keep the fractures open. In this case, a viscosifier is usually added to help transport the sand into the fractures. In low permeability carbonate aquifers, hydrofracturing may be carried out in combination with acidization in order to stimulate well yield. Further information on hydrofraccing is included in Box 5.7, which also refers to well stimulation by explosives.

Box 5.7   Well yield stimulation by explosives and hydrofraccing

In hard, fractured rock aquifers it has historically been common practice to attempt to enhance the yield of boreholes by the downhole use of explosives. The objective here is to create new fractures (or to increase the aperture of existing fractures) in the immediate vicinity of the borehole, which can tap into the wider aquifer fracture network. Typically, 'tamps' of sand are placed above the explosive charge to ensure that its energy is directed *out* into the formation rather than simply up the hole. The technique may sound crude, but the use of sequential and directional charges has undoubtedly honed the applicability of this seemingly blunt tool. We would like to emphasise here that *the use of explosives must only be undertaken by experienced, trained and licensed operatives.*

A more controlled and sophisticated approach to well yield stimulation in hard rock aquifers is that of hydraulic fracturing ('hydrofraccing' or 'hydraulic jacking'), a method that was pioneered in the petroleum industry. Here, a packer is placed at some depth in the borehole, and water is pumped into the section below the packer using a specially designed pumping system, to result in a very high (several MPa) pressure. Once the pressure exceeds the *in situ* tectonic and local stresses in the rock, small existing fractures can be 'jacked' open to increase their aperture. If no such fractures exist, new fractures can be initiated once the applied pressure exceeds the *in situ* stress and the tensile strength of the rock (Less and Andersen, 1994; Banks *et al.*, 1996). It is important that the pressure is maintained after 'jacking' or fracturing is initiated, in order to propagate the fracture some distance into the rock. 'Propping agents' such as sand/fine gravel grains or small beads can be suspended in the fraccing fluid, with the objective of propping open the fracture once the pressure is released.

Double-packer hydrofraccing systems can be used, in order to apply the water pressure to a specific zone of borehole wall. The double packer can then be systematically moved within the well to hydrofracture successive zones.

Explosives and hydrofraccing are applicable only in unlined, open boreholes. They both have the potential to create undesirable open fractures between the borehole and the surface (rendering the borehole vulnerable to surface contamination). In general, it is thus recommended that hydrofraccing packers should not be placed shallower than, say, 30 m from the surface.

Few statistics are published on the impact of explosives and hydrofraccing on water well yield. A Swedish study (Müllern and Eriksson, 1977; Fagerlind, 1979) of 60 boreholes in crystalline bedrock found that use of explosives (n = 30) resulted in yield increases of between 0 and 900 l hr^{-1} (median 65 l hr^{-1}), while hydraulic fracturing using single packer systems (n = 30) resulted in yield increases of

0 to 1880 l hr^{-1} (median 180 l hr^{-1}). This yield improvement represents the difference between an unusable borehole and one that is adequate for domestic or small farm supply. For comparison, the median yield of all Swedish hard rock boreholes is 600 l hr^{-1} (Gustafson, 2002). The majority of those boreholes selected for yield stimulation would have almost certainly had an initial yield significantly below this value.

Fagerlind (1982) cites another Swedish study of 182 hydrofracced boreholes from Stockholm County [Figure B5.7(i)]. The initial yield varied from 0 to 300 l hr^{-1} (median 5 l hr^{-1}), while the post-hydrofraccing yield had increased to a range of 15–1800 l hr^{-1} (median 225 l hr^{-1}). In the Norwegian Iddefjord Granite, a small study of seven boreholes demonstrated a median yield increase of 261 l hr^{-1} following hydrofraccing (Banks *et al.*, 1996). All of these studies used relatively primitive hydrofraccing equipment. With modern zoned hydrofraccing, using high pressure, high volume pumped systems, it is probable that even better enhancement statistics would be achieved.

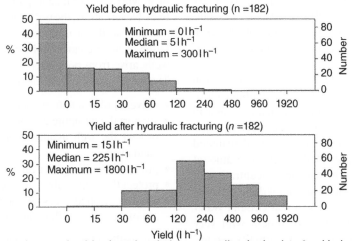

Figure B5.7(i)   *Distribution of yields of 182 boreholes in crystalline bedrock in Stockholm County, Sweden, before and after hydraulic fracturing. Modified after Fagerlind (1982), reproduced by permission*

5.9.5 Disinfecting the well

Drilling operations will unavoidably introduce bacteria into a well – from the soil zone, for example. After development and well testing, a disinfectant should be added to sterilize the well and to inhibit future biofouling problems. We should recognize, however, that while the introduction of bacteria from the soil may increase the risk of biofouling in the well, many bacterial genera live and can thrive in the groundwater environment. Thus, biofouling can (and probably will eventually) occur even if the most stringent disinfection rules are followed.

The most common disinfectants used are chlorine-based compounds such as chlorine gas, sodium hypochlorite and calcium hypochlorite. Chlorine gas is difficult and hazardous to handle, whilst calcium hypochlorite is not recommended for use in some groundwaters since insoluble calcium reaction products could lead to clogging problems in the well (Houben and Treskatis, 2007; American Society of Civil Engineers, 2014). Therefore, sodium hypochlorite is generally the preferred disinfection agent, although it has a shorter shelf life than calcium hypochlorite and so the latter may be more suitable for use in remote locations in developing countries where it is not easy to acquire fresh chemical supplies regularly (Smith and Comeskey, 2010). All chlorine compounds are strong oxidants, and care must be taken in their handling to avoid contact with skin, eyes, and so on.

Sodium hypochlorite is supplied in a solution containing between 5% and 12% available chlorine. It is added to the well and then the water in the well is agitated using a surge block or other method so that the chlorine is distributed throughout the length of the well. The amount added should be such that the concentration of chlorine in the well after mixing will be between $50\,mg\,l^{-1}$ and $200\,mg\,l^{-1}$ (National Ground Water Association, 2014); shock treatment with very high levels of chlorine is not normally recommended since this can lead to a sharp rise in water pH and the precipitation of carbonates and

other minerals (American Society of Civil Engineers, 2014). The chlorine solution should be left in the well for at least 24 hours and then removed by pumping. The chlorine-rich discharge water should not be discharged directly to streams or other surface waters. The well should be disinfected again after the production pump is installed and the headworks are complete.

Finally, we must be aware that while chlorine and other disinfectant compounds are effective against bacteria, they are less effective in eliminating other microorganisms such as viruses and *Cryptosporidium*. The best insurance for a clean, potable supply is to choose the well site carefully, and to construct the well and the wellhead properly so as to minimize the risk of pollution. When a well site is being chosen, the vulnerability of groundwater and the land use in the vicinity in the likely zone of contribution should be considered (Section 2.8).

5.10 Wellhead completion

The headworks on a water supply well are usually constructed as part of a civil engineering contract, along with the distribution, storage and any treatment works, rather than as part of the drilling contract. Nevertheless, the wellhead is a very important aspect of well construction, since many of the pollution or maintenance problems common in water wells arise from poor attention to the headworks. The purpose of the wellhead is to:

- prevent pollutants from entering the well;
- support the pumping equipment;
- protect the well against accidental damage or vandalism;
- protect the well against flooding;
- protect the well against freezing;
- allow controlled access to the well for monitoring of water level, flow rate and water quality.

Many wellheads have traditionally been completed in underground chambers, as these offer advantages in terms of minimizing visual impact and land take, and in providing security. However, wellheads that are below ground are more susceptible to flooding, and to the accumulation of water from leaking valves or pipework connections, which can be difficult to drain properly, and therefore are much more vulnerable to pollution than above-ground completions. In an assessment of the risk of *Cryptosporidium* contamination to groundwater supplies for an area in southeast England, headworks were included as one of the risk assessment categories, with the greatest risk being assigned to 'Headworks in outside chamber and/or below ground level, liable to flooding' and the lowest risk being 'Completely sealed, raised borehole cover, inside secure building' (Boak and Packman, 2001).

We therefore recommend here that, wherever possible, the headworks should be above ground. While constructing a building over the well provides good security, it can make well maintenance and rehabilitation operations difficult, since it does not allow easy rig access to the well. One solution is to place all the valves and control equipment in a building or above-ground chamber, but to keep the well outside (Figure 5.29). In this scenario it is essential to seal the wellhead properly, and to have secure fencing around the site. The headworks for

a production well equipped with an electric submersible pump would normally include:

- a sealed well cover, with a screened vent pipe and access for a water level monitoring dip tube;
- one or more air valves on the pump rising main, to let air enter or leave the pipework when it is being emptied or filled with water;
- a check or non-return valve to prevent water from rushing back down into the well when the pump is switched off;
- a flow meter;
- a gate valve for adjusting the flow;
- a washout pipe to allow pumping to waste and sampling, or for carrying out a pumping test;
- a control panel;
- some form of sampling tap or line for acquisition of water samples.

Sometimes it is necessary to install a well at a site that may flood occasionally – in arid regions, for example, where a well is sometimes located in a wadi bed that may be subject to flooding once every few years. In this case, the wellhead should be designed so as to protect the well, valves, pipework, flow meter and switchgear from flood damage and pollution (in as far as this is practicable), whilst keeping the well itself

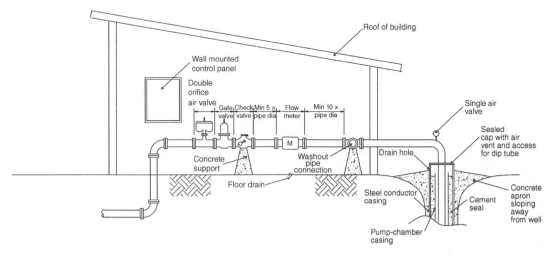

Figure 5.29 *Example of an above-ground wellhead completion for a production well equipped with a submersible pump*

Figure 5.30 *A wellhead arrangement at a wadi site in northern Oman which is susceptible to occasional flooding. The valves and other pipework are located in a gabion-protected concrete chamber, whilst the electric control panel is on top of the chamber (where the hydrogeologist is standing) to keep it as high as practicable above anticipated flood levels. The well, which has a watertight cover, is outside the chamber, so as to be accessible for maintenance. Photo by Bruce Misstear*

accessible for cleaning and other maintenance operations (Figure 5.30).

In North America, wellheads are often completed underground to avoid freezing problems. Nowadays, this can be done using a 'pitless adaptor', which attaches to the well casing below the ground freezing depth and provides a watertight right-angle connection between the pump rising main and the water supply distribution pipe (Sterrett, 2007).

Observation boreholes should also be completed above ground level, if this practicable (Aller *et al.*, 1991). Again the borehole should be fitted with a secure watertight cap, and the concrete apron surrounding the wellhead should have a slope so as to direct surface water away from the observation borehole (Figure 5.31). Where the

observation borehole needs to be finished below ground level, for example where it is exposed to traffic at the side of roads or in car parks, this should be done with a watertight manhole cover which is flush with the ground surface, and extends from the surface down into the concrete apron surrounding the well casing; the top of the casing should be proud of the concrete apron and be fitted with a secure cap.

It should be noted that some aquifers are especially susceptible to high natural concentrations of dissolved gases such as methane (perhaps derived from organic carbon or coal deposits in a nearby aquitard, for example) and carbon dioxide (especially limestones). In contaminated aquifers, the range of potentially dangerous volatile substances is wide. These gases and

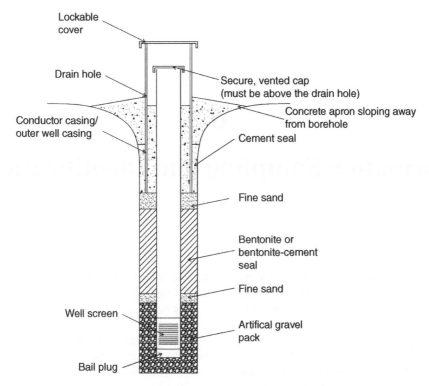

Lockable cover

Drain hole

Secure, vented cap
(must be above the drain hole)

Conductor casing/
outer well casing

Concrete apron sloping away
from borehole

Cement seal

Fine sand

Bentonite or
bentonite-cement
seal

Fine sand

Well screen

Artifical gravel
pack

Bail plug

Figure 5.31   *Example of an above-ground wellhead completion for an observation borehole*

volatiles may de-gas from the wellhead, especially during periods of low barometric pressure, to accumulate in underground chambers and manholes, or even in buildings constructed over the wellhead. Wellhead installations, manholes and buildings should thus be properly vented to prevent build-up of explosive or toxic gases. Persons should only enter the wellhead environment when necessary, and when properly equipped with appropriate safety equipment, and should monitor the gas composition of the air prior to entry.

6

Formation Sampling and Identification

The aim of this chapter is to demonstrate the type of information that can be gathered by a hydrogeologist or groundwater engineer while a well is being constructed, and the uses to which this information can be put.

A driller will usually produce a 'driller's log' that records his/her own observations on construction and hydrogeology (Box 5.2). The usefulness of this log will depend on the experience and competence of the driller. There are a number of reasons why a hydrogeologist should also be present during drilling (preferably continuously) to either supplement the driller's log or to produce his/her own definitive log (Box 6.1).

6.1 Observing the drilling process

The driller, geologist or engineer can learn a lot about the well's geology and groundwater potential by observing the drilling process. This is especially the case with rotary drilling, which is the main focus for the following text, but drilling observations are also important with other drilling methods. The three main types of drilling observations are:

a. The drilling penetration rate;
b. The drill action;
c. The behaviour of the drilling fluid.

Penetration rate. The drilling penetration rate provides a simple indication of the well's geology. In general, the harder the rock, the slower the penetration using rotary, percussion or auger drilling. This is of course a simplification, and there are no absolute penetration rates for individual drilling techniques in particular rock types: the penetration rate is influenced by many factors including the type of equipment, the well depth, the drill diameter, the bit type and the weight on the bit. There are also differences in penetration rate relating to the type of lithology as well as to its hardness; for example, rotary drilling tends to have a faster penetration in shales than sandstones, but may be slower in clays than sands and gravels because the clay can ball up on the drill bit.

Figure 6.1 gives an example of a drilling penetration log for a well in the Middle East, in which the penetration rates are recorded in minutes per metre. This well was drilled by the direct circulation rotary method through alluvium into

Water Wells and Boreholes, Second Edition. Bruce Misstear, David Banks and Lewis Clark.
© 2017 John Wiley & Sons Ltd. Published 2017 by John Wiley & Sons Ltd.

Box 6.1   What are the uses of a well log?

The purposes of a well log are to:

i. Assist in quality control and supervision of the drilling process, and to ensure the specifications of the materials being used during drilling. Is the well vertical? Is it being drilled at the specified diameter? Is grouting being carried out to an approved standard?

ii. Identify where water strikes are made during drilling, and how the well's water level changes during drilling. These observations may determine where a well screen or casing length should be installed, where a pump should be installed, or what the pumping regime of the well should be.

iii. Make a first assessment of water quality during drilling. Strong discolouration, odour or taste might indicate intervals of a well that should be cased out so that water of inferior quality does not enter the well. The progressive increase in salinity of groundwater during drilling may be an indication that drilling should not progress any deeper. (Note that, if a site is suspected to be contaminated, any member of a drilling crew should be mindful of health and safety and should use odour/taste tests with the utmost caution).

iv. Assess the geological succession being penetrated by the well or borehole. Where does the aquifer commence? Where does it finish? Where are the main water-bearing horizons or fracture zones?

v. Help decide where samples of formation (or even water) should be taken for more detailed analysis, and to quality control the collection of such samples.

vi. Characterize the aquifer horizons being penetrated. The degree of cementation of the aquifer may determine whether a well screen is necessary or not. The grain size and its distribution will determine the need for a gravel pack, and will control the design and sizing of a well screen. It may be that rapid confirmation of the aquifer grain size is required for a specially-designed well screen to be produced and delivered to the well site in a timely manner for installation.

vii. Provide information to regional or national environmental, hydrometric or geological authorities. Many nations have a legal requirement for drillers (or well owners) to submit a borehole or well log of construction and geology to a central authority. Some authorities may even require physical formation samples from the drilling process to be submitted. Although some drillers may regard this as a tedious burden, the databases containing such information are a tremendously valuable resource for hydrogeologists and planners. They enable the production of hydrogeological maps and assessments that can be used by individual consultants or drillers, often reducing the need for costly, site-specific surveys. We would argue that, even if legislation does not require it, every driller or consultant should archive hydrogeological information arising from a new well with a responsible state authority. An example of a regional groundwater atlas compiled largely on the basis of responsibly archived well information is that for the Faryab Province of northern Afghanistan (Banks, 2014c).

igneous bedrock. As is common in alluvial aquifers in the Middle East, the alluvium exhibited various degrees of cementation, with the amount of cement generally increasing towards the base of the alluvium. This is reflected in the slower penetration rate with increasing well depth. The drilling rate at the base of the alluvium is slightly slower than that in the bedrock below; hence the contact between the two geologies can be detected from the penetration log. The penetration log also

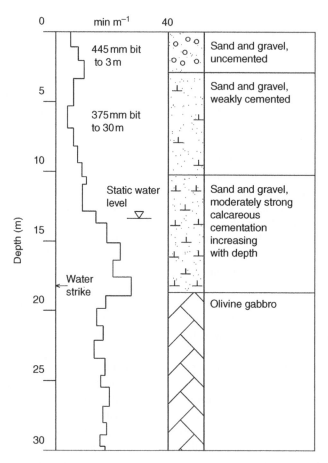

Figure 6.1 *Penetration log from a borehole drilled through alluvium and into underlying igneous rock, in the Middle East*

shows the influence of drill bit size, with a faster penetration rate being achieved after the changeover to a smaller bit at 3 m.

Drill action. The experienced driller is able to identify changes in lithology from the behaviour of the drill string. In percussion drilling the feel of the drill cable can provide clues to the geology, and the presence of loose sand is usually shown by sand heaving up into the temporary casing. In reverse circulation rotary drilling, if a drag bit is used, the drill action is generally much smoother in clays than in sands and gravels, especially if the latter contain large cobbles. Indeed, the change in

lithology from clay to gravel can often be detected by the onset of vibration in the drill string.

Another type of drill string behaviour is where the drill string falls under its own weight on encountering a large fracture. An extreme example of this was witnessed by one of the authors of this book whilst supervising the drilling of a deep well in karstified limestones in southern Oman. The borehole was being drilled by the down-the-hole hammer technique using air-foam as the circulation fluid (Section 5.2.4) when it presumably penetrated the roof of a large cavern, after which the drill string was observed to fall freely for approximately 10 m. Naturally, this resulted in a complete loss of drilling fluid circulation, and drilling could only continue once the cavern had been 'cased out'.

Drilling fluid. The detection of fractures, or even caverns, from the total loss of drilling fluid circulation in limestones is just one example of the importance of observing fluid behaviour. In drilling through granular formations using rotary mud-flush, the presence of permeable horizons can be indicated by a temporary loss of mud circulation as the mud filtrate enters the formation. This loss of circulation will cease once the filter cake develops on the borehole wall. Therefore, the mud level in the mud pits should always be monitored, and any falls in mud level noted. Again, the presence of a strongly artesian aquifer may be detected by dilution of the mud by aquifer water entering the mud column, until such time as the aquifer is 'mudded off'. The driller should always monitor mud density and viscosity to make sure these are appropriate for the borehole conditions (Section 5.2.2 and Box 5.4). It is also useful to monitor the pH and EC of the flushing fluid, whether this is mud, polymer or water, since these parameters can provide good indications of changes in groundwater quality as drilling proceeds.

Where a liquid drilling fluid is not used, such as in air-flush, down-the-hole hammer (DTH) or cable-tool percussion drilling, it is still important to observe the behaviour of the natural groundwater within the well. The first water strike may be an indicator of the level of the water table in unconfined permeable formations, of the top of a confined aquifer, or of yielding fractures in lower permeability hard-rock formations. It is often good practice to observe how the rest water level in the well changes as drilling progresses (for example, at the start of each drilling shift – see Box 5.2). A groundwater level that falls with increasing borehole depth may be an indicator of a downward vertical hydraulic head gradient (and *vice versa*). Water samples taken during drilling can give some indication of the evolution of water quality with depth (e.g., increasing salinity). One advantage of air-flush,

DTH or cable tool methods is the ability to acquire, in a cost-effective manner, some impression of any hydrochemical stratification in the aquifer.

6.1.1   Observing the drilling process in hard-rock aquifers

In crystalline rock aquifers, such as gneisses or granites, wells are typically drilled using air-flush down-the-hole hammer (DTH) techniques. Here, the drilling process returns to the surface a continuous stream of pulverized rock dust in a carrier fluid of compressed air. A specially designed collar can be attached to the well top (Figure 6.2) such that this dust can be collected in sample bags.

In hard-rock aquifers it is important to be able to recognize the following features:

- *Changes in lithology.* These can be recognized by changes in colour of rock dust, changes of drilling rate or changes in the size of the material returned to the surface. Changes in lithology can be confirmed by analysing the rock dust sample at a laboratory (for example, by X-Ray Diffraction (XRD) to identify the minerals present, or by X-Ray Fluorescence (XRF) to determine the bulk elemental composition of the sample).

- *Zones of fracturing or altered rock.* These can be recognized as intervals of faster drilling or zones where the bit can be heard/felt to catch on loose rock fragments. Zones of fracturing or alteration (at least in relatively shallow rock) will often have been subject to throughflow of oxygenated water in the geological past (even if they are not especially permeable today). Such throughflow may have oxidized the iron minerals in the rock (e.g., magnetite) to reddish ferric oxides. Zones of potentially transmissive fracturing *may* thus be recognized by a transition from greyish (fresh) bedrock cuttings, to reddish-brown coloration.

- *Zones of water inflow.* Major intervals of water inflow can be recognized by visible water emerging from the well along with drilling cuttings (and a first assessment of yield and water quality may be possible by collecting the ejected

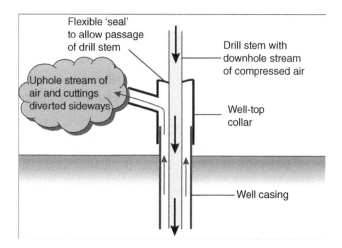

Figure 6.2 Schematic diagram of drilling collar for DTH drilling, enabling the relatively easy collection of cuttings and assessment of water flow at the wellhead

water in a bucket of known volume). More minor inflows will manifest themselves as damp cuttings, which will often emerge as larger clumps of material, rather than the fine dust characteristic of dry, fresh bedrock.

Figure 6.3 shows a typical field drilling log from a well in migmatite gneiss in Flatanger, Norway, illustrating these observations.

6.2   Collecting formation samples

Formation samples are needed to establish the lithological succession at a site and to assess the hydrogeological characteristics of the aquifers. The establishment of a lithological section is a particularly important part of exploratory drilling, but is also important in the construction of water wells and observation boreholes. The most common methods produce disturbed samples, which can be categorized as either bulk samples or representative samples (Section 6.3). These are used to identify the formation and also the grain-size distribution in aquifer material. They are not suitable for measuring aquifer properties. When samples are needed for laboratory tests of porosity and hydraulic conductivity, or even for better

formation identification, then undisturbed samples must be obtained. The collection of disturbed and undisturbed samples is considered in turn for the main drilling methods described in Chapter 5: percussion drilling, rotary drilling and auger drilling.

6.2.1   Disturbed formation sampling

Percussion (cable-tool) drilling: crystalline and consolidated rocks. Chiselling of hard rocks in the unsaturated zone produces a granular mixture of crushed rock, which may bear little resemblance to the original rock. The samples are removed by bailer or similar tool, and their identification will follow the same procedures as for the samples recovered from rotary drilling (Section 6.3).

The destruction of the rock fabric and the addition of drilling water mean that the bulk samples from bailed cuttings can only be used for broad lithological identification; features such as bedding, porosity or texture cannot usually be ascertained. The depth control of the sampling can be good, because the hole is bailed clean after each drilling period, but some 'contamination' can still arise from caving formations above.

Once the water table has been reached, water no longer has to be added. The water in the well is

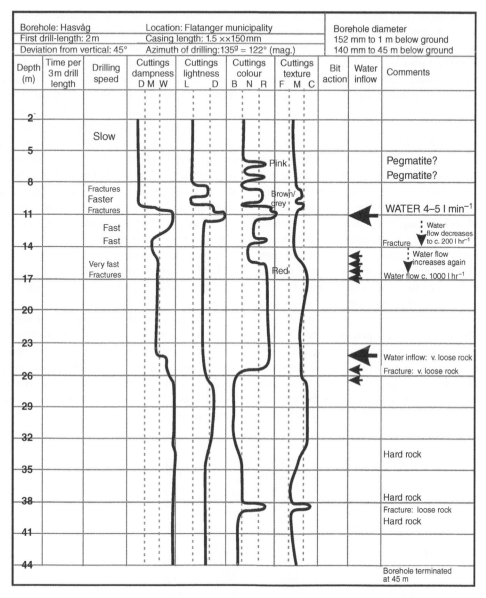

Figure 6.3 *Portion of a DTH drilling log from a borehole in Precambrian migmatite gneiss at Hasvåg, Flatanger, mid-Norway (modified after Banks & Mauring, 1993). Note that the left-hand column (although unused here) can be utilized to record the time taken for every 3 m drill-string section to be completed. Key: cutting dampness (D = dry, M = moist, W = wet); cutting lightness (L = light, D = dark); cuttings colour (B = black, N = neutral, R = red); cuttings texture (F = fine, M = medium, C = coarse). Reproduced by permission of Norges geologiske undersøkelse (Geological Survey of Norway)*

mixed but, as the drilling progresses, and the drilling water is progressively removed, the water remaining becomes closer to true groundwater. Bailed samples, therefore, can be used to give an indication of groundwater quality during exploration drilling when pumped sampling may not be possible.

Percussion (cable-tool) drilling: unconsolidated sediments. Sampling in unconsolidated soft formations is done by removing the formation in a shell, from ahead of the advancing temporary casing. The sample arrives at the surface as a slurry, as in hard rock drilling, but in this case the sample is merely a disaggregated sediment. In competent clays or silts, a clay cutter may be used and then reasonable samples may be recovered in the quadrants between the cutter's fins (Figure 6.4).

Percussion samples, because they are taken ahead of the temporary casing (thus preventing wall collapse or caving), generally have good depth control and, despite being disturbed, are representative of the formation lithology and to some extent of pore-water quality. Below the water table, however, pore-water quality control may be lost because of mixing through the vertical profile of the well and because, in unstable sands, heaving sands may fill the well from below, so contaminating the formation samples.

The bailer samples of disaggregated formation material present some problems in sample treatment. All sandy sediments contain a certain proportion of clay and silt, which can have a significant bearing on the hydrogeological character of the sediments; for example, a few percent of clay in a sand can reduce its hydraulic conductivity by an order of magnitude or more. The action of taking a sample in a water slurry can wash out much of the clay or silt fraction from the sand, so care must be taken to avoid underestimating the 'fines' content of a sample.

It is worth remembering that, where temporary casing is advanced down the borehole, only a small section of the formation at the base of the hole will be exposed at any time. This allows some opportunity to collect bailed or pumped water samples of the water that percolates into the well after episodes of drilling and bailing. These water samples will contain sediment and they will by no means be perfectly representative but, after filtration, they may be valuable as a first, approximate means of assessing water quality at a particular horizon (e.g., salinity).

Figure 6.4 *A clay cutter, used to extract samples of cohesive sediments encountered during cable-tool percussion drilling. Photo by Bruce Misstear*

Direct circulation rotary drilling. The drilling returns from direct circulation rotary drilling are a slurry of rock fragments suspended in a drilling fluid. The fragments may be broken chips of rock or disaggregated sediments, and may range from clays to chips several millimetres in size. Examination of the drilling returns and their use in producing a lithological log of the well, therefore, can be a complex task and requires expertise and experience on the part of the geologist or engineer on site.

Formation samples, after being cut by the drill bit, take a certain time to return to the surface. Thus, during continuous drilling, a sample is derived from a level above that at which the bit is located when the sample is collected. A correction has to be made to obtain a true sample depth. This depends on the velocity of drilling fluid up the hole, which can be calculated (assuming no circulation losses) from the size of the hole and drill stem, and the drilling fluid pumping rate. In deep wells, we should bear in mind that up-hole travel times may vary for different sized cuttings, potentially resulting in a degree of artificial sorting of samples according to grain size.

The cuttings in the drilling return will be contaminated by the drilling fluid as they are being cut. Additionally, during their travel to the surface, they can be contaminated with debris eroded from the borehole wall in overlying formations. This debris can represent a large proportion of the cuttings in any mud sample, so a lithological log based on cuttings should be recorded as a percentile log. The depth at which any particular lithology is first detected is taken as the top of the formation that the lithology represents.

Samples of cuttings can be collected from the returning drilling fluid, by passing the drilling returns over a container, such as a bucket or drum, set in the mud channel. This container acts as a small mud pit into which cuttings settle and, in continuous drilling, it is emptied every time the drilling has progressed for a fixed interval such as 1 m. The samples collected from the sampling container may contain drilling mud. The addition of a mud dispersant, such as polyphosphate or polyacrylamide, will help to disperse this mud and clean the sample. On the other hand, in direct circulation drilling, where a liquid drilling fluid has been used, loss of some of the finer fraction of the formation from the sample container is almost inevitable and the lithological log will suggest that the formation is coarser than it is in reality. This must be taken into account when interpreting the log, notably in selecting the screen and artificial gravel pack.

Alternatively, special filter screens, sieves and shakers are available that can be placed across the stream of drilling fluid emerging from the hole and used to separate out cuttings from the mud stream. The use of such 'sieving' methods will inevitably result in the loss of particles finer-grained than the mesh size (Figure 6.5).

The drilling returns, when mud, water or foam is used, are ultimately directed to the mud pits which are designed to allow the solid cuttings to settle out of the fluid before it is returned down the borehole (Figure 6.5). When mud control is lost and the drilling mud viscosity is too high, some cuttings will be returned to the well in the fluid circulation. These cuttings can severely contaminate the new cuttings and give an erroneous lithological log if the site geologist is not aware of the problem.

Sample mixing may occur at geological boundaries. Two examples of situations where sample mixing is likely are:

1. Drilling into a hard-rock formation immediately below a caving unconsolidated formation.
2. Drilling thinly bedded formations.

In these cases, the cutting samples are likely to represent two different formations. If sample mixing is suspected, we can check this by:

- Suspending the drill string about 0.5 m above the base of the well and continuing fluid circulation until the drilling fluid is free of cuttings.
- Drilling 1 m and collecting a sample which should now be representative of the formation being drilled.

Figure 6.5 *(a) Direct circulation rotary drilling in Faryab Province, Northern Afghanistan. Samples of alluvial sand and gravel were collected in a sieve (c) from the drilling mud and then stored in flasks clearly labelled with depth intervals (b). Fine material may be under-represented in the samples. (a)–(c) reproduced by permission of NORPLAN, Norway*

Rotary drilling using air flush (Section 5.2.2), possibly involving a top-hole or down-the-hole hammer (Section 5.2.4), avoids some of the problems encountered when drilling using liquid fluids. The air flush is not returned to the borehole, and therefore there is no cross-contamination by recirculated cuttings. While drilling with air through the unsaturated zone, the cuttings are not mixed with mud or water and therefore do not need cleaning before inspection (unless a foaming agent has been used to help lift the cuttings to the surface). Air-flush sampling, however, cannot avoid contamination of samples by caving formations, and drilling below the water table will produce a slurry of cuttings in native groundwater. The samples must be washed carefully, taking care to avoid losing the finer cuttings (Figure 6.6).

Figure 6.6 *Hydrogeologist David Ball washing limestone drill cuttings collected during air rotary drilling (using a down-the-hole hammer), County Laois, Ireland. Photo by Bruce Misstear*

When using air-flush drilling to investigate sites contaminated by hydrocarbons, oil traps must be used on the air line to avoid contamination of the samples and *in situ* formations by compressor oils (or alternatively, vegetable oils can be used as lubricants). It must also be recognized that in such a situation air flush will strip any volatile pollutants from the formation samples.

Reverse circulation rotary drilling. The samples derived from reverse circulation drilling are returned to the surface up the drill stem and do not come into contact with overlying formations. The samples, however, may be contaminated by caved debris carried down to the drill bit in water flowing down the annulus around the drill string. The velocity of the return flow with reverse circulation is high, so that the larger fragments of the formation can be returned to the surface. This is a great advantage when drilling and sampling gravel formations, because the coarser fractions can be recovered uncrushed and so sample identification is much easier. A problem with sampling from

reverse circulation rigs is caused by the high velocity of the returns, which emerge as a jet. The samples can be collected in a sieve or a bucket held in the jet discharge (Figure 6.7). However, loss of the finer fraction is almost inevitable.

Dual-wall reverse circulation drilling. The dual-walled reverse circulation technique employs a drill stem comprising an outer and an inner pipe, with an annular space between. The technique can be used to obtain good disturbed formation samples, especially using a rock-roller bit, although blade bits and down-the-hole hammer (DTH) bits are also commonly used (Strauss *et al.,* 1989). The drilling fluid, which is usually compressed air (especially with a DTH bit) or sometimes water, is pumped down the annular space between the outer and inner drill pipes, and the cuttings and fluid are returned to the surface up through the inner pipe. The drill bit is located only a few centimetres ahead of the drill pipes and hence the samples returned to the surface should be representative of the short section of formation just drilled (Aller *et al.*, 1991).

Figure 6.7   *Collecting formation samples from reverse circulation drilling in Sind province, Pakistan. Photo by Bruce Misstear*

Auger drilling. Drilling with a solid stem flight auger can be a good method for sampling shallow, unconsolidated formations. Samples from auger drilling are recovered as fragments from auger flights. With care, fairly representative samples of the formation can be collected. The sampling technique involves lowering the auger to the bottom of the hole and then spinning it at high rotation to clean debris out of the hole. The depth of the auger is recorded, and then the auger is advanced at a slow rotation speed for a fixed depth interval. Rotation is stopped and the augers retrieved by straight pull. The sample adhering to the auger flight comes from the fixed interval.

The weakness in solid stem auger drilling lies in the great difficulty in preventing cross-contamination. As the sample is pulled from the hole it is contaminated by the material above the sample interval. The method cannot be used for taking samples where absolute integrity of the sample is necessary. For undisturbed sampling, a hollow stem auger with a thin-walled sampler can be used (Section 6.2.2).

Storage of disturbed formation samples. The storage and clear labelling of formation samples is extremely important in order to avoid future confusion. With disturbed material, samples of about 0.5 to 1 kg in weight are taken from the well, although larger samples may be needed for laboratory sieve analysis where the formation material is coarse. Samples are normally placed initially into a sampling tray on site. This tray is divided into compartments, with each compartment containing the sample for a particular depth interval (Figure 6.8). The samples are thus readily accessible for inspection and description by the hydrogeologist or engineer on site. When the samples are ready to be removed from site, they should be stored in sealable plastic containers, either bags (Figure 6.9) or jars (Figure 6.5). Each sample container must be labelled with waterproof

Figure 6.8   *Disturbed formation samples being stored in sequential order of depth on site in a compartmentalized sample box. Reproduced by permission of Paul Ashley, Mott MacDonald Ltd*

Figure 6.9   *Disturbed formation samples placed in plastic bags. Photo by Bruce Misstear*

ink. The label should contain all relevant information and be clearly visible when the sample is picked up. The label should be durable, preferably plastic, to avoid being spoiled by the wet sample. It is advisable to double-bag samples, with the label being put between the two bags to keep it clean, clear and secure.

6.2.2   Undisturbed formation sampling

The action of drilling to obtain any sample will disturb that sample to some extent; 'undisturbed formation sample', therefore, is a relative term. The sampling methods described below solve the problem of obtaining truly undisturbed samples with various degrees of success, depending on the formation being drilled and whether or not drilling fluids have been used. British Standards Institution (2006) provides further guidance.

Percussion (cable-tool) drilling. The equipment usually used for obtaining undisturbed samples by percussion drilling is the general purpose open-tube sampler. The most common type is known as the U100 sampler (formerly the U4 sampler – 4 inch in Imperial measurements), comprising a 450 mm-long steel tube of 100 mm internal diameter to which a cutting shoe is screwed (Figure 6.10). The steel tube and shoe can be machined so that an aluminium liner can be inserted in the tube. The tube is then screwed to a

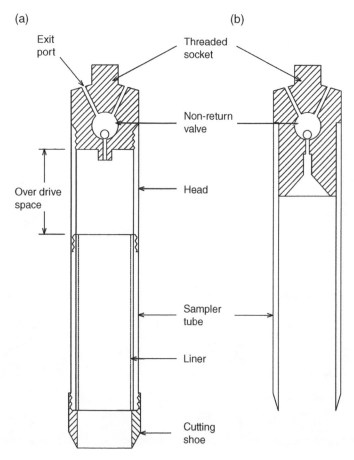

Figure 6.10   *(a) U100 tube sampler and (b) a thin-walled sampler*

U100 head, which in turn is attached to a slide hammer below a sinker bar.

The well to be sampled is cleaned out to the bottom with a bailer, and then the U100 assembly is lowered slowly to the base of the hole. Drilling then proceeds by blows to the slide hammer until 450 mm penetration has been achieved. The U100 assembly is lifted to the surface and the U100 tube is unscrewed. Either the liner or the tube itself is kept until the sample can be removed for examination. The full U100 tubes should be sealed with end-caps (typically rubber or alloy) and stored in either polythene bags or plastic tubing to avoid contamination or moisture loss.

The sample is extruded from the tube or liner by a manual or electric hydraulic ram. The sample appears as a core, but very commonly it is in three parts, the lower part (close to the cutting shoe) is virtually undisturbed, the middle part is fractured and the upper 100 mm or so have been disturbed by the percussion techniques. If the whole core is needed in an almost undisturbed state, then a double U100 can be used in which two U100 tubes are connected with a collar. Sampling is as for a normal U100 but the sampler is driven in up to 900 mm. The formation held in the cutting shoe and in the top tube is discarded and only that held in the bottom tube is retained for examination. In soft formations, sample recovery can be improved by incorporating spring clips or core lifters in the cutting shoe.

An advantage of the U100 sampling method is that relatively uncontaminated samples are obtained because they do not come into contact with drilling fluids except on their upper surface. This makes them particularly suitable in pollution investigations, although special precautions may be required (Figure 6.11). For example, when sampling for material polluted by organic chemicals, no organic material such as grease or oil (which could introduce secondary contamination) must be used on the U100 assembly. The assembly must be degreased before sampling begins and the samples must be sealed in airtight bags to avoid the loss of volatiles. In obtaining samples for microbiological analysis, even greater care must

Figure 6.11   *Flame cleaning of U100 tube sampler during an investigation of trace organic pollution, UK. Reproduced by permission of Paul Ashley, Mott MacDonald Ltd*

be taken to avoid sample contamination. The U100 tubes, liners and end-caps must be sterilized and kept in sterile bags before use. The entire U100 assembly has to be flame-cleaned before it is lowered to take each sample (Figure 6.11). The reader should seek specialist advice on degreasing and sterilization techniques.

One of the requirements for collecting undisturbed samples is that the mechanical disturbance caused by the sampler should be as small as possible. One measure of this disturbance is the *area ratio* of the sampler, which is the ratio of the volume of soil displaced by the sampler to the volume of sample. A U100 sampler has an area ratio of about 25-30%.

Another type of open-tube sampler, known as a thin-walled sampler (Figure 6.10), has an area ratio of only around 10%, and is used to obtain good-quality undisturbed samples in soft, cohesive sediments. The thin-walled sampler is pushed, rather than hammered, into the ground. The standard thin-walled sampler is much weaker than a U100 tube, and is therefore much more restricted in its use – it may not be suitable in very stiff clays or where large stones are present. A third type of open-tube sampler is the split-spoon or split-barrel type. This sampler has a high area ratio of around 100%, and the samples cannot properly be regarded as undisturbed. It is used for carrying out standard penetration tests in ground investigations [see, for example, *Eurocode 7: Geotechnical design – Ground investigation and testing* (British Standards Institution BS EN 1997-2, 2007) or American Society for Testing and Materials D1586 (ASTM, 2011)].

Other samplers that can be used with percussion drilling include:

- reinforced thin-walled samplers, such as the Vicksburg and Denison samplers (Shuter and Teasdale, 1989; Aller *et al.*, 1991; Clayton *et al.*, 1995), for obtaining undisturbed samples in stiffer sediments than would be possible with standard thin-walled samplers;
- stationary piston thin-walled sampler, used for soft non-cohesive sediments;
- continuous drive or Delft sampler (Begemann, 1974; Clayton *et al.*, 1995), designed for sampling very soft, non-cohesive sediments;
- compressed-air or Bishop sand sampler (Bishop, 1948; Clayton *et al.*, 1995), designed to collect samples of non-cohesive sands below the water table;
- window sampler, mainly suitable for collecting samples of cohesive sediments above the water table; the 'flow through sampler' is a type of window sampler.

Whilst most of the above methods are more commonly employed in site investigations for civil engineering works than in groundwater investigations, various direct push and drive sampling techniques are used in the construction of shallow observation boreholes and in investigations of contaminated land (Section 5.6).

Rotary drilling. The recovery of undisturbed formation samples by rotary drilling is by coring using direct circulation (Section 5.2.1). Equipment available for coring is versatile and variable because of its importance to the oil and mineral exploration industries. Coring is carried out by means of a special drill bit assembly known as the core barrel (Figure 6.12). The core barrel is a hollow steel tube made up of four parts: head, tube, reaming shell and core bit. The head is threaded to be compatible with the drill stem. The tube and reamer are both hollow tubes but the reamer has raised, diamond-studded plates on its face to clean and ream out the borehole wall behind the bit. The core bit is a steel tube with the bottom rim studded with diamonds or tungsten carbide inserts. Between the bit and the reamer is an internal rim of steel ribs called the core lifter, which retains the core in the barrel when it is being lifted from the hole.

In hard-rock formations a single-tube core barrel is sometimes used. The rotation of the rim of the core bit cuts a circular plug or core of rock, which passes up into the core barrel. The drilling fluid passes down the drill stem, between the core and the barrel, to cool the bit. This single-tube core barrel is suitable only for massive, strong, uniform rock because the drilling fluid will wash out all but the most cohesive cores.

The most commonly used core barrels are double-tube types (Figure 6.12), as these provide much better core recovery than the single-tube type. These core barrels have an outer and inner tube: the outer tube rotates with the drill string whereas the inner tube is mounted on a swivel so that it does not rotate. The drilling fluid passes between the two tubes and therefore does not wash away the core; only the bottom of the core is in contact with the drilling fluid.

In a triple-tube core barrel a liner, usually split, is incorporated in the inner tube and the core is extracted inside the liner. The addition of this liner

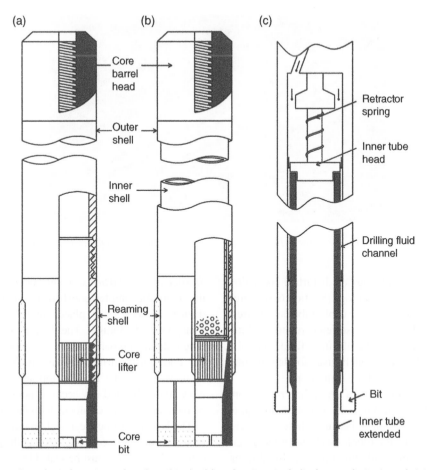

Figure 6.12 *Core barrels. (a) single-tube, (b) double-tube (internal discharge design) and (c) triple-tube (Mazier type)*

does not in itself increase core recovery; rather, it is designed to protect the core from the drilling fluid and to preserve it in good condition. An alternative to purpose-designed triple-tube core barrels is to add a plastic inner liner to a conventional double-tube core barrel.

In conventional coring, when the core barrel is full the entire drill string has to be removed from the hole in order to retrieve the core. With the alternative system of wire-line coring, the inner core barrel can be brought to the surface inside the drill pipe, thus avoiding the need to remove the drill string from the hole every time the barrel is filled (Figure 5.10).

The core is retrieved from the core barrel by setting the barrel on a clean rack and then removing the core bit, the core lifter and the reaming shell. The core then usually slides out if the barrel is tilted and tapped. If a plastic or split liner is used, then these can be pulled out with the core intact inside. In those cases where a core will not slide out, then the core has to be extruded, usually by a hydraulic ram to which the inner tube of the barrel can be attached.

Coring is relatively expensive compared with normal rotary drilling, but can provide valuable information for the hydrogeologist. A core gives unambiguous evidence of the lithology present at the sampling depth, and this will act as a control

Figure 6.13   *Storage of newly extracted cores in a core-box, landfill site, England. Photo by Bruce Misstear*

for comparison with logs of disturbed formation samples and geophysical logs (Section 6.4). Cores taken from double-tube or triple-tube core barrels using air flush will have suffered relatively little contamination from the drilling fluid, and the central part of the core can be used for determining the moisture content of the formation.

The diameter of cores of crystalline rocks from mineral exploration boreholes is commonly quite small – 20 to 50 mm – because these cores are taken principally for lithological and mineralogical identification, and any invasion by drilling fluid is of minor importance. Cores of sedimentary formations taken during groundwater investigations generally should be of larger diameter, preferably at least 75 mm. This larger diameter will give sufficient core material for laboratory tests on porosity and permeability.

Hollow-stem augers. The central part of the hollow-stem auger, as the name implies, is a hollow tube through which a core of sample can be taken using a thin-walled sampler pushed into the base of the hole beyond the auger bit. This method of sampling can only be used to investigate unconsolidated formations that are capable of being augered, but it can be used to take samples above and below the water table (the auger stem acts as casing to prevent the hole from collapsing; see Figure 5.20).

Storage of undisturbed samples. On retrieval from the open-tube sampler, core barrel or other sampling device, the cores are placed in a core box for transport and storage. The box is divided into longitudinal compartments wide enough to hold the core fairly firmly (Figure 6.13). The core sections have to be labelled top and bottom with relevant depths, and any sections not cored should be indicated by spacer blocks with their depths.

The boxes must be clean and, in cases where the moisture content is to be measured or the pore-water extracted, the cores must be sealed in sleeves of polythene to avoid evaporation. Cores of unconsolidated sediment should be extruded and kept in sleeves to keep them intact. Samples which are going to have their pore-waters analysed for organic contaminants must be chilled or frozen immediately to avoid biodegradation of the contaminants. Samples for microbiological examination similarly must be kept in sterile sleeves of polythene and stored in the dark at around 4 °C to avoid microbial growth (Section 8.5.3).

6.3   Description and analysis of drilling samples

During the drilling process, samples will have been collected at regular intervals. These may be, as we have seen, disturbed or 'bulk' samples, or they may be more or less undisturbed. Depending on the drilling and sampling process, the samples may or may not be representative of the aquifer's composition, texture or pore-water chemistry. An important initial step in sample description will be to consider these factors and make a preliminary assessment of how representative or undisturbed the sample is. In some cases, this will be obvious; in others, specialist techniques may be required. For example, coring of hard-rock aquifers has the potential to stress the rock fabric and a specialist may be able to distinguish primary, original fracturing from fracturing caused by the coring process.

If core is used to obtain pore-water for analysis, we need to be aware that any drilling fluid may invade the periphery of the core. However, with a core diameter of 75 mm or 100 mm, there is a good chance that the centre of the core will be undisturbed. In this case, only the centre of the core should be used for pore-water analysis to ensure, as far as possible, that contamination by drilling fluid is avoided. Additionally, some form of tracer can be added to the drilling/coring fluid to ascertain whether the sample's pore-water has been contaminated during sampling.

6.3.1   Characterizing disturbed samples

Bulk samples will typically be inspected in the field and (preferably) also in a laboratory in order to produce a likely interpretation of the geology penetrated by the well. On the basis of these samples, the limits of the aquifer horizons and aquitards will be identified. An assessment of the aquifer's likely hydrogeological properties may also be possible. The senses of sight and touch will be paramount. The use of standard procedures for sample description, such as those published by the British Standards Institution (2010a) and the American Society for Testing of Materials (ASTM, 2009b) is encouraged.

Samples taken at regular intervals should ideally be placed in a sample tray (Figure 6.8). This enables adjacent samples to be compared and small shifts in colour or texture to be identified. The

experienced geologist may, on the basis of the sample tray, be able immediately to identify stratigraphic units. While this is a hugely useful skill, it is important to distinguish such *interpretation* from *observation*. In other words, any geologist should record the basic raw data as well as his/her interpretation of it on any log. His or her interpretation *may* be wrong and it is important for a third party (or, indeed, for subsequent generations) to be able to check and, if necessary, reinterpret the data. The basic data to be recorded from bulk samples are listed in Table 6.1, and charts to aid in sample description are included in Figures 6.14 to 6.16.

6.3.2   Characterization of representative samples

The hydrogeologist or groundwater engineer will often select specific horizons (typically aquifer horizons where a well screen or gravel pack needs to be installed) for representative sampling. Disturbed samples may be adequately representative of grain-size distribution or lithology in some cases (typically with "dry" drilling methods such as cable-tool percussion, shell-and-auger or flight auger), provided sufficient sample can be recovered. Otherwise, separate representative samples will need to be taken, typically using a sample tube or core apparatus. On recovery, representative samples need to be transferred to sampling containers, which will typically either be robust bags or glass/plastic containers, which will be sealed until arrival at the laboratory. If the samples are to be subject to chemical analysis, the laboratory should be consulted to ensure that sample packing materials are sufficiently inert so as not to react with the sample.

Representative samples are often analysed for grain size distribution. The most common method of doing this is to pass an air-dried, representative sample of sediment through a stack of sieve trays. This is typically performed by selecting between five and eight differing sieve sizes, the coarsest mesh being selected such that it will not retain more than 20% of the sample. Large clasts (e.g., greater than 20 mm) are often removed from the

Table 6.1   *List of sample characteristics to be evaluated during logging of drill samples*

Characteristic	Observations	Comparators/Indicators
Colour	Be aware that subtle gradations can be important. For example, chalk is not always just 'white' – it can vary through buff to grey, and such distinctions have stratigraphic and hydrogeological significance. Also be aware that colour may depend on the moistness of the sample	As an aid to objectivity, a definitive colour chart (e.g., the Munsell Colour Chart) may be used as a comparator
Dryness	Is the sample dry, moist or wet/slurry?	
Odour	Is there a characteristic odour: for example, of rotting organic matter or of contamination such as fuels/solvents?	
Texture	Is the sample compact and dense, or light and friable? Is it granular or plastic? Can it be moulded or rolled?	
Hardness	Hardness of minerals or clasts. Can the fragment be scratched with a steel blade, or by a fingernail	Moh's Scale of Hardness. Quartz(ites) will not be scratched by a steel blade, whereas carbonates will
Degree of cementation	Degree of cementation of grains or any lithic clasts. What is mineralogy of cement?	Dilute hydrochloric acid may reveal carbonate cements (see below)
Characteristic mineralogy or lithology of clasts	Are grains visibly of quartz or of feldspar? Are rock fragments ('lithics') present and recognisable? Are flints present in Chalk samples? What is the mineralogy or petrology of the clasts: are they limestone or sandstone? Can the proportions of different minerals/lithic fragments be identified?	See Figure 6.14. Dilute hydrochloric acid will effervesce on contact with carbonates such as limestone, or with sandstones with carbonate cement. Pure quartzites will not effervesce with acid
Dominant grain size	See Table 6.2	Visible grains can be compared with a comparator diagram, grain sample card or grain sample set. Essentially, if grains can be seen by the naked eye, they are sand grade or coarser. Further guidance on distinguishing sands, silts and clays is provided in Table 6.3
Degree of sorting	A 'sand' will seldom be purely a sand. It will have a mixture of grain sizes: it may for example be a 'silty, slightly clayey fine-medium sand, with occasional coarser fragments, up to gravel grade'	Degree of sorting (i.e., uniformity of grain size) can be estimated from Figure 6.15
Degree of rounding and sphericity	Tabular clasts might be expected in formations with prominent thinly spaced bedding or cleavage planes, such as shales or slates	See Figure 6.16

Table 6.2 *Grain size classification of sedimentary clasts. It is important to note that there are differing systems in operation throughout the world. This table shows the Wentworth Classification, commonly used in the United States (modified by the U.S. Geological Survey, and cited by Driscoll, 1986), and the Norwegian Geotechnical Institute (NGI) classification (itself a modification of the Atterberg scheme, and cited by Selmer-Olsen, 1980)*

Wentworth/USGS Classification			Grain size (mm)	Grain size (mm)	NGI (Modified Atterberg) Classification	
Boulder			>25.6 cm	>20 cm		Coarse block
Cobble			6.4–25.6 cm	6–20 cm		Fine block
Gravel	Pebble	V. coarse gravel	3.2–6.4 cm	2–6 cm	Coarse gravel	
		Coarse gravel	1.6–3.2 cm	6–20 mm	Medium gravel	Gravel
		Medium gravel	8–16 mm	2–6 mm	Fine gravel	
		Fine gravel	4–8 mm			
	Granule	V. fine gravel	2–4 mm			
Sand		V. coarse sand	1–2 mm			Sand
		Coarse sand	0.5–1 mm	0.6–2 mm	Coarse sand	
		Medium sand	0.25–0.5 mm	0.2–0.6 mm	Medium sand	
		Fine sand	0.125–0.25 mm	0.06–0.2 mm	Fine sand	
		Very fine sand	0.063–0.125 mm			
Silt			0.004–0.063 mm	0.02–0.06 mm	Coarse silt	Silt
				0.006–0.02 mm	Medium silt	
				0.002–0.006 mm	Fine silt	
Clay			<0.004 mm	<0.002 mm		Clay

sample before sieving, as they can have a disproportionate influence on the grain size distribution by weight. The sieves are stacked on top of each other, with the coarsest mesh at the top, progressively finer meshes downwards and a residual collection pan (for fines) at the base. The sieve stack is shaken (often on a special vibrating machine) for at least five minutes. Thereafter the material retained in each mesh is weighed. Having checked that the sum of the masses of sediment in each sieve is the same as the original total sample, the grain size distribution can be plotted on a cumulative curve (Figure 6.17). In some cases, mechanical sieving of dry sediment may not suffice to define the silty or clayey fraction of a sediment, as clayey particles may tend to form clumps which will not pass through fine mesh sizes. To overcome this, wet sieving may be used, possibly in conjunction with some dispersing agent.

If a detailed description of grain size distribution in the clayey fraction is required, alternative methods must be employed. For example, the clayey fraction may be suspended in a solution of water (possibly with dispersing agent added) in 'velocity settling tubes'. Here, different grain sizes will settle from suspension at differing rates, the finest particles settling slowest. Some particles (typically < 1 µm) may not settle at all, if they are fine enough to be affected by Brownian motion to the extent that they remain in colloidal form.

The sieve and velocity settling tube methods have been described above, as they are conceptually easiest to understand, but alternative methods are available for grain size analysis, including:

- Photoanalysis: manual or automated analysis of grain sizes in an image.

Table 6.3 *Distinguishing between sands, silts and clays in the field (some observations based on Building Research Establishment, 1976)*

Grain Size	Field Tests
Sand	Individual grains visible to the naked eye (although fine sands only just visible). Feels gritty between fingers. Breaks down to individual grains when dry (if matrix not clayey)
Silt	Individual grains invisible to the naked eye, but may be visible under hand lens. When moist can be moulded but cannot be rolled out to form 'threads'. When moist, does not adhere to or smear out on skin as effectively as clay. When dry, will become powdery and readily brush off from skin. May feel slightly gritty between teeth
Clay	Individual grains invisible to the naked eye and under hand lens. Require microscopy to see grains. When moist, is plastic and can be moulded. Can also be rolled out to form "threads". When moist, has high adherence and can be smeared out on skin. When dry, hard to remove. No grittiness to skin or between teeth Five types: • very soft: exudes between fingers when squeezed; • soft: easily moulded with the fingers; • firm: moulded with strong finger pressure; • stiff: cannot be moulded by the fingers; • hard: brittle or tough.

- Laser diffraction: a laser beam shone through a suspension of sediment will be diffracted at differing angles depending on the effective diameter of the grains in suspension. The smaller the grain, the greater the diffraction angle. The method is especially suitable for particles in the sub-micrometre range.
- Sedimentation methods, such as the Imhoff cone (see Box 4.4). These are based on the principle of larger, heavier particles settling from a fluid quicker than small, light ones. A hydrometer may be applied ('Hydrometer method') to determine the change in density of the residual suspension, as various fractions settle out.
- Fluid elutriation methods: here, the sample is placed in a vertical tube containing an upwards-directed fluid flow (air or a liquid). As the upward fluid velocity increases and exceeds the particle settling velocity, progressively larger particles will be mobilized and eluted from the top of the cylinder.
- Acoustic (ultrasound attenuation) spectroscopy: here a spectrum of ultrasound is passed through a suspension of particles and the sound transmittance is plotted against frequency to obtain a spectrum, from which the distribution of particle masses can be estimated (McClements, 2006; Ali and Bandyopadhyay, 2013); however, additional data on the properties of the particles may be necessary for a good calibration.

It should be noted that all the above methods are based on subtly differing particle properties. For example, laser diffraction is determined essentially by particle size, sedimentation/elutriation by the ratio between weight and drag and acoustic attenuation by a number of factors including particle density.

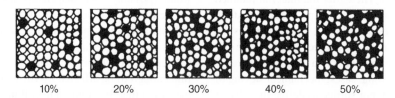

| 10% | 20% | 30% | 40% | 50% |

Figure 6.14 *Diagram to illustrate the appearance of different percentages of light and dark minerals in a grain sample. From Clark (1988)*

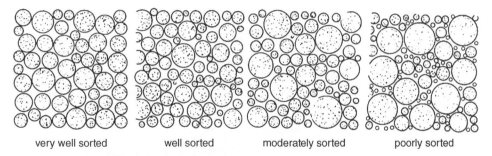

very well sorted   well sorted   moderately sorted   poorly sorted

Figure 6.15 *Diagram to illustrate differing degrees of sorting in a grain sample. Reproduced by permission from Tucker (1996). Copyright (1996) John Wiley & Sons Ltd*

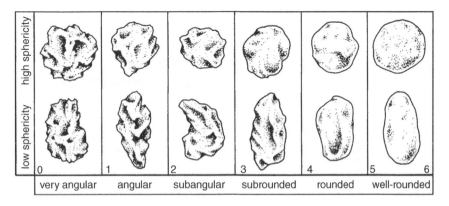

very angular   angular   subangular   subrounded   rounded   well-rounded

Figure 6.16 *Diagram to illustrate differing degrees of rounding and sphericity in a grain sample. Reproduced by permission from Tucker (1996). Copyright (1996) John Wiley & Sons Ltd*

Other analyses which may be carried out on representative samples are:

1. *Mineralogical composition.* This may be important if the aquifer contains, for example, sulphide minerals which may oxidize to release acid, sulphate and metals on dewatering and oxidation. It may also be important to identify buffering minerals such as carbonates if the aquifer is shallow and subject to acid rain or mine run-off. X-Ray Diffraction is the most common technique for semi-quantitatively determining the mineralogy present in a sample. However, other specific minerals such as sulphides or carbonates can be detected by pyrolysis and gas analysis, or by titration with acids.

2. *Determination of content of organic matter.* This may be important for assessing (i) whether shrinkage may occur on dewatering, or (ii) the potential for retardation of contaminants. We should take care to note whether results are cited as organic matter or as organic carbon. Analysis is usually by pyrolysis.

3. *Determination of cation exchange capacity, or sorption coefficient* for specific chemicals, which may be important for assessing the potential for retardation of contaminants. These determinations will typically be made by batch shaker or column experiments in the laboratory.

4. *Determination of geochemical composition,* which may be important if it is suspected that the aquifer matrix hosts a particular element that may be detrimental to water quality, such as arsenic or uranium. Bulk (total) geochemistry of rock materials is most readily performed

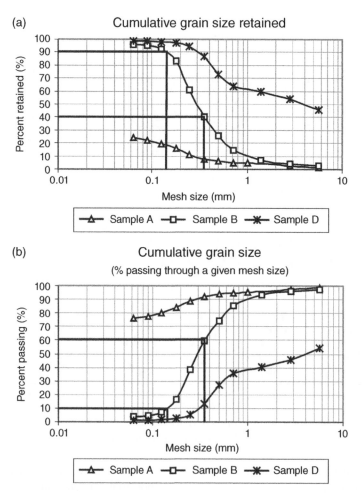

Figure 6.17 *Three examples of grain size distribution curves derived from wet sieving of sediments: sample (A) a boulder clay, (B) a glaciofluvial sand and (D) a glaciofluvial sandy gravel from Hertfordshire, UK. The grain size distribution is plotted as both (a) grain size retained and (b) grain size passing a given sieve mesh size. The D10 and D60 grain sizes are marked on both diagrams for sample B. Data from Banks (1984)*

by X-Ray Fluorescence (XRF), although this will not be suitable for all elements. The geochemistry of elements which can be readily dissolved from the aquifer materials is typically determined from extractions using water or acids of varying strength/temperature. Where extractive methods are used to determine the geochemistry of sediments, it is important to note that results will always depend on the extraction method. Thus, the following should always be noted and cited when presenting results: (i) the type and concentration of acid/solvent, (ii) the temperature of extraction, (iii) the duration of extraction and (iv) the extractant fluid/sample ratio. For example, in a recent study of salinity in sediments in Afghanistan, 20 g of the <2 mm fraction of the air-dried sediment was shaken for 1 hr at room temperature with 400 ml of distilled water, and then allowed to stand for 20 hours, before the supernatant water was extracted, filtered at 0.45 μm and analysed (Banks 2014c).

6.3.3   Characterization of undisturbed samples

Undisturbed samples will normally be taken using either a sample tube or a core barrel, in an attempt to preserve intact the fabric and intrinsic properties of the aquifer being sampled. The core sample will usually be protected in a core sleeve or tube and shipped carefully to a laboratory within the tube. In some cases, a core may be carefully cooled or even frozen to preserve the water chemistry within the core. Undisturbed samples, however carefully sampled, will usually be subject to at least three drawbacks:

- There is likely to be at least some disturbance, fracturing or contamination of the core during sampling. It is important for the analyst to be able to recognize such features in order to be able to make an assessment of the reliability of the results.
- A core sample comprises a tiny volume of the total aquifer system, and may thus be unrepresentative – especially in highly heterogeneous systems. For example, a permeability test carried out on a core of a crystalline rock formation or of a dual porosity aquifer (i.e., one with both intergranular and fracture permeability) will not be representative of the aquifer, as the volume of the core is not large enough to contain a representative fracture network.
- Most cores will be vertical sections through the aquifer and any determination of hydraulic conductivity will usually be of the formation's vertical hydraulic conductivity K_V. In most bedded aquifers, the laminar microstructure of bedding will mean that K_V will usually be significantly less than horizontal conductivity K_H, which is the parameter of most relevance for aquifer assessment.

Laboratory tests on undisturbed samples are typically of one of four types:

1. Geomechanical/geohydraulic analyses.
   - The sample's *dry bulk density* can be determined from the volume and mass of the undisturbed material following drying at 105 °C (British Standards Institution, 2014a).

- The *porosity* (*n*) can be deduced [Equation (6.1)] from the ratio of the dry bulk density (ρ_b) to the average density of the grains/material comprising the aquifer matrix (ρ_M):

$$n = 1 - \frac{\rho_b}{\rho_M} \qquad (6.1)$$

where ρ_M is typically 2.65 g cm^{-3} for quartz-dominated sediments and 2.7-2.8 g cm^{-3} for limestones and granite. Alternatively, porosity can be measured directly by saturating the dry core in water.

- Use of a porosimeter based on a non-wetting fluid such as mercury can also give information about the distribution of pore spaces in a sample (British Standards Institution, 2005a). The sample is saturated with mercury under gradually increasing applied pressure and mercury uptake is plotted against pressure. Progressively larger applied pressures are required to force the mercury into progressively smaller pore apertures. A broadly similar principle is applied by measuring progressive adsorption of low temperature gases (especially nitrogen) into a solid at increasing pressures (Sing, 1975).

- To determine *hydraulic conductivity*, the core is placed in a permeameter (Figure 6.18). A given head of water (*Δh*) is applied across a core of known area (*A*) and length (*L*), the flow of fluid through the core is measured and the permeability or hydraulic conductivity (*K*) is deduced from Darcy's Law (see Box 1.5). This type of test is called a *constant head test*. The alternative *falling head* method involves starting the test with an initial applied head and observing how this head decreases with time as the test progresses. The relevant equations are:

$$K = \frac{VL}{At\Delta h} = \frac{QL}{A\Delta h} \qquad \left(\text{constant head}\right) \quad (6.2a)$$

$$K = \frac{aL}{At}\ln\left(\frac{\Delta h_0}{\Delta h_t}\right) \qquad \left(\text{falling head}\right) \quad (6.2b)$$

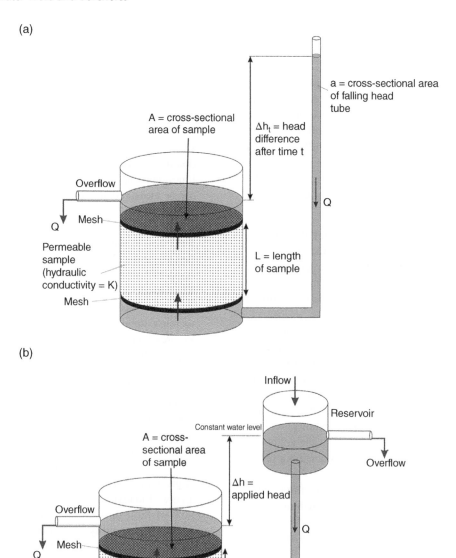

Figure 6.18   *The principle of a permeameter to determine the hydraulic conductivity of a formation sample or core. (a) The falling head permeameter; (b) the constant head permeameter. See Equations (6.2a,b)*

where *V* is the volume of water passing through the sample in time *t* (i.e., flow rate $Q = V/t$). For the falling head test, Δh_0 is the initial head difference across the sample, Δh_t is the head difference after time *t*, and *a* is the cross sectional area of the pipe or burette in which the falling head is measured. The permeameter may be a simple fixed wall permeameter (i.e., a rigid cylinder), whose ends may be compressed to ensure the sample is well compacted. Alternatively, a more complex, flexible wall permeameter can be used, where the entire sample can be pressurized to simulate the *in situ* effective stress (Daniel *et al.*, 1985). In a flexible wall apparatus, the risks of fluid leakage around the sides of the core or sample are much less. Permeability can also be determined using fluids other than water (including gases in a *gas permeameter*), using pressures rather than heads, and with a knowledge of the fluid's viscosity and density being required to convert to hydraulic conductivity.

- An alternative method to determine *hydraulic conductivity* employs a centrifuge to drive water through a core. The centrifuge can apply a driving force greater than gravity, and thus significant flows can be induced. Unlike a conventionally applied driving pressure (piston or high head of water), this does not necessarily imply that the sample becomes saturated, however. Thus, the method can be used to determine unsaturated hydraulic conductivity. The method is described by the American Society for Testing and Materials (2008).

2. Pore-water analyses. Pore-water from cores taken at various depths, or from various sections of core, can be extracted from the core by spinning in a centrifuge. The water can then be subjected to chemical or isotopic analysis and a hydrochemical profile of the pore-water composition in the unsaturated or saturated zone can be constructed. Tracers can be added to drilling fluids, which potentially allows any pore water contamination by such fluids to be identified. Pore water analysis has been used for, amongst other purposes:

- Assessing the progress of recharge water through an aquifer system (by dating the pore-water).
- Assessing the progress of nitrate-contaminated recharge through the unsaturated and saturated zones below agricultural areas (Foster *et al.*, 1985; Lawrence and Foster, 1986).
- Identifying zones of preferential contamination by industrial solvents (Misstear *et al.*, 1998a,b).

3. An extracted core can, of course, be subject to inspection and sub-sampling:

- Detailed *visual* examination to identify the aquifer fabric's bedding and microstructure, grain size variations, mineralogical variations and fracturing structure.
- Microscopic inspection of grain and pore size distributions in *petrological thin sections* can also be carried out to estimate porosity and, with a lower degree of certainty, hydraulic conductivity and specific yield (Younger, 1992).
- *Geophysical* techniques (tomographic techniques, radiometric profiling) can be used to map the core's internal structure, variations in clay content, porosity and fluid content.
- *Geochemical* sub-sampling or microprobing techniques can be applied to map, at a high level of resolution, the chemical composition of the core.

4. With the increasing popularity of geothermal and ground source heating and cooling schemes, representative samples can be analysed for thermal properties, such as thermal conductivity, volumetric heat capacity and thermal expansivity.

6.4   Downhole geophysical logging

Coring is an expensive procedure and will never be totally representative of *in situ* conditions. Geophysicists might thus argue that the best way of measuring *in situ* conditions in an aquifer or well is to place the measuring instruments *in* the aquifer. The large quantity of data that can be acquired by geophysical logging at relatively low cost means that it is a technique that should be considered for every newly drilled well.

Downhole geophysical logging involves the lowering of various sensors (housed in logging tools or *sondes*) into a well or borehole to map the physical properties of the borehole fluid, borehole construction or geological formations penetrated (sensors can even be mounted on a powered submarine to allow navigation into a horizontal adit). There are many reasons to carry out geophysical logging:

1. To collect information on the geological sequence penetrated by the well or borehole. If this is carried out in a mud-filled hole prior to installation of casing and well-screen, it can be invaluable in assisting with screen design and placement.
2. To determine the physical properties (porosity, bulk density, formation resistivity, fluid resistivity) of aquifer units.
3. To identify inflow horizons and fractures, and to quantify water inflows.
4. To assess and check on the integrity and construction of completed or existing wells.
5. To assess the changes in the well's fluid properties with depth.
6. To determine geothermal temperature gradient.
7. To correlate lithological and hydrogeological features between wells and boreholes.
8. To assist in diagnosing biofouling and other operational problems, thus aiding the maintenance of the well (Chapter 9).

The applications of geophysical logs, and the conditions under which they can be run, are summarized in Table 6.4. Further information on geophysical logging can be obtained from the literature (Robinson and Oliver, 1981; Digby, undated; Whittaker *et al.*, 1985; Schlumberger, 1989; Hurst *et al.*, 1992; Scott Keys, 1997; Hearst *et al.*, 2000; Keary *et al.*, 2002). The website of the Society of Petrophysicists and Well Log Analysts (SPWLA) is also highly informative. Naturally, during geophysical logging (which may take place

on an active drilling site, and may also involve combinations of electricity, water and highly stressed cables), the health and safety precautions summarized in Box 6.3 should be observed (see also Appendix 3 for guidance on the preparation of a project health and safety plan).

6.4.1   The geophysical logging package

Geophysical logging equipment typically consists of the following elements (Figures 6.19 to 6.21):

1. *One or more sondes*. A sonde is the downhole package of sensing equipment. It often looks like a slim stainless steel tube (Figure 6.21). Sondes may be able to be joined end to end to form a single long 'sonde stack' comprising several different instruments.
2. *A cable and winch*. The cable is typically a reinforced cable with an armoured screen that both supports the sonde and carries electrical signals between the sonde and the topside equipment, via the winch. The signals carried may be digital or analogue. The winch may be manually powered, but is usually electrically powered, with variable speed control and forward/reverse gears.
3. *A well-top pulley* to convey the sonde and cable smoothly downhole without abrasion. A depth-measuring device (such as an optical shaft encoder) will often be integrated into the pulley wheel, and a depth signal will be conveyed in a separate cable back to the topside console.
4. *A signal receiving and processing console*. In the simplest logging equipment this may merely be, for example, a Wheatstone Bridge to measure resistivity. Most consoles will perform some kind of data manipulation of an analogue or digital signal. In many modern logging systems, the console will often comprise an interface and a laptop computer, where signal manipulation can be controlled in a software environment. Such manipulation will typically involve (in very broad terms):
   • Association of the incoming data with a depth coordinate from the optical shaft encoder on the pulley or winch, and possibly

Table 6.4 *Applications of geophysical logs in wells and boreholes*

Type of log	Geophysical log	Cased/screened well	Uncased (open) well	Mud/water filled well	Dry section of well
Formation	Resistivity	No (except CCL)	Yes	Yes	No
	Self potential (SP)	No	Yes	Yes. Best in mud	No
	Electromagnetic induction	No (Yes in plastic)	Yes	Yes	Yes
	Sonic/Acoustic	No (except CBL)	Yes	Yes	Typically no
	Natural gamma	Yes	Yes	Yes	Yes
	Neutron	Yes	Yes	Yes	Yes
	Gamma-gamma	Yes	Yes	Yes	Yes
Structural	CCTV	Yes	Yes	Clean water only	Yes
	Caliper	Yes	Yes	Yes	Yes
	Casing collar locator (CCL)	Yes (metal)	No	Yes. Not used in mud	Yes
	Cement bond log (CBL)	Yes	No	Yes. Not used in mud	Yes
Fluid	Flow meter	Yes but of limited use	Yes	Water only	No
	Temperature	Yes but of limited use	Yes	Yes	No
	Differential temperature	Yes but of limited use	Yes	Yes	No
	Conductivity	Yes but of limited use	Yes	Yes	No
	Differential conductivity	Yes but of limited use	Yes	Yes	No

carrying out some kind of depth correction to account for the length of the sonde or the height of the well top above ground level.

- Averaging the signal over a user-specified distance or time interval (smoothing).
- Conversion of the raw signal to a meaningful output (millimetres of well diameter, counts per second of gamma radioactivity or ohm-m of resistivity). This will usually require a tool-specific calibration algorithm or a user-defined calibration.
- Output of the data to, for example, a paper scroll output (see Figure 6.22) and/or to a digital file on a laptop or digital storage medium.

5. *A vehicle* for transport of equipment. Basic lightweight logger packages are available that are designed to be carried on foot in a backpack. Most packages are, however, designed to be mounted and transported in a four-wheel drive vehicle.

6. *A power source*. This will often be a vehicle battery, which may need constant charging from the vehicle engine during running. For some tools a separate generator may be required.

A successful geophysical logging operation also requires experienced personnel. For most systems, two operators are recommended, one to operate

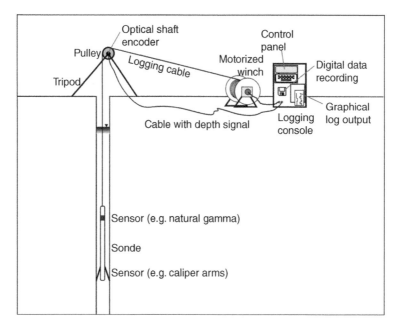

Figure 6.19   *Schematic diagram of a geophysical logging operation*

Figure 6.20   *Photograph of the geophysical logging of a Chalk well in Southern England. Note the chisel of the cable-tool percussion drilling rig. Note also the logging cable emerging from the front of the vehicle and running over a pulley into the well. Photo by David Banks*

(a)

(b)

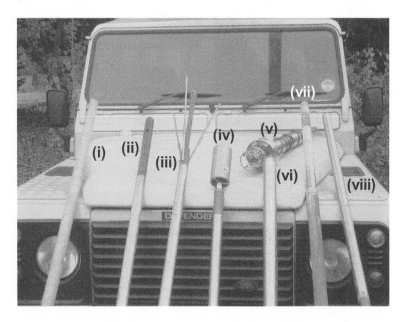

Figure 6.21   *(a) photograph of a geophysical logging vehicle, with interior processing consoles and sondes mounted on the roof. (b) photograph of an array of geophysical logging sondes: (i) electrical induction, (ii) fluid temperature/conductivity, (iii) gamma-caliper, (iv) impeller flowmeter, (v) CCTV, (vi) 1.25 litre depth sampler, (vii) normal resistivity, (viii) guard resistivity. Public domain material, provided by and reproduced with the permission of the Environment Agency of England & Wales (Thames Region)*

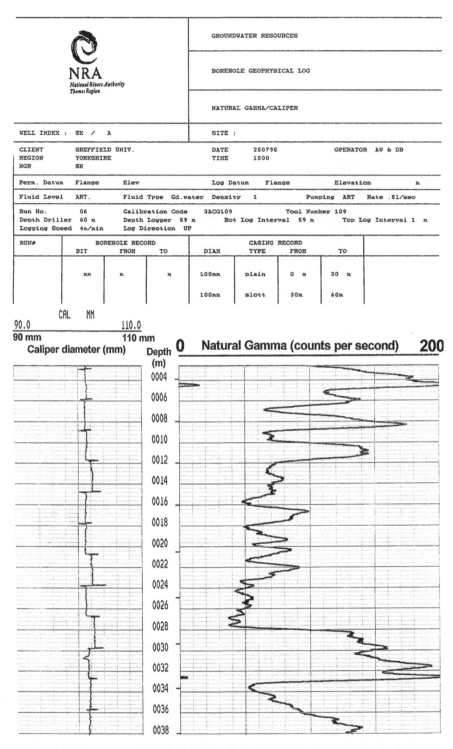

Figure 6.22 *Example of a paper scroll output from geophysical borehole logging of a borehole in the Coal Measures of Sheffield, South Yorkshire. The borehole has plain casing to 30 m and then slotted casing to full depth. The caliper log (left) shows the casing joints at regular intervals. The natural gamma log (right) shows alternating horizons of shales (high gamma signal) and sandstones/siltstones. Public domain information, provided by and reproduced with the permission of the Environment Agency of England & Wales (Thames Region)*

and monitor the console from the vehicle, the second to stand close to the well top and monitor the progress of the sonde. A single operator may all too easily become absorbed in the readout from the console, and fail to notice changes in tension in the cable (a sign of the sonde being trapped within the well), or even that the sonde has reached the surface. At unmanned well sites, two operators are also recommended for health and safety reasons (Box 6.3).

6.4.2 Organizing a geophysical logging mission

Before embarking on a geophysical logging mission, as much information should be acquired about the borehole or well and its current status as possible. This information will include the elements listed in Box 6.2 and a health and safety assessment (Box 6.3). Additionally, an assessment needs to be made of the risk posed to the logging sonde and cable by lowering it into the 'black hole' that is a well or borehole. The potential for a sonde to become entangled in any equipment or debris downhole is high.

Newly drilled, stable wells prior to pump installation represent the ideal condition for well logging. Modern wells or boreholes operated by responsible operators from which the pump and ancillary equipment have been removed might also be considered a relatively low-risk environment. Nevertheless, even here, experience suggests that such wells may contain old cables, lost pipes, cable clips, and even abandoned pumps! Thus, before logging commences, the *pre-logging checklist* (Box 6.2) should be carried out, and a closed circuit television (CCTV) inspection of the well should be considered.

Large diameter wells or boreholes can be geophysically logged with the pump or rising main still in place. Here it is imperative to know the pump depth and the diameter of the well, pump and rising main. It is also important to note that non-verticality in the well or rising main may lead to the position of the pump within the well section at depth not being the same as it appears at the surface. When logging below the pump it is preferable

to log through an access tube to ensure the sonde cable does not become entangled with the pump or its pipework and cables (see Figure 6.26 in Section 6.4.5).

In old boreholes or wells, there is always a strong likelihood that a previous operator may have discarded old equipment or other debris within the well. Geophysical logging should not be contemplated unless the well can be cleanly plumbed to its full depth and a CCTV survey has been carried out. We should be aware that even a television camera can become entangled with debris and so we need to be very careful about sending a CCTV sonde below a piece of visible debris in a well.

6.4.3 On arriving on site

On arriving on site, the first steps are to inspect the wellhead site, verify health and safety factors (Box 6.3), and carry out the pre-logging checklist (Box 6.2).

The order in which the geophysical logs will be run will depend on whether the well is newly-drilled and full of drilling mud, or completed and full of representative groundwater. In the former case, the main purpose of geophysical logging will be to locate aquifer and aquitard horizons, assess their properties and confirm the final well design, including the screen setting depths. Formation logs (electrical and radiometric logs) will typically be run at this stage.

In an existing groundwater-filled well, geophysical logs will typically be run in the following order:

1. CCTV log, if required, to ensure well is safe and unobstructed.
2. Fluid logs (temperature, conductivity, etc.). Ideally, these require an undisturbed column of water and thus are typically run in a downhole direction before any other sonde. If the well has been disturbed by previous plumbing or CCTV logging, it may be advisable to wait until the fluid column has re-stabilized.
3. Caliper logging, to check well construction and diameter. This is run in an uphole direction to

---

Box 6.2   Information to be collected prior to geophysically logging a well

When planning a geophysical logging mission, the following information should be collected:

1. Location of the well, any security clearance required and contact details of any on-site personnel.
2. Is the well top accessible to the logging vehicle? If not, can a cable realistically be run from the vehicle to the wellhead?
3. Is the well top aperture wide enough to accommodate the sonde, and low enough to accommodate the pulley and tripod?
4. What is the basic construction of the well/borehole? What is the depth, casing material and diameter? Does the diameter decrease with depth?
5. Is the well in use? What equipment is installed in the borehole? Is there a pump (what depth? What diameter?), a rising main (with or without protruding flanges? What diameter?), any electrical cables, pump rods?
6. What is the age of the well/borehole? Has it had more than one owner/operator? Can the current operator guarantee that the borehole is free from debris, rubbish or old equipment?
7. Are there known or suspected to be any issues of contamination at the site?

8. Has a health and safety assessment (Box 6.3) been carried out? How many operators are required?

Immediately before commencing the logging, the well top should be inspected and the following checks carried out:

9. Select the measuring datum. Will this be the well top or ground level? Relate all subsequent measurements back to this datum.
10. Establish when the well was last pumped/disturbed.
11. Measure borehole diameter at well top.
12. Note visible borehole construction and materials and any in-well installations (rising main, cables, etc.).
13. Inspect the upper part of the well, down to water level, using either (a) a powerful torch, or (b) a mirror to reflect a beam of sunlight down the well (highly effective).
14. Use an electrical dipper to measure water level below datum.
15. Use a plumb line (a metal weight on a graduated cable or tape) to confirm the well depth and to check the well is not obstructed. The plumb should reach the bottom smoothly, without snagging. Any snagging might suggest the presence of debris within the well.
16. If in doubt that the well is unobstructed, consider running a downhole CCTV survey.

---

ensure constant tension in the cable and better depth control.
4. Other sondes (gamma radiation and resistivity logs are the most common). These are usually run uphole.

A single sonde may contain combinations of sensors; for example, both the natural gamma and caliper tools (Figure 6.21b). It is becoming increasingly common for sondes to be 'stacked', such that a composite sonde can be compiled from various modules (gamma, caliper, fluid, electric) coupled together

in sequence, as required, usually with the fluid temperature/conductivity module lowermost in the stack.

6.4.4   Formation logs

An ever increasing array of different formation sondes is available for downhole application. Space does not permit a full description of all of these here. Discussion will be restricted to five of the most common types: the electrical resistivity and induction sondes, the passive (natural) gamma sonde, the acoustic sondes and the radioactive sondes.

Box 6.3   Health and safety at the wellhead

Before any geophysical logging, sampling or testing of a well, it is essential that a health and safety assessment is carried out as part of the overall health and safety plan for the well project (Appendix 3). The form of this assessment may be specified in national legislation and this book does not aim to provide detailed guidance on this topic. However, a health and safety assessment will usually include consideration of all or some of the following risk elements:

- risk of falling into open excavations, wellhead chambers or wells;
- exposure to possible contamination of soils or groundwater at site;
- heavy lifting;
- working at heights;
- working below unstable or dangerous machinery (including drilling rigs), buildings or natural features (quarry faces/cliffs);
- presence of inflammable or explosive materials (including the possibility of degassing of methane from excavations or wells);
- risk of electrocution from generators, pump cables or electrically powered logging tools;
- risk of working with rigs or other large conductors in the vicinity of high tension electricity cables (the rig does *not* need to be in contact with the cable to induce a high voltage in the rig);

- working machinery (turbines/generators) and pressurized equipment;
- working in confined/enclosed spaces;
- asphyxiation related to degassing of carbon dioxide from wells (particularly in carbonate strata and often as a result of a rapid fall in atmospheric pressure). This applies especially at wellheads in buildings, in sumps or chambers and in confined spaces;
- risk of drilling into electric cables, gas/oil/water pipes or other underground services;
- risks from working in the vicinity of highly tensioned drilling, logging or sampling cables.

As a result of the assessment, the health and safety plan will incorporate strategies for minimizing the risks arising from all the identified risk elements. For example, decisions will be made covering the number of operators required (there should be a minimum of two persons present on site), communication equipment, reporting protocols, working methods, safety and monitoring equipment. The plan should also describe the proposed response in the event of an accident or other emergency, including information on first aid facilities available at the site, the locations and contact details for the nearest hospitals, the procedures to follow in the event of a fire, etc.

Electrical resistivity logs. Electrical resistivity logs are run in fluid-filled boreholes where the fluid column electrically 'couples' the electrodes to the strata being measured. The fluid in question may be drilling fluid (mud), typically in uncased wells that have been newly drilled by rotary methods. Electrical resistivity logs may also be run in existing, completed wells, filled with natural formation water, but their use is restricted to unlined portions of the well. These types of sondes measure the electrical resistivity of the formations in the wall of the well or

borehole and are used to distinguish aquifer and aquitard horizons and to assess their properties.

The standard electrical resistivity logs [single point resistance, short normal, long normal, focused (laterolog) and self potential – see text below for descriptions] do not function in dry, cased or screened sections. All these logs will, however, identify the base of casing below water level.

Electrical resistivity is determined by a number of factors including: clay content, porosity, fluid column conductivity and pore water fluid conductivity, all of which are of interest in assessing aquifer properties. If appropriate geometric factors are taken into consideration, these intrinsic properties can often be deduced from the apparent resistivity measurements generated by electrical sondes (especially if used in conjunction with other tools such as the fluid conductivity and gamma sondes).

Often in hydrogeological studies, electrical tools are simply used to distinguish high resistivity formations (clay-poor, often good aquifers such as sandstones or limestones) from low resistivity formations (typically clays and other aquitards). In crystalline rock settings, the converse may apply: low resistivity may indicate a porous, fractured, weathered zone, whereas high resistivity may indicate intact rock.

Furthermore, in some aquifer units, stratigraphically characteristic electrical resistivity fingerprints may be identified that can allow wells to be stratigraphically correlated over distances of tens or even hundreds of kilometres. For example, the Middle (*New Pit*) Chalk of southern England has (amongst other features) a pair of low resistivity spikes which can be identified throughout the region (Figure 6.23).

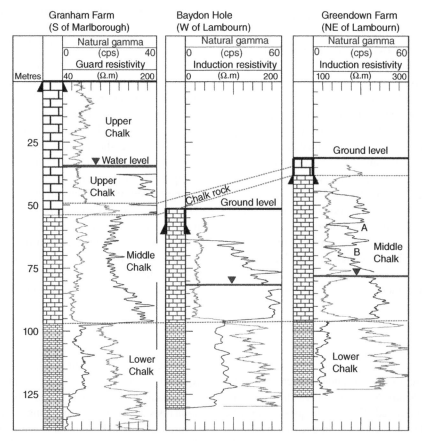

Figure 6.23 *Formation resistivity and natural gamma logs from three wells in the Chalk aquifer of the Berkshire/ Wiltshire Downs of southern England, aligned such that the transition from the Lower to Middle Chalk corresponds in all wells. Note the generally downwards increasing clay content of the Chalk (increasing gamma, decreasing resistivity), the glauconite content of the Chalk Rock hard-band (high gamma and high resistivity) and the two characteristic low-resistivity marly spikes (marked A and B) in the Middle Chalk. Public domain information, provided by and reproduced with the permission of the Environment Agency of England & Wales (Thames Region)*

Electrical resistivity logs are of several basic types, but all of them comprise an array of potential (voltage) and current electrodes. Most of the electrodes are downhole, but some configurations often include electrodes at the surface. Current is passed between two electrodes (AB), and potential (voltage) difference is measured between two electrodes (MN). An apparent resistivity (R_a) is then calculated (Box 6.4), which approximates to the true resistivity (R_t) of the formation. As well as the resistivity of the formation(s) along the current pathway, the apparent resistivity may be affected by the resistivity of the borehole fluid (the wider the well diameter, the greater the effect), the geometry of the well relative to the sonde and any resistance associated with the

---

Box 6.4   Quantitative interpretation of electrical logs

Electrical resistivity logging typically involves a measurement of the apparent electrical resistivity (R_a) of the water-saturated formation. This 'apparent' value needs to be corrected for borehole diameter and borehole fluid resistivity (and drilling fluid infiltration etc.) to yield a corrected value of the true formation resistivity (R_t). The formation resistivity in turn depends (Robinson and Oliver 1981) on:

- R_w – the resistivity of the formation water (this is the inverse of electrical conductivity, which is approximately proportional to water salinity or total dissolved solids). In groundwater-filled wells, this can be derived from fluid conductivity logs or water sampling.
- Degree of saturation (assume 100% below the water table).
- Porosity (n) and type of porosity.
- Resistivity of rock matrix.

Archie (1942) stated that the resistivity of a formation can be described by:

$$R_t = FR_w$$

where R_t is the 'true' formation resistivity and F is the *formation factor*, given by:

$$F = \frac{a}{n^m}$$

Here, a is a constant varying from 0.62 to 1 in unconsolidated sediments and hard rocks

respectively; m is also a constant varying from 2.2 (unconsolidated sediments) to 2 (hard rocks).

Electrical resistivity logs can be run in newly-drilled mud-filled wells, and in completed unlined portions of water-filled wells (and electromagnetic induction logs can be run in plastic-lined wells). In wells filled with formation water, R_t can be derived from R_a by applying corrections for (i) borehole diameter, derived from caliper logs, and (ii) fluid column resistivity derived from conductivity logs. If the fluid column consists of formation water of resistivity R_w from the aquifer of interest, then F can be calculated from:

$$F = \frac{R_t}{R_w}$$

and thus the aquifer porosity (n) can be derived.

In newly-drilled, mud-filled boreholes, R_w cannot be directly measured, so that F is estimated by:

$$F = \frac{R_{xo}}{R_{mf}}$$

where R_{xo} is the resistivity of the flushed zone; that is, the zone adjacent to the borehole wall, which has become saturated by mud filtrate (this can be measured using shallow-penetration sondes such as micro-resistivity) and R_{mf} is the resistivity of the mud filtrate itself (derived by knowing the mud resistivity, R_m). From the value of F, formation porosity can then be calculated.

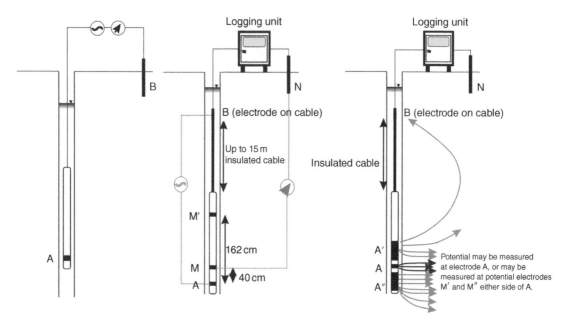

Figure 6.24 *Electrode configurations for the most common types of electrical resistivity sondes. From left to right: single point resistance; long/short normal; guard resistivity*

electrode/soil interface at the surface. The main resistivity log types (Figure 6.24) are:

1. *The single point resistance* (SPR) log. Here, resistance is measured between a single current/potential electrode on the sonde, and another current/potential electrode at the well top. The SPR log is difficult to use quantitatively, but has very good vertical resolution in narrow boreholes. In wider wells, its utility is more limited.
2. The *short normal* electrode configuration, where current passes between a current electrode A at the base of the sonde and a second electrode B, at the top of the sonde (*or*, sometimes, at the surface). Potential difference is measured between an electrode M on the sonde (at a distance of 16 in, or 406 mm, above current electrode A) and a second potential electrode N at the surface or sometimes higher on the sonde. This tool has relatively shallow penetration of the formation, but relatively good depth resolution. It is best used in moderately narrow wells.
3. The *long normal* electrode configuration, where the AM electrode spacing is 64 in, or

1626 mm. This has good penetration (due to wide electrode spacing) but poor depth resolution. It performs best in wide diameter wells.

4. The *focused electrical resistivity* (*laterolog* or *guard*) log. Here, two guard electrodes straddle the sonde's current electrode. Their electric field focuses the main electric field out into the formation in a narrow 'beam', thus achieving good penetration *and* good depth resolution.
5. The *microlog*. Here, the electrode array is very closely spaced (only a few cm apart) and mounted on a pad which is pressed by means of a sprung arm against the borehole wall. This achieves excellent vertical resolution. Penetration is, however, very shallow, and the log is affected by any filter cake on the borehole wall, or by the penetration of any drilling fluid (mud filtrate) into the formation.

There is a further common electrical log that does not measure resistivity but, rather, natural potential differences:

6. The *self potential* (SP) log passively measures natural potential differences set up in the earth's stratigraphy at junctions, for example between

mudstone horizons and sandstone horizons. These can be very difficult to interpret unambiguously, especially in freshwater situations. The SP log is best run in mud-filled holes, immediately following drilling.

Often a single tool may contain an array of electrodes allowing several configurations to be determined simultaneously in a single run – for example, SPR, short and long normal resistivity.

Electromagnetic induction log. The electromagnetic induction sonde involves passing a transient electrical signal through a *transmitting (induction) antenna.* This, in turn, induces a magnetic field in the adjacent geological formation and thence modest "eddy currents" in the formation. This secondary field then generates out-of-phase magnetic fields and thence electrical potentials in one or more receiver antennae, separated by a given distance(s) from the transmitting antenna (Moran and Kunz, 1962). Depending on the geometry of the antennae and borehole, and on the properties of the formation, the received signal can be processed (deconvoluted) to yield an apparent electrical resistivity of the formation (as in the pure electrical resistivity sondes described above). The great advantage of the electromagnetic induction sonde is that it can be used:

- in uncased dry sections of a borehole, and
- in plastic cased sections.

The technique does not work in metal-cased boreholes, nor does it perform especially well in highly saline (conductive) situations.

A specialized type of passive electromagnetic sonde is the *casing collar locator (CCL)*, which is specifically designed for steel-cased wells. Here, currents are induced in coils set within magnetic fields when the sonde passes any massive conducting object such as the chunk of metal that is the collar or joint between strings of casing.

Passive (natural) gamma log. All geological materials contain naturally radioactive nuclides that emit gamma (γ) radiation (in addition to alpha and beta radiation) upon decay. The most commonly occurring such nuclides are potassium-40 (^{40}K), and the uranium and thorium radioisotopes. ^{40}K occurs alongside non-radioactive potassium in many minerals including: most clays, alkali feldspars, many micas, glauconite and sylvite. The occurrence of natural γ radiation is usually regarded as a good indicator of the clay content of a sedimentary rock sequence (although it may also indicate other lithologies such as arkoses, glauconite-rich beds, and uranium/potassium-rich horizons).

The passive gamma sonde essentially comprises a γ-radiation detector such as a scintillation crystal (e.g., sodium iodide) which releases a flash of light when struck by a γ-photon. This is amplified by a photomultiplier into a detectable signal and is sent as an electrical or digital pulse to the logging console. The sonde is highly versatile, functioning in cased and uncased holes, both above and below the water table. The gamma sonde is run slowly up the hole (typically 2-4 m min^{-1}), as it needs a certain amount of time to aggregate enough counts to give a stable average. The slower the line speed, the better the resolution of a passive gamma log. The resulting log enables sandstones to be clearly distinguished from shales and clays. It is also able to reveal stratigraphic gradations: for example, the southern English Chalk (Figure 6.23) shows a decrease in clay (and ^{40}K) content up through the Lower Chalk and a significant decrease from the Lower to Middle Chalk. If the well diameter is known, it may be possible to use certain corrections to interpret gamma logs semi-quantitatively to provide a rough indication of the clay content of the lithology.

The natural gamma log also responds to other factors, which are important to recognize for the correct log interpretation:

1. The γ-radiation count depends on well diameter. Thus a sudden decrease in γ-signal can be due simply to an increase in well diameter.
2. γ-radiation is attenuated by water. Thus a sudden increase in γ-counts can be due to the sonde emerging from the water level into air.

3. γ-radiation is attenuated by casing. Thus a sudden decrease in γ-counts can be due to the sonde transiting from open hole to a cased section.

4. γ-radiation is generated by ^{40}K in the clay content of bentonite grouts. Thus, a sudden increase in γ-counts can be due to the sonde passing into a zone of grouted casing.

5. Some sandstones may be rich in glauconite, K-mica or K-feldspar, yielding high γ-signals. Conversely some clays may be relatively poor in potassium.

It is possible to obtain spectroscopic gamma sondes that will identify gamma photons of specific frequencies (or energies), allowing one to distinguish between radiation due to the differing source nuclides (^{40}K, U, Th).

Active neutron and gamma-gamma sondes. While the natural gamma sonde is merely a passive detector of natural radiation, the neutron and gamma-gamma sondes contain radioactive sources. They will thus be subject to stringent licensing and health and safety legislation in many countries. Indeed a number of nations will simply not permit their use in potable groundwater environments. Furthermore, great care (and, most likely, training and certification) will be required in handling such sondes.

The neutron sonde contains an emitter of fast neutrons (typically a pellet of ^{241}Am and Be) and a detector. Neutrons are slowed and backscattered by particles of a similar size: in practice, this usually means the hydrogen atoms in water molecules. Thus, the higher the water content of the formation, the higher the backscattering and the greater the slow neutron signal detected by the sonde. The water content of a saturated formation is, in most cases, closely related to its porosity.

In most aquifer formations, with appropriate matrix and borehole corrections, the neutron log signal can be converted to porosity directly. In clays and marls, the direct relationship breaks down, however, and in fractured rock, the log can at best only be used qualitatively.

Above the water table, the neutron log cannot readily be used quantitatively to determine porosity (as not all voids are water filled), but it can still be a useful qualitative tool to indicate water content. The main pitfall in the application of neutron logging is the potential presence of other hydrogen-bearing materials (for example, hydrocarbons, or water bound in minerals).

The gamma-gamma (γ-γ) log uses a similar concept. It contains a gamma emitter (typically ^{137}Cs or ^{60}Co), and a gamma detector to measure backscattered radiation. The rate at which the gamma radiation is backscattered is related to the bulk density of the materials in the borehole walls. Thus, the sonde can be calibrated to estimate formation bulk density. Well geometry needs to be taken into account when interpreting results. Some sondes have two or more detectors at different spacings to automatically calculate and compensate for well geometry.

Sonic (acoustic) logs. Sonic logging sondes are essentially the downhole analogue of seismic refraction surface geophysical techniques. The sonde normally comprises a sonically isolated acoustic transmitter and two receivers. The transmitter must not be too close to a receiver, otherwise it will be difficult to distinguish the signal passing through the rock of the borehole wall from a signal passing through the borehole fluid. Distances between the sender and nearest receiver are usually between 0.9 and 1.5 m, and the two receivers are usually about 0.3 m apart (Keary *et al.*, 2002). The further apart the sender and receivers, the better the penetration and the more reliable the average signal, but the poorer the resolution and the more susceptible to interference from, for example, caving of the borehole wall. The receivers typically detect the compressional p-wave, although s-wave (shear wave) detectors are available. The sonde measures the velocity of sound waves through the material of the borehole wall. Sonic logging can be performed for at least three purposes:

1. Stratigraphic identification. Different lithologies have different acoustic velocities (see Table 6.5).

Table 6.5 *Sonic matrix and fluid travel times (after Digby, undated). Reproduced by permission of Adrian J. Digby*

Material	Matrix velocity (m s^{-1})	Matrix travel time t$_m$ (μs m^{-1})	Fluid	Fluid velocity (m s^{-1})	Fluid travel time t$_f$ (μs m^{-1})
Mudstone	1750–5100	196–570	Fresh water	1460	685
Sandstone	5500–6000	167–182	Saline water	≤1675	≥600
Limestone	6400–7000	143–156			
Dolomite	7000	143			
Steel casing	5330	187			
PVC casing	2350	395			

2. Estimation of porosity (n) by:

$$n_{sonic} = \frac{\left(t - t_m\right)}{\left(t_f - t_m\right)} \qquad (6.3)$$

where t is the measured transit time (usually in μs m^{-1}), t_m is the transit time of the rock matrix component and t_f is the transit time of the interstitial fluid.

3. To check the integrity of a grout seal behind a casing. The *cement bond log* is a sonic log whose depth of penetration is adjusted such that it monitors the degree of energy loss along the casing. This is, in turn, related to the amount of casing in contact with cement grout.

6.4.5 Fluid logs

Fluid conductivity and temperature. The most common fluid logs are those which determine the temperature and electrical conductivity (EC) of the fluid at a given depth in a well using a small temperature sensor (e.g., a thermistor or resistance temperature detector – RTD) and a conductivity cell (Michalski, 1989). The sonde is typically run first down the borehole at a modest speed, in order to minimize disturbance of the fluid column and to allow time for the sensor to thermally equilibrate. Disturbance of the fluid column can be minimized and logging speed increased, by using fibre optic temperature sensors (Förster *et al.*, 1997): these utilize the temperature-dependence of Raman backscattering of laser light. In a non-pumping well, with no axial fluid flow, the temperature log will typically reveal an upper zone of seasonal perturbation in water temperature, below which the groundwater is not affected by seasonal fluctuation, and where the temperature increases by around 1 to 3° C per 100 m, reflecting the typical global geothermal gradient. The reader should, however, be aware that, in urban or industrial environments, downward heat leakage from houses or industrial processes can perturb the natural geothermal gradient for depths of several tens of metres (Banks, 2012; Westaway *et al.* 2015).

The conductivity log may or may not show an increase in conductivity with depth. Occasionally, a sharp increase at a given horizon may reflect the fact that the well has penetrated a deep-seated saline water body. Here we should recognize that, due to buoyancy effects and the possibility of flow in the well bore, the depth of saline water in the well *may* not exactly correspond to the position of the saline interface in the aquifer.

If the well is in an aquifer with significant vertical head gradients, natural flow may occur along the well axis (Figure 3.15). In the flowing section of the well, the fluid temperature will usually be rather constant (typically higher if the inflow is upwards from deep in the well, lower if downward from a shallow inflow horizon). The flowing section will be bounded by inflow and outflow horizons, which will normally show as sharp or diffuse "steps" in the log. Similarly, electrical conductivity will tend to be homogenized in regions of active axial flow. In heterogeneous, fissured or fractured aquifers, sharp jumps or spikes in the temperature or EC logs may represent inflows (or outflows) of groundwater from discrete horizons or fissures containing water of slightly differing temperature or conductivity to that in the well column.

It is worth noting that, as well as advective axial fluid flow, convection also can have an effect on temperature and conductivity logs. It has been shown that convection flow cells can develop in boreholes as narrow as 50 mm with geothermal gradients as low as 0.03 °C m^{-1} (Krige, 1939; Vroblesky *et al.*, 2006). This clearly has the potential to homogenize the fluid properties in the well (Sammel, 1968). Generally, the greater the temperature gradient, the less the viscosity and the wider the well, the more likely convection cells are to develop (Gretener, 1967). In a vertical tube, this occurs when the Rayleigh Number (*Ra*) exceeds a critical value:

$$Ra = \left(\frac{gr^4}{\mu\alpha}\right)\frac{\partial\rho}{\partial z} \qquad (6.4)$$

or, translated into terms of temperature-driven density differences:

$$Ra = \left(\frac{g\beta r^4}{v\alpha}\right)\frac{\partial T}{dz} \qquad (6.5)$$

where ρ is fluid density (kg m^{-3}), T is absolute temperature (K), g is the acceleration due to gravity (9.81 m s^{-2}), r is the tube radius (m), μ is dynamic viscosity (kg m^{-1} s^{-1}), v is kinematic viscosity $= \mu/\rho$ (m^2 s^{-1}), β is the fluid's isobaric thermal expansion coefficient (K^{-1}) and α is the fluid's thermal diffusivity (m^2 s^{-1}).

The critical *Ra* depends on the contrast between the thermal properties of the fluid and the borehole wall and varies between 69 (when the borehole wall/casing has a very small thermal conductivity) and 216 (when it has a very high conductivity – Cussler, 2009; Berthold, 2010). Krige (1939) developed this concept by combining this criterion with the adiabatic lapse rate, to assert a critical temperature gradient for free convection to commence:

$$\text{Critical gradient}\left(°\text{C m}^{-1}\right) = \frac{g\beta T}{c_p} + \frac{Ra_{crit}\alpha v}{g\beta r^4} \qquad (6.6)$$

where T is absolute temperature (K) and c_p is the fluid heat capacity at constant pressure (J kg^{-1} K^{-1}).

In newly drilled wells, elevated temperatures in cased sections may be due to the heat released by the exothermic setting reaction of many grouts (especially cement-based grouts). Also, a short zone of high conductivity may be encountered at the base of newly drilled or newly acidized wells, representing residual drilling slurry or acidization residue.

Fluid logs other than EC and temperature are available: for example, pH, Eh or ion specific electrodes can be attached to logging sondes to return in situ information about the variation of water chemistry with well depth. However, such electrodes may have a limited life, require frequent re-calibration and are more susceptible to drift, interference and poisoning than the simple EC and temperature sensors.

Fluid logs can be especially useful when run immediately before and during a pumping test. Significant inflows will often be identified as steps in the logs (see Figure 6.25) or by the differences between the non-pumping and pumping logs. Fluid logs often also incorporate *differential temperature* and *differential conductivity* traces (Table 6.4), which are simply the calculated rate of change of temperature/conductivity with depth (°C m^{-1} or µS cm^{-1} m^{-1}).

Flow velocity log. The other main type of fluid log is the flow velocity log. Several different sondes are available to measure uphole or downhole axial fluid flow velocity in wells, including:

- The *heat pulse log*. This is designed to be used statically in a well. A central electrical heating element is placed equidistantly between two closely-spaced temperature sensors. A short pulse of current is passed through the element, releasing a pulse of heat to the water column. This travels with fluid flow either up or down in the well and reaches either the uphole or downhole thermistor after a given time. From the time interval after the heat pulse is detected, the flow velocity can be calculated.

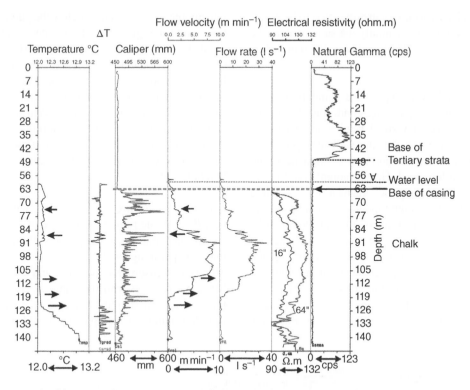

Figure 6.25 *Fluid logging of a non-pumping Chalk well (SU56/18B) at Bishop's Green pumping station, Southern England. The logged borehole was located c. 5m from an adjacent pumping well, with the pump located in the base of the well. This induced a downhole flow in the logged well, in which the flowmeter log was run in the uphole direction. Flow logs are corrected for line-speed. The water is entering the well through fissures below the base of the casing, in particular in the zone from approximately 74m to 91m (left hand arrows) and largely leaving via fissures between 108 and 119m (right hand arrows). The main fissures may also be seen on the caliper log. Note the relatively constant temperature down to 119m, confirming this interpretation, with a sharp rise below 120m (indicating warm water immobile below the flowing zone). The natural gamma log indicates clearly the base of Tertiary deposits overlying the Chalk at about 48m. ΔT is differential temperature – the calculated rate of change of temperature. Public domain information, provided by and reproduced with the permission of the Environment Agency of England & Wales (Thames Region)*

- The *impeller flowmeter*. This is simply a sensitive propeller which is positioned with its axis of rotation parallel with the well's axis. Flow passing through it causes it to rotate at a given rate, sending a proportionate signal back to the surface console. Good impellers can measure fluid flows down to 1-2 m min⁻¹. The impeller can be calibrated in a known static hole by winching it up and down through the water column at a range of known speeds. Calibration varies slightly with hole diameter. The impeller can be used as a static tool to measure flow velocity by positioning it at different depths in the hole and averaging the reading over a time interval. Alternatively, the impeller can be run at a constant line speed. In this case, one must remember to subtract (or add, depending on the logging direction) the line speed from the measured flow velocity to yield the true flow velocity.

- *Electromagnetic flowmeters*. These work on the principle that a conductor (in our case, water) passing through a magnetic field generates a small electrical current in proportion to its velocity. It is used in essentially the same way as the impeller but (at least in theory) requires less calibration.

Many wells will experience some fluid flow, even in their non-pumping condition, due to the likely presence of vertical head gradients in the aquifer and the fact that the well may provide a short circuit between horizons of high head and horizons of low head. Head gradients (and hence flows) will typically be downward in recharge areas and areas of high topography, and upward in discharge areas and areas of low topography (Box 1.3). Often (but not always) such natural flows are too slow to measure, and flow logs are thus most often employed in pumping wells or (ideally) in wells subject to artesian overflow. For example, if a well (Figure 6.26) of diameter 450 mm pumps at a rate of $20\,s^{-1}$, the average flow rate uphole a short distance below the pump chamber would be given by flow rate $(0.02\,m^3\,s^{-1})$ divided by area of well $(0.16\,m^2)$, or $0.13\,m\,s^{-1} = 7.5\,m\,min^{-1}$. If an impeller flowmeter log was run downhole at a line-speed of $4\,m\,min^{-1}$, the measured velocity below the pump chamber would be $11.5\,m\,min^{-1}$, reducing to $4\,m\,min^{-1}$ at the base of the hole, below the lowest inflow horizon. Step changes or gradual changes in flow velocity signify discrete or diffuse flow horizons in the borehole wall.

Logging in pumping wells or wells with pumps installed. While running any sonde in a well with an installed pump (and especially fluid or flow logs in pumping wells), great care must be exercised to ensure that the sonde and cable do not become entangled with the pump or rising main. Firstly, the well should be free of debris and any installed pumping equipment should be tidily

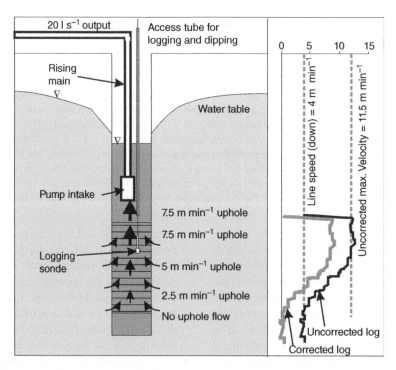

Figure 6.26   *Illustration of logging in a well of 450 mm diameter, pumped at $20\,s^{-1}$. This results in an uphole flow velocity of $7.5\,m\,min^{-1}$. The logging sonde is being run downhole at $4.0\,m\,min^{-1}$*

arranged in the well; for example, electrical cables should be attached to the rising main or enclosed in a dedicated duct, and not left floating loose in the well. The condition of the well should ideally be checked prior to logging using CCTV. In order to log above the pump, the diameter of the well must be great enough to allow free access for the tool without nearing the rising main. To log below the pump, an access or 'guide' pipe should be used (Figure 6.26) to at least 2 m below the pump.

6.4.6   Well construction logs

Caliper log. The caliper sonde typically has three sprung arms which push outwards against the borehole wall and track its contours. The extent of the arms is converted to a signal sent to the console and thereafter into a well diameter in mm. The caliper log can be run in a wet or dry well, but it must be empty of pump, rising main and any other downhole equipment. It allows the user to identify the length and diameter of the casing, which will typically appear as a straight line of fixed diameter, possibly with small blips representing casing joints or welds. Well screen may be distinguished from plain casing by small outward deviations at regular intervals representing slots. Open-hole sections will typically be represented by more irregular lines. Weaker horizons (e.g., shales) may be washed out during drilling and therefore be represented by zones of larger diameter. Fissures or fractures in limestones or crystalline rock may be visible as sharp outward deviations in the caliper trace. A selection of these features is visible in the sample logs shown in Figure 6.27.

CCTV log. The Closed Circuit Television (CCTV) sonde is typically used to inspect the well construction and to identify any debris, equipment or irregularity in the well. Most modern CCTV units allow colour inspection and employ light sensitive chips (CCDs) to return a digital image. Some modern cameras have a single, remotely-controlled, 3-axis, omni-directional lens. Older models may have a choice of two lenses: (i) forward view which provides a view down the axis of the well, and (ii) side view, which employs a rotatable mirror at 45° to give a view of the borehole wall. As well as inspecting the construction and integrity of the well's casing and filter (Figure 6.28), the CCTV may allow the identification of biofouling of well screens (often recognisable on colour CCTV as characteristic reddish brown slime). Some geological features might be identifiable due to contrasts in colour or texture (e.g., flint beds in Chalk), as will fractures or fissure horizons in limestones or crystalline rocks. CCTV has even been used to identify the presence of macro fauna, such as leeches and cave-dwelling shrimps in water wells in quasi-karstic Chalk aquifers of southern Britain (Waters & Banks, 1997).

ROV surveys. Remotely operated vehicles (ROVs) are becoming increasingly used to inspect larger diameter wells and shafts. These are, in essence, miniature propeller-driven submarines, connected to the surface operator by a cable for power and control. An ROV is typically equipped with a CCTV camera, appropriate lighting and, sometimes, with other sensors or even sampling equipment. An ROV is especially useful for inspecting tunnels or adits leading horizontally away from large-diameter hand-dug water wells (such as those that were constructed in the Chalk in London during the nineteenth century). The smaller ROVs available are around 0.4 × 0.25 × 0.25 m in dimension.

Other construction logs. Two other types of log that can be used to investigate the integrity of the well structure are the cement bond log and the casing collar locator. These are described in Section 6.4.4, under the headings 'Sonic logs' and 'Electromagnetic induction logs', respectively.

6.5   Downhole geophysical imaging

With increasing computer processing power, sondes have been developed which do not simply return a single signal to a console for a printout as

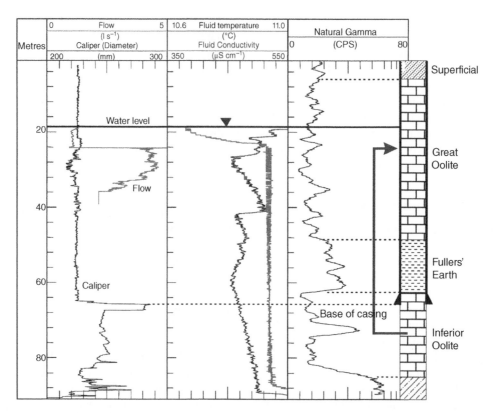

Figure 6.27 *Geophysical logs from a site in the Jurassic Limestones of the English Cotswolds. The well was constructed (and cased out) through the Great Oolite limestone aquifer and the Fullers' Earth clayey aquitard to abstract water from the underlying Inferior Oolite limestone. The fluid and flowmeter logs indicate, however, that even in non-pumping condition, water from the Inferior Oolite is flowing up the well to exit via a breach in the casing at around 24 m depth to the Great Oolite (see arrow on geological column). Public domain information, provided by and reproduced with the permission of the Environment Agency of England & Wales (Thames Region)*

a line trace. Imaging sondes now contain large arrays of sensors, which allow a processor to build up a 'picture' of the borehole or well wall. There are obviously many possible combinations of sensors, so our discussion here will be restricted to a few types only. In all the imaging techniques discussed below, planar features (such as fractures or lithological boundaries) will appear as lines on the flat (unwrapped) image (Figures 6.29 and 6.30). Horizontal features will appear as straight horizontal lines, while dipping features will appear as sine curves, whose degree of sinuosity increases as dip increases (thereby allowing dip to be calculated).

The *optical imager* is essentially a digital CCTV device coupled with an integral magnetometer and inclinometer (to orientate the sonde). This tool, in contrast to a CCTV that simply returns an instantaneous shot of a portion of well, can store data to build up a composite, oriented image of the borehole wall. An example is shown in Figure 6.29.

The *acoustic televiewer* is a sonic tool and is essentially a form of sonar. An ultrasound signal is sent out from a rotating transceiver as a series of pulses, and the return of their echoes from the borehole wall is detected by the transceiver. Intensive signal processing allows an image to be

(a)

(b)

(c)

114.20

(d)

140.66

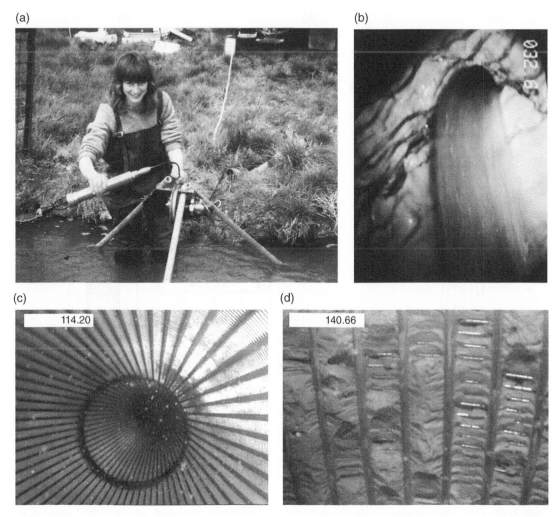

Figure 6.28   *A selection of CCTV images, showing (a) preparation for CCTV logging a submerged overflowing borehole in the Jurassic limestones of southern England, (b) a side view of a cascading Chalk fissure (approximately 4 to 5 cm in aperture) above the water level in a pumped well, (c) a forward view image of a continuous-slot well screen at 114.2 m depth in a Lower Greensand well, southern England, (d) clogged well-screen in the same borehole at 140.7 m depth. Public domain information, provided by and reproduced with the permission of the Environment Agency of England & Wales (Thames Region)*

built of the borehole wall that is particularly well-suited to detecting fracture planes. As with the optical imager, the fracture orientation (dip and strike) can readily be determined (Figure 6.30). Unlike the optical imager, the acoustic televiewer's use is not restricted to clean water wells, but can also be used in turbid water or mud-filled holes.

Electrical imaging tools are also available under names such as *dipmeter* or *formation microscanner* (FMS). These are typically variants of the electrical resistivity microsonde tool (see above), but instead of employing a single resistivity electrode micro-array (pad) pressed against the borehole wall, they employ a large number of pads, spaced around the (orientated) sonde. The electrodes measure the

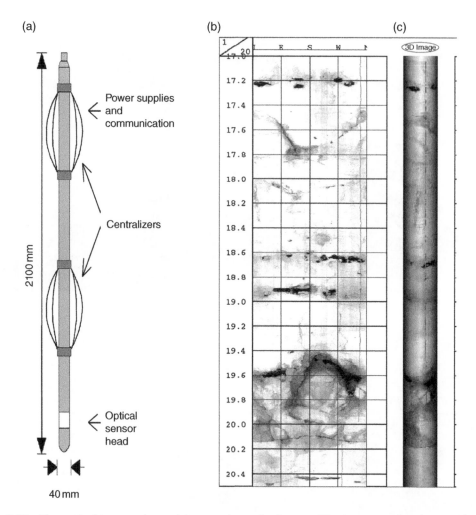

Figure 6.29   *The optical imager, shown (a) as a schematic diagram. The image can be shown either as an oriented (note points of compass at top of image) unwrapped image (b), in this case of a Chalk borehole, where fractures appear as sinusoidal lines, or as a conceptualized three-dimensional core (c). Reproduced by permission of European Geophysical Services, Shrewsbury, UK*

detailed resistivity profile at the point of contact with the borehole wall. Fractures and changes in lithology will be represented by resistivity anomalies. If a similar anomaly is detected by all the pads around the borehole wall, it is likely to be a continuous feature, such as a fracture or bedding plane, and can be identified as such on the final image. The dipmeter tools typically have fewer electrode arrays (four to eight) than the more sophisticated tools and produce only a partial image, which

suffices, nevertheless, to identify and calculate the dip of bedding and prominent fracture features.

6.6   Distributed (fibre-optic) temperature sensing (DTS)

A new technology is rapidly being developed which allows permanent (or temporary) real-time monitoring of temperature along the length of a

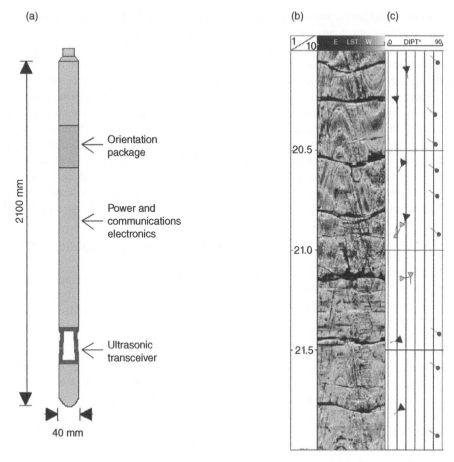

Figure 6.30 *A schematic diagram of an acoustic televiewer tool (a) and an example of an acoustic televiewer image, shown as an unwrapped image, where fractures appear as sinusoidal lines (b). The image has been interpreted in a fracture dip log (c) where fracture azimuth is shown by the direction of the tails of the symbols, and fracture dip (°) is shown by the lateral position of the symbol on the diagram (horizontal fractures on left, vertical on right). Reproduced by permission of European Geophysical Services, Shrewsbury, UK*

well, borehole, lake, watercourse or soil profile. It involves installing a fibre optic cable down the length of the borehole (the cable can be up to several km in length, so it allows monitoring of really deep geothermal boreholes).

A laser signal is sent down the fibre. At every point along the cable, a small fraction of the laser light is back-scattered and can be detected at the top of the cable. The distance to the point of back-scattering can be calculated from the two-way travel time of the light (laser) signal (Hurtig *et al.*,

1995, 1996, 1997). In fact, the laser light is back-scattered by several different mechanisms. The main backscattering component is termed the "Rayleigh" component and has the same wavelength as the main laser pulse. In addition, there are two Raman backscattering components, with wavelengths slightly greater and less than the incident wavelength. These two Raman components are known as "Stokes" and "Anti-Stokes". It turns out that the ratio between the "Stokes" and "Anti-Stokes" backscattering is temperature-dependent. Thus, measuring this ratio in the backscattered signal as a function of time allows the temperature profile along the entire length of the fibre to be deduced; the technical term for the technique is Optical Time Domain Reflectometry (OTDR). The cable itself is not hugely expensive, but the laser source and receiver can cost many thousands of pounds.

The applications of this will become immediately apparent to a hydrogeologist. Instead of winding a fluid temperature sonde up and down a borehole (Section 6.4.5), an installed optical fibre will immediately return a borehole fluid temperature profile, potentially allowing flowing fractures to be identified. The evolution of the temperature profile as pumping commences can be followed in real time. Yamano and Goto (2005) describe the installation of a fibre optic DTS in a deep geothermal reinjection well. Temperatures monitored during injection of cooler water clearly revealed a transmissive horizon at 540 m depth.

If hot or cold water is injected into an observation or reinjection borehole, its dispersion in the injection borehole can be monitored, as can its breakthrough in a nearby pumping well. Macfarlane *et al.* (2002) describe a thermal tracer experiment in a well doublet system in Kansas, United States (see Section 4.8). Hot water (73 °C) was introduced at $0.69 l s^{-1}$ into an injection well some 13.2 m from an abstraction well (pumped at $3.8 l s^{-1}$). The breakthrough of this warm water was monitored by fibre optic DTS systems installed in the wells.

As an alternative to using a one-off heat pulse flowmeter measurement (Section 6.4.5), a heat pulse can be generated (or injected as a slug of hot water) at a given point in a borehole and its progress up or down the borehole (and its dilution or removal by flowing fractures) tracked through time. This is exactly the approach taken by Leaf *et al.* (2012) who plotted temperature as coloured pixels in a depth vs. time diagram to graphically show the velocity of axial water movement along the axis of a borehole in a fractured rock aquifer. Locations of transmissive fractures were very apparent. Read *et al.* (2013) carried out thermal dilution tests and borehole-to-borehole thermal

tracer tests in fractured rock aquifers, employing similar graphical techniques.

Read *et al.* (2014) developed a technique whereby a fibre optic DTS was encased in a steel armouring and lowered down the axis of a drilled well. Electricity was passed through the armouring, which generated (via electrical resistance) a constant heat power per metre of cable. The equilibrium temperature of the armoured cable depends on the rate at which the generated heat is advected away from the cable by groundwater flow along the borehole. Thus, the temperature profile of the heated armoured cable allows the flow profile of the borehole to be deduced.

6.7 Preparing a composite well log

When drilling, sample logging and analysis and geophysical logging have been performed, a composite log can be compiled. A composite log usually comprises:

- a header, containing basic information on the well: name, grid reference, index number, date of drilling, depth, construction, lengths of casing, rest water level, and so on.
- the log itself, which is made up of a number of columns, where various observations or parameters are plotted against depth down the page.

The following columns of information should be shown on all composite logs:

- A graphical column, showing the lithology (using standardized symbols where possible), and construction details such as length of casing, position of screen, etc.
- A column containing a lithological description of the formation (based on raw observational data). For example: 'Sticky, somewhat clayey, very fine grained, pale grey coloured, soft LIMESTONE, without flints'.
- A column containing hydrological information, such as water strikes/inflows, rest water level at commencement of drilling every day, rest water level on completion of drilling, and so forth. Note that, if rest water level is measured at the

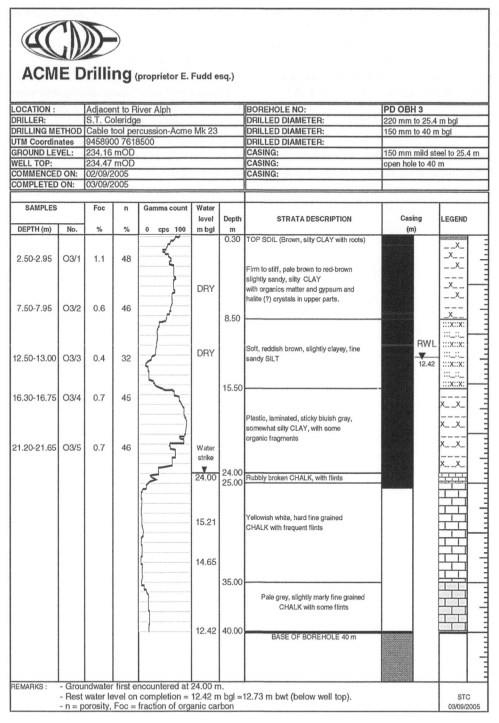

Figure 6.31   *Typical composite geological log, including a gamma log from the borehole and laboratory-determined porosity and organic carbon values from samples*

commencement of drilling every day and declines gradually as drilling progresses, this can be indicative of a downward vertical head gradient (and *vice versa*).

- A column containing a stratigraphic interpretation: for example, 'Lower (*Zig-Zag*) Chalk'.

The following optional columns may also be included:

- A column containing drilling observations, such as drilling penetration rate (Figure 6.1), colour/ texture of cuttings, loss of drilling fluid circulation, etc.
- Columns containing graphical representations of geophysical logs
- Columns containing geomechanical data (e.g., porosity, bulk density, permeability determinations).
- Columns containing results of mineralogical or geochemical analyses.

The composite log can take a number of formats, of which only one possible example is shown in Figure 6.31.

7

Well and Borehole Testing

What is the purpose of test pumping water wells? In this chapter we will examine five objectives for test pumping, and then consider in more detail how the data collected can be analysed to provide meaningful answers. The five objectives are:

1. To confirm the yield, efficiency and performance of the well.
2. To investigate water quality.
3. To assess whether the abstraction can be sustained in terms of yield and quality.
4. To identify potential environmental impacts.
5. To characterize the aquifer properties (such as transmissivity, hydraulic conductivity and storage).

Numerous national or regional standards for pumping tests exist. Of relevance to the United Kingdom and many other countries are those published jointly by the British Standards Institution and the International Standards Organization (BS ISO 14686: 2003, BS ISO 21413: 2005 and BS EN ISO 22282–4:2012). The original British Standard BS 6316: 1983, Code of practice for test pumping water wells (now superseded by BS ISO 14686: 2003), was influential in the development of the Australian Standard on test pumping (AS 2368–1990). In the United States, relevant standards include those published by the American Society for Testing and Materials (2010b and 2010c – that is, ASTM D4043–96(2010) E1 and ASTM D5092–04(2010)E1 especially) plus the National Ground Water Association standard on well construction (ANSI/NGWA-01–14).

7.1 Objectives of test pumping

7.1.1 Well performance

The first and foremost objective of test pumping, from the point of view of the target audience of this book, is to confirm that the well is performing as expected. Questions to be answered may include:

- Does the well provide the required yield?
- How does drawdown vary with pumping rate?
- What is the drawdown for the optimal yield and the peak yield and how can the energy efficiency of pumping be optimized?
- Is the well abstracting efficiently or are there large well losses resulting ultimately in excessive pumping costs (see Chapter 4)?
- Have any yield stimulation techniques been effective?
- Has the yield deteriorated over time due to clogging (chemical incrustation, biofouling) of well screen, gravel pack or aquifer formation (see Chapter 9)?

Water Wells and Boreholes, Second Edition. Bruce Misstear, David Banks and Lewis Clark.
© 2017 John Wiley & Sons Ltd. Published 2017 by John Wiley & Sons Ltd.

To answer such questions, some form of relatively short-term yield testing during drilling will often be carried out (Section 5.2.6), followed by a test which assesses the performance of the well under differing pumping conditions. This is typically achieved via a step test. For this purpose a large monitoring network is not required, and discharge and drawdown observations in the pumped well will usually suffice. However, if the efficiency of the screen or gravel pack needs to be assessed, pie-zometers should be placed just outside the well – within the gravel pack, for example (Box 3.3). If there is a major difference in pumped water level between well and gravel pack, there may be signifi-cant losses in energy as water flows into and within the well screen (see Section 9.2.1 and Figure 9.9).

If external observation boreholes are available, the drawdowns due to step testing may be able to be interpreted by standard analysis methods (for exam-ple, Thiem, Theis or Cooper-Jacob – Sections 7.4.4 and 7.4.5) to give values of aquifer parameters.

7.1.2 Water quality

Here, the questions to be addressed may include:

- Is the water quality acceptable for the purpose intended? Will it comply with legislation regard-ing drinking water (see European Commission, 1998; World Health Organization, 2011)?
- Does it vary with rainfall events or seasonally (Box 8.1)? If so, this may be an indicator of vul-nerability to contamination from the surface;
- Does it vary with pumping rate?
- Does it evolve with time of pumping?

The pumped well must be adequately clearance-pumped or 'purged' (Box 8.6) before any sampling commences, to remove any residues of drilling cuttings, drilling fluid or well stimulation residues. It has already been noted (Section 5.2.6) that water quality in freshly drilled wells in crystalline rock can be affected for some considerable time by the presence of fresh drilling cuttings and rock surfaces in the well.

Sampling can take place during step testing to ascertain whether water quality changes rapidly with yield. In this case, on-line physico-chemical readings and samples can be taken during and, especially, towards the end of each step. During long-term testing, samples may also be taken at regular intervals to identify any seasonal or long term changes in water quality. In Figure 7.1, it can be seen that water quality pumped at Verdal, Central Norway (Hilmo *et al.*, 1992) changed from a fresh calcium bicarbonate water to a rather saline sodium chloride water during the course of around six months' test pumping, due to leakage of pore water from overlying marine clays.

If the objective of the sampling is to identify seasonal changes in water chemistry, test pumping and/or systematic sampling ideally should extend throughout the year's main dry season and the main recharge event. If the objective of the sampling is to identify transient water quality effects (for example, low electrical conductivity, high colour, microbiological contamination asso-ciated with rainfall, snow melt or other recharge events), there will need to be either:

- continuous monitoring of an indicator parameter (such as low electrical conductivity or low pH as an indicator of rapid recharge of rainfall or snow melt), or
- a high density of sampling around the event in question.

Sampling from the pumped well will often be suf-ficient to address the above issues. However, in some circumstances – for example, if progressive saline intrusion or migration of contaminants is suspected – sampling from a network of observation boreholes *may* also be desirable.

The topic of groundwater quality sampling will not be dealt with further in this chapter, but will be considered in some detail in Chapter 8.

7.1.3 Sustainability

The term sustainability is here used to indicate whether a well can maintain a given yield and water quality in the long term, without leading to unac-ceptable environmental impact. Short-term testing *may* be adequate to predict the sustainability of an

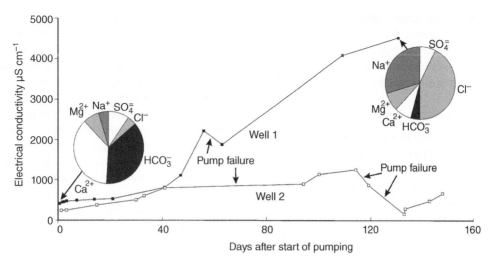

Figure 7.1 *Changes in chemistry during test pumping of two shallow drilled wells in glaciofluvial sands and gravel in Verdal, Central Norway (after Hilmo et al. 1992). The lines show increasing electrical conductivity (note the temporary decreases following pump failure). The pie diagrams show the major ion composition of the pumped water (in meq l⁻¹ proportions) at the start and end of the test. Reproduced by permission of Norges geologiske undersøkelse (Geological Survey of Norway)*

abstraction, especially if the abstraction is small and the aquifer is relatively uniform. However, unless one has good knowledge and experience of the aquifer in question, or unless one has an extraordinarily good conceptual model of the aquifer, it will seldom be possible to confirm the sustainability of a major abstraction on the basis of a short-term test. Even the long-term behaviour and yield of a modestly-yielding well in a marginal crystalline rock aquifer can be difficult to guarantee with any certainty. Despite the neat formulae and assurances of theoretical hydrogeologists, many aquifers are just too complex to predict from theoretical principles. In order to answer questions such as:

- Do the groundwater heads within the cone of depression caused by the abstraction stabilize, or continue to decline over time?
- How do the groundwater levels and/or well yields respond to recharge events?

some form of empirical long-term testing will be needed (the "suck it and see" approach). Whatever testing programme is adopted before commissioning, monitoring of the well or well field in operation is also essential (Chapter 9). It should also be recognized that, at a distance from the pumped abstraction, hydraulic effects can take a significant time to develop (the speed of propagation of an effect is broadly related to the aquifer's diffusivity – the ratio between transmissivity and storage T/S) – this will have implications for the duration of monitoring (including after pumping has ceased) and the certainty with which any event can unequivocally be ascribed to the pumping test (Bredehoeft, 2011).

The question of test duration is difficult to answer: for major abstractions it is tempting to recommend that a test should span at least the main dry season and the major recharge period. Long term testing can be an expensive commitment; pragmatically, longer testing will be justified if (a) the size of the abstraction and the capital investment are large, (b) the conceptual understanding is poor and (c) there is significant risk of serious adverse environmental impact (see below). If the water produced during testing can be utilized for water supply or an economic purpose during long-term testing, this can reduce the net economic costs.

A more comprehensive observation network is likely to be required for an assessment of the sustainability of larger abstractions. This may contain a number of elements in addition to observation boreholes and is discussed further in Section 7.2.5.

7.1.4   Environmental impacts

The question of negative environmental impact of an abstraction will often be related to that of sustainable yield (Section 7.1.3). A groundwater abstraction will alter flow patterns and will draw down groundwater heads within its cone of depression. This drawdown may adversely affect neighbouring abstractions or valued environmental features. It is therefore necessary to identify:

- Whether the abstraction affects other nearby wells/boreholes, spring flows, aflaj/qanats (Banks and Soldal, 2002)?
- Whether the abstraction affects groundwater-dependent features such as wetlands or baseflow-fed streams?

As with the question of sustainability, in the absence of a good conceptual model answers to these questions may only found after empirical long-term pumping (Bredehoeft, 2011; Kelly *et al.*, 2013).

7.1.5   Aquifer properties

We will usually also wish to know something about the characteristics of the aquifer. Obtaining estimates for transmissivity and storage may be important, especially for larger abstractions, as they enable us to answer questions like:

- How will the cone of depression evolve with time?
- How will water levels and flows in the aquifer respond to a new abstraction?
- How can wellhead protection areas be delineated (Section 2.8.2)?

Values of aquifer parameters can be placed in analytical or numerical models of the aquifer, which can then be used to make predictions that can answer some of these questions. Even for minor abstractions, which will impart a minimal stress on the aquifer, the derivation of values of aquifer parameters will provide inputs to the body of knowledge held by national geological surveys and regulatory authorities. This will assist in the management of the aquifer, to the benefit of all current and future users.

Aquifer properties can be estimated from short-term testing and from step tests, but are probably best ascertained from a medium-term constant rate drawdown and recovery test. Drawdown readings from the pumped well *may* suffice to derive values of transmissivity, but readings from one or more observation boreholes will be required to obtain trustworthy values of storage.

7.2   Planning a well pumping test

7.2.1   Before starting

A pumping test involves a considerable investment in equipment, time, non-productive water abstraction and power. To ensure that this investment is not wasted, a period of planning needs to take place before test-pumping commences. A number of issues need to be clarified, which may include those listed in Box 7.1.

Water features survey. Prior to commencing a pumping test on any well intended to become a significant permanent abstraction (and preferably prior to commencing drilling), the hydrogeologist or engineer should carry out a 'water features survey' of the area likely to be affected by the new abstraction. The aim of this is to ensure that any other abstractions (which might be derogated by the new abstraction) or valued environmental features are identified and monitored. It is, in effect, a concise hydrological environmental impact assessment. The water features survey should aim to identify:

1. Wetlands or similar habitats that may depend on the maintenance of a high groundwater levels.
2. Springs and streams/rivers that may depend on a significant component of groundwater-derived baseflow.

3. Lakes and ponds that may enjoy a degree of hydraulic continuity with the pumped aquifer.
4. Wells, boreholes, springs or aflaj (qanats) that are used by other abstractors or by livestock, or where rights to abstraction might exist (even if the sources are not actively in use).

The following types of information should be collected on the above sources:

- location and elevation (for this purpose, an accurate map and/or a GPS will be valuable);
- water flow or water level at time of survey;
- weather conditions at time of survey;

Box 7.1   Before commencing a pumping test

Before commencing a pumping test, the following issues need to be clarified:

1. What questions are we trying to answer with the pumping test?
2. Is there a conceptual model of the aquifer, to predict the likely impacts of the pumping test? How can this model be most efficiently tested and refined?
3. Have the abstraction wells been fully developed and clearance pumped?
4. What is the approximate water quality to be yielded by the wells?
5. From the preliminary pumping, what are the approximate yield and drawdown expected from the wells?
6. In the case of proposed abstraction wells, what are the desired operational pumping rates and regimes for the wells?
7. What permissions or consents are needed to carry out test pumping?
8. What neighbouring abstractions might be affected by the pumping well?
9. What hydraulic impacts might arise from the pumping well? Reduction in flows of springs, rivers, wetlands? Land subsidence?
10. Are there other potential (non-hydraulic) environmental impacts from the test: noise, fumes, incompatibility of discharge water with recipient?
11. Where will the pumped water be discharged so as not to interfere hydraulically with the test, and so as not to cause any risk of flooding or degradation of water quality?
12. How long does the pumping test need to last? What will be the regime (see above)?
13. When should the pumping test ideally take place?
14. How will the pumping test be analysed? Are any of the underlying theoretical assumptions violated? What data are required for analysis? What data frequency is required?
15. What will be the source of power for the pumping test – generators, mains electricity?
16. Is the site vulnerable to vandalism, flooding or freezing? Will the test pumping set-up need to be continuously manned?
17. What equipment resources are needed for the pumping test: data loggers, sampling equipment, sensors, pipeline, pumps, flow gauges? Is continuous monitoring required, or will sporadic measurements suffice? Which approach is most cost-effective?
18. What human resources are needed?
19. What analytical resources are needed to characterize water quality? Frequent/periodic sampling? Continuous in-line sensors?
20. What health and safety (H&S) issues are relevant to the pumping test for site operators and the general public (Box 6.3)? Is there a H&S plan (Appendix 3)?

- visual appearance/photographic record;
- construction details, depth, pump installation;
- apparent geological/hydrogeological setting;
- current use; approximate daily abstraction;
- name and address of land owner and/or user;
- is the abstraction licensed or does the owner/user enjoy other legally-protected rights to continued abstraction?
- is the habitat or wetland officially recognized by any environmental agency or nature conservancy body and does it enjoy protection under law?
- is the feature accessible and suitable for monitoring during the test pumping?
- what are the potential impacts on these water features from the proposed well test-pumping, according to our best current conceptual model?

Background data collection. As well as thorough planning, the collection of adequate background data is usually a prerequisite for valid interpretation of pumping tests. For example, if groundwater levels are not monitored prior to the pumping test, it may not be possible to tell whether a decline in water level during the test is (a) drawdown due to

abstraction, or (b) a purely natural recession due to lack of recharge in a dry spell. Figure 7.2 shows a groundwater level hydrograph for a pumping test in a limestone aquifer of southern England. If background water level monitoring (both before and after the test) had not been carried out, it would not have been possible to correct drawdown measurements for the natural recession which was superimposed on the pumping response.

Thus, before the pumping test commences, a period of background monitoring needs to take place. The duration of this background monitoring will depend both on climatic and practical considerations, but a period of two weeks before the test is considered a minimum by the British Standards Institution (2003), and a similar period afterwards is probably appropriate. Frequency of background monitoring will depend on the 'wavelength' of natural fluctuations in water level or stream and spring flow, but typically a relatively low monitoring frequency (daily measurements) will suffice. Background monitoring will be required of water levels in the wells to be pumped, and of the various components of the observation network enumerated in Section 7.2.5 (including rainfall and

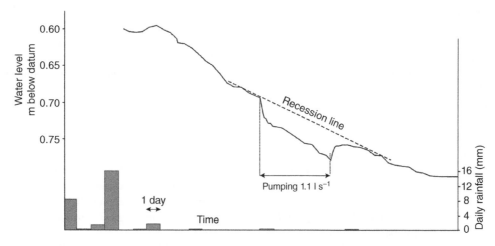

Figure 7.2 *A pumping test hydrograph from a well in the Corallian limestone aquifer in southern England. Note the natural groundwater recession taking place in the aquifer prior to and during the test. Brisk recessions are typical for such high transmissivity and low storage aquifers, but may not be as significant for other hydrogeological contexts. Public domain information, provided by and reproduced with the permission of the Environment Agency of England & Wales (Thames Region)*

possibly barometric pressure). Background monitoring of water quality may also be beneficial.

7.2.2   When to test pump

The best time to carry out a pumping test will depend on climate, practical considerations and the urgency with which the source needs to be commissioned. All other factors being equal, it is good to test the well or well field under 'worst case' conditions. For example, we might wish:

- to carry out well-performance (short term) testing during the dry season or season of low recharge;
- to initiate a long term test such that it straddles the latter part of the dry or low-recharge season, when natural groundwater levels are at their lowest (to evaluate sustainability), and also the period of main recharge (to identify any water quality issues associated with such an event, and to ascertain how the aquifer responds to recharge).

For example, in central Scandinavia and northern Canada, we might choose to carry out short term well-performance testing towards the end of winter (low recharge, due to frozen surface conditions precluding rainfall infiltration), and initiate a long-term test to encompass the recharge event associated with snow-melt and continuing into the summer period of low recharge. On the other hand, winter conditions may pose insurmountable practical problems with access or freezing of pipes, rendering this ideal plan unrealistic.

In general, it is best to avoid commencing pumping or recovery at the same time as a major rainfall event, as it can then be difficult to distinguish drawdown/recovery effects related to pumping from recharge effects related to rainfall. Short to medium-term weather forecasts can thus assist in planning the timing of test pumping.

As another example, in an emergency refugee camp situation, only a relatively brief and immediate period of test pumping may be defensible, due to the urgent need to commission the source. Given that this is unlikely to be adequate to allow a proper assessment of source sustainability, it would then be necessary to (i) monitor the well's performance, the aquifer drawdown and the groundwater quality during the first year of operation, and (ii) develop contingency plans for other solutions in case of source failure.

7.2.3   Consents and permissions

In many countries, some form of consent or licence needs to be obtained from the relevant environmental or water agency before drilling, testing and commissioning a well. This consent may specify the minimum requirements for test pumping – types of test, duration, monitoring, and collection of data. It may also require some form of water features survey (Section 7.2.1) to identify nearby abstractions, springs, streams or habitats that may be impacted by the pumping. The test pumping requirement will need to be fulfilled, documented, analysed and filed with the relevant agency before any groundwater supply can be licensed and commissioned.

Furthermore, it may also be necessary to obtain some kind of consent before any water from the clearance pumping or test pumping of a well can be discharged to an external recipient. This is potentially a difficult issue: the water cannot necessarily be returned to the aquifer via a nearby injection borehole, pit or soakaway, as this would be a form of artificial aquifer recharge and would hydraulically interfere with the test. Similarly, if the water is discharged to a nearby dry valley, gully, stream or river, care needs to be taken to ensure that the excess water would not cause additional infiltration from the watercourse to the aquifer, disturbing the pumping test. Suitable recipients for discharge of test pumping water might include:

- a sealed sewerage system removing the water from the well's catchment;
- a surface water feature outside the area of influence of the pumping test;
- a surface water feature that is not in hydraulic continuity with the pumped aquifer;
- a surface water feature whose existing flow or volume is so large that the addition of pumped

water would make only a negligible difference to water level, and therefore to interactions with the aquifer (as in the discharge to the River Thames for the Gatehampton test illustrated in Figure 7.3);

- a soakaway or injection well draining to a hydraulically separate aquifer.

It is also important to consider whether the quality of the pumped water is compatible with any existing recipient and its wildlife. For example:

- the pumped water may contain saline or slightly acidic residue from acidization of a limestone well;
- the pumped water may contain suspended particles, residual mud or drilling cuttings;
- the natural groundwater quality may be very different – more saline, poorer in dissolved oxygen or richer in, for example, iron or hydrogen sulphide – from the surface water, posing a risk to aquatic species.

Before discharging pumped water from a well to the natural environment, consent from the relevant environmental authority will usually need to be obtained. If discharging water to a public sewer, the permission of the sewerage utility will be required. If no legislation is in place controlling discharge of pumped water, an internal environmental impact assessment should be carried out.

7.2.4   Equipment

The equipment required for a pumping test will not be discussed in great detail here. However, the basic resources outlined in Box 7.2 will usually need to be in place.

Prior to test pumping in earnest, some form of *equipment test* may be carried out. This short period of pumping should aim to ensure that all equipment (especially pumping, pump-regulation and discharge monitoring equipment) is functioning as it

Figure 7.3 *Two large submersible pumps (and two rising mains) are installed for test pumping of this abstraction well in the Chalk aquifer at Gatehampton, southern England. The pumped water is discharged to the River Thames. Photo by David Banks*

Box 7.2   Equipment and resources required for test pumping

This will include (but may not be limited to):

- A reliable source of power: usually a generator, a motor-driven pump or mains electricity.
- Pump(s): positive displacement pumps will be preferable from the point of view of maintaining a constant pumping rate when the water level is falling, but electric submersible pumps have the advantage of convenience (Section 3.7).
- Means of regulating flow (valve-work, pump regulator).
- Discharge pipe.
- In some cases, in-line or down-hole continuous water quality monitoring sensors.
- Water sampling equipment: wellhead tap, equipment for on-site measurements (e.g., pH, temperature, conductivity, alkalinity), appropriate flasks, preservatives, filters, coolboxes, packaging, and so forth.
- Meteorological monitoring equipment (rain gauge) if local meteorological data are not readily available from other sources.

- Access tube or dip tubes in pumped well(s) for unimpeded dipping of water level or installation of water level or quality sensors. BSI (2003) recommends that the access tube should extend to at least 2 m below the pump intake.
- Water level monitoring equipment (typically a manual dipper tape or pressure transducer/data logger).
- Flow measuring equipment (flow meter, weir tank, orifice weir).
- Accurate watch or stopwatch.
- Accommodation for operatives.
- Prepared paper field sheets or computer spreadsheets for recording manual readings of time, pumping rate and drawdown.
- Communication (two-way radios or mobile phones – ideally hands-free types). These can be very valuable for communication between the pump-test supervisor, the person measuring discharge and the pump operator, for example.

---

should. This can also be used as a *dimension pumping test* to gain a first impression of drawdown for a given yield.

Water level monitoring.   There are three basic types of equipment for monitoring water levels during a pumping test (Banks, 2004; British Standards Institution, 2005b):

i. *An electrical dipping tape.* This comprises a graduated inelastic tape with two embedded and sheathed electrical wires, connected to a battery at the surface, and to a metallic weighted contact on the downhole end of the tape (Figure 7.4). On touching the water surface, a circuit is completed causing a buzzer to sound and/or a light to flash. The dipper should have an accuracy of at worst 10 mm and preferably 5 mm. By virtue of its simplicity, a

dipper should *always* be employed during a pumping test, sometimes as the primary means of collecting water level data but, in any case, as a means of regularly checking and calibrating data acquired by other mechanical or electronic means.

ii. *A float-operated chart recorder* (Figure 7.5). This comprises a graduated paper chart wrapped around a geared, rotating cylinder. An indented tape is suspended over a wheel connected to the cylinder. On one end of the tape is a float on the surface of the water, on the other a counterbalancing weight. As the water level rises and falls, the tape moves, rotating the chart. A pen is driven by a clockwork or electric motor across the chart, directly recording the water hydrograph on the graduated paper. Different gearings can result in different

Figure 7.4   *Measuring water level in an observation well in Dublin using electrical dipping tapes ('dippers').*
Photo by Bruce Misstear

scales on the chart, and in daily, weekly or monthly records. At their best, these chart recorders can be very reliable and robust. Such rotating chart recorders can also be attached to optical shaft encoders to generate an electronic signal that can be stored in a data logger or telemetered to a control centre.

iii. *An electronic sensor.* In the industrialized world, it is becoming increasingly common to use downhole electronic pressure sensors, connected to a data logger, to record groundwater levels. These sensors are emplaced at a known depth below the water surface (Figure 2.21). The sensor contains a transducer (a device that converts one kind of signal to another – i.e., pressure to electrical) to measure pressure – this might be a piezoresistive strain gauge, a piezoelectric crystal, ceramic or membrane. In many devices, the parameter recorded is total pressure – that is, the sum of water pressure and atmospheric pressure – above the sensor. Thus, to convert the signal to water level below well

top, several calculation steps must be applied (Schlumberger, 2014; Figure 7.6):

- The atmospheric pressure must be subtracted from the total pressure. This is most conveniently done by installing a separate barometric pressure sensor at the surface near the well head. (In some cases, the pressure sensor may be automatically compensated for atmospheric pressure by means of an open air line to the surface).

- The water pressure must be converted to water head, sometimes by making a density correction (important if the water is brackish or saline).

- The water head must be subtracted from the known logger depth below well top.

Modern downhole pressure sensors may also incorporate sensors for temperature, electrical conductivity or other parameters. The downhole sensor(s) may be connected by an electronic cable to a data logger at the surface, from which data may be telemetered, or sent

(a)

(b)

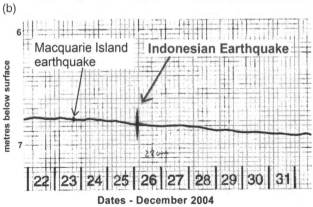

Dates - December 2004

Figure 7.5 (a) A float-operated clockwork chart recorder for groundwater level monitoring, mounted on top of an observation borehole in the Chalk aquifer at the Gatehampton site, southern England. A chart recorder such as this will have produced the hydrograph shown in Figure 7.26. Photo by David Banks. (b) A chart from a well at Knocktopher Manor, Republic of Ireland, showing a water level movement caused by the Indonesian earthquake of December 2004. Modified after Tedd et al. (2012), Quarterly Journal of Engineering Geology and Hydrogeology © Geological Society of London (2012)

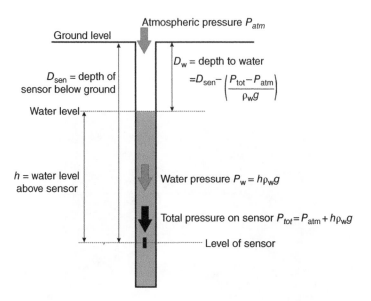

Figure 7.6   *The measurement of water level using a downhole pressure transducer. The measured pressure often needs to be corrected for atmospheric pressure (P_{atm}), converted from water pressure P_w to water head (h) using water density (ρ_w) and then subtracted from the sensor depth (D_{sen}) to yield a depth to water (D_w)*

via the internet, to a remote monitoring station. Alternatively, the downhole sensor capsule may have an inbuilt data logger sealed within (Figure 2.21). This means that the sensor unit can simply be suspended in the well on a steel wire and recovered to the surface every few weeks or months to download the data to a portable computer.

Pressure sensors may be prone to drift (and thus need to be periodically checked or calibrated). Most providers offer a variety of sensors, each tailored to a specific maximum range in water level – for example, to 1.5 m range or 100 m range. One should select a sensor with a range a little greater than the maximum anticipated variation in water level. Modern downhole pressure sensors have an accuracy of around 0.1–0.2% and a resolution of 0.02% (for example, a sensor rated for 10 m of water level range may have an accuracy of 1–2 cm and a resolution of 2 mm).

These types of monitoring equipment may be used in pumping wells, observation boreholes, lakes or rivers. In societies where labour is cheap, and computers are sparse, manual or simple mechanical means of water level monitoring are likely to be the most reliable and appropriate. Where labour is expensive and computing more commonplace, electronic solutions are more attractive, but, even here, periodic visits will be necessary to perform calibration and manual checking.

In pumping wells or in any surface water, all three types of water level monitoring equipment are typically placed in some form of stilling tube, to dampen the effects of turbulence due to pumping (wells) or to wind, eddies or even passing boats (surface water).

Monitoring of pumped discharge. Despite its deceptively simple-sounding nature, this task can be one of the most problematic, and more than one method of discharge measurement is advisable to lend confidence to the data. Small groundwater abstractions can be accurately (although not continuously) measured by the use of a bucket of known volume and a stopwatch (Figure 7.7a). For

(a)

(b)

Figure 7.7   *(a) The Norwegian hydrogeologist Steinar Skjeseth uses a watch and bucket of known capacity to determine the yield from a pumping test of a well. After Banks and Robins (2002) and reproduced by permission of Norges geologiske undersøkelse (Geological Survey of Norway). (b) A V-notch weir tank used during test-pumping in Malaysia. The height of water flowing over the weir can be related, with good accuracy, to the flow. The weir tank contains baffles to calm the discharge from the well. Photo reproduced by permission of David Ball*

larger abstractions, measurement of pumped discharge can be achieved by one of the following methods:

i. An in-line *flow meter*; this may be one of several types: (a) an impeller meter, whose rate of rotation is proportional to flow; (b) an ultrasonic flowmeter, based on the speed or Doppler effect of sound in flowing water; (c) an electromagnetic meter, based on the principle that a flowing conductor (water) in a magnetic field will generate a field of electrical potential difference; or (d) a venturi/orifice flow meter, based on the pressure differential across a constriction/orifice plate within the pipe. Most flow meters will only function adequately over a given range of flow and, in any case, the meter *must* be adequately calibrated prior to installation. It must also be installed correctly and, for many types, this implies that a certain length of straight pipe (of diameter compatible with the meter) needs to be installed before and after the meter. Several types of meter may be amenable to data logging and telemetry.

ii. A *weir tank* (Figure 7.7b). This is a rectangular steel tank (British Standards Institution, 2003), often resting above the ground on supports (horizontality is of prime importance). Pumped water enters stilling chambers at the head of the tank. From there, the water passes to the tank's main chamber, from which it should flow in laminar fashion over a (removable and replaceable) weir slot dimensioned for the pumping rate in question. For low pumping rates, v-notch weirs are preferable; for high rates, rectangular weir slots are used. For each type of weir slot, a calibration curve exists relating upstream head (in the main tank) to discharge. Provided the water level in the main tank can be measured accurately, the discharge can be determined with a high degree of precision. The water level in the main tank is often measured by a float-operated chart recorder, a pressure sensor, or by a manual hook gauge.

iii. A calibrated sharp-edged *orifice plate* (British Standards Institution, 2003) at the end of the discharge pipe.

iv. *Purdue trajectory method.* The trajectory of water exiting a horizontal pipe of internal diameter d at full bore can also be employed to estimate the discharge Q (Bos, 1989). If x is the horizontal distance the water travels (m), y is the vertical drop (m) and g is the acceleration due to gravity (m s^{-2}), then:

$$Q = \frac{1.1\pi}{4} d^2 \sqrt{\frac{gx^2}{2y}}$$
$$= 1.91 d^2 \sqrt{x^2 / y} \text{ in m}^3\text{s}^{-1} \tag{7.1}$$

v. *Vertical pipe water jet method.* For a pipe or artesian borehole of diameter d (m) discharging "fountain flow" vertically, the height of the water column (Δh in m) above the pipe rim is related to discharge (Q) by (Bos, 1989):

$$Q = 3.15 d^{1.99} \Delta h^{0.53} \text{ in m}^3\text{s}^{-1}, \text{if } \Delta h > 1.4d \tag{7.2}$$

$$Q = 5.47 d^{1.25} \Delta h^{1.35} \text{ in m}^3\text{s}^{-1}, \text{if } \Delta h < 0.37d \tag{7.3}$$

As drawdown develops in the pumped well following pump switch-on, the pump rate will tend to diminish (due to the pump's head-discharge relationship – see Section 3.7). Thus, excellent communication is required between the person monitoring discharge rate and the person regulating the pump in the early stages of a test, to ensure the constant discharge that is a prerequisite for much interpretation of test data.

7.2.5 The observation network

The observation network required for a pumping test will depend on the nature and objectives of the test, the size and importance of the abstraction, and the economic resources available. An observation network may include one or more of the elements in Table 7.1.

Table 7.1 *Elements of an observation network for different pumping test objectives*

Objective	Type of test	Elements of observation network
Well performance	Short term variable rate step testing	• Water levels in abstraction well(s)
Aquifer properties	Short to medium term constant rate testing and recovery	The above, and: • Water levels in purpose drilled observation boreholes at short-medium distance (typically 10s of metres) from abstraction well and/or: • Water levels in any convenient existing (preferably disused) well or borehole • In some cases, barometric pressure and tidal fluctuations
Long-term sustainability	Medium to long term constant rate testing	The above, and: • Water levels in purpose drilled or existing observation boreholes at medium-long distance (typically 100s to 1000s of metres) from abstraction well • Meteorological data, especially rainfall (if not available from public meteorological networks) • Water levels and flows in rivers, lakes, springs and streams that might interact with (provide recharge to, or obtain baseflow from) groundwater
Impact on environment and other abstractions	Medium to long term constant rate testing	The above, and: • Flows or water levels (and possibly water quality) in lakes, rivers, ponds, streams, springs, wetlands or bogs, identified by the *water features survey* and deemed to be potentially impacted by the planned abstraction • Water levels in dedicated observation wells/boreholes located at environmentally sensitive spring or wetland features • Visual inspection or photography of other environmentally sensitive features not deemed amenable to quantitative monitoring • Flows, water levels (and possibly water quality) in springs, aflaj (qanats) or other wells with abstraction rights, deemed to be at risk from the planned abstraction • Accurate surveying of ground levels (if abstraction-related ground subsidence is an issue)
Water quality	Any of the above	• Water quality in any of the above (especially if changes in water quality due to abstraction are suspected)

Aquifer testing: short- to medium-term constant rate tests. While estimates of transmissivity (T) can be derived purely from drawdown measurements in the abstraction well during short-term testing (step tests – see Section 7.3.2), better estimates will be obtained if drawdown data are also available from one or more dedicated observation boreholes situated at known distances from the pumping well. To derive an accurate value of storage (S), data from an observation borehole during a constant rate test are regarded as essential.

The design of observation boreholes is considered in Section 3.5. Their number and position will depend on the size of the abstraction and the geometry and properties of the aquifer. In a homogeneous and laterally isotropic aquifer, data from a single observation borehole are often adequate to derive values of T and S. The optimum distance of the observation borehole from the abstraction well needs to be such that a measurable degree of drawdown will be produced in the observation borehole. This distance can be calculated using, for example, Thiem's formula – discussed in Section 7.4.4 – provided some estimate can be made of T from the aquifer lithology and thickness, or from other data available in the region. The rate at which drawdown effects migrate will also depend on aquifer storage S. In confined aquifers the drawdown will migrate faster than in an unconfined situation. For a constant rate pumping test of several days' duration, the British Standards Institution BSI ISO 14686:2003 recommends that, for an unconfined aquifer, observation boreholes should be 25 to 60 m away, depending on transmissivity. For a confined, high transmissivity aquifer, observation boreholes could be 100 to 200 m away.

The British Standards Institution (2003) recommends an ideal of at least four observation boreholes spaced at geometrically increasing intervals, along two lines from the pumping well, at right-angles to each other. It is recognized, however, that one or two observation boreholes will be more economically realistic and will suffice in many cases.

More than one observation borehole can be used (a) to increase confidence in estimates of T and S; or (b) if the aquifer is laterally anisotropic or heterogeneous. Greater drawdowns in one direction (east-west, say) might suggest the existence of an anisotropic element in the aquifer: for example, a highly permeable buried channel deposit with an east-west orientation (Figure 7.8). Where hydrogeological barriers (such as impermeable faults) or recharge features (rivers or other surface water features) exist, observation boreholes located between the abstraction well and the fault/river will provide data that may allow the hydrogeological role of the feature to be clarified.

Observation boreholes can be purpose-drilled (Section 3.5). However, to save money, any pre-existing wells or boreholes (especially if disused) can be pressed into service for monitoring, provided they are at a suitable location and of suitable construction and depth. Caution needs to be

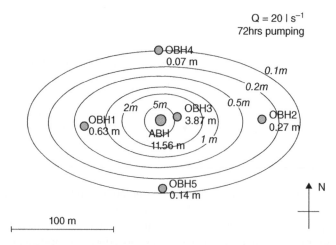

Figure 7.8 *Plan of a well field showing drawdown contours (in m) caused by pumping a single abstraction well (ABH) in an alluvial aquifer. The measured drawdowns in the five observation boreholes (OBH1–5) suggest the presence of aquifer anisotropy (for example, the possible existence of an east-west high transmissivity feature running through the area, perhaps related to a buried alluvial channel)*

exercised if these are still in use, as any pumping will disturb water level measurements.

In general, observation boreholes for the purpose of collecting data to characterize an aquifer's hydraulic properties should be of a similar depth and/or emplaced in the same aquifer horizon(s) as the pumping well. Exceptions to this guideline occur if the aquifer is multilayered and we wish to investigate responses in individual aquifer horizons or in adjacent aquifer units (British Standards Institution, 2003), or if we specifically need to investigate the aquifer's connection with a surface water feature.

Barometric and tidal effects. Monitoring of groundwater levels in wells can be subject to fluctuations due to a range of natural or man-made phenomena:

- Fluctuations in barometric pressure (Ferris *et al.*, 1962; Rasmussen and Crawford, 1997). If the atmospheric pressure in a well falls, and if that change in atmospheric pressure is not efficiently transmitted through overlying sediments or rocks to the aquifer, then the pressure in the aquifer will force the water slightly higher in the well (in looser terms, the drop in atmospheric pressure 'sucks' the water further up the well bore).
- In coastal wells, tidal effects (either direct infiltration or pressure loading) due to the sea (Turner *et al.*, 1996). A "Tidal Efficiency" (TE) can be defined as the ratio between the fluctuation in well water level to the fluctuation in tide water level (both corrected for density), within the range 0–100% (Ferris *et al.*, 1962).
- Earth tides. The land's surface rises and falls with the moon's and sun's gravity by up to several tens of cm. This causes decompression and compression of aquifers, and results in changes in water levels in boreholes, sometimes of a few cm (Ferris *et al.*, 1962; Bredehoeft, 1967). Confined aquifers and thick, low-porosity unconfined aquifers are most susceptible to earth tide effects. The effect was first observed in a coal mine at Duchcov, Bohemia (modern Czech Republic) by Klönne (1880).

- Loading and compression of aquifers due to passage of heavy vehicles (freight trains – King, 1892);
- Changes in river/canal water levels, for example, due to locking.

In the case of barometric and tidal effects, these should either be monitored directly or, where possible, publicly-available data on these phenomena should be collected from meteorological/coast-guard authorities.

A perfectly confined aquifer would exhibit the greatest barometric effects (100% "barometric efficiency"), while a perfectly open shallow unconfined aquifer would exhibit the lowest effects. Indeed, barometric efficiency has been used as an indicator of aquifer vulnerability to pollution (Hussein *et al.*, 2013) and even of the success of hydrofraccing (Burbey and Zhang, 2010). Barometric efficiency can be used to estimate aquifer specific storage and aquifer compressibility (Ferris *et al.*, 1962; Acworth and Brain, 2008). Many unconfined aquifers do show a barometric effect, however (Hare and Morse, 1997).

$$\text{Barometric efficiency}\left(\text{BE}\right)=\frac{\Delta h \rho_w g}{\Delta P_{atm}}\times100\% \quad (7.4)$$

where Δh is the well water level (m) change in response to a ΔP_{atm} change in atmospheric pressure (Pa), ρ_w is water density (kg m^{-3}) and g is acceleration due to gravity (m s^{-2}). For an ideal confined aquifer, the sum of barometric and tidal efficiencies should be 100% (i.e., a well in an aquifer overlain by a perfectly rigid confining layer should show a perfect barometric response, but no tidal loading effect; Ferris *et al.*, 1962).

Sustainability/long-term testing. The objective of longer-term testing is typically to ascertain whether the abstraction can be sustained. Such testing needs to identify whether the cone of depression of the well becomes unacceptably large or does not stabilize, whether the abstraction exceeds the available recharge to the aquifer unit, or whether the abstraction induces recharge from surface water features or overlying aquifer units. The positioning of an observation network for this purpose will depend on the prevailing conceptual model of the aquifer system: the observation network will be designed to test and challenge this conceptual model.

In two key papers for hydrogeologists, Theis (1940) and Bredehoeft *et al.* (1982) demonstrated that a cone of depression continues to expand until it has either induced a quantity of recharge or captured a quantity of groundwater discharge equal to the amount abstracted (or a combination thereof). Cones of depression around major abstractions can thus be very large and networks of observation boreholes are likely to be larger than for short-medium term testing, extending to distances of several km in some instances. With major abstractions that approach or exceed the available resources of an aquifer, a cone of depression may continue to expand laterally at some distance from the abstraction even when it has apparently stabilized in observation boreholes near to it. Especially in arid regions, cones of depression may continue to develop laterally even after an abstraction has ceased to pump, as groundwater from distant regions flows in to fill the cone of depression near the abstraction centre.

In addition, the observation network for assessment of the sustainability of large abstractions will likely include monitoring of (a) rainfall and other climatic data, (b) surface water features which may interact with the aquifer, (c) groundwater heads in the vicinity of surface water features, to ascertain the degree to which the aquifer is interacting with surface water, and (d) groundwater heads in laterally or vertically adjacent aquifer units.

Environmental impact/long-term testing. A water features survey carried out prior to drilling and testing may have identified other abstractions (springs, wells, boreholes, qanats) or environmentally sensitive hydrological features (ponds, lakes, springs, wetlands, baseflow-fed streams) that might be adversely impacted by drawdowns associated with a new planned abstraction. A long term test designed to demonstrate the magnitude of such an impact will probably have to undertake monitoring of flows or water levels (or both), and even sometimes water quality, directly in these features. Alternatively, observation boreholes located between the abstraction and the threatened feature *may* be adequate to deduce the magnitude of likely impact.

A period of background monitoring before and after the test (see Section 7.2.1) is essential to be able to identify impacts which can be ascribed to the test pumping. Also, when monitoring the impact of groundwater levels on a surface feature such as a wetland, careful thought needs to be given to the depth of any observation borehole.

In some cases, monitoring of the ground surface may also be necessary to identify any subsidence issues related to groundwater abstraction. Subsidence may be reversible, due to linear poro-elastic compression and expansion of an aquifer matrix. It may, however, be non-linear and may be irreversible, especially in clayey or organic rich strata whose physical and hydraulic properties are affected by compaction and dewatering. A number of well-known regional subsidence effects have been related to groundwater abstraction, including Shanghai (Banks, 2012a), Mexico City, Venice and several localities in the United States (Bawden *et al.*, 2003). Modelling approaches are available to simulate these various effects in relation to groundwater abstraction (Rivera *et al.*, 1991; Lebbe, 1995).

Vandal-proofing. During long dark nights of test pumping, huddled around a generator, an experienced groundwater engineer may be able to tell disturbing tales of chart recorders being used for shotgun practice in the UK, of pressure transducers being eaten by polar bears on Spitsbergen or of observation borehole casing being uprooted for 'domestic recycling' in Ethiopia. Observation boreholes should always be immediately equipped with lockable well caps (Figure 5.31), and more complex installations should be protected by robust and lockable chambers or cabinets. They, and the data they collect, represent an investment that, once lost, is difficult to replace. No cabinet is wholly proof against human or animal interference, however, and monitoring installations should therefore be made to look as uninteresting as possible, in order to discourage unwanted attention (albeit they do need to be findable by the hydrogeologist, even when vegetation has grown up or the ground is covered by snow!)

7.2.6   Recording of data

The primary data required for the interpretation of any pumping test are:

- discharge rate of pumping well (Q);
- water level in the abstraction well or observation borehole (h);
- time of observation – absolute time and time since start of test (t);
- location of observation, and distance from pumping well (r).

Before the test commences, a set of field observation sheets (one for each monitoring point) should be prepared for pumping test observers. These will be used to record manual determinations of the above data (and may be used to calibrate data collected continuously by other mechanical or electronic means). Figure 7.9 suggests the format of a pumping test field data sheet, with recommended intervals of observation for a constant rate test.

The variation in frequency of monitoring shown in Figure 7.9 reflects the fact that water levels in wells tend to fall rapidly soon after pump start-up, but later settle down to a much slower rate of decline, sometimes even stabilizing altogether. This is also why many methods of test pumping analysis require the time data to be plotted on a logarithmic scale. Pumping rates may also vary with drawdown, as the falling water level means the pump has to work harder to lift the same quantity of water, although in most tests the pumping rate will be regulated (either by means of an electrical frequency control or a wellhead valve) to achieve as constant a discharge as possible. Similarly, just as measurements will be taken most frequently soon after the start of pumping, so the frequency needs to be highest immediately after switch-off in a recovery test.

Datum. One common source of confusion when monitoring water levels is that the observer has not recorded where he/she is measuring water level *from* (i.e., the measuring datum). Is the water level below the well top (bwt) or below ground level (bgl)? It is important always to note the datum being used, and to be consistent (see the unambiguous statement of the location of the datum in Figure 7.9). If possible, the datum point should be physically marked on the wellhead with paint or an indelible marker. Also the height of the datum above or below ground level should be recorded. Note that wellheads may be modified during or after testing: well tops may be sawn off or placed below ground level in a chamber. A photographic record of the well top should be taken at the time of the pumping test, so that subsequent observers can ascertain whether the wellhead configuration has changed.

Often, the well will be accurately surveyed to the nearest cm. Again it is important to make sure that the surveyor knows what well datum is being used, and that he/she records the location of the point that has been surveyed.

Well numbering. In a large pumping test with many observation points, there is also huge scope for confusion as regards *which wells are being used for pumping and which for monitoring*. Before commencing the test, it is useful to issue each member of the test pump crew with a map unambiguously identifying each well with a unique number. If possible, the well top should be physically marked with its well number.

Metadata. In many societies and sectors of industry, quality assurance and documentation of a train of evidence are becoming increasingly important. It is therefore essential to not only produce good data, but also to demonstrate *how* those data were produced. A pumping test

Pumping Test:	**Sleepyville New Waterworks, Sept. 2005**			
Abstraction well	ABH 2	**Observation well**	OBH 6	
Nominal rate	30 l s⁻¹	**Obs. well depth**	80 m	
Type of test	72hr drawdown	**Well datum**	Casing flange top	
Abs. well depth	85 m	**Datum elevation**	32.61	**m above sea level**
Abstraction or	~~Recovery~~	**Datum above ground**	0.51	**m above/below**
Time of start	09:30 GMT	**Ground level elevation**	32.10	**m above sea level**
Date of start	2/9/05	**Initial water level**	9.60	**m below datum**
Time of stop	09:30 GMT	**Distance from abs. well**	70	**m**
Date of stop	5/9/05	**Obs well grid reference**	SU 7272 8150	
Test supervisor	W. Ham	**Type of flow metering**	In line impeller (Acme mark 1.52)	

Time after start	Water level	Discharge		Comment
	m below datum	**mm on hook gauge at weir tank**	**l s⁻¹**	*Samples taken etc.*
0 s	9.60			
10 s	9.60			
20 s	9.60			
30 s	9.60			
40 s	9.60			
50 s	9.60			
1 min	9.60		30.5	
every 10 s to				
2 min	9.60			
2½ min	9.60			
3 min	9.60		30.5	
3½ min	9.60			
every 30 s to				
5 min	9.60			
6 min	9.60		30.5	
every minute to				
15 min	9.61			
every 5 min to				
1 hr	9.65		29.4	
every 10 min to				
2 hrs	9.81		29.2	
every 30 min to				
5 hrs	10.25		29.8	
6 hrs	10.36			
7 hrs	10.42		29.6	
every hour to				
72 hrs	12.20		30.1	

Figure 7.9 *Illustration of a possible format for a field data sheet for recording pumping test information from an observation well. The observation intervals are similar to those suggested by Kruseman and de Ridder (Kruseman et al., 1990)*

supervisor will thus need to document so-called *metadata*: information which informs about the quality of the primary data. Such metadata may include (this is not a comprehensive list):

- the name and contact details of any observer or sampler;
- the type and serial number of any flowmeter, chart recorder, transducer;
- calibration curves/certificates for flowmeters, transducers, pH electrodes and so on;
- filtration, preservation and storage conditions of any water sample (see Chapter 8);
- chain of custody documentation for water samples (see Chapter 8);
- weather conditions when making observations;
- levels and locations of measuring datum points.

7.3 Types of pumping test

7.3.1 Dimension pumping

It has earlier been noted [Equation (1.21)] that the steady-state drawdown (s_w) in a well, pumping at rate Q, in a confined aquifer of transmissivity T, is given by:

$$s_w = \frac{Q}{2\pi T} \ln \frac{r_e}{r_w} \qquad (7.5)$$

or, using the Logan approximation [Equation (1.23)]:

$$s_w = \frac{1.22Q}{T} \qquad (7.6)$$

where r_e is the radius of influence of the pumping well and r_w is the radius of the pumping well. These equations have been derived on the assumptions that the aquifer is confined, infinite, homogeneous, isotropic, fully penetrated by the pumping well, of uniform thickness and with an initially horizontal potentiometric surface. The equations can also be regarded as useful in unconfined aquifers where the drawdown is small relative to the aquifer thickness.

Provided that the well is operating efficiently (i.e., head losses in the well screen and gravel pack,

and head losses due to turbulence or non-linear conditions near the pumping well, are low – see Section 4.5), it will be seen that there is an approximately proportional relationship between:

a. drawdown and discharge $s_w \propto Q$
b. transmissivity and specific capacity $T \propto \dfrac{Q}{s_w}$

Thus, in theory, a single measurement of 'equilibrium' drawdown for a given yield is sufficient to (a) make a first estimate of transmissivity of the aquifer, and (b) estimate the approximate steady-state drawdown for a given pumping rate. Therefore, prior to embarking on any pumping test, it is important (either during clearance pumping or during an *equipment test* or a dedicated *dimension pumping test* of at least two hours – Section 7.2.4) to gather some information on the drawdown for a given yield or range of yields. This will enable an estimate to be made of the maximum yield of the well that will result in the maximum acceptable drawdown (which *may* be governed by the elevation of the top of the well screen or main yielding fracture, but will be ultimately constrained by the level of the pump intake). With this information, we can design the *step test*.

7.3.2 The step test

The step test (Clark, 1977; Karami and Younger, 2002) is designed to provide measurements of drawdown in the pumping well for a range of yields. There will usually be at least four steps of rates Q_1 to Q_4, and it is often suggested that these should be at:

$$Q_1 = \frac{Q_{des}}{3}, Q_2 = \frac{2Q_{des}}{3}, Q_3 = Q_{des} \text{ and } Q_4 = \frac{4Q_{des}}{3}$$

where Q_{des} is the expected design pumping rate of the well under operational conditions. The steps thus encompass the range of realistic operating pumping rates of the well.

All steps should ideally be of the same duration. This will depend on site-specific conditions, but should be such that the drawdown has approximately stabilized by the end of the step. A step of 100 minutes or two hours is typical, however. Step tests progress from low to high discharge rate and

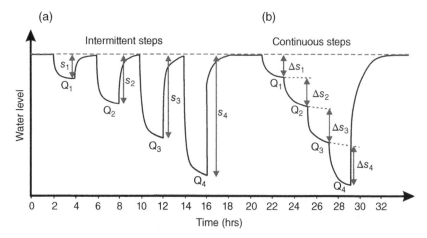

Figure 7.10 *Schematic diagrams of drawdown vs. time in a step test, illustrating the difference between (a) intermittent steps and (b) consecutive steps. After Banks and Robins (2002) and reproduced by permission of Norges geologiske undersøkelse (Geological Survey of Norway)*

are typically carried out in one of two different sequences:

i. *Intermittent steps*, where the drawdown is allowed to recover after each step. The period of recovery will be at least the same as the duration of the pumping step. A typical such test is shown in Figure 7.10(a), where each step of two hours is followed by a recovery of two hours, yielding a test regime of sixteen hours total. The advantage of this regime is ease of interpretation, the disadvantage is the duration (two working days).

ii. *Consecutive steps*, where each step is followed immediately by an increase in pumping rate to that of the next step. Thus, in Figure 7.10(b), four steps of two hours are followed by a recovery (total test period about 10 or 11 hours). The advantage of this test is that it can be completed in a shorter duration (one working day), but the results can be more difficult to interpret.

7.3.3 Medium to long-term (constant rate) test

The step test is normally followed by a period of recovery, such that the aquifer approximately returns to pre-pumping conditions. This recovery is likely to be at least one day, following step testing of eight hours' duration.

Typically, some form of constant rate testing will then follow. Constant rate testing will usually be designed to ascertain:

1. The hydraulic properties of the aquifer.
2. Whether the operational rate and drawdown can be sustained in a stable condition over a protracted period, or whether yield drops off (or drawdown increases continuously) with time.
3. Whether the abstraction could be responsible for reducing water availability in other wells, boreholes, aflaj, springs, or environmentally or economically significant hydrological water features such as lakes, streams or wetlands.
4. Whether water quality changes during the duration of the test.

A decline in performance (declining yield/increasing drawdown) of a well during a pumping test may be related to a number of factors:

• the cone of depression expanding to encounter the edge of the aquifer or a zone of lower aquifer transmissivity;
• high permeability layers or fractures being dewatered within the region of the pumping well;
• interference with other abstractions.

Ultimately, however, the failure of the yield and drawdown of an abstraction to approach a steady state condition will be due to the well's cone of depression either (a) failing to intercept an equal quantity of discharge that otherwise would have left the aquifer, or (b) failing to induce an equal quantity of additional recharge, or a combination of both (Theis, 1940; Bredehoeft *et al.*, 1982).

The duration of constant rate testing will depend on the size and importance of the well field development, the environmental sensitivity of the aquifer, whether there are sustainability issues to address, and the requirements of the regulatory authority. However, constant rate testing will usually last at least one day and commonly up to ten days, depending on discharge rate and the potential for delayed yield (Section 7.6.3). The BSI ISO 14686:2003 suggests that the minimum duration of a constant rate test should be 1 day for pumping rates up to $500 \, \text{m}^3 \, \text{d}^{-1}$, increasing to 4 days for pumping rates of $1000–3000 \, \text{m}^3 \, \text{d}^{-1}$, and to 10 days for rates above $5000 \, \text{m}^3 \, \text{d}^{-1}$. These durations are usually adequate to allow enough data to be collected for derivation of values for aquifer properties. Indeed, the first few hours of data will often be the most useful for this purpose and intensive data collection during this interval will be required.

Where there are complicated aquifer boundary conditions, issues of sustainability or environmental impact to address, a theoretical approach based on aquifer parameters may not be adequate, and a more empirical approach will be required. In such cases, and especially in the case of major public water supply or industrial abstractions, constant rate testing may continue over several weeks or months. However, during such periods, a relatively sparse data collection frequency may suffice. (Monitoring of well performance should continue during the operational lifetime of the well: this will confirm the reliable yield of the well, and also will help identify clogging or other problems affecting the well system performance – see Chapter 9).

The pump discharge rate for constant rate testing will be related to the anticipated operational production yield of the well. For example, if a well is designed to be pumped every day for 10 hours at $50 \, \text{l} \, \text{s}^{-1}$, we might choose firstly to carry out a constant rate test at $50 \, \text{l} \, \text{s}^{-1}$ of (at least) 10 hours duration, followed by a full recovery, to demonstrate that

the peak discharge can be sustained over 10 hours without unacceptable drawdown, *and* that full recovery takes place before the next pumping cycle starts 14 hours later. Thereafter, we might carry out a longer-term test (of several weeks/months duration) at a constant rate of $20.8 \, \text{l} \, \text{s}^{-1}$ (i.e., the continuous rate equivalent to pumping at $50 \, \text{l} \, \text{s}^{-1}$ for 10 hours a day), to demonstrate that the *average* yield can be sustained without adverse environmental impact.

7.3.4   Recovery test

When pumping ceases, the water level in a pumping well or observation borehole will usually begin to recover towards its rest water level. This rebound will occur quickly at first, and more slowly as the rest water level is approached. In fact, the recovery curve is often (under ideal conditions) the inverse of the drawdown curve under pumping conditions. It can be analysed using essentially the same methods (Section 7.4.6). The recovery curve can even be easier to analyse than the drawdown curve, as it will not be subject to the inevitable variations in pumping rate which often affect the early stages of a drawdown test.

One practical consideration is to be careful that, following pump switch off, water does not run back down the rising main, through the pump and back into the well. This can be avoided by ensuring that the pump string is fitted with a non-return valve at the foot of the rising main.

7.4   Analysis of test pumping data from single wells

7.4.1   Fundamentals

In carrying out a pumping test, we are primarily trying to ascertain if the well performs as expected, and in an efficient manner. Secondly, however, we are trying to find out about a complex system (an aquifer) by stressing it to see how it responds.

Before embarking on a pumping test, it is important to have a conceptual model of how the aquifer system behaves (and this conceptual model may be formalized as an analytical or numerical computer model). By carrying out a pumping test, this conceptual model can be tested and modified as necessary.

The pumping test stresses the aquifer by removing a given volume or given rate of groundwater. The aquifer will react to try and achieve a new equilibrium condition (if such a condition exists), and this will involve changes in heads and water levels in the aquifer with time (see Table 7.2 for terminology). The primary data to be collected during a pumping test relate to:

- the stress imposed: the pumping rate (Q) at a given location and time;
- the response induced: the changes in head (h) at various distances (r) from the pumping well at various times (t) after start of pumping.

7.4.2   The misuse of test pumping analysis

Various mathematical techniques (test pumping analyses) can be employed to interrogate the test data with the aim of learning more about the aquifer and, in particular, its properties such as transmissivity (T) and storage (S). These parameters are necessary for predicting aquifer behaviour, and thus for managing and protecting groundwater resources. However, real aquifers are complex, and mathematical models are only simplified versions of this reality. Thus, almost all techniques for analysing pumping tests make a lot of simplifying assumptions. Typically, most techniques assume that:

- The aquifer is confined.
- The well diameter is small, and well storage can be neglected.
- The aquifer is of infinite areal extent or at least significantly bigger than the area of the cone of depression.
- The aquifer is homogeneous, isotropic and of constant thickness.
- The potentiometric surface is approximately horizontal prior to the test.
- The pumping rate (Q) is constant.
- The pumped well fully penetrates the entire thickness of the aquifer.
- Groundwater flow is Darcian (laminar).
- Water is released from aquifer storage instantaneously with declining head.

There may additionally be other assumptions specific to each analytical method. These are listed

systematically in manuals such as that of Kruseman *et al.* (1990). However, real pumping tests do not satisfy most of these assumptions. The validity of the answers derived from the analyses depends on how closely the hydrogeological setting of the pumping test approximates the ideal situation. Before embarking on any test pumping analysis, it is essential to consider all the assumptions that apply to the analysis method. If the real situation deviates significantly from the ideal assumptions, we need to consider whether there might be other, more specialized methods of analysis which can tackle specific conditions. There are methods, for example, which allow for partial penetration of the aquifer, non-infinite aquifers (aquifers bounded by low permeability barriers or recharge features – see Section 7.6), fractured aquifers, finite well storage, delayed release of water from storage in unconfined aquifers (see Section 7.6.3) and non-constant pumping rates (Kruseman *et al.*, 1990).

If the real situation does not adequately fit any of the available analytical methods, it may be necessary to simulate the pumping test in some form of computer-based numerical model, which allows for a greater degree of aquifer complexity. Such an approach is costly and time-consuming, however, and may not be justified in the context of a minor abstraction.

Finally, we need to be careful in the use of computer programs which are designed to aid interpretation of pumping test analyses. They are attractive, as they allow ready manipulation of data and facilitate curve-fitting techniques (such as the Theis technique). They have a number of drawbacks, however:

- They may not warn the user that some of the fundamental assumptions of the method are being violated.
- They leave the user to choose the method of analysis. This may be fine for experienced pump test analysts, but the inexperienced practitioner is left with many possible methods for analysis and only limited guidance towards the correct one (if any).
- The automated curve-fitting techniques applied in some software packages may not be appropriate for many techniques of analysis.
- They may allow a user to draw a straight line (for example, in the Cooper-Jacob

Table 7.2 *Notation and terminology used in Chapter 7. The first section covers properties of the aquifer, the second properties of the well and the third other properties*

Symbol	Description	Units (metric)
K	Hydraulic conductivity	m d^{-1} or m s^{-1}
T	Transmissivity	m^2 d^{-1} or m^2 s^{-1}
T_a	Apparent transmissivity (fractured-rock aquifer)	m^2 d^{-1} or m^2 s^{-1}
S	Storage coefficient	Dimensionless
L	Leakage factor	m
$ß$	Drainage factor (Neuman analysis)	Dimensionless
R	Hydraulic resistance of a leaky layer	d or s
h	Head (relative to arbitrary datum)	m
h_1	Head in observation borehole no. 1	m
Δh_0	Initial displacement of water level in well during slug test	m
Δh	Displacement of water level in well at time t during slug test	m
Δh_{atm}	Well water level change in response to a ΔP_{atm} change in atmospheric pressure	m
H	Groundwater level (relative to base of aquifer)	m
H_w	Groundwater level in pumped well (relative to base of aquifer)	m
H_1	Groundwater level in observation borehole no. 1 (relative to base of aquifer)	m
H_0	Static groundwater level (relative to base of aquifer)	m
Q	Pumped discharge rate	m^3 d^{-1} or m^3 s^{-1}
Q_i	Pumped discharge rate during step number i of a step test	m^3 d^{-1} or m^3 s^{-1}
$q=Q/s_w$	Specific capacity	m^2 d^{-1} or m^2 s^{-1}
B	Linear well/aquifer losses	d m^{-2} or s m^{-2}
C	Non-linear well/aquifer losses	d^2 m^{-5} or s^2 m^{-5} *
s	Drawdown	m
s_w	Drawdown in pumping well	m
s_{w1}	Drawdown in pumping well during step 1 of a step test	m
s_1	Drawdown in observation borehole no. 1	m
Δs	Change in drawdown per log cycle of t or r	m
s_{max}	Maximum drawdown at end of pumping test or step	m
r	Radial distance from pumping well	m
r_1	Radial distance of observation borehole no. 1 from pumped well	m
r_i	Distance of observation borehole from image well	m
r_b	Distance of pumped well from impermeable/recharge boundary	m
r_r	Ratio of distance to image well and distance to real well	Dimensionless
r_w	Radius of pumping well	m
r_c	Radius of casing where water level fluctuation occurs	m
r_{wf}	Effective radius of well filter in slug test	m
r_e	Radius of cone of depression	m
L	Length of response zone (packer testing)	m
L_i	Length of zone of intake (slug testing)	m
s''	Residual drawdown during recovery test	m
Q''	Discharge rate prior to recovery test	m^3 d^{-1} or m^3 s^{-1}
s_p	Drawdown at point of inflection in a Hantush-type analysis	m
Δs_p	Gradient of s vs. $\log_{10} t$ curve at point of inflection	m
F	Geometric factor for well/piezometer in slug testing	Dimensionless
t	Time since start of pumping	day or s
t_0	Intercept on time axis, where $s=0$, in Cooper-Jacob approximation	day or s
t''	Time since start of recovery	day or s

(Continued)

Table 7.2   (Continued)

Symbol	Description	Units (metric)
t_p	Time coordinate of point of inflection in a Hantush-type analysis	day or s
Δt_i	Duration of step i in a step test	day or s
Δt_{step}	Duration of step in step test	day or s
T_0	Time taken for 63% of recovery to take place during slug test	s
ρ, ρ_w	Density (typically of fluid or of water)	kg m^{-3}
α	Coefficient of proportionality between specific capacity and T_a	Dimensionless
$W(u)$	Theis Well Function (a function of u)	Dimensionless
u	Variable in Theis analysis, $u = r^2 S/4Tt$	Dimensionless
$F(u_w, \gamma)$	Well function for large diameter wells	Dimensionless
γ	The ratio $r_w^2 S/r_c^2$ for large diameter wells	Dimensionless
K_o	Modified zero order Bessel function of the second kind	Dimensionless
N	Flow dimensionality	Dimensionless

* Assumes the non-linear term is proportional to the square of the pumped discharge.
Note that other systems of units can be used, but in test pump analyses, consistency of units is of paramount importance

method: Section 7.4.5) through a data set which manifestly was never meant to fit on a straight line, and to derive values of T and S therefrom – a problem also evident in some manual interpretations (Misstear, 2001).

Computer programs can be highly useful tools, but they should not be allowed to reduce the application of common sense and sound analytical thought.

7.4.3   Well performance – the step test

The step test (see Section 7.3.2 and Figure 7.10) enables four (or more) coordinates (Q_1, s_{w1})… (Q_4, s_{w4}) to be identified on a curve of discharge against drawdown (Clark 1977; Box 7.3), where $s_{w1}…s_{w4}$ are drawdowns (for time Δt_{step}) in the pumping well from steps 1–4 and where Δt_{step} is the step duration. If a regime has been chosen using intermittent steps, $s_{w1}…s_{w4}$ can be deduced simply by subtracting the final water level from the water level at the beginning of each step. If consecutive steps have been used, $s_{w1}…s_{w4}$ can only be derived after correcting for any continued drawdown accruing from the effects of previous steps (see Figure 7.10).

If the well satisfies all the assumptions made in connection with Equations (7.5) and (7.6), and if it is operating wholly efficiently with no turbulence and no head losses at the well-aquifer interface, the curve will be a straight line with a gradient proportional to transmissivity.

In reality, however, the points $(Q_1, s_{w1})…(Q_4, s_{w4})$ usually define a convex curve, such that the ratio Q/s_w decreases with increasing Q (Box 7.3). This is because of additional non-linear head losses at the well/gravel pack/aquifer interface. Many hydrogeologists identify two types of head loss associated with a pumping well in an aquifer (Section 4.5):

i. *Aquifer losses*. These are typically linear and in inverse proportion to hydraulic conductivity, if non-turbulent Darcian flow predominates. Aquifer losses *can* also be non-linear: for example, if turbulent flow occurs in the aquifer in the immediate vicinity of the well due to increased flow rate, or within fractures or fissures.

ii. *Well losses*. These may be linear or non-linear, and may be due to additional resistance imposed by the well screen, frictional losses in the gravel pack, resistance associated with any seepage face (transition from saturated to unsaturated conditions at the borehole wall), or turbulent flow associated with any of these components or within the well bore itself (Parsons, 1994).

Jacob (1946) suggested that, in many circumstances:

$$s_w = BQ + CQ^2 \qquad (4.2;7.7)$$

where B and C are constants. Thus,

$$\frac{s_w}{Q} = B + CQ \qquad (7.8)$$

Box 7.3   Example of a step-drawdown test analysis

A step test was carried out comprising four two-hour steps. The following data were obtained for yield (Q) and corresponding drawdown (s_w) in the pumping well:

Step	Q ($l\,s^{-1}$)	s_w (m)	Q/s_w ($m^2\,d^{-1}$)
Rest	0	0	
1	14.7	1.43	888
2	31.5	3.46	787
3	44.4	5.41	709
4	57.6	8.90	559

The data are plotted on the yield-drawdown graph shown in Figure B7.3(i), where the dashed line shows the yield that might have been predicted based solely on the first data point, before well losses become significant [or, in an unconfined aquifer, before the seepage face effect becomes significant – Equation (7.20)].

The Hantush-Bierschenk plot of s_w/Q vs. Q is shown in Figure B7.3(ii). The first three data points fall on a straight line, giving values of $B=0.001$ d m^{-2} and $C=1\times10^{-7}$ d^2 m^{-5}. The well efficiency drops from 89% at the first step to

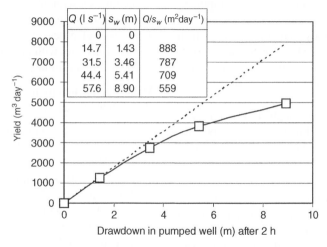

Figure B7.3(i)   *Yield–drawdown graph for step test example*

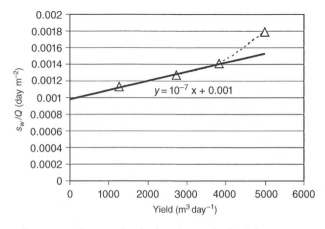

Figure B7.3(ii)   *Hantush-Bierschenk plot of specific drawdown against well yield*

56% at the fourth. The fourth point falls above the straight line (dashed curve) showing that it does not fit the Hantush-Bierschenk theory.

The inverse of *B* is the specific capacity at negligible yield, and equates to 1000 m² d⁻¹. An initial estimate of transmissivity (by the Logan approximation) can then be calculated from $1.22 \times 1000 \, \text{m}^2 \, \text{d}^{-1} = 1220 \, \text{m}^2 \, \text{d}^{-1}$. (A further estimate of *T* would be made from the constant discharge test data).

Hantush (1964) and Bierschenk (1963) noted that, if s_w/Q (known as specific drawdown, the inverse of specific capacity) is plotted against yield (*Q*), as in Box 7.3, a straight line should result, whose gradient is equal to *C*, and whose intercept on the s_w/Q axis is *B*. Some authors argue that *B* equates to aquifer loss (and is thus inversely proportional to *T*) and *C* to well loss, but this is something of a simplification. *B* represents linear head losses (which are likely to be predominantly aquifer losses, but which may contain well losses also; see Section 4.5). *C* represents the non-linear component, which also includes both well and aquifer losses, although it is typically dominated by well losses.

Other researchers (Rorabaugh, 1953) argue that the quadratic relation of Equation (7.7) is too simplistic, and that the true relationship is a power law:

$$s_w = BQ + CQ^n \qquad (7.9)$$

where $n > 1$. If we assume that $n = 2$ and that the term *B* is dominated by linear aquifer losses, we can also define well efficiency (E_w), the ratio between drawdown in an ideally functioning well and the actual drawdown:

$$E_w = \left(\frac{BQ}{s_w}\right)100\% = \left(\frac{BQ}{BQ + CQ^2}\right)100\%$$

$$(7.10)$$

The efficiency will decrease with increasing pumping rate. The efficiency of a production well can also be examined in terms of the change in operating specific capacity over time (Section 9.2).

In summary, the step test can be used to:

1. Define a practical tool (a curve of *Q* vs. s_w) to predict the drawdown for a range of yields spanning the probable operating range of the well, and thus determine whether the well is likely to produce the design yield for an acceptable drawdown value. It is important to remember, however, that the drawdown will increase with increasing pumping time, and that the points on the step test curve *are only strictly valid for pumping duration* Δt_{step}.
2. Assess at what depth the pump should be situated for subsequent testing.
3. Assess the magnitude of non-linear head losses.
4. Assess well efficiency and the rate at which efficiency declines with increasing pumping rate.
5. Make a first estimate of aquifer transmissivity.

Many other, generally more sophisticated, methods are available for analysis of step tests, including:

1. The Thiem (distance-drawdown) approach, provided that several observation boreholes are available.
2. The Eden-Hazel/Ehlig-Economides method (Eden and Hazel, 1973; Ehlig-Economides *et al.*, 1994), which is based on the Cooper-Jacob approximation of the Theis equation [Equation (7.25)], combined with Equation (7.7) to account for well losses (see Clark, 1977 and Kruseman *et al.*, 1990).
3. Cooper-Jacob, Theis (or similar) time-variant analyses, which can be applied to the drawdown or recovery portions of individual steps. Birsoy and Summers (1980) applied the Cooper-Jacob formula to water level recovery following consecutive steps of pumping, but modified the discharge during the last step (Q_n) using a function $\Psi_{t(n)}$ to take account of the different pumping rates and durations in previous steps ($\Psi_{t(n)}$ is calculated from these rates and durations; see Kruseman *et al.*, 1990):

$$s'' = \frac{2.30 Q_n}{4\pi T} \log_{10}\left\{\Psi_{t(n)}\left(\frac{t - t_n}{t - t'_n}\right)\right\} \quad (7.11)$$

where s'' is the residual drawdown (see Section 7.4.6); t=time since start of pumping; t_n=time of start of nth (last) step; and t'_n=time of end of nth step.

4. Karami and Younger (2002) proposed a method for normalizing the various steps to a standard yield, and also for graphically separating aquifer and well-loss components. They also argue that this method of analysis can provide valuable information about aquifer heterogeneity.

7.4.4 Steady state analyses

By 'steady-state' conditions, we mean a situation where a well is being pumped, but where water levels in the pumped and observation wells have reached a steady state and are no longer falling.

The Thiem solution. As has been seen in Chapter 1, Adolf Thiem (based on earlier work by Jules Dupuit – Box 7.4) derived the following equation for variation of head (h_1, h_2) with radial distance (r_1, r_2) from a pumping well under steady-state conditions in a confined aquifer. The points r_1 and r_2 can refer to two observation wells:

$$h_2 - h_1 = \frac{Q}{2\pi T} \ln \frac{r_2}{r_1} \qquad (7.12)$$

or, in terms of drawdown (s_1 and s_2) in the two wells:

$$s_1 - s_2 = \frac{Q}{2\pi T} \ln \frac{r_2}{r_1} \qquad (7.13)$$

A value of transmissivity can therefore be derived from any pair of observation boreholes at distances r_1 and r_2 from the pumping well. Alternatively, the pumping well can be used as one of the pair, by setting $s_1 = s_w$ (the drawdown in the pumped well) and $r_1 = r_w$ (the radius of the pumped well). This may yield inferior results, however, if well losses or turbulent flow in the vicinity of the pumping well lead to excessive drawdowns.

The Thiem solution is a solution to the more general differential equation for steady-state radial flow to a well:

$$-2\pi r T \frac{dh}{dr} = Q \qquad (7.14)$$

and thus:

$$\frac{dh}{dr} = \frac{-Q}{2\pi r T} \text{ or } \frac{ds}{dr} = \frac{Q}{2\pi r T} \qquad (7.15)$$

In Equation (7.15), Q is a negative quantity because water is flowing in the opposite direction to increasing r (although the negative sign is often omitted for convenience). The general solution to this is:

$$s = \frac{Q}{2\pi T} \ln r + c = \frac{2.30Q}{2\pi T} \log_{10} r + c \qquad (7.16)$$

where c is an arbitrary constant of integration. Hence, if there are several observation boreholes at a variety of distances (r) from the pumped well, drawdown (s) can be plotted on the y-axis against $log_{10}(r)$ on the x-axis. For an ideal aquifer, the relationship will be a straight line of gradient $\Delta s = \frac{2.30Q}{2\pi T}$, where Δs is the change in drawdown per $\log_{10}$ cycle of distance. Therefore:

$$T = \frac{2.30Q}{2\pi\Delta s} \qquad (7.17)$$

For the Thiem method to be truly applicable, the assumptions listed in Section 7.4.2 need to be valid. Also, Thiem developed this relationship assuming steady state flow conditions. Strictly speaking, this is impossible in an ideal confined aquifer as the cone of depression continues to expand *ad infinitum* with time. In practical terms, however, the Thiem method can be used where the heads have approximately stabilized and where the rate of change of drawdown is low. More interestingly, it will be seen that the Thiem relationship still retains some degree of validity under transient conditions (see Section 7.4.5 on the Cooper-Jacob method).

If some of the necessary assumptions are violated (for example, if the aquifer is anisotropic, of varying thickness or heterogeneous), different pairs of wells will yield different values of T. By plotting these values of T on a sketch of the well network, some idea of any systematic variation in aquifer properties can be obtained. For example, on Figure 7.8, the high values of T in an alluvial aquifer in the east-west direction might suggest the presence of a gravel-filled buried channel in that

---

Box 7.4   Jules Dupuit (1804–1866), Adolf Thiem (1836–1908) and Günther Thiem (1875–1959)

Arsène Jules-Emile Juvenal Dupuit was a Frenchman, born on 18th May 1804 at Fossano in the Italian Piedmont (at that time, part of Napoleon's empire). Like Henry Darcy (Box 1.5), he was educated at the School of Bridges and Roads in Paris (1824–1827). He commenced his career as Chief Engineer for the Sarthe region of France and gained experience in the fields of road maintenance and flood management. In 1850 he was called to Paris as Chief Engineer, where he began to work on issues of water supply and hydraulics. Later in life he also developed a considerable reputation as an economist, writing on topics such as the theory of marginal utility and the measurement of utility of public services (Ekelund and Hebert, 1999).

In contrast to Darcy's practical, empirical approach, Dupuit was more of a theoretician. Much of his work concerned the flow of water through channels (Dupuit, 1863; Narasimhan, 1998). Indeed, he attempted to calculate the flow of groundwater through many pore channels, by comparing this to the flow of water through pipes (for which proven formulae already existed). In 1863 he published a thesis on the axially symmetric groundwater flow towards a well in a porous medium. He theoretically confirmed Darcy's Law and derived his own calculation for groundwater flow to wells (Q) in confined and unconfined aquifers by considering the water heads at an 'undisturbed' radius of influence. Today, the formula for the unconfined case can be written thus:

$$Q = \frac{\pi K \left(H_e^2 - H_w^2 \right)}{\ln {r_e}/{r_w}}$$

where H_e and H_w are the elevation of the water table above the aquifer base at the edge of the cone of depression and at the well, respectively; and r_e and r_w are the radii of the cone of depression and of the well, respectively. Dupuit made a lot of assumptions (e.g., that the water table was effectively flat, that groundwater flow was

effectively horizontal and that there was no seepage face at the well), he did not test his formula empirically and did not seem unduly disturbed by how one actually defines r_e. Dupuit died in Paris on 5th September 1866. He left a large written opus, but the work of most interest to hydrogeologists and engineers is that of 1863:

> Dupuit, J. (1863). *Études théoretiques et pratiques sur le mouvement des eaux dans les canaux découverts et à travers les terrains perméables.* 2nd Edn, Dunod, *Paris.*

Adolf Thiem and later his son, *Günther Thiem*, carried out pioneering studies on the flow of groundwater to wells. Adolf Thiem was a municipal engineer largely based in the German state of Saxony. He consulted widely in Germany and abroad on water supply problems. According to Narasimhan (1998) and Grombach *et al.* (2000), as early as 1870 he independently derived and published radial flow formulae similar to Dupuit's. However, unlike Dupuit, he put forward ideas as to how to circumvent the problem of what r_e (the radius of influence or the radius of the cone of depression) actually means. Adolf is clearly credited by Weyrauch (1914) with Equation (7.13), using two drawdowns at two specific observation wells (rather than a nebulous 'radius of influence'). It appears that Thiem developed the approach during planning of groundwater supplies for the cities of Prague and Leipzig. Thiem also published work on using the diffusion of saline tracers as a means for estimating groundwater flow velocity. Thiem was one of the earliest workers to use the term 'groundwater' ('Grundwasser'). Adolf Thiem's major works are considered to be:

> Thiem A (1870) Die Ergiebigkeit artesischer Bohrlöcher, Schachtbrunnen und Filtergallerien. *Journal für Gasbeleuchtung und Wasserversorgung* 14, 450–467.

> Thiem A (1887). Verfahren zur Messung natürlicher Grundwassergeschwindigkeiten. *Polytechnisches Notizblatt* 42: 229–232.

Günther Thiem focused on field observations, carrying out extensive pumping tests to quantify aquifer properties (Simmons, 2008). It is Günther Thiem who is often credited (Logan, 1964) with the development the 'Thiem Equation (1.19)' but, in fact, it seems (Grombach *et al.*, 2000) that he further developed pumping test approaches based on his father's equation for radial flow to a well, to enable values of hydraulic conductivity to be derived:

Thiem GA (1906) *Hydrologische Methoden.* Gebhardt, Leipzig

The practice of referring to the steady-state radial flow equations as the Dupuit-Thiem formulae leaves the issue of attribution of the mathematics suitably vague!

orientation (although there are plenty of other plausible explanations).

Steady state flow in unconfined aquifers. Although the Thiem method explicitly assumes confined conditions, the method can be used with caution in unconfined aquifers provided a number of assumptions are made, the chief of these being that: *the aquifer thickness is large compared with the observed drawdowns.* In this case, transmissivity can be regarded as being independent of drawdown, and Equation (7.12) is still approximately valid.

If, however, this assumption cannot be made, T should be regarded as a variable dependent on drawdown or head, and the Dupuit assumptions should be applied (see Sections 1.3.2 and Box 7.4), giving:

$$-2\pi r K H \frac{dH}{dr} = Q \quad \text{or} \quad \frac{dH}{dr} = \frac{-Q}{2\pi r K H} \quad (7.18)$$

where K is hydraulic conductivity, and H is elevation of water table above the base of the aquifer (i.e., saturated thickness of aquifer).

Integrating between limits representing two observation boreholes at distances r_1 and r_2 from the pumping well, and with water table elevations H_1 and H_2 respectively, gives:

$$H_2^2 - H_1^2 = \frac{Q}{\pi K} \ln\left(\frac{r_2}{r_1}\right) \quad \text{or}$$

$$K = \frac{Q}{\pi\left(H_2^2 - H_1^2\right)} \ln\left(\frac{r_2}{r_1}\right) \quad (7.19a)$$

Therefore, for any pair of wells, where 'equilibrium' drawdown and radial distance from pumping well are known, we can derive a value of K.

In the case where $H_1 = H_w$, the water level in a pumped well of radius r_w, and $H_2 = H_0$, the undisturbed water table elevation at a certain effective radius of influence r_e, then we can estimate the well yield:

$$Q = \frac{K\pi\left(H_0^2 - H_w^2\right)}{\ln\left(r_e / r_w\right)} \quad (7.19b)$$

However, Rushton (2006) and Chenaf and Chapuis (2007) both note that Equations (7.19a and 7.19b) cannot be reliably used to predict water table levels in an aquifer. This is, fundamentally, because the Dupuit approach assumes (a) that the water level H_w in the well is the same as that in the aquifer (H_{aqw}) immediately outside the well and (b) that flow in the aquifer is quasi-horizontal. As drawdowns increase, these become increasingly poor assumptions. If vertical flow is taken into account, a *seepage face* is predicted in the well, where $H_{aqw} > H_w$ and water flows from the saturated aquifer outside the well into a portion of the well above the pumping water level. At first sight, the practical implications of this appear very similar to well efficiency – there appears to be an additional borehole hydraulic resistance impeding flow into the well – but in fact a seepage face can occur in a 100% efficient well. The seepage face is due to the vertical components of the hydraulics of the *aquifer*. Several analytical approximations have been proposed to predict the height of the seepage face. Rushton (2006) cites the following:

$$H_{aqw} = H_0 - \frac{0.6\left(H_0^2 - H_w^2\right)}{H_0 \ln\left(r_e / r_w\right)} \ln\left(\frac{r_e}{0.1 H_0}\right)$$

$$(7.20)$$

where H_0 is the rest water level. Here, we see that the height of the seepage face is not directly dependent on hydraulic conductivity. The seepage face becomes thicker the further the water level is drawn down in the well (Figure 7.11).

7.4.5   Time-variant analysis

To derive a value of S from a pumping test (in addition to T), analysis of time-variant (or non-steady state) drawdown data from one or more non-pumped observation boreholes is required.

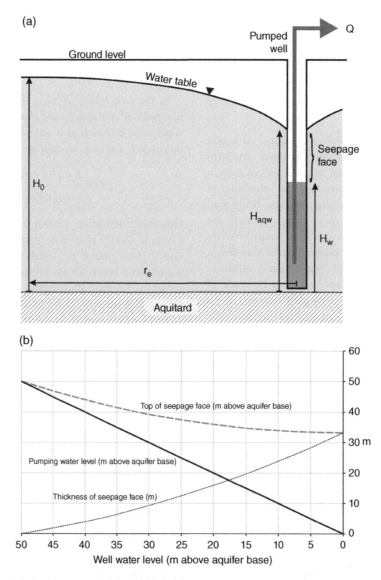

Figure 7.11   (a) Schematic section of the development of a seepage face in an unconfined aquifer, as outlined in Rushton's (2006) Equation (7.20). (b) The use of the Equation (7.20) to estimate the seepage face's dependence on pumping water level H_w (x axis) using Rushton's example of a well of radius r_w 0.4m, with a rest water level H_0 of 50m at a radius r_e of 126.5m. Towards the right, the estimate becomes somewhat unrealistic as water disappears from the well!

Non-steady state flow in a confined aquifer: Theis method. In 1935, CV Theis published a famous equation for transient flow to a pumping well (Box 7.5). Essentially, it was a solution to Equations (1.10 and 7.21), which related the head (*h*) in an aquifer of transmissivity (*T*) and storage coefficient (*S*), at a radial distance (*r*) from the pumping well, to the time (*t*) after pumping commences (at rate *Q*):

$$\frac{\partial^2 h}{\partial r^2} + \frac{1}{r}\frac{\partial h}{\partial r} = \frac{S}{T}\frac{\partial h}{\partial t} \qquad (7.21)$$

Theis's solution was not trivial. It turns out to be best described by the following equations (7.22–7.24):

$$s = \frac{Q}{4\pi T} W(u) \qquad (7.22)$$

where *s* is the drawdown at distance *r* from the well (typically measured in an observation borehole) at time *t;* and *W(u)* is a function known as the *Theis well function*:

Box 7.5   Charles Theis (1900–1987)

Charles Vernon Theis was born in the U.S. state of Kentucky on March 27th 1900. He commenced his study of Civil Engineering at the University of Cincinnati in 1917 and gained experience in carpentry, construction and surveying work. After gaining his degree in 1922, Theis was offered a post as an assistant at the same University's geology department. During this time, he also found summer work with the Kentucky Geological Survey and found his original engineering interests moving towards matters geological. In fact, in June 1927, Theis was given a post as Junior Geologist at the U.S. Geological Survey (USGS), based in Moab, Utah. In 1929, having gained his PhD in geology from the University of Cincinnati, he moved on again to the U.S. Army Corps of Engineers where he worked on dam sites and with meteorological measurements.

In 1930 he returned to the USGS, this time to their Ground Water Division, where he spent the remainder of his career. During his first four years with USGS, Theis worked extensively in the arid climate of New Mexico with irrigation wells. He soon realized that existing methods (such as the Thiem equation) were not wholly adequate and could not effectively deal with situations where a state of equilibrium had not been reached. He also identified flaws in the Dupuit–Thiem conceptualization of a cone of depression with a finite radius; namely that this required an implausible 'edge' in the potentiometric surface at the radius of influence. Theis drew upon analogies with heat flow theory. He asked his old university friend Clarence Lubin if a non-equilibrium solution had been found to the problem of radial heat conduction towards a heat sink, which might be analogous to groundwater flow to a well. Lubin suggested that existing equations from thermal theory (published in H.S. Carslaw's 1921 book, 'Introduction to the Mathematical Theory of the Conduction of Heat in Solids' – Banks, 2014b) could be applied to groundwater flow. Theis published the non-equilibrium groundwater solution in his 1935 paper (of which Lubin was offered co-authorship, but modestly refused):

> Theis, C.V. (1935). The relation between the lowering of the piezometric surface and the rate and duration of discharge of a well using ground-water storage. *Transactions of the American Geophysical Union* 16: 519–524.

Theis enjoyed many further years with the USGS, continuing to work on the hydrogeology of New Mexico as well as becoming involved in numerous other issues, including those of artificial recharge, aquifer inhomogeneity and anisotropy, and radioactive waste disposal (White and Clebsch, 1994).

$$W(u) = -0.5772 - \ln u + u$$
$$-\frac{u^2}{2.2!} + \frac{u^3}{3.3!} - \frac{u^4}{4.4!} + \frac{u^5}{5.5!} - .... \quad (7.23)$$

$$\text{and} \quad u = \frac{r^2 S}{4Tt} \quad (7.24)$$

The polynomial term $W(u)$ converges quickly and can usually be calculated by considering the first, say, five power terms (after checking that adequate convergence has occurred), allowing manual calculation of drawdown for a given r, T, S and t.

In a pumping test situation, s, r and t are normally known and we wish to find T and S. This is not a simple task, and cannot readily be solved analytically. The most tractable solution is the curve-matching technique, according to the procedure described in Box 7.6. This is a neat method, and several software packages are commercially available to perform curve matching and subsequent calculation automatically, or in a computer-screen environment. However, we should take care with our analysis: it is sometimes possible to convince oneself that a "match" exists even when one does not, and thus to derive

---

Box 7.6   Applying the Theis method by curve matching

The Theis method can be applied to the hypothetical data in Figure 7.12, from an observation borehole 50 m from a pumping well, abstracting 4320 m³ d⁻¹ of groundwater.

  i. Drawdown s from a given observation borehole at distance r from the pumped well is plotted (y-axis) against t/r^2 (x-axis) on translucent log-log graph paper.

  ii. If Theis's assumptions (in Section 7.4.2) are fulfilled, this should define a convex-upward curve of characteristic shape (Figure B7.6(i)).

 iii. The *Theis type curve* is prepared by plotting values of $W(u)$ (y-axis) against corresponding values of $1/u$ (x-axis) on similar log-log paper [Figure B7.6(ii)]. Tables of such data are included in many standard hydrogeological texts (including Driscoll, 1986;

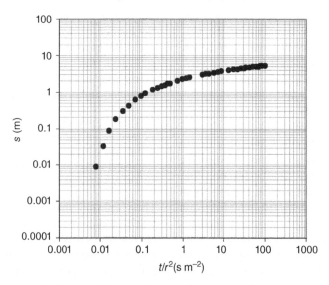

Figure B7.6(i)   *Plot of s vs. t/r² for Theis analysis*

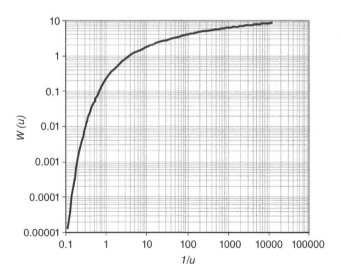

Figure B7.6(ii) *Theis type curve*

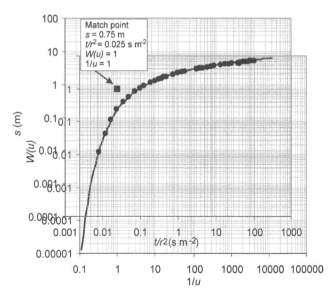

Figure B7.6(iii) *Matching of Theis type curve to data plot*

Kruseman *et al.*, 1990), but they may be readily calculated from Equation (7.23) using a spreadsheet program, by considering a limited number of terms in the polynomial expansion.

iv. The real data (on translucent paper) are superimposed on the Type Curve (at the same scale) and manipulated until a match is obtained, as demonstrated in Figure B7.6(iii).

v. An arbitrary match point is determined on the two superimposed sheets of paper, and its s, t/r^2, $W(u)$ and $1/u$ coordinates are obtained. In this case, $s=0.75\,\text{m}$, $t/r^2 = 0.025\,\text{s}\,\text{m}^{-2} = 2.9\times10^{-7}\,\text{d}\,\text{m}^{-2}$, $W(u)=1$ and $1/u=1$.

vi. A value of T can then be derived from Equation (7.22), using the co-ordinates of the match-point:

$$T = \frac{Q}{4\pi s}W(u) = 4320\,\text{m}^3\text{d}^{-1}$$
$$\times 1/(4.\pi\times0.75\,\text{m}) = 458\,\text{m}^2\text{d}^{-1}$$

vii. And a value of S can be obtained from Equation (7.24):

$$S = \frac{4Ttu}{r^2} = \frac{4T\left(\dfrac{t}{r^2}\right)}{\left(\dfrac{1}{u}\right)} = 4\times458\,\text{m}^2\text{d}^{-1}$$
$$\times 2.9\times10^{-7}\,\text{d}\,\text{m}^{-2}/1 = 0.0005$$

The transmissivity is thus around $460\,\text{m}^2\,\text{d}^{-1}$ and the dimensionless storage coefficient 0.05%, suggestive of a rather productive, confined aquifer.

(a)

Observation well 7, Banksville, Ruritania				
Distance from pumping well (r) = 50 m				
Pumping rate = 50 l s⁻¹ = 4320 m³ day⁻¹				
t	t	WL	s	t/r^2
(s)	(h)	(m bwt)	(m)	(s m⁻²)
0	0.000	3.21	0.00	0
20	0.006	3.22	0.01	0.008
30	0.008	3.24	0.03	0.012
40	0.011	3.30	0.09	0.016
60	0.017	3.39	0.18	0.024
90	0.025	3.51	0.30	0.036
120	0.033	3.64	0.43	0.048
180	0.050	3.82	0.61	0.072
240	0.067	3.99	0.78	0.096
300	0.083	4.12	0.91	0.12
450	0.125	4.35	1.14	0.18
600	0.167	4.54	1.33	0.24
750	0.208	4.67	1.46	0.3
900	0.250	4.79	1.58	0.36
1050	0.292	4.91	1.70	0.42
1200	0.333	4.98	1.77	0.48
1800	0.500	5.23	2.02	0.72
2400	0.667	5.44	2.23	0.96
3000	0.833	5.58	2.37	1.2
3600	1.0	5.71	2.50	1.44
7500	2.1	6.22	3.01	3
9000	2.5	6.33	3.12	3.6
10800	3.0	6.46	3.25	4.32
14400	4.0	6.66	3.45	5.76
18000	5.0	6.82	3.61	7.2
21600	6.0	6.93	3.72	8.64
etc.	etc.	etc.	etc.	etc.

(b)

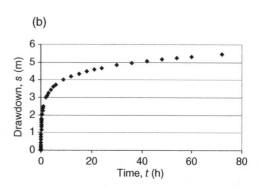

Figure 7.12 *A typical set of data resulting from a pumping test (a), plotted as columns of time, water level (below well top), and drawdown. The parameter t/r² is used in Theis analysis (Box 7.6). The data are plotted (b) as a linear drawdown vs. time diagram. Note the characteristic shape, with rapid drawdown at early time and a tendency towards stabilization of drawdown at later time*

values of T and S, and even express these values to several decimal places. We should carefully examine the log time versus log drawdown curve to see whether the match is real, or if it deviates significantly from the type curve. If it does deviate, then it means that the assumptions underlying this method of analysis are not being satisfied (see Section 7.4.2).

The Cooper-Jacob approximation. Cooper and Jacob (1946) attempted to simplify Theis's equation, such that a solution using a single sheet of semi-log graph paper was possible. They looked at Equations (7.22–7.24) and noted that, for large values of t and small values of r, the function u is small, and that for values of $u < 0.01$, Equation (7.22) can be approximated to:

$$s = \frac{Q}{4\pi T}\left[-0.5772 - \ln\left(\frac{r^2 S}{4Tt}\right)\right]$$
$$= \frac{2.30Q}{4\pi T}\log_{10}\left(\frac{2.25Tt}{r^2 S}\right) \qquad (7.25)$$

$$\text{or} \quad s = \frac{2.30Q}{4\pi T}\log_{10}\left(\frac{2.25T}{r^2 S}\right) + \frac{2.30Q}{4\pi T}\log_{10} t$$
$$(7.26)$$

Even with a less stringent criterion of $u < 0.05$, the errors are still very small (about 2% according to Kruseman *et al.*, 1990). Thus, if drawdown (s) is plotted on the y-axis, against $log_{10}t$ on the x-axis (Box 7.7), the result is a straight line with gradient:

$$\Delta s = \frac{2.30Q}{4\pi T} \qquad (7.27)$$

where Δs is the change in drawdown per $\log_{10}$ cycle of t. Finding the intercept on the x-axis ($t = t_o$ at $s = 0$) then, from Equation (7.25), it follows that:

$$0 = \log_{10}\left(\frac{2.25Tt_o}{r^2 S}\right) \qquad (7.28)$$

and therefore $S = \left(\frac{2.25Tt_o}{r^2}\right) \qquad (7.29)$

The major advantage of this method is its simplicity. It only requires semi-log graph paper or a standard software spreadsheet package. It suffers from the same limitations and assumptions as the Theis method, however, and one major assumption in addition: *the Cooper-Jacob approximation is only valid for situations where u is small (say, u < 0.05); that is, where r is small and/or t is large.* In practice, however, this condition is often satisfied (and at small values of t, the time-drawdown response is often swamped by well-bore storage effects in any case).

It should be added that, while data from observation boreholes (Figure 7.12) usually produce superior test pump analyses, the Theis or Cooper-Jacob procedures *can* be applied to data from the pumping well, by setting $s = s_w$ and $r = r_w$ to derive a value of T. Values of S derived from data analysis from the pumping well are almost always unrealistic, however, and should be disregarded.

Finally, it is interesting to note that the Cooper-Jacob solution [Equation (7.25)] can be re-written as:

$$s = \frac{2.30Q}{4\pi T}\log_{10}\left(\frac{2.25Tt}{r^2 S}\right)$$
$$= \frac{2.30Q}{4\pi T}\log_{10}\left(\frac{2.25Tt}{S}\right) - \frac{2.30Q}{2\pi T}\log_{10} r$$
$$(7.30)$$

For any given time t, drawdown data from a set of observation boreholes at varying distance r can be plotted on semi-log graph paper to give a straight line relationship where the gradient Δs (change in s per log cycle of distance) is described by:

$$\Delta s = \frac{2.30Q}{2\pi T} \qquad (7.31)$$

This is identical to Thiem's steady state solution (Section 7.4.4) and allows the shape of the cone of depression of a pumping well to be approximated at any time t following commencement of pumping (provided u is small).

7.4.6 Analysis of recovery tests

At the end of a step test or constant rate test, when the pump is switched off, the water level in an ideal well in an ideal aquifer (see Section 7.4.2 for the 'ideal aquifer' assumptions of pumping test analysis) will recover, quickly at first, then slower

Box 7.7   Applying the Cooper-Jacob approximation

The Cooper-Jacob method can be demonstrated again using the data from Figure 7.12, derived from an observation borehole 50 m from a pumping well, abstracting 4320 m³ d⁻¹ groundwater.

i. Firstly, drawdown (linear axis) is plotted versus time (logarithmic axis) on semi-log graph paper (Figure B7.7(i)). The early data do not fall on a straight line, as the Cooper-Jacob approximation is not valid at low values of t.

ii. The later data do fall on a straight line of gradient Δs = 1.56 m per log cycle, and T can be found from:

$$T = \frac{2.30Q}{4\pi\Delta s} = 2.30 \times 4320 \, \text{m}^3\text{d}^{-1} / \left(4 \times \pi \times 1.56\,\text{m}\right) = 507\,\text{m}^2\text{d}^{-1}$$

iii. To obtain a value for the storage coefficient S, the intercept t_o on the time axis should be noted and converted to consistent units. In this case, $t_o = 0.025\,\text{hrs} = 0.00104\,\text{days}$, and S is given by:

$$S = \left(\frac{2.25Tt_o}{r^2}\right) = 2.25 \times 507\,\text{m}^2\text{d}^{-1}$$
$$\times 0.00104\,\text{d} / \left(50\,\text{m}\right)^2 = 0.0005$$

iv. Finally, we need to check that the Cooper-Jacob approximation is valid. If the criterion for validity is:

$$u = \frac{r^2S}{4Tt} < 0.05, \quad \text{then } t > \frac{r^2S}{0.2T}$$
$$= \left(50\,\text{m}\right)^2 \times 0.0005 / \left(0.2 \times 507\,\text{m}^2\text{d}^{-1}\right)$$

or, $t > 0.012$ days or 0.3 hrs. As the majority of the data used to draw the straight line are at time $t > 0.3$ hrs, the use of the Cooper-Jacob approximation should be valid in this example.

Comparing these values $(T = 510\,\text{m}^2\text{d}^{-1},$ $S = 0.05\%)$ with those derived by the Theis method (Box 7.6), small differences are apparent. These are possibly due to the subjective nature of curve matching, as well as the simplifying assumptions of the Cooper-Jacob method.

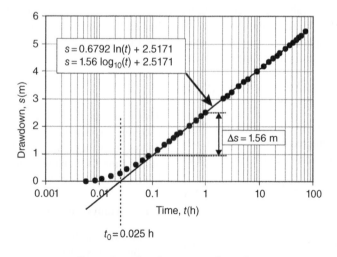

Figure B7.7(i)   Cooper-Jacob analysis

and then asymptotically towards the static water level. The shape of the recovery curve will approximately be the inverse of the Theis Type Curve (Box 7.6). If, prior to pump shut-down, the water level had stabilized at a drawdown s_{max}, the *recovery* at time *t'* after pump shut-down can be defined as the maximum drawdown (s_{max} at *t'*=0) less the residual drawdown (*s"*):

$$\text{Recovery} = s_{max} - s'' \qquad (7.32)$$

In this case, either the Theis or Cooper-Jacob methods (the latter subject to constraints regarding *u*) can be applied directly, by substituting ($s_{max} - s''$) for *s*, and *t'* for *t*. Where discharge during the pumping period varied – for example, in a step test with consecutive steps – the discharge term will need to be modified using one of several methods [see Section 7.4.3 and Equation (7.11)].

Where the water level has not fully stabilized prior to pump shutdown, but is still slowly declining (Figure 7.13), the recovery must be related back to the virtual drawdown curve (s_v) *as it would have continued had pumping not ceased*. In other words:

$$\text{Recovery} = (s_v - s'') = \frac{Q}{4\pi T} W(u) - s'' \qquad (7.33)$$

where $u = \dfrac{r^2 S}{4Tt}$ and *t*=time since *start* of pumping.

We can then apply the Theis or Cooper-Jacob methods, substituting ($s_v - s''$) for *s* and *t'* for *t*.

The Theis recovery method. Theis used the Cooper-Jacob approximation and devised his own elegant method for recovery analysis. On the basis of the arguments above, he stated that:

$$s'' = \frac{Q}{4\pi T} W(u) - \text{Recovery}$$
$$= \frac{Q}{4\pi T} W(u) - \frac{Q''}{4\pi T} W(u'') \qquad (7.34)$$

where $u = \dfrac{r^2 S}{4Tt}$ and $u'' = \dfrac{r^2 S}{4Tt''}$

Then, by assuming $Q=Q''$, and that *u* and *u"* are both sufficiently small (<0.05), the Cooper-Jacob approximation can be applied:

$$s'' = \frac{2.30Q}{4\pi T} \log_{10}\left(\frac{2.25T}{r^2 S}\right) + \frac{2.30Q}{4\pi T}\log_{10} t$$
$$- \left[\frac{2.30Q}{4\pi T}\log_{10}\left(\frac{2.25T}{r^2 S}\right) + \frac{2.30Q}{4\pi T}\log_{10} t''\right]$$

$$(7.35)$$

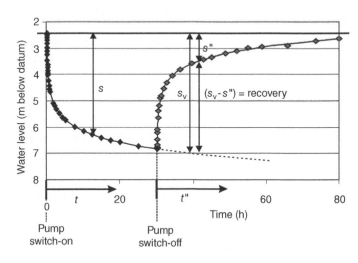

Figure 7.13   *A recovery test. To analyse the test, water level recovery (= $s_v - s''$) can be plotted against time since pump switch-off (t"), using the Theis, Cooper-Jacob or similar methods. In this diagram, s is drawdown, s_v is virtual drawdown (the drawdown if the pumping had continued), s" is residual drawdown and t is time since pump-start*

giving:

$$s" = \frac{2.30Q}{4\pi T} \log_{10} \frac{t}{t"} \qquad (7.36)$$

T can then be derived from the gradient of a plot of $s"$ versus $\log_{10}(t/t")$. No value of S is obtained, but if the line does not pass through the origin, this indicates that the values of S during storage and recovery apparently differ.

7.5   Multiple wells

7.5.1   Steady state analysis of multiple pumping wells

In situations where more than a single well is pumping during a pumping test, the drawdown observed in an observation borehole can be found by the *principle of superposition* (Section 2.9), which is valid both for steady-state and time-variant conditions in confined aquifers. It is also valid for unconfined aquifers, provided that drawdown is modest relative to total aquifer thickness. The principle states that the total drawdown at any given point is found by summing the drawdown effects from each pumping well. If an equilibrium condition has been reached, the Thiem equation can be applied:

$$s = s_{wA} - \frac{Q_A}{2\pi T} \ln \frac{r_A}{r_{wA}} + s_{wB} - \frac{Q_B}{2\pi T} \ln \frac{r_B}{r_{wB}}$$
$$+ s_{wC} - \frac{Q_C}{2\pi T} \ln \frac{r_C}{r_{wC}} + \dots \qquad (7.37)$$

For a system of three pumping wells:

$$s = s_{wA} + s_{wB} + s_{wC}$$
$$- \frac{1}{2\pi T} \left(Q_A \ln \frac{r_A}{r_{wA}} + Q_B \ln \frac{r_B}{r_{wB}} + Q_C \ln \frac{r_C}{r_{wC}} \right) \qquad (7.38)$$

where s is drawdown at the observation point; s_{wA}, s_{wB}, s_{wC} are equilibrium drawdowns in pumping wells A, B and C; Q_A, Q_B, Q_C are pumping rates in wells A, B and C; r_A, r_B, r_C are distances of the observation point from pumping wells A, B and C; r_{wA}, r_{wB}, r_{wC} are radii of pumping wells A, B and C.

This allows an approximate solution to be found for T (see also Section 2.9). Application of Equations (7.37) and (7.38) assumes that there are negligible well losses in the pumping wells, or that the well drawdown values s_{wA}, s_{wB}, s_{wC} have been corrected for well losses.

7.5.2   Time-variant analysis of multiple wells

The principle of superposition can also be applied to non-equilibrium conditions in a group of pumping wells A, B, C, and so on, by applying the Theis method:

$$s = \frac{Q_A}{4\pi T} W(u_A) + \frac{Q_B}{4\pi T} W(u_B) + \frac{Q_C}{4\pi T} W(u_C) + \dots \qquad (7.39)$$

where s is total drawdown in the observation borehole; Q_A, Q_B, Q_C are pumping rates from wells A, B and C; r_A, r_B, r_C are distances of the observation point from pumping wells A, B and C; t_A, t_B, t_C are times since pumping commenced in wells A, B and C; $u_A = \dfrac{r_A^2 S}{4 T t_A}$ and so on.

It is worth adding that a very common application of Equation (7.39) is in the design of multiple-well dewatering systems, using well-points or drilled boreholes. Here, it is common practice for all wells to commence pumping simultaneously (they are often connected to the same electrical or hydraulic manifold). Hence t_A, t_B, and so on, are represented by a single t, and the equation can be solved in a spreadsheet [see Equation (7.41) below].

7.5.3   Application of the Cooper-Jacob approximation to multiple wells

If u_A, u_B, u_C are all small (e.g., <0.05), and if all wells started pumping at the same time, then the Cooper-Jacob approximation can be applied:

$$s = \frac{2.30Q_A}{4\pi T} \log_{10} \left(\frac{2.25T}{r_A^2 S} \right) + \frac{2.30Q_A}{4\pi T} \log_{10} t$$
$$+ \frac{2.30Q_B}{4\pi T} \log_{10} \left(\frac{2.25T}{r_B^2 S} \right) + \frac{2.30Q_B}{4\pi T} \log_{10} t + \dots \qquad (7.40)$$

If s is plotted against $\log_{10}t$, a straight line should be obtained, with gradient:

$$\Delta s = \frac{2.30(Q_A + Q_B + Q_C + ...)}{4\pi T} \quad (7.41)$$

T can then be calculated. On such a Cooper-Jacob plot, a linear gradient in proportion with Q and $1/T$ is achieved only after a certain time, corresponding with the time at which $u = r^2 S/4Tt$ becomes sufficiently small. Thus, if two wells are pumping, at different distances from an observation borehole, the effect of the nearest well (A) will achieve linearity first and an initial quasi-linear segment will be observed with a gradient approximating to:

$$\Delta s = \frac{2.30(Q_A)}{4\pi T} \quad (7.42)$$

When the further pumping well (B) achieves linearity, the gradient will increase to:

$$\Delta s = \frac{2.30(Q_A + Q_B)}{4\pi T} \quad (7.43)$$

In other words, *two* or more quasi-linear segments may be observed on a Cooper-Jacob plot involving more than one pumping well (Figure 7.14).

7.6 The shape of the yield-drawdown curve: Deviations from the ideal response

Often, drawdown-time data will not exactly fit a Theis type curve or a Cooper-Jacob straight line. However, by examining the characteristic shapes of log-log and linear-log drawdown vs. time plots (Renard *et al.*, 2009), some indication can often be obtained as to *how* the aquifer is behaving non-ideally, and which assumptions (Section 7.4.2) are being violated. Figure 7.14 shows a number of characteristic responses for various aquifer situations, and a few of these are discussed in Sections 7.6.1 to 7.6.3. In some aquifer systems, 'poroelasticity' and 'subsidence' of the aquifer and aquitard material can lead to unusual pumping test

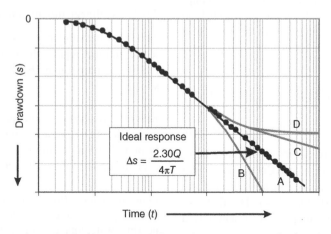

Figure 7.14 *Deviations from an ideal response as shown by a Cooper-Jacob plot. Curve A shows the ideal aquifer response, where $\Delta s = 2.30\ Q/4\pi T$. Curve B shows a steepening of the drawdown response: this could be due to (i) the cone of depression expanding into a region of lower transmissivity, (ii) interference with another well, where $\Delta s = 2.30\ (Q_A + Q_B)/4\pi T$, or (iii) the cone of depression encountering an impermeable boundary, where $\Delta s = 2.30\ Q/2\pi T$. Curve C shows a flattening trend of drawdown with time, and could be due to (iv) the cone of depression expanding into a region of elevated transmissivity or (v) some degree of recharge entering the aquifer. In Curve D, the drawdown trend stabilizes at a constant level ($\Delta s = 0$); this could be due to (vi) induced leakage from an adjacent aquifer that eventually balances the abstraction rate or (vii) induced recharge from a fully penetrating recharge boundary that eventually balances the abstraction rate*

responses, including the so-called 'Noordbergum effect' (Verruijt, 2013; see Section 7.6.4). For most pumping tests, well storage is considered negligible, but this is not the case with large diameter wells in a low transmissivity aquifer (Section 7.6.5). In addition to classic log-log and linear-log plots of drawdown versus time, several authors have found it useful to plot the logarithmic derivative of drawdown ($ds/dln(t)$) against time as a diagnostic plot (Bourdet *et al.*, 1983, 1989; Spane and Wurstner, 1993; Renard *et al.*, 2009), as discussed in Section 7.6.6.

7.6.1   A non-infinite aquifer: Presence of an impermeable barrier

If the aquifer is not infinite, but bounded at some distance by an impermeable feature such as a fault, up-throwing low permeability strata against the aquifer, the cone of drawdown cannot expand laterally in this direction. Instead of expanding it will deepen. It can be shown that the effect of such a boundary is analogous to a mirror: its effect can be simulated by imagining a virtual *image well* pumping at an identical rate, at the same distance behind the barrier as the pumping well is 'in front' of it (Figure 7.15). This then becomes a multiple well problem.

$$s = \frac{Q}{4\pi T}\left(W\left(u\right)+W\left(u_i\right)\right) \qquad (7.44)$$

where $u_i = \frac{r_i^2 S}{4Tt}$ and r_i is the distance of the point under consideration (e.g., an observation borehole) from the image well. On a Cooper-Jacob plot on $\log_{10}$ graph paper, two linear segments are sometimes observed (Figure 7.14). The first corresponds to the real pumping well and has a gradient:

$$\Delta s = \frac{2.30Q}{4\pi T} \qquad (7.45)$$

while the latter segment represents both the image and the real well and has a gradient:

$$\Delta s = \frac{2.30Q}{2\pi T} \qquad (7.46)$$

Todd and Mays (2005) point out that if such a two-segment curve with a clear inflection point can be observed, we can identify the times t_r and t_i necessary for the real well and the image well to produce identical (arbitrary) drawdowns s_A. Todd and Mays (2005) then suggest that:

$$\frac{t_r}{t_i} = \frac{r^2}{r_i^2} \qquad (7.47)$$

This allows the locus of the image well and thus the impermeable boundary to be constrained, but data from more than one observation borehole will be required to pinpoint it exactly. If the observation point is the (real) pumped well, the drawdown (s_w) will be given by:

$$s_w = \frac{Q}{4\pi T}\left(W\left(u_w\right)+W\left(u_i\right)\right) \qquad (7.48)$$

where $u_i = \frac{r_b^2 S}{Tt}$ and $u_w = \frac{r_w^2 S}{4Tt}$ and r_b is the distance from the pumped well to the barrier (= $r_i/2$) and r_w is the radius of the pumped well.

7.6.2   Recharge during a pumping test

The Theis and similar methods assume no recharge to the 'infinite' aquifer during the pumping test. In reality, recharge may be derived from several sources:

Rainfall recharge. Sporadic *rainfall recharge* events during a long-term pumping test may result in periods of apparently less steep drawdown, or even increases in groundwater level.

Induced recharge from a watercourse: the recharge boundary. Recharge may also be induced from nearby watercourses (or natural discharge of baseflow may be captured or suppressed). This is, in essence, the opposite of the hydraulic barrier considered above. Indeed, a fully penetrating river, in hydraulic continuity with the aquifer, can also be considered as a kind of hydraulic 'mirror'. However, instead of reflecting the cone of depression, causing it to deepen, it supports it, causing it to be shallower than it otherwise would have been. Indeed, it can be

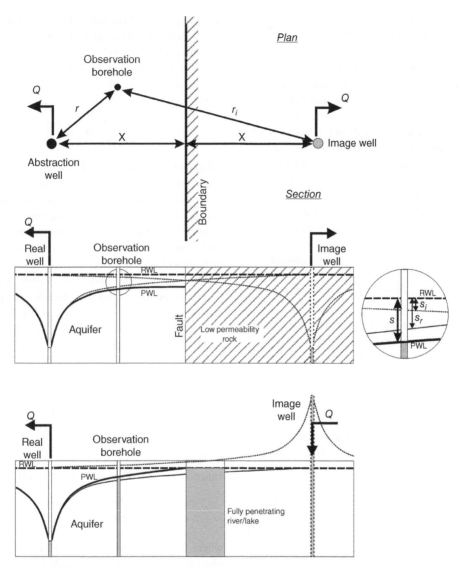

Figure 7.15 *The use of 'image wells' or 'virtual wells' to simulate the effects of impermeable boundaries or recharge boundaries. RWL = rest water level, PWL = pumping water level*

represented also by an image well, but in this case the image well has a negative abstraction rate $(-Q)$, injecting water into the aquifer, rather than abstracting it (Figure 7.15). In this case, the drawdown in an observation borehole is given by [compare with Equation (7.44)]:

$$s = \frac{Q}{4\pi T}\left(W(u) - W(u_i)\right) \quad (7.49)$$

We can define a ratio $r_r = \dfrac{r_i}{r}$ $\quad (7.50)$

And, as $u_i = \dfrac{r_i^2 S}{4Tt}$, then:

$$u_i = r_r^2 u \quad (7.51)$$

For either a barrier boundary or a recharge boundary, it follows that:

$$s = \frac{Q}{4\pi T}\left(W(u) \pm W\left(r_r^2 u\right)\right) \qquad (7.52)$$

where the operator is positive for a barrier boundary and negative for a recharge boundary. Tables of values of $(W(u) \pm W(r_r^2 u))$ are published (for example, in Kruseman *et al.*, 1990), allowing a type curve to be constructed and T and S to be deduced by curve matching. However, the location of the barrier or recharge boundary (r and r_i) needs to be known beforehand to yield a solution.

If a semi-log plot (Cooper-Jacob-type) of s vs. $\log_{10}(t)$ is produced, where an ideal recharge boundary is active, then, by analogy with the barrier boundary (above), one would expect to see an initial linear segment of approximate gradient:

$$\Delta s \approx \frac{2.30Q}{4\pi T} \qquad (7.53)$$

followed by a change in slope towards gradient ≈ 0 (i.e., horizontal) as the effect of the image well cancels out that of the real well (Figure 7.14). The maximum drawdown (s_{max}) will be given by:

$$s_{max} \approx \frac{Q}{2\pi T} \ln r_r \qquad (7.54)$$

where $r_r = r_i/r$ (or if the observation point is the pumped well, $r_r = 2r_b/r_w$, where r_b is the distance from the pumped well to the recharge boundary and r_w is the radius of the pumped well).

As with the recharge boundary, methods of analysis of the semi-log curve are available, such as the *Hantush inflection point method*, described by Kruseman *et al.* (1990).

Recharge via a leaky layer from an adjacent aquifer. In many cases, a confined aquifer may not be wholly confined. The overlying or underlying strata will never be completely impermeable and may respond to drawdown in the aquifer by vertical leakage into the aquifer. As drawdown in the aquifer increases, the quantity of vertical leakage increases. This will have the effect of reducing the drawdown at any given time, relative to the Theis solution, and flattening any curve of drawdown vs. time. This often leads to a horizontal gradient at a drawdown

(s_{max}) when the additional induced vertical leakage rate balances the abstraction (Figure 7.14).

Walton (1960, 1962) corrected the Theis well function for leakage, such that it is not just a function of u, but also of r/L, where L is a leakage factor:

$$s = \frac{Q}{4\pi T} W\left(u, \frac{r}{L}\right) \qquad (7.55)$$

Tabulated data for $W(u, r/L)$ exist (see, for example, Kruseman *et al.*, 1990), allowing type curves to be constructed on log-log paper and the equation to be solved by curve-matching techniques (Figure 7.16). In practice, curve matching can be a tricky process and a unique solution may be difficult to find.

Alternatively, the *Hantush (1956) inflection point method* may be used, by locating the point of inflection on a semi-log plot of s vs. $\log_{10} t$. The full mathematical solution will not be given here, but the following relations hold:

$$s_{max} = 2s_p = \frac{Q}{2\pi T} K_o\left(\frac{r}{L}\right) \qquad (7.56)$$

where $K_o()$ is a Bessel function, and s_p is the drawdown at the point of inflection. This helps us to locate the point of inflection, and also allows T to be found if s_{max} is known from a number of observation boreholes (De Glee, 1930; 1951). Also:

$$2.30 \frac{s_p}{\Delta s_p} = e^{r/L} K_o\left(\frac{r}{L}\right) \qquad (7.57)$$

where Δs_p is the slope of the curve on the semi-log plot (that is, the change in drawdown per log cycle of time) at the point of inflection. From this, (r/L), and thus L, can be derived, and:

$$\Delta s_p = \frac{2.30Q}{4\pi T} e^{-r/L} \qquad (7.58)$$

from which T, the transmissivity of the aquifer, can be derived. The storage coefficient S can be found from:

$$L = \frac{2Tt_p}{rS} \qquad (7.59)$$

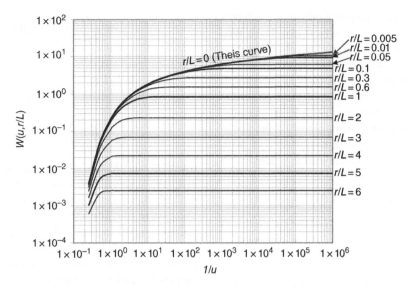

Figure 7.16 *Family of Walton (1960) type curves for W(u,r/L) versus 1/u, for different values of r/L. These can be used for curve matching log-log plots of s versus t/r² for pumping tests in aquifers affected by leakage from adjacent strata. The curve for r/L=0 corresponds to the Theis type curve (Box 7.6)*

where t_p is the time coordinate of the point of inflection. The hydraulic resistance (R) of the leaky layer can be determined from:

$$R = \frac{L^2}{T} \qquad (7.60)$$

7.6.3 Unconfined aquifers: Delayed yield

As has been noted in Section 1.2.2, there are two types of aquifer storage:

1. Elastic storage, due to the fact that water itself and the aquifer matrix can expand and contract slightly under changing head conditions. Storage coefficients due to elastic storage are very small. When head changes, these elastic responses occur almost immediately, such that changes in head propagate very quickly in aquifers dominated by elastic storage, such as confined aquifers.
2. Drainable storage, due to water draining out of (or filling up) pore spaces near the water table in response to changes in head. This type of storage occurs in unconfined aquifers and is much larger than elastic storage. However,

release of water from drainable storage *may* not take place immediately, especially if the material in the vicinity of the water table is relatively fine-grained.

In fact, in some unconfined aquifers, one of the fundamental assumptions (Section 7.4.2) may be violated: namely that water is not released instantaneously from storage with changing head. Earlier (now partially superseded – see below) conventional models of the response of some unconfined or leaky confined aquifers, have postulated two forms of storage response:

i. An *early time storage* (S_A), which releases water almost instantaneously with declining head.
ii. A drainable storage (S_y) releasing a delayed yield at later time. This is essentially the *specific yield* (see Section 1.2.2).

In medium-coarse sand it has been suggested that a minimum pumping time of 4 hours is often required to ensure that any delayed yield is fully detected, but for fine sands more than 30 hrs may be needed and for silty materials, in excess of 100 hrs is advised (see Todd and Mays, 2005).

The delayed yield response may thus produce a log(s) vs. log(t) curve on which two Theis type responses can be seen, with a relatively flat section in between (similar to the type curves in Figure 7.17). Boulton (1963) modified Theis's well function to describe this delayed yield phenomenon. Subsequently, Neuman (1972, 1975) proposed a similar solution, but more rigorously expressed in physical parameters. For situations where $S_Y/S_A > 10$, Neuman proposed a modified well function:

$$s = \frac{Q}{4\pi K_H D} W\left(u_A, u_B, \beta\right) \qquad (7.61)$$

where the subscripts A and B refer to the early and late Theis-like portions of the curve, respectively. For the early elastic response:

$$s = \frac{Q}{4\pi K_H D} W\left(u_A, \beta\right) \qquad (7.62)$$

where $u_A = \dfrac{r^2 S_A}{4K_H Dt}$ and D is aquifer thickness. Similarly, for the late delayed yield response:

$$s = \frac{Q}{4\pi K_H D} W\left(u_B, \beta\right) \qquad (7.63)$$

where $u_B = \dfrac{r^2 S_Y}{4K_H Dt}$

Neuman's parameter β is given by:

$$\beta = \frac{r^2 K_V}{D^2 K_H} \qquad (7.64)$$

where K_V and K_H are the vertical and horizontal hydraulic conductivity values, respectively. This method, therefore, allows for aquifer anisotropy. The system of equations can be solved by first of all matching the early time data to the left hand end of one of a family of Neuman type curves (Figure 7.17). By matching co-ordinates, this enables T, β and S_A to be found. Then, the late time data are matched to the right hand side of the type curve for the same value of β, allowing T to be confirmed and S_y to be derived. The method may tend to underestimate S_Y and overestimate elastic

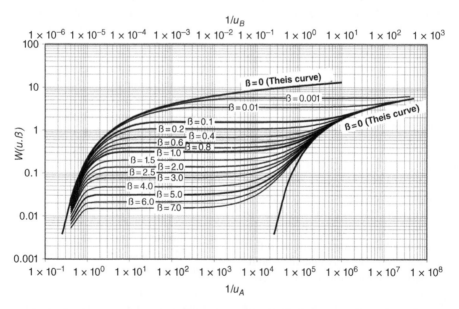

Figure 7.17 *Family of Neuman type curves for W(u_A, u_B, ß) versus 1/u_A and 1/u_B for different values of ß. These can be used for curve matching log-log plots of s versus t/r² for pumping tests in unconfined aquifers affected by delayed yield. Early time data are matched to the left hand side of one of the type curves, while later data are matched to the right hand side of the same type curve. The curve for ß=0 corresponds to the Theis type curve (Box 7.6)*

storativity (if we have assumed that the early response S_A is related to elastic storativity) because drainage from the unsaturated zone and capillary fringe above the water table is not explicitly considered (Kruseman *et al.*, 1990). Indeed, Nwankwor *et al.* (1992) and Akindunni and Gillham (1992) conducted a series of detailed field studies of the response to pumping in the capillary fringe and unsaturated zone, supported by modelling exercises. They demonstrated that the earlier Boulton (1963) and Neuman (1972, 1975) conceptualizations were oversimplified and analysis according to these methods led, in their case, to an apparent S_Y of 5–8%, compared with volume balance and laboratory determinations of 25–30%. Transmissivity/hydraulic conductivity estimates were not found to be greatly affected by the new conceptualization, however.

7.6.4 Poroelasticity, subsidence and the 'Noordbergum Effect'

We have already encountered (Section 7.6.3 and elsewhere) the concept of elastic storage (Wang, 2000). It is typically assumed that pumping causes a drop in pressure in aquifers, leading to a slight elastic expansion of water and a compression of the solid aquifer skeleton, potentially leading to aquifer 'subsidence'. Such a conceptual model can allow estimation of vertical poroelastic subsidence in response to abstraction (Lebbe, 1995). However, observation has indicated that the response of heads in aquifer/aquitard systems to pumping-induced poroelastic compression and expansion are not straightforward (Berg *et al.*, 2011). This is often because horizontal deformation effects in super- or subjacent aquifers and aquitards are not taken into account. One type of classic response was observed in the Dutch village of Noordbergum, in northern Friesland, where heads were observed to rise temporarily in a stratum following commencement of pumping in a stratigraphically deeper aquifer (Verruijt, 2013). The opposite to the 'Noordbergum effect' is the 'Rhade effect', where water levels drop briefly in response to cessation of pumping (Langguth and Treskatis, 1989).

A number of possible causative factors have been postulated to explain these effects but, currently, it is usually ascribed to deformation of the overlying aquitard/aquifer sequence in response to pumping. Typically, pumping a well will cause a compression (sinking) of the aquifer itself, which will in turn induce a slight vertical extension, but also a modest horizontal compression in the overlying aquifer/aquitard system. It is this horizontal compression that produces the "Noordbergum" rise in water levels in overlying strata. Hsieh (1996) and Berg *et al.* (2011) used a modelling approach to reproduce the Noordbergum effect and compare it with more conventional analytical solutions (Neuman and Witherspoon, 1969, 1972). Rodrigues (1983) also explored the possibility of correcting pumping test data for the Noordbergum effect in applications of the Neuman and Witherspoon analytical solutions. The range of possible effects is complex, however, and occasionally extensional effects dominate in adjacent strata, especially close to the pumping well, leading to an apparent *fall* in groundwater levels. These deformation-related water level fluctuations are usually modest in magnitude (a few cm or tens of cm) and brief in duration (but not always!) – Hsieh (1996).

7.6.5 Large diameter wells

We will recall (Section 7.4.2) that one of the fundamental assumptions for pumping test analysis using Theis-like solutions was that the well is of negligible diameter – and thus that the water stored in the well bore is insignificant compared to the pumped quantities. For a high-yielding 200 mm borehole, penetrating 80 m below the water table, and pumping 20 L/s, this is a good assumption (well storage = 2.51 m^3, or just 2 minutes' worth of pumping), but for a wider, dug well in a poorly transmissive aquifer, the amount of water stored in the well is not negligible.

Let us imagine that a well has a radius r_w in the aquifer and a radius r_c ($\geq r_w$) above the aquifer. Let us assume that the water level drawdown takes place wholly in the portion with radius r_c.

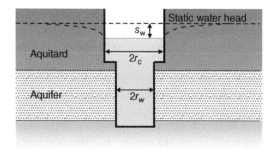

Figure 7.18 *A large diameter well, with a radius r_w in the aquifer and r_c above the aquifer in the zone where drawdown occurs. The drawdown in the well is termed s_w. See text and Papadopulos and Cooper (1967) for further explanation*

(Figure 7.18) If *no* water enters the well from the surrounding formation, then pumping the well at a rate Q will simply result in a linear decline in water level with time:

$$\frac{ds}{dt} = \frac{Q}{\pi r_c^2} \qquad (7.65)$$

Thus, a linear section at the start of a plot of drawdown s versus time t (or log s versus log t) can simply indicate a well-bore storage effect.

Papadopulos and Cooper (1967) found a good solution for analysis of pumped wells of large diameter. They derived expressions for the drawdown s in the aquifer and also for drawdown s_w in the pumped well itself (assuming negligible well losses). The expression they obtained was:

$$s_w = \frac{Q}{4\pi T} F\left(u_w, \gamma\right) \qquad (7.66)$$

where $u_w = \dfrac{r_w^2 S}{4Tt}$, $\gamma = \dfrac{r_w^2 S}{r_c^2}$ and

$$F\left(u_w, \gamma\right) = \frac{32\gamma^2}{\pi^2} \int_0^\infty \frac{\left(1 - \exp\left(-\omega^2 / 4u_w\right)\right)}{\omega^3 \Delta(\omega)} d\omega \qquad (7.67)$$

The term ω in this equation is a dummy variable for integration. The term $\Delta(\omega)$ involves monstrous Bessel functions and, frankly, the maths is getting

too complicated already. However, the expression is Theis-like and can be calculated. At values of $t > 250\, r_c^2/T$, the $F(u_w, \gamma)$ approximates to the Theis function $W(u_w)$. At low values of time,

$$s_w \to \frac{Q}{4\pi T}\left(\frac{\gamma}{u_w}\right) = \frac{Qt}{\pi r_c^2} \qquad (7.68)$$

which is our straight-line well-bore storage response (Equation 7.65). Barker (1985) has applied a similar conceptual model to large diameter wells in fractured aquifers.

7.6.6 Diagnostic plots

At the start of Section 7.6, we mentioned the concept of diagnostic plots (Renard *et al.*, 2009). These can be of three types:

a. Type 1 – log s versus log t
b. Type 2 – s versus log t
c. Type 3 – ds/dlog$t$ versus log$t$ (the differential drawdown/time plot). ds/dlog$t$ can be plotted on either a linear or a logarithmic y-axis (Figure 7.19 uses a logarithmic y-axis).

Characteristic responses are summarized below and are shown in Figure 7.19:

- *Classic radial flow.* The Type 1 plot is the classic Theis curve. The Type 2 plot tends towards a straight line of constant positive gradient (Cooper-Jacob solution). The Type 3 curve tends towards a flat (horizontal) line at a value of ds/dln$t = (Q/4\pi T)$ or ds/dlog$_{10}t = (2.30Q/4\pi T)$ as the value of u becomes small (Equation 7.27).
- *Impermeable boundary.* The Type 2 plot initially follows the Cooper-Jacob response, and then the straight line gradient doubles. The Type 3 curve has two flat (horizontal) line segments at values of ds/dlog$_{10}t = (2.30Q/4\pi T)$ and ds/dlog$_{10}t = (2.30Q/2\pi T)$.
- *Multiple wells.* The Type 2 plot initially follows the Cooper-Jacob response, and then the straight line gradient increases progressively in proportion to the abstracted quantity as the response "sees" new pumped wells. The Type 3 curve has

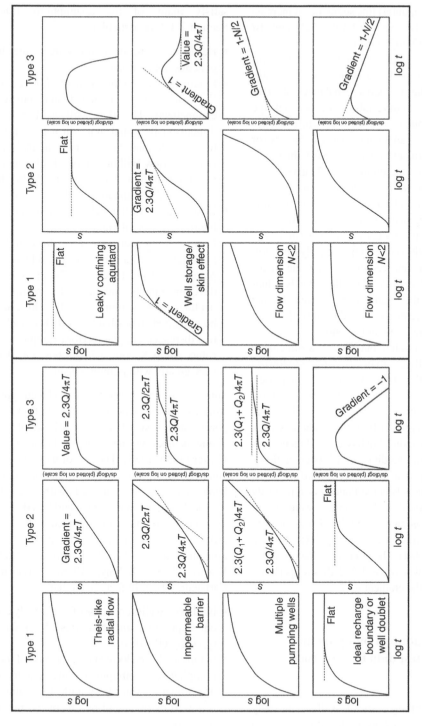

Figure 7.19 Families of diagnostic drawdown-time curves for pumping tests, following the concept of Renard et al. (2009). The Type 1 curves are log s versus log t curves, Type 2 are s versus log t and Type 3 are differential log(ds/dlog t) versus log t curves. s is drawdown and t is time, while all the logarithms are base 10 in this diagram

several flat (horizontal) line segments at increasing values of $ds/d\log_{10}t = (2.30Q_1/4\pi T)$ and $ds/d\log_{10}t = (2.30(Q_1+Q_2)/4\pi T)$ and so on.

- *Ideal recharge boundary, or leaky aquitard response.* The Type 1 and Type 2 curves eventually become flat as recharge balances abstraction. The Type 3 curve falls towards a flat line value of zero (but, of course, never quite reaches it!). In the case of an ideal recharge boundary (or an abstraction-injection well doublet), the gradient of the falling line on the Type 3 curve is -1.
- *Well-bore storage response in large-diameter well and/or skin effect.* The Type 1 curve has an initial straight line segment with a gradient of unity, which then reverts gradually to a Theis curve. The Type 3 curve exhibits a "hump" at early time, followed by a flat line response at a value of $ds/d\log_{10}t = (2.30Q/4\pi T)$. A "skin effect" tends to hinder flow from the aquifer into the well, forcing flow to be preferentially derived from well storage and thus eliciting a similar response.
- If the flow is not radial, but has a flow dimension N (Barker, 1988) less than 2 (i.e., tending toward linear flow), the Type 3 response increases, while the Type 2 plot become steeper. The gradient of the derivative plot on a log scale is $1-N/2$, thus for perfectly linear flow, the gradient is $+0.5$.
- If the flow is not radial, but has a flow dimension N greater than 2 (i.e., tending toward spherical flow), the Type 3 response falls away towards 0 as the Type 1 and 2 plots become flatter. For a perfectly spherical flow, the drawdown tends towards a constant value of $Q/4\pi Kr$, where K is hydraulic conductivity and r is distance from the point abstraction Q (Thomson, 1884). The gradient of the derivative plot on a log scale tends to $1-N/2$, thus for perfectly spherical flow, the gradient is -0.5.

7.7 Interpretation of pumping and recovery test data in hard-rock aquifers

Analyses of pumping tests of wells (especially low-yielding ones) in hard-rock aquifers do not satisfy many of the assumptions listed in Section 7.4.2. In particular:

- Hard fractured-rock aquifers are neither homogeneous nor isotropic.
- Fractures are seldom exactly horizontal and flow in fractures is not planar but usually channelized. The flow will neither be two-dimensionally radial, nor will it be constrained to a single one-dimensional channel. It will typically flow within a network of intersecting fracture planes or channels and will thus have a *fractional* flow dimension (Black, 1994).
- If flows are low, the well storage is not negligible relative to well yield.
- Low yielding wells are very difficult to test-pump at a constant rate. If variable discharge pumps are used (Section 3.7), the drawdown develops rapidly and hence the pumping rate declines correspondingly rapidly. This effect can be mitigated somewhat by using low-yield positive displacement pumps (for example, a helical rotary pump; Section 3.7.4), whose rate does not vary greatly with pumping head.

Some hydrogeologists have produced powerful work that allows general solutions to flow fields in porous and fissured aquifers. Those of an academic inclination can refer to classic papers by Barker (1985, 1988). The rest of us without mathematical superpowers will, in the context of this book, have to be satisfied with the simpler approaches adopted by Czech and Swedish hydrogeologists, who have long utilized well yield as a proxy for 'order of magnitude' of local aquifer transmissivity in hard-rock terrain (Jetel and Kràsny, 1968; Kràsny, 1975; Carlsson and Carlstedt, 1977). In fact, given that:

- hard-rock wells tend to be of a relatively consistent depth in many countries (50–100 m, and typically 60–80 m);
- pumps tend to be located near the base of the well;
- pumping or test pumping regimes tend to involve drawdown of water level almost to pump level, followed by pump cut-out, recovery, pump cut-in and renewed pumping;

it follows that cited yields can be taken as roughly proportional to specific capacity, as drawdown is relatively consistent. As we have seen, the Logan approximation (Sections 1.3.3 and 7.3.1) suggests that T is proportional to specific capacity. Banks (1992b) examined worldwide practice for relating apparent transmissivity (T_a) to specific capacity ($q=Q/s_w$) in hard-rock aquifers, and concluded that:

$$T_a = q/\alpha \qquad (7.69)$$

where α is a constant. He also noted that α typically had a value of about 0.9 in most published relations (Table 7.3; note Logan's approximation produces a value of 0.82 for α). This approach was revisited by Banks *et al.* (2010), who recommended the use of $\alpha=0.7\pm0.2$, for large databases of drilled wells in Scandinavian hard rock terrain. They found the distribution of well yields to be remarkably consistent in Norway, Sweden and Finland, largely irrespective of lithology, with a median yield of 600–700 l hr^{-1}, an apparent transmissivity of around 0.56 m^2 d^{-1} and a bulk hydraulic conductivity of around 1×10^{-7} m s^{-1}.

MacDonald *et al.* (2002, 2005) propose a simple bailer test for low permeability terrain in Africa whereby a given number of bailers-full of water are removed from a drilled well in a period of 10 minutes. The bailer is usually a 75 mm diameter bucket, about 1 m long, with a capacity therefore of about 4.4 l. The time is measured for the water level in the well to achieve 50% and 75% recovery and a look-up table is used to determine whether the well is 'successful' or not. For example, if water is bailed from a 125 mm diameter drilled well at 17.5 l min^{-1} for a ten minute period (40 standard bailers-full), the well is deemed to be a success if (i) the maximum drawdown does not exceed 7.1 m, (ii) 50% recovery is achieved within 9 minutes and (iii) 75% recovery is achieved within 21 minutes. A 'successful well' is defined as one that can supply 250 people with at least 25 l per person per day throughout a 6 month dry season, without the drawdown exceeding 15 m. This success criterion is believed to correspond approximately to a transmissivity of 1 m^2 d^{-1}.

7.7.1   High yielding hard-rock wells

If the hard rock well in question has a relatively high yield, it may be possible to carry out step, constant-rate and recovery testing as described in Sections 7.2–7.3. Standard test pump analysis methods (Section 7.4) can then be applied and it will normally be possible to characterize well performance adequately and to derive a value of T. However, it is necessary to be aware of violations

Table 7.3   *Published values of α, the constant of proportionality between apparent transmissivity (T_a) and well specific capacity (q) in the formula $T=q/\alpha$ in hard rock aquifers (modified after Banks, 1992b)*

References	Type of test	Value of α
Huntley *et al.* (1992)	Empirical short term testing of whole wells in fractured aquifers	2
Moye (1967), Banks (1972)	Lugeon testing of short (2 m) borehole sections in fractured aquifers	1.32–1.46
Carlsson & Carlstedt (1977)	Short term testing of whole wells in fractured aquifers (theory)	0.9–1.1
Kràsny (1975)	Short term testing of whole wells in fractured aquifers	0.91
Banks (1992b)	Typical value	0.9
Banks *et al.* (2010)	Recommended values for large databases of hard-rock wells in Scandinavia	0.7 ± 0.2
Carlsson & Carlstedt (1977)	Practical, short term testing of whole wells in fractured aquifers	0.84
Logan approximation (1964)	Testing of wells in porous aquifers	$(1.22)^{-1}=0.82$

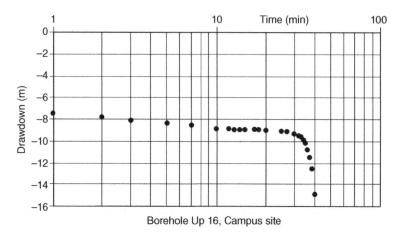

Figure 7.20 *Decline in available yield following dewatering of a yielding fracture (at a level corresponding to 9–10 m drawdown) in a South African well. Reproduced by permission from Van Tonder et al. (1998) 'Estimation of the sustainable yield…'. Copyright (1998) the Institute for Groundwater Studies, University of the Free State, South Africa*

of assumptions concerning homogeneity, isotropy and radial flow, and to question what the derived value of T actually means in such a situation. Indeed, authors such as Banks (1992b) have encouraged the use of terms such as *apparent transmissivity T_a*, for such situations.

In South African hard-rock wells, it is common practice to try to maintain operational pumping water levels above the level of the main yielding fracture. Once the main yielding fracture becomes partially dewatered, yields drop off rapidly, resulting in dramatic declines in well efficiency (Figure 7.20). Thus, test pumping is aimed at trying to identify an optimal rate of discharge Q such that drawdown does not fall below the main inflow level after a pumping time t_{max} without any recharge to replenish the aquifer. The time t_{max} is chosen to match the duration of a plausible drought (which may be 3–5 years in some parts of southern Africa). Standard methods such as Theis can be used to this end, although Van Tonder *et al.* (1998) recommend the more conservative *flow characteristics* (FC) approach. The main differences between the FC and Theis approaches are:

- The FC analysis uses differential drawdown/time data (i.e., *ds/dt* plotted against *t* – see Section 7.6.6).

This allows more accurate identification of, for example, well-bore storage effects, boundary effects and recharge, *but* presupposes good control of a constant rate of pumping.
- Boundary effects (e.g., the edge of a highly transmissive fracture zone) are considered.
- Uncertainty in aquifer parameters is explicitly considered.

7.7.2 Low-yielding hard-rock wells

If the well yield is low (perhaps a few hundreds of l hr^{-1}), it is unlikely to be practical to undertake a satisfactory constant rate test. Here, some form of *constant drawdown test* or *rapid drawdown/recovery test* may be the best option.

A constant drawdown test is usually carried out where the capacity of the pump is much larger than the capacity of the well. Essentially, the water level is dropped rapidly to the level of the pump intake until it sucks air (an old or dedicated test pump should be used for this). The well is then pumped continuously and the discharge monitored (using a bucket and stopwatch; Figure 7.7a) for this constant drawdown. The discharge will typically decrease with time as storage is depleted and as a zone of drawdown develops in the aquifer. Alternatively, the water level can be held within a constrained range of levels by

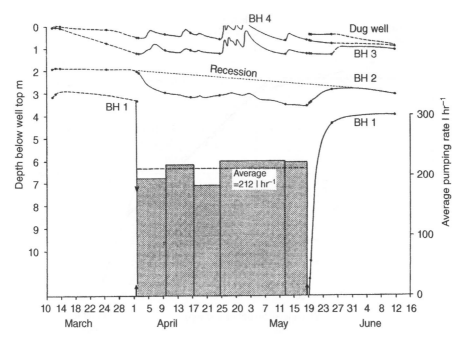

Figure 7.21 *A constant drawdown test (the water level is controlled by pump cut-in and cut-out switches) performed at a well field in the Iddefjord Granite at Hvaler, Norway. BH 1 was pumped at c. 212 l hr⁻¹ throughout the test. BHs 2, 3, and 4 were observation wells. A nearby dug well in superficial alluvium was also monitored. After Banks et al. (1993c) and reproduced by permission of Norges geologiske undersøkelse (Geological Survey of Norway)*

water-level sensitive cut-in and cut-out switches at different depths in the well (as in Figure 7.21).

Recovery tests can also be carried out after water has been expelled or pumped from a well, for example, at the end of a brief constant drawdown test. Banks (1992b) examined recovery curves from such short-term recovery tests, and noted that it was possible to derive individual specific capacities and transmissivities for specific fracture horizons. Imagine a well with three yielding fractures (Figure 7.22) with flow rates Q_1, Q_2 and Q_3 at elevations ε_1, ε_2 and ε_3, all above the 'drawn down' water level. At the start of recovery, the rate of rise of water level is given by:

$$Q_{tot} = \frac{dh}{dt}\pi r_w^2 = Q_1 + Q_2 + Q_3 \qquad (7.70)$$

where h is the water level in the well and r_w the radius of the well. The flow from each fracture is proportional to:

- the head difference between the ends of the fracture: $h_{aq} - \varepsilon_1$, where h_{aq} is the static groundwater head at some distance from the well in the aquifer, and ε_1 the elevation of the fracture in the pumping well, and
- a term loosely designated as *fracture specific capacity*, C_1.

In other words:

$$Q_1 = C_1\left(h_{aq} - \varepsilon_1\right) \qquad (7.71)$$

and $Q_{tot} = \dfrac{dh}{dt}\pi r_w^2 = C_1\left(h_{aq} - \varepsilon_1\right)$
$$+ C_2\left(h_{aq} - \varepsilon_2\right) + C_3\left(h_{aq} - \varepsilon_3\right) \qquad (7.72)$$

The initial inflow rate is independent of h, the water level in the well (Section 4 of the recovery curve in Figure 7.22). However, when Fracture 3 becomes submerged, the flow rate from that fracture decreases and becomes dependent on h:

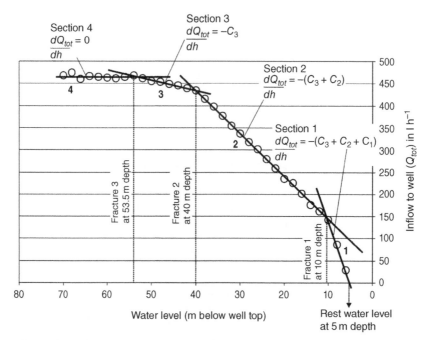

Figure 7.22 *The analysis of a recovery test in a poorly-yielding well, fed by three fractures, using the method of Banks (1992b). The total inflow Q is calculated from the well diameter and rate of recovery. The 'kinks' in the curve represent each fracture becoming successively submerged*

$$Q_3 = C_3 \left(h_{aq} - h \right) \qquad (7.73)$$

and therefore:

$$Q_{tot} = \frac{dh}{dt} \pi r_w^2 = C_1 \left(h_{aq} - \varepsilon_1 \right) + C_2 \left(h_{aq} - \varepsilon_2 \right)$$
$$+ C_3 \left(h_{aq} - h \right)$$

$$(7.74)$$

The gradient of the recovery curve abruptly changes from horizontal to a gradient (Section 3 of the recovery curve in Figure 7.22) of:

$$\frac{dQ_{tot}}{dh} = -C_3 \qquad (7.75)$$

Similarly, as Fracture 2 becomes submerged (Section 2):

$$Q_{tot} = \frac{dh}{dt} \pi r_w^2 = C_1 \left(h_{aq} - \varepsilon_1 \right) + C_2 \left(h_{aq} - h \right)$$
$$+ C_3 \left(h_{aq} - h \right)$$

$$(7.76)$$

and $$\frac{dQ_{tot}}{dh} = -\left(C_2 + C_3 \right) \qquad (7.77)$$

and so on. The Banks (1992b) method enables us to:

- locate the depths of yielding fractures from recovery test information;
- assess their relative specific capacities and hence their apparent transmissivities from $T_1 = \dfrac{C_1}{\alpha}$ [see Equation (7.69)].

De Lange and Van Tonder (2000) have also developed a method for estimating T_a and thickness of fracture zones in South Africa from early specific capacity data, using a somewhat more sophisticated Logan-type relationship (Box 7.8).

7.7.3   Sustainable yield of hard-rock wells

The problem of how to derive a value of sustainable yield from a relatively short term test pumping

---

Box 7.8 Use of short-time specific capacity data to deduce properties of fracture zones in hard-rock aquifers

De Lange and Van Tonder (2000) suggested that the drawdown (s) and corresponding yield (Q) after one minute of pumping can allow a rapid characterization of the properties of the fracture or fracture zone feeding a well or borehole. While such a method does not provide any information about the wider aquifer or fracture network, it can be useful in helping define well-head protection areas based on relatively short travel times. De Lange and Van Tonder (2000) propose that, in general, where $\dfrac{Q}{s} > 170$ m² d⁻¹ after 1 minute, this indicates an extensive feeder fracture, whereas $\dfrac{Q}{s} < 86$ m² d⁻¹ indicates limited fracture extent. They also propose a modified Logan-type equation to relate specific capacity to short-time apparent transmissivity (T_a):

$$\log_{10} T_a = \frac{\left(\log_{10}\left[\dfrac{1.20Q}{s} \right] + \log_{10}\left[0.174Q \right] \right)}{2}$$

where T_a is in m² d⁻¹, Q is the yield after 1 minute in m³ d⁻¹ and s is the corresponding drawdown in the pumped well in m. The specific storage (S_s) of a fracture zone can be estimated from geomechanical principles:

$$S_s \approx \rho_w g \left(\alpha + n\beta \right)$$

where $\rho_w g$ is the specific weight of water (9804 N m⁻³), n is the porosity of the fracture zone, α is the compressibility of rock and β is the compressibility of water (4.74 × 10⁻¹⁰ m² N⁻¹). If, on the basis of tracer tests and inverse modelling, n is set at 0.13 and α is 5.56 × 10⁻⁹ m² N⁻¹, then this gives a value of $S_s = 6 × 10⁻⁵$ m⁻¹.

In their methodology, De Lange and Van Tonder (2000) assume relatively high-yielding wells and a well-conducted constant yield test allowing meaningful Q/s determinations after 1 minute's pumping. In low yielding wells, well storage effects would need to be considered.

---

of a hard-rock well is a difficult one to solve. Certainly, the yields and specific capacities derived from short-term testing (above) will often tend to overestimate the long-term sustainable yields of hard-rock wells.

Murray (1997) notes that, in South Africa, it is common practice to set the operational yield (for a 12 hour per day pump cycle) at a value of 60–65% of the yield resulting in the water level being drawn down to the pump intake. Murray argues, however, that this is a highly arbitrary method, and that an informed decision should be made on the basis of an understanding of whether sustainable yield is limited by, for example:

- decreasing bulk aquifer transmissivity at some distance from the pumping well (the edge of a transmissive fracture zone);

- limited groundwater storage in the aquifer;
- limited recharge to the aquifer.

We have already noted (above) the Van Tonder *et al.* (1998) flow characteristics approach to identifying the yield which ensures that the main inflow horizon does not become dewatered within a period t_{max}, the duration of a typical drought.

Finally, we can fall back on the 'suck it and see' approach – the long term testing of a well to ascertain whether well yield decreases with time. Figure 7.21 shows a hydrograph for a long term, constant drawdown test of Borehole 1 in granite at Hvaler, Norway. There is a drawdown response in observation Borehole 2, but not in 3 or 4, confirming the anisotropic and heterogeneous nature of hard-rock aquifers.

7.8   Single borehole tests: slug tests

This section will deal exclusively with tests that are carried out in single wells or boreholes to obtain values of aquifer parameters. Their advantages include their short duration and (in some cases) their ease of application. Their main disadvantage is that, while a full pumping test will investigate the entire aquifer volume between the pumping well and the observation borehole, a single borehole test will typically only stress a small volume of aquifer immediately around the borehole. Therefore, the derived values may not be representative of the aquifer as a whole.

7.8.1   Slug tests

The slug test is a short term, transient (time-variant) test. It is often carried out in piezometers of narrow diameter and with a relatively small section open to the aquifer. Measurements from slug tests can be useful in investigation boreholes (before the installation of a full scale well field, for example) or in piezometers which form part of the observation network of a pumping test. As a slug test only investigates the volume of aquifer immediately around the borehole, any gravel pack or open spaces behind the screen or casing may influence the analysis and result.

In a slug test, the borehole water level is instantaneously raised or dropped by one of three methods:

- removing a known volume of water (e.g., by a bailer);
- adding a known volume of water to the borehole;
- rapidly emplacing (or removing) a metal object (or "slug") of known volume on a line below the water level, thus displacing an equal volume of water upwards (or downwards; USGS, 2011).

The decay or recovery of the water level back towards the static condition is observed as a function of time. In many low to moderate permeability materials, the recovery of the water level is steady and gradual, with no oscillations – an 'overdamped' response. In some highly permeable aquifers, an 'underdamped' response is observed, where the

water level oscillates up and down several times in response to the slug – this can be more difficult to analyse (Butler *et al.*, 2003). Many analysis methods are available (Cooper *et al.*, 1967; Oosterbaan and Nijland, 1994; Hyder *et al.*, 1994), not all of which assume full penetration of the aquifer by the borehole or piezometer. We will merely present two of the most established analysis techniques here, although most depend on the plotting of a normalized head displacement $\Delta h/\Delta h_0$ versus time t.

The Hvorslev (1951) method can be applied to partially- or fully-penetrating boreholes, but only in saturated strata. The method can be applied to a positive or negative displacement of water level (Δh_0) from the static water level at time $t=0$ (Figure 7.23). Thereafter, readings of the water level Δh (again relative to static water level) are taken systematically at various times (t) after the perturbation was performed. It is assumed that the rate of exchange of water between aquifer and borehole casing is proportional to the formation's hydraulic conductivity (K) and the excess head of water (Δh) above the static water level. Furthermore the rate of change of water level is proportional to the volumetric rate of exchange, but also inversely proportional to the cross-sectional area of the borehole casing. In other words:

$$-\frac{d(\Delta h)}{dt} = \frac{KF\Delta h}{\pi r_c^2} \tag{7.78}$$

where F is a factor depending on the borehole or piezometer geometry and r_c is the radius of the borehole casing within which the water level perturbation takes place. Integrating between two given times results in:

$$K = -\frac{\pi r_c^2}{F(t_2 - t_1)} \ln\left(\frac{\Delta h_2}{\Delta h_1}\right) \tag{7.79}$$

If $\Delta h = \Delta h_0$ at $t=0$ (Figure 7.23), then:

$$K = -\frac{\pi r_c^2}{Ft} \ln\left(\frac{\Delta h}{\Delta h_0}\right) \tag{7.80}$$

The ratio ($\Delta h/\Delta h_0$) is calculated for each observation, then plotted on log-linear paper against t

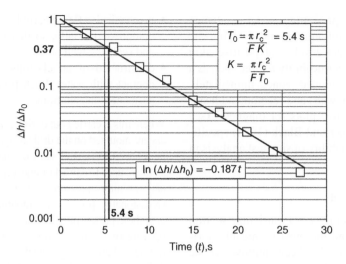

$$T_0 = \frac{\pi\, r_c^2}{F\,K} = 5.4\ \text{s}$$

$$K = \frac{\pi\, r_c^2}{F\, T_0}$$

$$\ln(\Delta h/\Delta h_0) = -0.187\,t$$

Figure 7.23 *Schematic representation of a 'slug test' analysis using the Hvorslev method*

(Figure 7.23), a procedure which should result in a straight line plot. K is found from the gradient of the plot. Alternatively, the time (T_0) taken for 63% of recovery is found (that is, the time t at which $(\Delta h/\Delta h_0) = 0.37$, and thus $\ln(\Delta h/\Delta h_0) = -1$; Figure 7.23). Then:

$$T_0 = \frac{\pi r_c^2}{FK} \qquad (7.81)$$

As mentioned above, the factor F depends of the geometry of the borehole or piezometer in relation to the aquifer. In the most common formulation, however, if the intake length of the borehole/piezometer filter section (L_i) is more than eight times its effective radius (r_{wf}), including gravel pack (if present), then:

$$F = \frac{2\pi L_i}{\ln\!\left(\dfrac{L_i}{r_{wf}}\right)} \qquad (7.82)$$

$$\text{and } K = \frac{r_c^2 \ln\!\left(\dfrac{L_i}{r_{wf}}\right)}{2L_i T_0} \qquad (7.83)$$

The Bouwer and Rice (1976; Bouwer, 1989) approach can be used in partially- or fully-penetrating boreholes in unconfined aquifers (but can also,

with caution, be applied to confined aquifers). It derives the equation:

$$K = \frac{r_c^2 \ln\!\left(r_e/r_{wf}\right)}{2L_i}\frac{1}{t}\ln\!\left(\frac{\Delta h_0}{\Delta h}\right) \qquad (7.84)$$

where r_e is the effective radius of the test's hydraulic influence and $\ln(r_e/r_{wf})$ is effectively a geometric factor found from nomograms; r_{wf} is the effective borehole radius in the formation, including filter pack; and r_c is the casing radius, which can be corrected to take account of water level changes within a gravel pack of porosity n (Bouwer, 1989):

Corrected equivalent casing radius

$$= \sqrt{\left([1-n]r_c^2 + n r_{wf}^2\right)} \qquad (7.85)$$

Bouwer (1989) notes that, when there is a strong contrast in permeability between, for example, a high K gravel pack and a low K formation, two straight line segments (corresponding to the two permeability regions) may be seen on a plot of $\log\Delta h$ versus t. Such skin effects (see Section 4.5) can be especially important in low permeability formations (Moench and Hsieh, 1985). Yang and Gates (1997) distinguish negative skin effects (i.e., an apparent low permeability resistance at the borehole wall preventing water exchange between borehole and aquifer, and delaying the recovery of the water level) and positive skin effects (e.g., a high permeability region, such as a gravel pack or developed zone, immediately beyond the borehole wall) and recommend head-displacement derivative plots for diagnosis (see Section 7.6.6). Barker and Black (1983) specifically consider the viability of slug test analysis in fractured aquifers.

7.8.2 Packer testing

Sometimes we wish to test a particular section of an aquifer, to find the most permeable horizons. One method that has been employed in the oil industry has been the 'bottom-up casing perforation' method. Here, a string of plain steel casing was set to the bottom of the borehole. A special tool was then used at the base of the hole to blow perforations into the casing, opening it to the formation. The fluid pressure, chemistry and flow rate could then be tested. Following testing, the base of the borehole was pumped full of cement and a new section perforated for testing. This was a somewhat primitive (but effective) method, which resulted in a borehole being progressively filled with cement. The use of so-called 'packer testing' avoids this limitation.

Packer testing involves the isolation and testing of specific sections of a borehole, and can be employed in open (unscreened) sections, or can be used in cased boreholes where the section to be tested is perforated by a special tool as described above. In the water industry, packer testing is of particular interest in fractured aquifers, multilayered aquifers or aquifers where hydraulic conductivity varies greatly with depth.

Packer testing can employ a single packer, allowing the section between the packer and the base of the hole to be tested. Alternatively, two packers can be employed to isolate a specific intermediate borehole section. Packers are typically:

- hydraulically inflated by gas or fluid to form a tight seal against the borehole wall;

- mechanical: a flexible, rubber-like packer can be compressed axially, so that it expands radially and seals the well bore;
- constructed of an elastomer that expands on contact with water. Such packers are typically permanent installations.

Where this is carried out during drilling, and the packers are mounted on the drill string, the technique may be referred to as drill stem testing. Ports in the drill stem are used to extract or inject fluid, collect samples and monitor pressures in the isolated section. Occasionally, packers are mounted on a dedicated downhole packer pipe, rather than the drill stem of a rig. Lugeon testing (Section 5.2.6) is a form of packer testing, where a single packer is employed during drilling (Lancaster-Jones, 1975; Houlsby, 1976).

In dual-packer testing, two packers are mounted on the test pipe/drill stem and a specific section of the borehole is tested. The complete packer/drill stem/port system is referred to as a *bottom-hole assembly (BHA)*. Several types of hydraulic test are available: flow extraction tests, flow injection tests and pressure build-up tests. These are all analysed by variations of the Theis radial flow method described in Section 7.4. The only major difference is that water pressure is typically plotted against time, rather than groundwater drawdown or head. Remember:

$$\text{Pressure}(\text{Pa}) = \text{Freshwater head}(\text{m})$$
$$\times 9810 \left(\text{kg m}^{-2}\text{s}^{-2} \right) \quad (7.86\text{a})$$

$$\text{Pressure}(\text{Pa}) = \text{Fluid head}(\text{m}) \times \rho g \quad (7.86\text{b})$$

where ρ = fluid density (kg m^{-3}) and g = 9.81 m s^{-2}.

Packer testing can be subdivided into flowing well drill-stem testing (FDST), where formation fluid pressure is artesian and fluid can overflow through the drill stem to the surface, and nonflowing tests (NFDST). The most common forms of test include (see Earlougher, 1977):

1. Production tests, where an artesian packered aquifer is allowed to flow freely and discharge at the surface (FDST). Alternatively water can be pumped from the packered section at a

constant rate. These are approximately equivalent to constant head and constant yield tests respectively (although there will be neither constant flow nor constant pressure during the period when fluid is flowing up the drill stem – a kind of well-bore storage effect).

2. Injection tests, where water is injected to the formation at a constant rate or constant pressure.

3. Stepped or phased injection tests, where injection of water is carried out during short-term (tens of minutes to hours) steps of increasing pressure, sometimes followed by steps of reducing pressure (Brassington and Walthall, 1985). This allows a plot of flow vs. pressure to be constructed. British Standards Institution (2003) describes the procedure and recommends five steps of constant pressure of duration 15 minutes, with injected volumes being monitored.

4. A 'shut-in' or 'pressure build-up' test, in an artesian formation, where a packered section is allowed to discharge fluid through the drill stem. The valve is then closed in the drill stem, isolating the packed section. The rate of pressure build-up back to formation pressure is monitored. This is analogous to a recovery test (Section 7.4.6) and the typical 'Horner (1951) plot' used by hydrocarbon engineers is analogous to the Theis recovery technique.

5. Transient 'pulse' tests, where a pressure pulse is applied and then allowed to decay, allowing information to be derived about the storage properties of the formation (via inverse modelling) and potentially also the *in-situ* stresses within the formation (Banks *et al.*, 1995).

If a steady injection rate Q and a steady excess head Δh are established in a packer test in a vertically unbounded aquifer, the following general relationship is valid (Bliss and Rushton, 1984):

$$Q = \frac{2\pi L K \Delta h}{\ln\left(\frac{L}{r_w}\sqrt{\frac{K_r}{K_z}}\right)} \qquad (7.87)$$

where L is the packer response section length (>10 times the borehole diameter), r_w is the well radius in

the open section, K is the hydraulic conductivity and $\sqrt{K_r / K_z}$ is an anisotropy ratio based on radial (K_r) and vertical (K_z) conductivities. For a constrained (vertically-bounded) aquifer, where radial flow dominates, a radial steady state solution is more appropriate [see Equations (7.5) and (7.6)].

The typical results from packer testing include: the water head/pressure in the section of interest, the water chemistry and temperature, the average hydraulic conductivity in the section of interest and, often, information about storage, skin effects and boundary conditions. It is worth noting that, if very high pressure differentials are applied, the apparent hydraulic conductivity obtained from discharge tests can be significantly lower than those from injection tests, as the injection fluid pressure can open fracture apertures or pore spaces, leading to artificially elevated calculated values. Of course, at even higher fluid pressures, hydraulic fracturing can be induced within the packered section.

In some deep geothermal wells, permanent bottom-hole assemblies (BHAs) up to 1500 m long are being installed, with up to 30 packered sections of 50 m, each with its own automated port from the stem, to allow 30 distinct sections of the aquifer to be independently hydraulically fractured or tested in any order (Meier *et al.*, 2015).

7.9 Tracer tests

The concept of tracers is introduced in Chapter 2. A tracer is simply a substance (or even simply a temperature or heat signal) that is characteristic of a given source of water. Tracers can be:

- *Natural*: For example, a given ^{2}H or ^{18}O isotopic signature characterizing a source of rainfall recharge,
- *Ambient artificial*: For example, a given pharmaceutical characteristic of sewage effluent in a watercourse (e.g., Bones *et al.*, 2007). Alternatively, ambient concentrations of certain substances in rainfall, allowing recharge to be approximately dated, such as tritium (derived

from former fusion bomb tests), methyl tertiary butyl ether (MTBE), sulphur hexafluoride (SF_6) and chlorinated fluorocarbons (CFCs) – see Cook and Solomon (1997); Plummer and Friedman (1999); Darling and Talbot (2003); Darling *et al.* (2003, 2012).

- *Deliberately added tracers*: These can be added to sinkholes, surface waters, groundwater (via injection into wells) and generally used to 'trace' what happens to the water and where it ends up.

Deliberately added tracers can be:

- organic dyes such as fluorescein or rhodamine (Banks *et al.*, 1995);
- non-radioactive (or in some cases, even radioactive) isotopes (Williams *et al.*, 1998);
- small particles or biological agents: DNA fragments, bacteriophages or spores such as lycopodium powder (Cronin and Pedley, 2002). These can typically be used qualitatively to prove a hydraulic connection but can be difficult to use quantitatively;
- inorganic salts: typically, conservative, unreactive, highly soluble salts of lithium, sodium or potassium (often chlorides or bromides);
- heat (a temperature signal – Stonestrom and Constantz, 2003; Anderson, 2005).

As the addition of tracers to natural water is often restricted or regulated (e.g., Environment Agency, 2012a) and given that there are health concerns over the use of some organic dyes, there is an increasing tendency to use simple salts, particulates or heat as environmental tracers.

One method of using a tracer in non-pumping wells is simply to disperse a known quantity of the tracer (e.g., salt) in the water column of a well (known volume). The rate at which the known initial tracer concentration (C_0) declines can be used to deduce the natural groundwater flow through the well. This is called a Single Borehole Dilution Test (Maurice *et al.*, 2011, 2012). If the vertical distribution of the tracer is uniform and there are no vertical flows, then the Darcy velocity (groundwater flux, v_D) in the aquifer can be estimated (Lewis *et al.*, 1966; Pitrak *et al.*, 2007):

$$\ln\left(\frac{C}{C_0}\right) = -\frac{4v_D t}{\pi r_w} \quad \text{or} \quad \log_{10}\left(\frac{C}{C_0}\right) = -0.553\frac{v_D t}{r_w}$$

$$(7.88)$$

where C is the concentration at time t and r_w is the well radius. v_D can thus readily be found by plotting $\log_{10}C$ versus t.

Depth-specific sampling or electrical conductivity logging (or temperature logging if heat is the tracer) can also be employed to pinpoint the locations at which fresh water is entering or tracer is leaving the well bore. In pumping tests, this technique can be used in (non-pumped) observation boreholes near the main pumping well (Figure 7.24; Mathias *et al.*, 2007). As pumping usually increases the lateral flow of water through nearby wells, this technique can be used to identify the main transmissive fissure or aquifer horizons near the pumping well. If the point of tracer addition is close enough to the pumping well, the tracer should eventually appear in the pumping well itself. The time at which the tracer 'breaks through' to the pumping well allows the groundwater flow velocity under pumping conditions to be estimated, while the shape of the breakthrough curve can indicate the nature of solute transport mechanisms (Mathias *et al.*, 2007) – for example, a long 'tail' on the breakthrough curve can indicate diffusion of tracer into the matrix. It should be noted that if the solute interacts with the aquifer matrix (e.g., by sorption) it will be somewhat retarded relative to groundwater flow and its arrival may tend to underestimate true groundwater flow velocities.

One circumstance where tracer breakthrough times are particularly important is in the scenario of the geothermal well doublet (see Section 4.8). A tracer can be added to the injection well and monitored in the abstraction well. Heat can be used as a tracer in this context, but it should be remembered that heat is also retarded relative to groundwater flow and conservative tracers, as heat is sorbed into the aquifer mineral matrix. The retardation factor is related to the volumetric heat capacity of the aquifer (Banks, 2009b, 2011, 2012a; Barker, 2012).

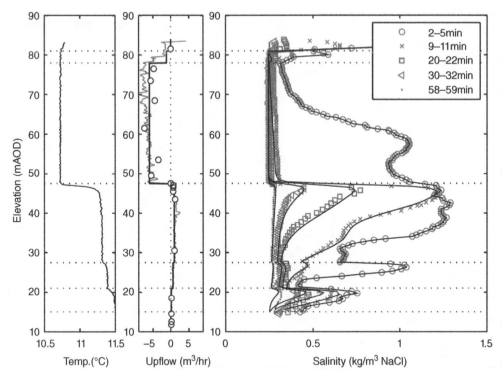

Figure 7.24 *Results of geophysical logging, with injection of saline tracer, in an observation borehole located 54 m from a pumped abstraction well, pumping at c. 67 ls⁻¹, in the English Chalk aquifer. The left hand column shows the groundwater temperature, the middle column the upward flow in the borehole estimated from an impeller flowmeter (grey line), a heat pulse flowmeter (circles), and numerical inversion of the EC logs (thick black line). The right hand column shows the fluid electrical conductivity (EC) logs recorded at various times after the salt tracer injection. The solid lines shown in the right column are synthetic logs generated from a numerical model (progressive dilution of tracer). The horizontal dotted lines across all the plots show the elevations of the flow horizons used for numerical inversion of the EC logs. After Mathias et al. (2007) and reproduced with permission of © the American Geophysical Union (2007) and John Wiley & Sons*

7.10   Geophysical logging during pumping tests

We have already encountered geophysical borehole logging in Section 6.4. During a pumping test, axial flow meters and heat pulse flow tools can be used in pumping wells (and nearby observation wells) to determine the flow up (or down) the borehole at any location. This assists in identifying the location of transmissive horizons or fractures that are contributing groundwater flow (Figure 7.24). In some fractured or fissured aquifers, flowing fractures above the pumping water level in the well can be directly observed by CCTV (Figure 6.28b).

Where saline or heat tracers are injected into nearby wells, fluid conductivity and temperature logs can be applied in the injection or observation wells to monitor the rapidity of tracer dilution and the locations at which the tracer is being removed (i.e., flowing horizons). In the pumping well, the same fluid conductivity and temperature logs can identify the horizons at which the tracer or heat is entering the well (see Figure 7.24; Mathias *et al.*, 2007).

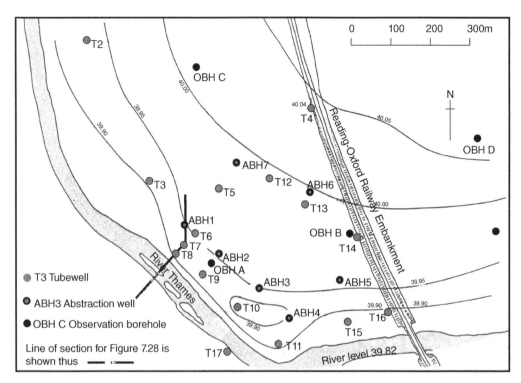

Figure 7.25 *The Gatehampton well field, showing contours on static groundwater level (in m above sea level) on 19 September 1986 (after Robinson et al., 1987). Public domain information, provided by and reproduced with the permission of the Environment Agency of England & Wales (Thames Region)*

In deep, wide diameter shafts (for example, mine shafts), tracer can be injected at a given point in the shaft and its slow progress towards a pump elsewhere in the shaft can be monitored by fluid logging or by conductivity or temperature sensors spaced at discrete depths along the shaft. Thus, the flow profile of the shaft can be deduced.

7.11   Test pumping a major well field: the Gatehampton case study

In 1986, England's largest groundwater source, at Gatehampton, near Goring-on-Thames, Oxfordshire, was test-pumped (Robinson *et al.,* 1987). The site lies on the flood plain of the River Thames, where several metres of alluvial sand and gravel overlie the Upper Cretaceous Chalk aquifer. The static water table varies from 1.5 to 6 m below ground level across the site. The pumping test network (Figure 7.25) consisted of:

- Seven, 70–80 m deep, 740 mm diameter abstraction wells (ABH 1 to 7), completed in the Chalk. Some of these were equipped with two large electrical submersible pumps in each well (Figure 7.3).
- Four, 60 m deep, 200 mm diameter observation boreholes (OBH A to D), completed in the Chalk and equipped with continuous float-operated chart recorders (Figure 7.5a).
- Seventeen shallow well-point tubewells in the alluvial gravel materials (T1-T17).
- Two existing Chalk observation boreholes (part of a regional network).
- Several existing private wells.
- A station to record the stage (level) of the River Thames.
- Rainfall station.

The test regime involved the following elements:

- Individual step testing (4×2 hr steps) of the seven abstraction wells. Wells 4 and 7 and wells 2 and 5 were tested in pairs, as they were so far apart that interference was judged minimal.
- Individual 3-day constant rate testing of the seven abstraction wells. Paired testing was carried out for some wells as above.
- Group testing of all seven wells, over a 4-month period, to ascertain sustainability and environmental effects. The total discharge was stepped up at intervals throughout the test, culminating in a total of around $110\,000\,\text{m}^3\,\text{d}^{-1}$ ($=1270\,\text{l}\,\text{s}^{-1}$).

For the step testing and three-day testing, almost all drawdown took place within the first 15 minutes of pumping, the drawdown-time curve becoming relatively flat thereafter. This was interpreted as the effect of a recharge boundary (the River Thames) in close proximity (see Section 7.6.2). The best performing well (ABH 4) yielded $15\,100\,\text{m}^3\,\text{d}^{-1}$ ($175\,\text{l}\,\text{s}^{-1}$), with only 3.45 m drawdown after 3 days. This led to great optimism that the site was capable of yielding huge volumes of groundwater for long periods, the yield being sustained by infiltration of surface water from the River Thames.

The subsequent group test demonstrated how short-term test interpretations can be over-optimistic. In contrast to the short term tests, the group test demonstrated that, at a high collective rate, stable drawdowns were not achieved. Drawdowns continued to increase, with a significant break in slope occurring after some 56 days, around 14th–15th November (Figure 7.26). It has been suggested that this might be characteristic of some form of hydraulic barrier (caused by, for example, the transmissivity of the Chalk declining rapidly away from the valley of the Thames). Alternatively it might reflect the dewatering of much of the high specific yield alluvial gravels or the uppermost, highly transmissive horizons of the Chalk.

Moreover, the cone of depression continued to expand during the group test, and eventually a cone of depression began to develop on the southwest side of the River Thames, opposite to the well field (Figures 7.27 and 7.28). This indicated that the Thames was not a perfect recharge boundary (perhaps not surprising, as it does not fully penetrate the aquifer) and that the site was drawing water from the Chalk aquifer *underneath* the Thames. This also suggested that the Thames was maybe not in such good hydraulic continuity with the Chalk as was originally supposed (initial groundwater modelling studies suggested <10% of

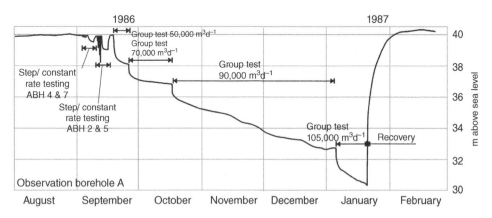

Figure 7.26 *Hydrograph of groundwater level in Observation Well A during testing of the Gatehampton source (after Robinson et al., 1987). Public domain information, provided by and reproduced with the permission of the Environment Agency of England & Wales (Thames Region)*

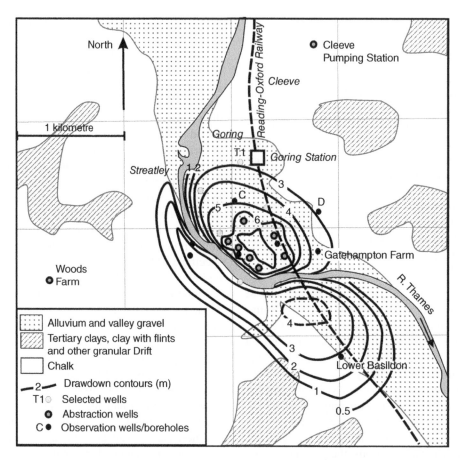

Figure 7.27 *Regional map (with simplified geology) of the Gatehampton area, showing contours on drawdown (m) on 19 January 1987 (relative to 19 September 1986) (after Robinson et al., 1987). Public domain information, provided by and reproduced with the permission of the Environment Agency of England & Wales (Thames Region)*

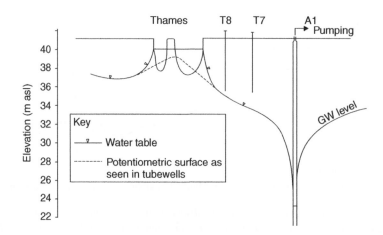

Figure 7.28 *Schematic cross-section below the Gatehampton site on 19 January 1987, demonstrating the development of a cone of depression on the southwest bank of the Thames (after Robinson et al., 1987). Public domain information, provided by and reproduced with the permission of the Environment Agency of England & Wales (Thames Region)*

the yield was derived from the River). Furthermore, hydrochemical sampling of the abstracted water during the test failed to demonstrate any clear ingress of large quantities of Thames water to the Chalk. Clearly, although the site was still a hugely valuable groundwater resource, it would not provide the almost limitless supply of 'filtered Thames water' that was originally envisaged. The site would have to be managed with caution.

Of course, the pumping test provides only part of an early stage in the understanding of a new well field, and as more data have been collected, the site has become better understood. In due course, Younger (1989) proposed a model of Gatehampton that implied that the Chalk did not have a high degree of hydraulic continuity with the Thames. Indeed, the river bed sediments were suggested to be of rather low vertical hydraulic conductivity. Younger argued that the high well yields at the site, and the flat drawdown vs. time responses during short term testing, were due to extremely high transmissivities in the Chalk below the valley of the Thames. This was supposed to be a consequence of the Thames having acted as a perennial *talik* – a zone of unfrozen ground, with upwelling groundwater flow in an otherwise permafrosted terrain, during periglacial conditions in the Devensian period, at which time fissure apertures would have been enhanced by intense dissolution. In effect, the Gatehampton site could thus be regarded as having tapped a huge underground drain (a zone of highly transmissive fissured Chalk), rather than feeding off induced recharge from the Thames.

The debate continues, however, and conceptual models are still being refined. Indeed, subsequent groundwater modelling investigations by Thames Water suggested a greater contribution from the River Thames to the Gatehampton abstraction, up to around 25% at abstraction rates of 86 000 to 105 000 m^3 d^{-1}.

7.12 Record-keeping

Following a pumping test, it is usual to generate a report. For detailed guidance on reporting, the reader is referred to British Standards Institution (2003). Briefly, it is suggested here that such a report should contain the following basic elements:

i. Executive summary.
ii. Introduction and objective of the testing programme.
iii. Brief characterization of the region: topography, climate, hydrology, geology and hydrogeology.
iv. Results of water features survey.
v. Description of well field and monitoring network.
vi. Map or plan of well field and monitoring network, with clear annotation and identification.
vii. Illustrative hydrogeological cross-section through well field.
viii. Description of test pumping regime.
ix. Brief description of results.
x. Hydrograph for entire test period including water levels in all monitoring points, surface water flows (if applicable), pumped discharge, rainfall (and, if necessary, barometric pressure and tidal fluctuations).
xi. Map or plan showing extent of drawdown at key points during the test.
xii. Discussion of water quality results, from both on-site and laboratory analyses (Chapter 8).
xiii. Brief discussion of interpretation of results, including:
   • a justification for the chosen pumping test analysis method;
   • the degree to which fundamental assumptions are satisfied or violated;
   • uncertainty attached to input data and to their interpretation.
xiv. Conclusions and recommendations.
xv. Annexes.

The Annexes should include:

i. Construction details of wells and boreholes.
ii. Calibration curves and specifications for monitoring equipment.
iii. Other quality assurance data and metadata.

iv. Listings of raw data from monitoring points (typically on digital media or via a reference to an internet-based data repository).

v. Hydrographs from individual monitoring points.

vi. Pumping test analysis plots and calculations.

vii. Water quality analysis sheets and graphs.

Much of the Annex material could be submitted on a digital medium. Whatever format is adopted, it is essential to document the raw data, and to document the means of analysis used in the data interpretation. In the future, another hydrogeologist may wish to revisit the results, and to add his or her own insights.

8

Groundwater Sampling and Analysis

The aim of this chapter is to give an introduction to good practice for water quality sampling and analysis of groundwater from wells and boreholes, both in the testing phase and the production phase. Many countries have overriding national standards for water quality, sampling and analysis and it is important for the practitioner to become familiar with these; for example: SFT (1991) guidance for Norway; the British Standards Institution (2008, 2009, 2010b, 2012b, 2014b) for the United Kingdom, and corresponding ISO/EN numbers for the remainder of the European Union. Additionally, the following standards and guidelines should be of interest to the reader:

1. The United Nations organizations publish guidelines for acceptable limits of various microbiological and chemical parameters in drinking water (WHO, 2011; see Appendix 4), and also guidance on sampling (Chapman, 1996).
2. The US Geological Survey (Wilde, 2011; US Geological Survey, 2015) publishes comprehensive advice on collection of samples and water quality data, available as an online Field Manual.
3. Supplementary advice can be obtained via publications of the Illinois State Water Survey

(Barcelona *et al.*, 1985) and the US Environmental Protection Agency (Puls and Barcelona, 1996). The US Environmental Protection Agency also produces standard documents for sampling, preservation and analysis of waters, for example, US Environmental Protection Agency (1999a, 2004, 2006) on total recoverable elements, general groundwater sampling and low-flow sampling techniques, respectively. Other publications also offer useful advice (Bartram and Ballance, 1996; UNESCO/WHO/UNEP, 1996; Mather, 1997; Yeskis and Zavala, 2002; and Appelo and Postma, 2005).

Before going into the details of groundwater sampling procedures, we would like to stress to the reader that thorough planning of a sampling campaign is a crucial element for success. A risk assessment and health and safety assessment should be carried out to identify and mitigate both risks to human health and risks to the success of the mission (see Box 6.3 and Appendix 3). Basic information concerning the condition of the well and sample location should be gathered, many elements of which will be similar to those in Box 6.2.

Water Wells and Boreholes, Second Edition. Bruce Misstear, David Banks and Lewis Clark.
© 2017 John Wiley & Sons Ltd. Published 2017 by John Wiley & Sons Ltd.

Table 8.1   *Checklist of accessories for use in groundwater sampling*

Essential	Agreement with laboratory on sample delivery and analysis
	Special instructions from laboratory
	Sample list and QA documentation
	Map
	Dipper tape (and/or tape measure and/or plumb line)
	Bailer/pump and sufficient cable/rising main
	Portable generator (if required)
	Sample labels (preferably pre-fixed to flasks)
	Waterproof marker pen and pencil
	Field notebook
	Deionized water
	Deionized water flask with 'squirtable' nozzle
	Thermometer/temperature meter
	Field meters (pH, E.C., etc.) with spare batteries
	Calibration solutions
	Powder-free latex gloves
	Any necessary field kits
	All necessary flasks and sample bottles (and spares)
	Sufficient syringes and filter units (and spares)
	Securely packed preservative agents (if not pre-filled)
	Coolbox with bubble wrap packing
	Ice packs/coolant packs
	Kitchen towel
	First aid kit
	All necessary personal protective equipment
	Means of communication
Useful/Optional	Portable fridge (runs off car battery)
	Portable VOC/gas detector
	Wellhead 'poncho': large waterproof groundsheet with hole cut in centre to fit over wellhead. This provides a dry clean base around the wellhead upon which to work
	Alternatively, small portable table
	Large fishing umbrella/small camping shelter (in case of rain)
	Torch and mirror (to inspect well using reflected sunlight)
	Cigarette lighter/gas torch to sterilize taps
	Bucket (graduated or of known volume)
	Watch (ideally a stop watch), to measure flow
	Tool-kit
	Digital camera
	Global positioning system (GPS) device
	Sealable freezer bags
	Geochemical grade sample bags (if soils or solids to be sampled)
	Wet wipes/alcohol-based gel

Access routes, transport and communication routines will need to be assessed and permissions from landowners obtained. Agreements will need to be made with laboratories regarding sampling routines (and a documented sampling plan developed), analyses required, delivery to and reception at the laboratory. Quality assurance documents will need to be written and, of course, equipment will need to be prepared, cleaned and calibrated (Table 8.1).

8.1   Water quality parameters and sampling objectives

The purpose of the monitoring or sampling must be clearly defined, as this will be crucial in constraining the programme design, the parameters to be analysed and the spatial and temporal frequency of sampling. The purpose may be for initial investigation, screening, contamination delineation, regulatory compliance, hydrometric monitoring, operational monitoring or for research.

When sampling a groundwater, there is often a temptation to sample only those parameters required by law or those which are directly relevant to a specific pollution problem. This limited approach to sampling is not recommended here: we contend that the hydrogeologist or engineer should *always* consider analysing what are referred to in this chapter as the *master variables* and the major ions on a representative selection of samples from a site. The reasons for this are:

- These parameters dominate the water chemistry and may influence how other elements or species behave. For example, it is of limited use to know the arsenic concentration in a water sample, if nothing is known about the redox and pH conditions and the concentrations of other dissolved elements such as iron. These will determine how the arsenic behaves in solution.
- They will also provide a hydrochemical context for the groundwater: they will provide information on how the water's chemistry has evolved and possibly even give some clues as to its residence time. This hydrochemical context can be interpreted by an experienced hydrochemist, who may then be able to provide plausible explanations for the occurrence of parameters of more direct interest.
- The major variables allow quality controls and reality checks to be made on the analytical results. For example, the concentrations of major ions enable an ion balance error to be calculated (Section 8.1.3). If this error greatly exceeds ±5%, it suggests that either (a) the analytical result is wrong or (b) that there are other parameters (perhaps due to gross pollution)

which are affecting the ion balance. As another example, if pH is neutral yet concentrations of aluminium are high, this might suggest that either (a) the result is wrong, as aluminium is not substantially soluble under neutral pH conditions, or (b) that the sample was not adequately filtered during sampling, allowing particulate or colloidal matter to enter the sample.

8.1.1   Master variables

The so-called master variables are normally regarded as:

- *pH*. This is a measure of the acidity of the solution. In other words, it is a measure of the activity of protons (or hydrogen ions $H^+_{(aq)}$) in solution:

$$pH = -\log_{10}\left[H^+\right] \qquad (8.1)$$

where $[H^+]$ is the activity of hydrogen ions in solution in mol l^{-1}.

- *pe*. This is a measure of how reducing or oxidizing a solution is. It is a measure of the activity of electrons in the solution:

$$pe = -\log_{10}\left[e^-\right] \qquad (8.2)$$

where $[e^-]$ is the activity of electrons. Very often, electron activity is expressed as a variable *Eh*, the so-called redox potential of the solution (in volts):

$$Eh = \frac{2.3RT^o}{F}pe \qquad (8.3)$$

where R is the gas constant (8.3143 J K^{-1} mol^{-1}), T^o is the temperature (in degrees Kelvin) and F is the Faraday constant (96 487 C mol^{-1}).

8.1.2   Main physicochemical parameters

There are other physicochemical parameters that may be readily measured, usually by field techniques. *Temperature* may be measured simply with a conventional thermometer or thermistor, and can be a useful parameter for distinguishing true groundwater from surface water seepage (true groundwater will usually approximate to annual average air temperature, whereas surface waters will more closely follow seasonal variations).

Electrical conductivity (EC) is usually measured in μS cm^{-1} (microSiemens per centimetre) and is a user-friendly indicator of how many dissolved ions there are in solution. As most of the dissolved content of natural groundwater is in ionic form, EC can be empirically related to *total dissolved solids* (TDS). For a typical fresh groundwater, dominated by calcium and bicarbonate, the following approximate relationship pertains at 25 °C (Appelo and Postma, 2005):

$$\sum anions\left(meq\,l^{-1}\right)$$
$$= \sum cations\left(meq\,l^{-1}\right) \approx \frac{EC}{100} \qquad (8.4)$$

where EC is in μS cm^{-1}. This relationship is valid up to around 1500 μS cm^{-1}. Alternatively, the following approximate relationship can be used:

$$TDS\left(mg\,l^{-1}\right) \approx EC\left(\mu S\ cm^{-1}\right) \times F \quad (8.5)$$

where the factor F typically varies between 0.55 and 0.75 (depending on ionic species present and total salinity – Hem, 1985). A value of around 0.64 is often used as an average. Appelo and Postma (2005) suggest that, theoretically, a 1 mmol l^{-1} solution of calcium bicarbonate (162 mg l^{-1}) has a conductivity of c. 208 μS cm^{-1} at 25 °C ($F=0.78$), while a 1 mmol l^{-1} solution of sodium chloride (58 mg l^{-1}) has a conductivity of 126 μS cm^{-1} ($F=0.46$). *TDS* can also be determined in the laboratory by evaporating the water to dryness and weighing the residue. Here, it is important to note that the mass of the residue may depend on the temperature of heating used by the laboratory. Excessive heating may cause some salts (for example, gypsum) to lose their water of crystallization, or cause bicarbonates or carbonates to be converted to oxides:

$$CaSO_4.2H_2O \rightarrow CaSO_4 + 2H_2O \uparrow$$

$$Ca^{2+} + 2HCO_3^- \rightarrow CaCO_3 + H_2O \uparrow$$
$$+ CO_2 \uparrow \ ^- \rightarrow CaO + CO_2 \uparrow$$

Dissolved oxygen (O_2) may be determined with an electrode and is a good indicator of the water's redox state. Water in equilibrium with atmospheric air contains 8.3 mg l^{-1} dissolved O_2 at 25 °C, increasing to 12.8 mg l^{-1} at 5 °C (Appelo and Postma, 2005). Dissolved oxygen can also be determined in a laboratory, provided that the dissolved oxygen content is immobilized in the field by addition of alkaline manganese (II) and iodide (the Winkler, 1888, method – see Section 8.5.2).

Dissolved carbon dioxide CO_2 (or carbonic acid $H_2O + CO_2 = H_2CO_3$) can be estimated by titration of the water to a pH of 8.2–8.3 (phenolphthalein end-point) using a strong base. At this pH, the carbonic acid will have been converted to bicarbonate (HCO_3^-). This technique assumes that carbon dioxide is the only significant acidic species in the water and it will not work if there are significant quantities of dissolved iron, manganese or other species that produce acid on oxidation and hydrolysis.

8.1.3   Major ions

The major ions in groundwater are commonly regarded as sodium, potassium, calcium, magnesium, chloride, (bi)carbonate, sulphate and nitrate (and occasionally ammonium). Table 8.2 provides an overview of the chemical notation, ionic masses and equivalent masses for these ions.

Alkalinity is the sum of alkaline species present in a solution. Alkalinity can be defined in a number of ways, but is broadly the amount of acid (in meq) which needs to be added to a litre of water to bring the pH down to a reference value (often taken to be pH $\approx$ 4.3, at which point effectively all of the bicarbonate and carbonate present will have been converted to carbonic acid). Hence the alkalinity (in meq l^{-1}) is approximately equal to the amount of carbonate and bicarbonate present in the water in meq l^{-1}.

Table 8.2 *Properties of selected ionic species in groundwater*

Ion	Notation	Ionic mass (g mol^{-1})	Equivalent mass (g eq^{-1} or mg meq^{-1})
Sodium	Na^+	22.99	22.99
Calcium	Ca^{2+}	40.08	20.04
Magnesium	Mg^{2+}	24.31	12.15
Potassium	K^+	39.10	39.10
Bicarbonate	HCO_3^-	61.02	61.02
Carbonate	CO_3^{2-}	60.01	30.00
Chloride	Cl^-	35.45	35.45
Sulphate	SO_4^{2-}	96.06	48.03
Nitrate	NO_3^-	62.00	62.00
Ammonium	NH_4^+	18.04	18.04
Ferrous iron	Fe^{2+}	55.85	27.93
Fluoride	F^-	19.00	19.00
Manganese (II)	Mn^{2+}	54.94	27.47
Hydrogen	H^+	1.008	1.008
Hydroxide	OH^-	17.01	17.01

Notes:
To convert mg l^{-1} to meq l^{-1}, divide by equivalent mass (g eq^{-1} or mg meq^{-1}).
To convert alkalinity cited as mg l^{-1} $CaCO_3$ to meq l^{-1}, divide by 50.04 mg meq^{-1}.
To convert mg l^{-1} NH_4^+–N (ammonium as nitrogen) or mg l^{-1} NO_3^-–N (nitrate as nitrogen) to meq l^{-1}, divide by 14.01 mg meq^{-1}.
To convert nitrate cited as mg l^{-1} N to mg l^{-1} NO_3^-, multiply by 62.00/14.01 = 4.43.

To calculate an ionic balance, the major ionic species are converted to meq l^{-1} (Table 8.2):

$$\text{Sum cations} \left(\text{meq}\,l^{-1} \right)$$
$$= \frac{\left(Na^+ \right)}{22.99} + \frac{2\left(Ca^{2+} \right)}{40.08} + \frac{2\left(Mg^{2+} \right)}{24.31} + \frac{\left(K^+ \right)}{39.1} \quad (8.6)$$

$$\text{Sum anions} \left(\text{meq}\,l^{-1} \right)$$
$$= \frac{\left(Cl^- \right)}{35.45} + \frac{2\left(SO_4^{2-} \right)}{96.06} + \frac{\left(NO_3^- \right)}{62.00} + \text{Alkalinity} \quad (8.7)$$

where ( ) refers to a concentration in mg l^{-1}, and alkalinity is already in meq l^{-1}.

The ion or charge balance error (*IBE* or *CBE* – Fritz, 1994), which should be <±5%, is then given by:

$$IBE = \frac{\left[\text{sum cations} - \text{sum anions} \right]}{\left[\text{sum cations} + \text{sum anions} \right]} \times 100\%$$

$$(8.8)$$

8.1.4 Drinking water

The chemical, radiological, microbiological, physicochemical and aesthetic parameters that should be analysed in drinking water will usually be defined by national drinking water regulations or, failing that, by referring to an international guideline, such as that of the WHO (2011, see Appendix 4).

Any given set of drinking water guidelines may not include every parameter of health significance. Examples of chemical parameters that should be considered for analysis in certain geological environments, even though they are not contained in some sets of national regulations, include:

- uranium and radon, especially in crystalline rock terrains, and particularly in acidic igneous or metamorphic rocks (see Box 2.9);
- thallium and beryllium (Box 2.10).

In addition to the parameters regulated for health or aesthetic reasons, waterworks feeding major community supplies will also be monitored on a regular basis for parameters that might indicate any rapid change in source water quality (such as a rapid recharge event), affect or reflect the efficiency of treatment or affect the suitability of the water for entry to the distribution network. The World Health Organization (2011) suggests the regular operational monitoring of factors such as wellhead structural integrity, chlorine (or other disinfectant residual) and turbidity as a minimum. The World Health Organization (2011) further suggests that the final choice of other operational parameters will be network-specific and will reflect the hazards to that network: other operational parameters may include colour, electrical conductivity (which, together with turbidity, can be indicators of rapid recharge events – see Box 8.1), UV absorbency, pH, redox potential, dissolved oxygen, heterotrophic plate count and organic carbon (high concentrations of which can

Box 8.1   Groundwater chemistry as a guide to vulnerability

In the Bergen region of Norway, the hydrochemical behaviour of groundwater from five drilled wells in crystalline bedrock was studied throughout the year. In three of the wells, water chemistry remained stable, while in the other two, it showed significant seasonal fluctuation (Figure B8.1(i)).

Periods of rainfall and recharge coincided with episodes of low pH, and low concentrations of lithologically-derived elements such as fluoride and radon. The wells simultaneously experienced elevated colour (derived from organic acids leached from peaty top soils), and elevated concentrations of pH-dependent (or organically-bound) elements such as iron and aluminium. In these wells, the 'true' groundwater, characterized by elevated pH, alkalinity and concentrations of lithologically-derived elements, is being periodically diluted by rapid influxes of low-pH, organic-rich recharge water. This recharge clearly has a low residence time as its low pH has not been neutralized by the geological environment and its organic content has not been biodegraded. The wells that respond rapidly to recharge events may thus be regarded as vulnerable to surface contamination: the recharge may be entering the wells via short, open fracture pathways, or via inadequate wellhead sealing.

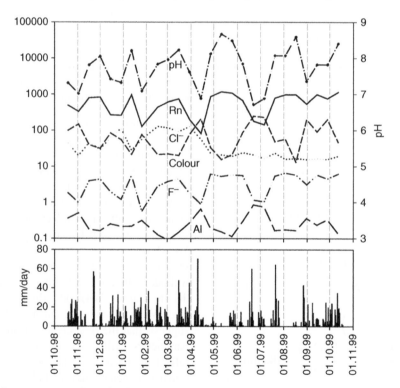

Figure B8.1(i)   *Variation in fluoride, aluminium and chloride (all mg l⁻¹), colour (mg l⁻¹ Pt), pH and radon (Bq l⁻¹) in a vulnerable borehole in western Norway (note logarithmic scale), with daily precipitation (which may fall as snow in winter) shown as a histogram. Reproduced by permission of Norges geologiske undersøkelse (Geological Survey of Norway)*

In summary, wells that are vulnerable to potential contamination during recharge events (at least in temperate regions of the world) may be characterized by limited episodes of:

- low pH and alkalinity;
- elevated colour and turbidity;
- low concentrations of lithologically-derived parameters;
- elevated concentrations of surface-derived parameters (such as nitrate);
- elevated concentrations of pH-dependent metals.

enhance biofilm growth in distribution systems). Monitoring of water pressure in piped supplies can also confirm whether periods of underpressure, and thus potential for ingress of contaminated soil water into the pipe system, are occurring.

For groundwaters being distributed through piped systems, the water in the distribution network will also be subject to a regular programme of monitoring and validation for indicators as to the effectiveness of treatment, such as chlorine residual, faecal indicator bacteria and, in some cases, concentrations of chemicals added in the treatment process.

Groundwaters being sold as spring waters or mineral waters will usually be subject to additional compliance criteria and monitoring schedules, which will be defined in national or international legislation. Microbiological compliance may be more rigid than for non-bottled potable water, although chemical criteria may be similar or less rigorous, sometimes permitting quite high concentrations of some elements which may be deemed desirable in mineral waters. In England, the relevant piece of legislation is *The Natural Mineral Water, Spring Water and Bottled Drinking Water (England) Regulations* of 2007, amended in 2010 and 2011. In the European Union, mineral waters are regulated by Directive 80/777/EEC, which has been subsequently modified and amended by Directives 96/70/EC and 2003/40/EC (European Commission 1980, 1996, 2003).

8.1.5 Water for agricultural and industrial purposes

The use of water for irrigation was considered in Section 2.7.4. Here, the important parameters to determine are: the major cations and anions [allowing the sodium adsorption ratio to be calculated – Equation (2.5)], the total dissolved solids or electrical conductivity, alkalinity and any specific phytotoxic compounds or elements (such as boron). It may also be advantageous to gain some information concerning the loading of nutrient parameters, especially nitrogen (nitrate and ammonium), base cations and sulphate, in the water, so that these may be taken into account when calculating fertilizer or lime applications.

As regards industrial usage, the quality of water required will depend on the industrial processes in question. However, for processes that involve heating or boiling water, permanent (sulphate) and temporary (carbonate) hardness will be important parameters to measure or calculate. In the food industry, any process water may be subject to standards of microbiological purity and may also have to satisfy drinking water standards for physicochemical parameters.

8.1.6 Pollution-related parameters

If there is suspicion of any past or present polluting practice or industry in the vicinity of the abstraction, specific parameters that may be derived from this pollution source will need to be considered (although many will already be regulated by national drinking water standards or international guidelines). Table 8.3 provides an overview of potential groundwater contaminants associated with particular industries or activities.

In considering potential contaminants, it is important to remember also that some contaminants biodegrade to other compounds which may be toxic or regulated by national standards. For example, the solvent trichloroethene (TCE) may degrade to vinyl chloride.

Table 8.3 *Main groundwater contaminants associated with specific industrial and other human activities (the list is far from exhaustive)*

Activity	Contaminant
Agriculture	Nitrate, ammonium, potassium, phosphorus (from fertilizer, manure)
	Nitrate (from deep ploughing, ploughing of virgin grassland)
	Microbes, for example, coliforms, *Cryptosporidium, Giardia* (manure, pasture)
	Pesticides
	Uranium (apatite-based fertilizers)
Airports	Hydrocarbons, de-icing fluids from aircraft and runways (e.g. urea, glycols: usually have high BOD[a], urea degrades to release ammonium), solvents, herbicides
Aluminium production	Fluoride, PAH[b]
Coal processing (gasworks, coking works)	Hydrocarbons[a], phenols, PAH[b], metals, metalloids, cyanide, ammoniacal nitrogen
Chlor-alkali production	Mercury (from liquid electrodes)
Dentists, older laboratories	Mercury, waste chemicals
Dry cleaning	Chlorinated solvents
Fertilizer production	Acid, sulphate, metals/metalloids (from sulphuric acid production)
	Phosphorus, radium, radon, uranium, fluoride (from apatite processing)
	Ammonium, nitrate, potassium
	Anthrax (from bone processing)
	Hydrocarbons[a] and solvents (production of packaging)
Gas (offshore) processing	Alcohols (especially methanol)
Harbours, ports	Hydrocarbons[a], chlorinated solvents, metals, metalloids, PAH (e.g. in dumped dredgings), PCB[c] (from old paints), anti-fouling agents (organic tin compounds)
Landfill	Nitrate, ammonium, base cations, chloride, sodium, potassium, carboxylic acids, BOD, COD[a], microbes
Military bases	Hydrocarbons[a], solvents, rocket/aircraft fuels, explosive and weapons residues (lead, nitrate)
Mining	Acid, sulphate, metals and metalloids (from coal/sulphide ores), saline formation water (deep formation water may also contain, for example, ammonium, barium, radium), turbidity, nitrate from explosive residues (if used)
Oil production	Hydrocarbons[a], drilling fluids, saline formation water, radioactive scale (although this is generally poorly soluble)
Oil processing and plastics production	Hydrocarbons[a], PAH[b], chlorinated solvents. (Sulphate, acid, heavy tars, PAH from re-refining of lubricating oils)
Petrol stations	Hydrocarbons[a], MTBE[d] (iron, manganese, alkalinity, low dissolved oxygen, resulting from biodegradation)
Power generation, heat pumps	Temperature (if coolant water or waste heat disposed to ground). Anti-freeze from leaking heat exchangers (typically glycols). Some power stations also release acidic gases (SO_2) and base cations to the atmosphere, which, via rainfall, *may* impact on water quality. Fly ash can generate alkaline leachates, rich in certain metalloids and elements forming mobile oxy-anions (e.g. boron, arsenic, selenium)
Quarrying	Turbidity, nitrate (from explosive residues), drilling fluids, hydrocarbon spills

Table 8.3   *(Continued)*

Activity	Contaminant
Sewage, municipal waste water	Faecal microbes, COD, BOD[a], chloride, nitrate, ammonium, surfactants, phosphorus, hydrocarbons, pharmaceuticals
Textiles	BOD[a], chloride, metals, surfactants, solvents, carbon disulphide
Timber, pulp and paper	BOD, COD[a], phenols, solvents, organic carbon, sulphate, sulphide, organic mercury compounds. Creosote, copper, chromium, arsenic from timber impregnation
Transport (railways, roads)	Hydrocarbons[a], PAH[b], MTBE[d], herbicides, road salt, other de-icing agents
Vehicle, metals industry, engineering, workshops	Many potential contaminants: especially chlorinated and other solvents, hydrocarbons[a]

BOD, biological oxygen demand; COD, chemical oxygen demand; MTBE, methyl tertiary butyl ether; PAH, polycyclic aromatic hydrocarbons; PCB, polychlorinated biphenyls.

[a] The biological or chemical oxidative degradation of organic compounds (hydrocarbons, de-icing fluids, BOD, COD) will typically consume oxygen and release CO_2, leading to acidic, reducing conditions. This will, in turn, result in reductive dissolution and mobilization of iron and manganese, and in intensified hydrolysis of carbonates and silicates, releasing base cations and alkalinity to the groundwater.

[b] PAH: A range of relatively persistent, potentially toxic organic compounds, based on linked carbon ring structures (e.g., naphthalene). They can be important soil contaminants at many sites (smelting works, coking works) but the most toxic PAH compounds tend not to be especially soluble in groundwater.

[c] PCB: Environmentally persistent carcinogens that were used in certain types of paints and in oils in electrical transformers.

[d] MTBE: A volatile, soluble and mobile ether, used as an additive to petrol (especially unleaded). Can be an important groundwater contaminant and also occurs at low levels in rainfall in industrialized countries. (Burgess *et al.*, 1998).

8.1.7   Indicator parameters

The potential number of groundwater quality parameters to be monitored in a new abstraction is enormous. National regulations will normally recommend the frequency at which water supplies need to be sampled (and which analytes should be measured). Often, the frequency will depend on the size of the population served (Table 8.4a). Any abstraction intended for public consumption should be sampled for the full suite of regulated parameters and any other parameters of concern, both initially and at intervals throughout its operation. However, it is not uncommon to measure:

- certain parameters indicative of good water quality, either continuously (at major abstractions in more wealthy parts of the world) or at frequent intervals;
- a fuller (and much more costly) suite of determinands at less frequent intervals to verify the conclusions drawn from the indicator parameters.

In terms of water chemistry, the most valuable indicator parameters include pH, electrical conductivity (EC), temperature, dissolved oxygen,

Table 8.4(a)   *The World Health Organization (World Health Organization, 2011) recommended minimum sample numbers for faecal indicator testing in distribution systems (parameters such as chlorine, turbidity and pH should be tested more frequently as part of operational verification monitoring)*

Population	Total number of samples per year
Piped supplies to:	
<5000 people	12 (i.e., monthly)
5000–100 000	12 per 5000 head of population (i.e., 12 to 240)
100 000–500 000	12 per 10 000 head of population, plus an additional 120 samples (i.e., 240 to 720)
>500 000	12 per 50 000 head of population, plus an additional 600 samples
Point sources	Progressive sampling of all sources over 3- to 5-year cycles (maximum)

Reproduced by permission of the *World Health Organization*

Table 8.4(b)   The quality of drinking water systems based on compliance with performance and safety targets for the absence of E. coli in potable water (after World Health Organization, 2011)

Quality of water system	Proportion (%) of samples negative for E. coli (WHO recommend zero organisms per 100 ml sample)		
	Population <5000	Population 5000–100 000	Population >100 000
A (formerly "Excellent")	90	95	99
B (formerly "Good")	80	90	95
C (formerly "Fair")	70	85	90
D (formerly "Poor")	60	80	85

Reproduced by permission of the *World Health Organization*

colour and turbidity. These can all be measured in the field using portable meters and comparators or, more rigorously, in the laboratory. Rapid variations in these parameters may indicate that the groundwater quality is unstable and responding to sporadic recharge events or seasonal events. Indeed, Norwegian experiences suggest that drilled wells in crystalline rock that are inadequately sealed in the casing annulus, or which have very short fracture flow pathways to the surface, respond rapidly to snowmelt or rainfall events. This is shown by lowering of pH, EC and alkalinity and increases in colour (due to leaching of humic and fulvic acids from organic-rich, peaty soils) and turbidity (Box 8.1). At a lower level of sampling frequency, major cations and anions and alkalinity can prove valuable, and relatively cheap, indicator analyses of gross groundwater chemistry variations.

To be able to use indicator parameters effectively in a public supply water well, the detailed water chemistry of the well ideally should be observed throughout a yearly cycle, by sampling on at least a monthly basis during the first year of pumping. During this period, it should be possible to establish which individual parameters correlate effectively with indicator parameters (and hence can be adequately represented by indicator parameters) and which do not. Based on these observations, it should be possible to define a schedule for regular monitoring of a limited selection of indicator parameters and for possibly less frequent monitoring of individual regulated parameters.

8.1.8   Microbiological quality and indicator parameters

In the industrialized world, citizens have become used to drinking water that has either been disinfected at a water treatment works, or been bottled from a safe source under sterile conditions. It should not be forgotten, however, that most of the world's population does not enjoy this luxury, and has no alternative but to consume water that may be vulnerable to pollution by microbes. Even in the industrial world, there are still significant and fatal outbreaks of disease due to treatment-resistant protozoa (Box 8.2), dangerous strains of *E. coli* (Box 8.3), and the airborne *Legionella*. Indeed, it has been estimated that 6.5 million illnesses annually in the United States are caused by microbiological contamination of groundwater (Reynolds *et al.*, 2008), whilst faecal contamination of a single shallow well at Walkerton in Canada resulted in 2,300 people becoming seriously ill, with seven fatalities (Hrudey *et al.*, 2002).

The term *microbes* applies to a variety of organisms. Many may occur naturally in the groundwater environment. Others may be related to faecal or other anthropogenic pollution and a few of

Box 8.2　Protozoa: *Cryptosporidium* and *Giardia*

Protozoa are a subkingdom of single-celled animals that are larger and more complex than bacteria. They range in size from macroscopic to submicroscopic and include amoebae. Some species of amoeba cause severe gastrointestinal disease: for example, *Entamoeba histolytica* can cause *amoebiasis* (the notorious amoebic dysentery).

In the context of groundwater supply, two particular protozoa (*Giardia* and *Cryptosporidium*) have attracted much attention, due to their ability to encyst. The cysts or oocysts are able to survive for prolonged periods (weeks or months) in the water environment, to resist treatment by chlorination and to cause gastrointestinal disorders. Symptoms are usually self-limiting in healthy individuals, but both parasites pose a severe risk to immuno-compromised individuals such as AIDS sufferers. Although, with both parasites, person-to-person contact is the most important transmission route, fewer than ten cysts or oocysts are required to infect humans, so that contaminated water is also a significant infection pathway. Raw sewage can contain tens of thousands of cysts per litre.

Giardia lamblia (also known as *G. duodenalis* or *G. intestinalis*) occurs as an intestinal parasite in domestic and wild animals. The parasite lives internally as a flagellate phase, but is periodically shed in faeces as a thick-walled cyst of diameter 8–12 µm, which can be transmitted to water supplies. Human infestation by *Giardia* is manifested as *giardiasis* and is characterized by severe diarrhoea, abdominal cramps and weight loss. Vomiting, chills, headache and fever can occur in more serious cases. The cysts are more resistant than bacteria to chlorination, with a contact time of 25–30 minutes at a free chlorine residual of 1 mg l^{-1} required to result in 90% inactivation (World Health Organization, 2011).

Cryptosporidium parvum lives as an intracellular parasite, occurring primarily in herd animals such as cows, goats and sheep, and also wild deer. Thick-walled oocysts of diameter 4–6 µm are shed in faeces, which can contaminate water supplies, especially during run-off and recharge events on pasture land following heavy rainfall or snowmelt. *Cryptosporidium* can infect humans resulting in diarrhoea (which is normally self-limiting), nausea, vomiting and fever. Although the impacts on human health and wellbeing are our primary concern, the socio-economic costs are also worth noting. The cost of illness associated with a cryptosporidiosis outbreak in 1993 in Milwaukee (USA), which affected around 400 000 people, is reported to have been about 96 million US$ (World Health Organization, 2011). The oocysts are very resistant to chlorination (even more so than *Giardia*) and are too small to be efficiently removed by conventional sand filtration. UV disinfection, if properly designed, can inactivate oocysts. Alternatively, they can be excluded by membrane filtration.

The World Health Organization (2011) sets no specific guideline values for protozoa in water, but instead recommends a risk-based approach. It does, however, recommend some guideline values for faecal indicator bacteria in potable water (which, at least in untreated water, may also be surrogate indicators of the presence of faecal protozoa). In treated water, however, the absence of faecal coliforms cannot be used to imply the absence of *Giardia* and *Cryptosporidium*, due to the protozoa's resistance to treatment. In treated waters, more resistant faecal bacteria (such as intestinal *enterococci*), coliphages or *Clostridium perfringens* spores may be more appropriate as faecal indicators.

Box 8.3 Bacteria: Total coliforms, faecal coliforms and *E. coli*

The kingdom *Monera* consists of what are commonly called bacteria: microscopic, one-celled prokaryotic organisms. Most fall into one of several classes based on shape: the rod-shaped bacteria (*-bacillus*), the spherical (*-coccus*), the spiral (*-spirillum*) and the curved, comma-shaped (*Vibrio*). Bacteria may be *aerobic* (requiring the presence of free oxygen), *obligate anaerobes* (to whom oxygen is toxic) or *facultative anaerobes* (can grow in the presence or absence of oxygen).

There are a great many bacterial species and most are not pathogenic to humans (thus, *total heterotrophic plate counts* are not necessarily good indicators of microbiological water quality). Many of the bacterial and protozoan species which cause gastrointestinal disease in humans are associated with a faecal transmission route. Thus, we tend to gauge water quality by the presence of *faecal indicator bacteria*. The most common faecal indicators are the *faecal coliform* bacteria, but others such as *faecal streptococci* are also used. We should, however, bear in mind that pathogenic microbes do exist which are not faecally derived, but which can infect the lungs by inhalation as aerosols (*Legionella*) or other organs via exposure during washing or bathing (World Health Organization, 2011).

Coliform bacteria are a group of related species. They may be *vegetative coliforms*, occurring in vegetation, sediment, soil or insects, or they may be *faecal coliforms*, occurring in the gastrointestinal tracts and faeces of humans and animals. Many coliforms are not pathogenic and around

60-90% of *total coliforms* are *faecal coliforms*. Thus the presence of total coliforms is not a wholly reliable indicator of faecal contamination.

Faecal coliforms can be analysed and distinguished by the fact that they may be cultivated by 24 hours incubation on a lactose medium at around 44 °C, resulting in fermentation and production of gas. *Faecal coliforms* (essentially synonymous with *thermotolerant coliforms*) are regarded as a good indicator of faecal contamination, and include genera such as *Enterobacter, Klebsiella and Citrobacter.*

Over 90% of faecal coliforms are typically of the genus *Escherichia*. One particular species of *Escherichia* is *E. coli*, which is also used as a faecal indicator in water analysis. *E. coli* thrives in the intestinal tract of humans and warm-blooded animals, where it is generally harmless (although it can cause infection if it enters other organs). *E. coli* occurs in several strains, a few of which can cause gastrointestinal disease. Two strains of enterohaemorrhagic *E. coli* (EHEC), O157:H7 and O111, can be fatal to humans. Both *faecal coliforms* and *E. coli* should normally be absent from potable water.

The World Health Organization (2011) has not gone down the difficult path of setting a wide variety of guidelines for individual microbes in potable water. Instead, WHO promotes a risk-based approach, coupled to water safety plans. The only guidelines offered are that *faecal coliforms* and/or *E. coli* (as faecal indicators) should be absent from a 100 ml sample of potable water.

---

these may be pathogenic to humans. Microbes that may be found in water include:

- helminths: multicellular worms (including roundworms, tapeworms and flukes) that are parasitic in vertebrate intestinal tracts;
- protozoa: single celled organisms, such as amoebae, *Cryptosporidium* or *Giardia* (Box 8.2);

- bacteria: simpler, single-celled organisms (see Box 8.3); tetanus, cholera and typhoid are diseases caused by bacteria;
- viruses: these are not cells at all, but tiny replicating clusters of nucleic acids; polio, hepatitis and foot-and-mouth disease are examples of diseases caused by viruses.

To analyse all of these on a regular basis would be very time-consuming and practically impossible in all but the best-equipped laboratories. However, many waterborne microorganisms that are pathogenic to humans are derived from faecal matter and thus exhibit a degree of co-variation. Therefore, it is normal practice to analyse for characteristic indicator parameters of faecal contamination. Three of the most common indicators are (i) thermotolerant (or faecal) coliforms (ii) faecal streptococci and (iii) *E. coli* (Box 8.3). The risk associated with various occurrences of faecal coliforms in water supplies of various sizes has been assessed by the WHO [2011; Table 8.4(b)].

Another microbiological indicator parameter that is commonly used is the total heterotrophic plate count (THPC). The heterotrophic plate count is essentially the total number of heterotrophic organisms (bacteria and some fungi) that can be grown on a rich growth medium, with incubation at a given temperature for a given time (typically 20–37 °C for several hours to 7 days, depending on the method). As most bacteria and fungi are not pathogenic and not faecally derived, a high THPC is not necessarily a health concern and THPC should *not* be used as an indicator for pathogenic bacterial contamination (WHO, 2011). Sporadic, high THPCs may be indicators of biofilm growth in a well or pipeline (see Chapter 9), with pieces of biofilm occasionally sloughing off and giving rise to high bacterial counts. Consistently high THPCs may also be indicative of poor system integrity (for example, due to ingress of surface water or sediment to the well or pipeline), while high THPC after major rainfall or snowmelt events may indicate vulnerability to surface contamination. THPC can also be used as an indicator of efficiency of disinfection of a supply, as viable bacteria should not be present in an adequately disinfected supply.

Finally, *Clostridium perfringens* is a faecal bacterium which is highly resistant to conventional treatment and can thus be a good indicator of risk of contamination by *Giardia* or *Cryptosporidium* protozoa (Box 8.2).

8.2 Field determinations

To analyse some parameters, it is not necessary to transport the samples to the laboratory; rather, the determinations can be undertaken in the field using portable meters (Figure 8.1), comparators, and field kits (titrations, spectrophotometry). Such determinations may be less accurate and more poorly controlled than laboratory analyses, but there are often good reasons to prefer them. Where the number of samples and their importance warrant it, it may be possible to bring a mobile field laboratory to site, allowing the rapid determination of parameters (including unstable ones) with a high degree of quality assurance.

8.2.1 The purpose of field determinations

Field determinations may be necessary for two main reasons:

1. The results are needed immediately. For example, determinations of pH and turbidity are required immediately to assess levels of chlorination necessary for water treatment. A field reading of electrical conductivity (EC) will provide prior information to a laboratory on the salinity of a water sample and the most appropriate method of analysis;
2. Some hydrochemical parameters are unstable and may change during storage and transport to a laboratory.

Parameters which are unstable following sampling include obvious examples such as temperature, dissolved oxygen, redox potential and CO_2 content, which will tend to change rapidly when exposed to ambient atmospheric conditions.

Under some circumstances, pH and alkalinity can also change during storage, due to possible degassing of carbon dioxide, oxidation and hydrolysis of iron or precipitation of calcite. Generally, if samples are not extremely acidic or alkaline, are free of hydrolysable metals and are delivered promptly, representative laboratory analyses are usually possible. However, field determinations are cheap, accurate and simple, and are strongly recommended.

Figure 8.1 *Magne Ødegård and Ola Magne Sæther of the Geological Survey of Norway demonstrate their portable truck-mounted sampling 'laboratory'. The open case contains a throughflow cell and a multi-meter to determine common physicochemical parameters (temperature, pH, EC, etc.). To the right of the case is a spool carrying a small diameter hose and cable attached to a lightweight electrical submersible pump*

Some compounds are sensitive to biodegradation, volatilization or redox changes, including sulphide, nitrate, nitrite and ammonium. Field kits are available to determine these approximately on site. Alternatively, some form of suitable preservation (simple cooling may be adequate for nitrate, or chemical preservation for other parameters) and rapid transport to a laboratory may be acceptable.

8.2.2 Downhole sondes and throughflow cells

The geosphere (constant temperature, basic and reducing) is a very different geochemical environment from the atmosphere (fluctuating temperature, acidic and oxidizing). Thus, when groundwater samples are brought to the surface, some parameters tend to change very rapidly on exposure to the atmosphere and good readings can

be difficult to obtain. Therefore, we often try to minimize exposure to the atmosphere either by (i) monitoring certain parameters downhole or (ii) pumping water through a sealed throughflow cell, within which the appropriate measurements are taken.

Even before a well has been drilled, it is possible to obtain *in-situ* information about an aquifer by using cone penetrometer testing (CPT). The cone penetrometer is essentially a steel cone that can be pushed under pressure into unconsolidated sediments to depths of some 40–50 m under favourable conditions. It has traditionally been used to determine the geomechanical properties of the sediments encountered, but nowadays it is possible to obtain penetrometers where downhole sensors can be employed to measure temperature, electrical conductivity, pH, radioactivity and even concentrations of hydrocarbons (using a downhole laser and determining UV fluorescence) in pore waters. CPT can also be employed to extract depth-specific groundwater samples at successive depth intervals.

If there is access to an existing well, a number of parameters can be determined by downhole geophysical logging (see Chapter 6) or monitored continuously by installing a downhole electronic sonde containing a package of physicochemical sensors. These can include a temperature sensor, a water level transducer, an electrical conductivity cell, and pH and dissolved oxygen electrodes. They may also include specific ion electrodes or specific sensors based on relatively new technology, such as optical sensors (where a light signal is used to investigate a film of chemically sensitive material whose optical properties change in relation to the concentration of a given parameter in solution).

If water is pumped from a well to the surface, certain unstable parameters (especially Eh and dissolved oxygen) should be determined by means of a throughflow cell (an example of which is shown in Figure 8.2). The inlet and outlet are arranged such that the cell is always filled with water and there is a constant flow of fresh groundwater past the various sensors. The sensors typically include the following:

- a thermometer or thermistor;
- a calibrated pH electrode;
- a redox electrode or dissolved oxygen probe;
- an electrical conductivity electrode.

Data from a downhole sonde or from the sensors in a throughflow cell may be recorded manually or transmitted to and stored on a data logger.

8.2.3   Field kits for other parameters

A wide array of field kits (Figure 8.3) is available for determination of various chemical and microbiological parameters on site. These are typically of five types:

i. *Ion specific electrodes*. The pH electrode is a form of ion specific electrode for hydrogen ions. Others are available – for fluoride, for example.
ii. *Titration kits*. Here a reagent is added to the sample from a graduated pipette, to react with

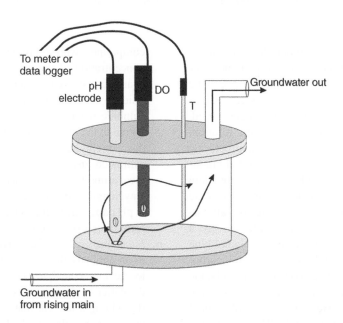

Figure 8.2   *A typical throughflow cell. DO, dissolved oxygen; T, temperature*

Figure 8.3 *Portable water testing kits being used in Muscat. Photo by Bruce Misstear*

the analyte in question. When the reaction has gone to completion, some form of indicator typically undergoes a colour change. The quantity of the analyte in question can be calculated from the volume of reagent used. A good example is the alkalinity titration, where a strong acid (such as hydrochloric acid) of known concentration is added gradually to the water sample. A phenolphthalein indicator undergoes a colour change when the pH reaches about 8.2, indicating that all carbonate alkalinity has reacted. Alternatively, a different indicator (such as bromophenol blue or methyl orange) will indicate when bicarbonate alkalinity has been consumed (that is, the total alkalinity) at an end point of pH ≈ 4.2–4.3.

iii. *Colorimetric ki*ts. Here some form of reagent is added to the water sample. This reacts with the analyte in question to form a coloured compound, whose colour is related to the concentration of the analyte in the water. An example is the determination of dissolved sulphide by the methylene blue method, where sulphide reacts with dimethyl-p-phenylenediamine in the presence of ferric chloride to produce methylene blue. The colour can be estimated using a comparator chart, or by measuring the absorbance of light at given wavelengths in a field spectrophotometer. A pH paper strip (litmus paper) is also a simple form of colorimetric test.

iv. *Immunoassay kits.* These involve the application of an antibody that is specially tailored to bind to the target analyte. The bound analyte can thereafter be determined either by colorimetric (by adding a reagent that reacts with enzymes produced by the antibody) or fluorometric means. Immunoassay techniques have been developed for pesticides, polycyclic aromatic hydrocarbons (PAH), polychlorinated biphenyls (PCB), explosives, fuel hydrocarbons, some volatile and semi-volatile organic compounds (VOC and SVOC) and some metals.

v. *Field incubators*, to determine simple microbiological parameters (these are discussed further in Section 8.2.4, below).

When considering field kits, it is important to be aware of the following limitations of some field methods (which should be documented in operating instructions):

- many field kits and ion specific electrodes will be subject to interferences with other species that may be present in solution;
- many field kits are only valid for given concentration ranges of the parameter in question, or for given ranges of salinity;
- colorimetric or spectrophotometric kits may be subject to interference if the water sample is turbid or coloured (from the presence of organic acids, for example); some spectrophotometric kits are able to compensate for this by comparing to a blank (i.e., a water sample where the reagent has not been added);
- electrodes may also be poisoned by certain compounds and they may degrade with time; many electrodes (such as pH) require regular field calibration and careful rinsing between samples to avoid cross-contamination.

It is advisable, where possible, to carry out some form of comparison of field determinations with a laboratory determination under a more controlled environment.

8.2.4   Emergency water supply

When supplying water in an emergency situation, such as to a refugee camp (Box 8.4), a permanent laboratory may not be accessible. Decisions regarding water quality and treatment may need to be made immediately on the basis of field determinations. For short-term emergencies, microbiological quality will often be the immediate concern (as, in fact, it will be in most public water supply

operations). Some form of chlorination will usually be undertaken. As the amount of chlorine required will depend on turbidity and pH, these two parameters need to be determined in the field:

1. *Turbidity*. This is most easily determined by adding water gradually to a long Perspex or glass tube until a marking on the base of the tube becomes invisible when viewed through the water column. The length of the water column for this to occur can be related to turbidity.
2. *pH*. This can be determined by a pH meter, or by a simple comparator kit where an indicator tablet is added to the water sample, resulting in a specific colour related to the pH of the water.

For chlorination of drinking water, chlorine (often as a solution of calcium hypochlorite – Box 8.5) is added to the raw water sample, with the aim of obtaining a particular free chlorine residual (typically 0.2 to $0.5 \, mg \, l^{-1}$) after 30 minutes' contact time. To ascertain the appropriate chlorine addition, differing amounts of hypochlorite are added to an array of flasks of water and, after 30 minutes, the free chlorine residual is determined in each flask in order to ascertain the correct initial chlorine dosage. Free chlorine residual can be determined by a comparator kit, whereby the treated water is added to a transparent plastic cell. An indicator tablet (DPD1) is added to the cell resulting in a pink colour, whose intensity can be related to chlorine residual via a comparator chart.

The microbiological quality of raw or treated water can be verified in the field by the use of a portable incubator, such as that contained in the Oxfam-Delagua (2000) portable laboratory. This equipment works on the following methodology:

i. A given quantity of water sample is passed through a fine filter mesh (trapping any coliform bacteria).
ii. The mesh is laid on a Petri dish containing culture medium, thus inoculating it with bacteria.
iii. The Petri dish is incubated at a stable temperature (37 °C for total coliforms, 44 °C for faecal coliforms) for 16–18 hours.
iv. The colonies of a specific colour (coliforms) are counted.

Box 8.4   Water quality and emergency water supply

When a natural disaster or a war strikes, large numbers of people are often suddenly displaced to refugee camps or settlements. In such situations, it is widely accepted (World Health Organization, 2011) that the most important factors in preventing gastroenteric diseases (such as cholera) and skin diseases are:

- Maintaining an adequate *quantity* of available water (even if not of pristine quality) to permit hand-washing and bodily, kitchen and household hygiene and ensuring that water is readily accessible to all
- Providing an adequate number of *sanitary latrine* facilities.

These two principles are clearly described in the international SPHERE (2011) standards, and emergency workers promote key indicators such as the availability of at least 15 litres water per person per day and a maximum of twenty people using each latrine. With priority being given to water quantity and sanitation, issues such as water quality are often forgotten. Yet, there is a clear SPHERE standard for water quality (which is broadly similar to specific World Health Organization (2011) recommendations for emergency provision of water):

> *"Water is palatable, and of sufficient quality to be drunk and used for cooking and personal and domestic hygiene without causing risk to health."*

There are several key actions and indicators for assessing whether this standard is being met (SPHERE, 2011). These include (among others):

- Rapid sanitary surveys are undertaken and water safety plans for the source implemented.
- Steps are taken to minimize post-delivery contamination. (It is of little value providing clean water if this is put into dirty containers).
- For piped water supplies (or for all water supplies at times of risk or presence of diarrhoea epidemic), water is treated with a disinfectant so that there is a free chlorine residual at the tap of 0.5 mg l^{-1} and turbidity is below 5 NTU.

(The chlorine residual in delivered water provides some protection against possible contamination of water as it is being transported in jerry cans or stored in the household environment).

- There should be no faecal coliforms per 100 ml at the point of delivery/use.
- People drink water from a protected or treated source in preference to other readily available water sources.
- No negative health effect is detected due to short-term use of water contaminated by chemical (including carry-over of treatment chemicals) or radiological sources, and assessment shows no significant probability of such an effect.

In other words, the water must be of adequate microbiological quality and palatable (otherwise people may choose to use alternative, better-tasting, but possibly less safe sources). The water also must have no acute or short-term adverse health effects. Of course, this assumes that the settlement or camp is a temporary phenomenon. Many refugee camps in the world persist for many years or even decades. In such situations, chronic long-term exposure to parameters such as fluoride or arsenic may become important (World Health Organization, 2011) and the normal World Health Organization Guidelines will become more relevant than emergency standards such as SPHERE (2011).

One parameter of particular concern in emergency situations is nitrate. SPHERE (2011) does suggest (presumably mainly in regard to microbiological quality, rather than nitrate) that "soakaway" type pit or trench latrines should terminate at least 1.5 m above the water table and should be at least 30 m from a groundwater source (depending on local geology). However, while such a recommendation is likely to be satisfactory for a single well and a single latrine, it may not be sufficiently conservative in a camp where several wells are located amongst high densities of family pit latrines, all potentially leaching nitrate and ammonium to the water table.

Bottle-fed infants with existing gastro-intestinal disorders are especially vulnerable to nitrate toxicity (methaemoglobinaemia; see Section 2.7.3): a potential situation in a refugee camp. Monitoring of groundwater quality for nitrate in such a situation is highly recommended and, if necessary, provision of low-nitrate bottled water for infants should be considered.

Box 8.5   Chlorination of potable water supplies

Chlorination kills most (but not all) pathogenic organisms by destructive oxidation of their organic/cellular material (some cysts of pathogenic protozoa, such as *Giardia* and *Cryptosporidium*, some viruses and some eggs are resistant to chlorination). However, chlorine is also consumed by any oxidizable matter in the water – not just by microbes – thus the presence of oxidizable matter (e.g., organic matter, ferrous iron, etc.) will use up the chlorine. Chlorination effectiveness is limited:

- if particles are present in the water – these should be removed (by flocculation and settlement or filtration) prior to chlorination to give a turbidity <5 NTU (although chlorination will have some effect up to 20 NTU);
- at high pH (pH>8).

Generally speaking, if chlorination is employed, enough chlorine should be added to kill microbes, oxidize any other substances *and* to leave a free chlorine residual of around 0.5 mg l^{-1} after 30 minutes. The free chlorine residual at the point of use should be around 0.2 to 0.5 mg l^{-1} (the free chlorine residual continues to provide protection during distribution of the water in a pipe network, jerry cans etc.). To achieve this, it is often necessary to add the equivalent of 1 to 5 mg l^{-1} chlorine equivalent (0.1 to 0.5 ml of 1% chlorine/hypochlorite solution per litre of water); the World Health Organization (2011) implies that chlorine concentrations of <5 mg l^{-1} do not present a significant health threat.

Chlorine is normally supplied as a solution of calcium or sodium hypochlorite or as calcium hypochlorite tablets/powder. Hypochlorite anions (and hypochlorous acid) are released on dissolution in water, which are able to oxidize organic matter (CH_2O), leaving harmless chloride in solution:

$$Ca(ClO)_2 + 2H_2O = Ca^{++} + 2ClO^-$$
$$+ 2H_2O = Ca^{++} + 2HClO + 2OH^-$$

$$Ca^{++} + 2ClO^- + CH_2O$$
$$= Ca^{++} + 2Cl^- + CO_2 + H_2O$$

The strength of hypochlorite solutions is measured by comparing with chlorine gas (Cl_2). 1 g of a 100% "active chlorine" product has the same oxidizing power as 1 g of chlorine gas. Quite coincidentally, pure calcium hypochlorite powder has a strength of almost 100%, while "High Test Hypochlorite" (HTH) granules typically have a strength of 70% available chlorine. Further advice on the use of water treatment chemicals can be found in NSF/ANSI (2012).

In emergency situations, it is common to dose potable water using a "mother solution" of hypochlorite of strength 1%. This solution can be prepared by adding 10 g of pure calcium hypochlorite or 14–15 g HTH granules to 1 l of water. Thus, 1 g of the solution contains 0.01 g (1%) of active chlorine. For effective disinfection, contact times of between 30 minutes (in warm water) to 60 minutes (cold water) are recommended.

For guidance on disinfecting the well, the reader is referred to Section 5.9.5. Note that sodium hypochlorite is preferred to calcium hypochlorite for well applications, as the latter can lead to clogging problems in some groundwaters.

The collection and handling of samples for microbiological analysis is discussed in Section 8.5.3.

8.3 Collecting water samples from production wells

8.3.1 The sample line

All production wells should be installed with the following wellhead equipment: an access tube for water level determination, a flowmeter to record discharge, and a wellhead tap for sampling. Water samples from production wells should be taken from a sample line attached to the wellhead tap. Samples should not be taken downstream of the throughflow cell: the sample line and the throughflow cell should ideally be connected to different branches of a line leading from the wellhead connection. The sample line should be thoroughly purged before sampling (Box 8.6). Thereafter, sampling line flow rates (which are not necessarily the same as well pumping rates) should be <0.5 l min^{-1} for most sample flasks (USGS, 2015). For volatile organic compounds (VOCs), very low flow rates of <0.1 l min^{-1} are recommended (Barcelona *et al.*, 1985). When sampling, contact between the sample line and the flask should be avoided, as should turbulence and exposure of the sample to air. Sampling for VOCs requires particular attention (and will be discussed below), with minimization of aeration/turbulence and special sampling vials. For some other parameters, such as carbon isotopes (^{14}C), large sample volumes may be required (Figure 8.4).

The choice of material for the tap, sample line and other sampling materials is complex, but the construction should be such that it will not "contaminate" the sample. For sampling of organic compounds, stainless steel or glass are probably the preferred materials, although a fluorocarbon such as polytetrafluoroethene (PTFE) is often utilized, being relatively, though not wholly, chemically inert. The choice of material can depend on

the analyte in question, however, and for some organic compounds, uPVC might be preferable.

8.3.2 When to sample: well testing

When designing a sampling programme to be carried out during a pumping test, it is important to remember the following:

- The presence of drilling fluids, slurry, cuttings, hydrofraccing water or acidization residue will impact on the sampled water quality. Before sampling commences, thorough clearance pumping must be carried out to ensure that true groundwater is being sampled. In some environments, especially crystalline-rock aquifers, the presence of freshly exposed drilling surfaces and residual cuttings may continue to influence groundwater quality for some months after drilling, resulting in elevated concentrations of some lithologically-derived parameters (Banks *et al.*, 1992b; 1993b).
- Water quality can vary with pumping rate or abstraction pattern in a well field.
- Water quality can change with time during prolonged pumping, as the well begins, for example, to draw on groundwater from remote parts of the aquifer, from greater depths, from induced infiltration of surface waters or from saline water bodies (see Figure 7.1).

To ascertain the possible dependence of water quality on pumping rate, samples can be taken during step testing (Section 7.3.2). During step testing, readings of physicochemical parameters (temperature, pH, electrical conductivity, Eh, dissolved oxygen) should be monitored throughout each step. Water samples and corresponding on-site readings should be taken at the end of each step.

During long-term testing (Section 7.3.3), master variables and physicochemical parameters should be monitored at frequent intervals throughout the test. The frequency with which water samples will be taken for more extensive analysis will depend on the hydrogeologist's conceptual understanding of the aquifer and on the issues of concern with regard

Box 8.6   Purging a well prior to sampling

Before taking a water sample from a well, it is important to ensure that the water is not 'stagnant water' that has been standing for a period in the well bore and casing. Therefore, before sampling, it is normal practice to purge water from the well and draw in 'representative' groundwater from the portion of the aquifer of interest. However, the act of pumping has the potential to disrupt any vertical stratification of water quality in the aquifer. Pumping rates and drawdowns should be kept as low as practically possible during purging, avoiding turbulence and disruption of any bottom sediments in the well. The rate of purging should, in general, be lower than the rate of post-completion well-development, but greater than the rate of sampling (Barcelona *et al.*, 1985; BSI, 2009).

There are many rules of thumb available to assess when the well has been purged sufficiently, of which the following are most common:

i. *Hydraulic analysis.* Analytical approaches, such as that of Papadopulos and Cooper (1967), can be used to calculate the proportions of water derived from well storage and from the aquifer after any given time of pumping, provided that the storage and hydraulic conductivity of the aquifer are known (Barcelona *et al.*, 1985). One could, for example, regard the well as adequately purged when 95% of the pumped water is calculated as aquifer-derived.

ii. *Well volume approach.* Typically, three to five volumes of water should be removed (by pumping or bailing) before commencing sampling (Puls and Barcelona, 1996; British Standards Institution, 2009; US Geological Survey, 2015). The well volume is calculated by:

$$\text{Well volume} = \pi r^2 H_w$$

where r is the internal well radius and H_w is the height of the water column in the well (or the section affected by pumping).

iii. *Parameter stabilization.* Pumping of the well continues until field hydrochemical indicators such as temperature, pH, electrical conductivity (EC), dissolved oxygen (DO) or redox potential become approximately stable. If planning a long-term monitoring programme, some sampling (and later analysis) during purging may be desirable, to confirm that parameters of interest are stabilizing satisfactorily. Criteria for 'stability' are given in Table B8.6(i).

In contaminant investigations, high purging rates run the risk of causing large lateral and vertical hydraulic head gradients in the aquifer near the well. These may induce 'unnatural' groundwater flows that can potentially disturb the distribution of water quality in the aquifer near the well. High purging rates may also generate large volumes of contaminated waste water that must be responsibly disposed of. Thus, many authors (e.g., Puls and Barcelona, 1996; Yeskis and Zavala, 2002) recommend *micro-purging* or *low-stress* purging. Here, the aim is not to empty a column of stagnant water from the well; rather, the aim is to place the pump within or at the top of the well screen to induce laminar flow from the flow horizons in the aquifer to the pump as immediately as possible. Thus, only the portion of the well containing the flow path of water from the aquifer to the pump intake is "purged". With micro-purging, the pumping rate should be such as to result in minimal drawdown (<0.1 m according to Yeskis and Zavala, 2002). A minimum volume v should initially be purged from the well and then monitoring of field parameters should commence in a throughflow cell; the parameters should be recorded after each successive volume v has been pumped. Here, v is defined as the total volume of the pump, the pumping and sample line and the throughflow cell. Sampling can commence when three successive sets of readings meet the stability criteria recommended by Yeskis and Zavala (2002) in Table B8.6(i). It should be noted that where there is a large vertical hydraulic gradient across the well screen (or open section of borehole), then low-stress pumping may not overcome the natural vertical flows within the well, and hence the sample will reflect the water quality of the main water inflow zone or zones (McMillan *et al.*, 2014).

In conclusion, no single recipe for purging can be recommended. Indeed, BSI (2009, 2010b) suggests a variety of approaches depending on aquifer properties and well construction, including recommendations for low permeability formations.

Table B8.6(i)   *Criteria for 'stability' during pumping (purging)*

Parameter		Stabilization criteria		
Source	Yeskis and Zavala (2002)[1]	USGS (2015)[2]	BSI (2009)	Barcelona *et al.* (1985)[3]
pH	±0.1	±0.1		±10%
Turbidity	±10%[4]	±10%		
Dissolved oxygen	±0.3 mg l⁻¹	±0.3 mg l⁻¹		
Electrical conductivity	±3%	±3%[6] ±5%[7]	±10%[5]	±10%
Redox potential	±10 mV			±10%
Temperature		±0.2 °C[8] ±0.5 °C[9]		±10%

[1] Over three successive well volumes
[2] Over five successive measurements
[3] Over two successive well volumes or two successive increments of 10% x the total purge volume
[4] When turbidity is greater than 10 NTU. Ideally turbidity should stabilize at <10 NTU
[5] Applies to any concentration parameter (mass per unit volume)
[6] For EC>100 µS cm⁻¹
[7] For EC<100 µS cm⁻¹
[8] For measurement by thermistor
[9] For measurement by liquid glass thermometer

to water quality. In general, it is desirable to carry out long-term testing of major groundwater sources to span (a) the major recharge event(s) of the year and (b) the dry season. Water quality sampling should also aim to characterize these seasonal episodes (for example, to ascertain if rapid recharge mechanisms pose a water quality risk; Box 8.1). Unscheduled water quality samples may also be taken during long-term test pumping in response to significant fluctuations in monitored physicochemical parameters (pH, temperature, and so forth).

8.3.3   When to sample: production wells

During water production from a well, sampling frequency will typically depend on national legislation or standards and also on the size of population served by the abstraction. Table 8.4(a) summarizes the WHO (2011) guidelines for minimum sampling frequency for faecal indicator bacteria for waterworks (chlorine, turbidity, pH will be tested more frequently). Inevitably, economic and access issues will also come into play, and the WHO (2011) suggests that small-scale, remote communal supplies (hand-pumps, small pumped supplies) should be visited, sampled and subjected to sanitary inspection at least every 3–5 years as part of an on-going programme.

Typically, routine water quality monitoring will focus on a limited number of hydrochemical and microbiological indicator parameters (e.g.,

Figure 8.4 *For some parameters, very large sample volumes are required. Here, Bruce Misstear prepares water samples for ^{14}C dating*

temperature, dissolved oxygen, pH, electrical conductivity, faecal coliforms, possibly colour, turbidity, major ions). Other regulated parameters will be sampled at less frequent intervals. This philosophy assumes that indicator parameters adequately reflect potential variations in a wide range of regulated parameters. Therefore, the test pumping period and the initial phase of production should be used to:

- characterize the water quality and its seasonal variation;
- verify that that indicator parameters are representative of most regulated parameters;
- identify any regulated parameters or other parameters of health concern that do not correlate very well with indicator parameters and which hence require a specific monitoring schedule.

Although ideally data for at least one year are required to characterize the water quality variation in a groundwater abstraction, a pumping test seldom has so long a duration. Thus, the verification of water quality will continue throughout the first year after the well is taken into production. Sampling frequency during this period will depend on national legislation and standards and on the population served, but a minimum of one sample per month is recommended during this phase.

8.4   Collecting water samples from observation boreholes

8.4.1   Preparation for sampling

It may be necessary to sample an observation borehole as part of a pumping test or during the operational monitoring of a well field (for example, to verify that a saline water or pollution front is not migrating towards the abstraction). Alternatively, sampling from an observation borehole may be part of a groundwater contamination investigation or regional water quality survey. Sampling from an observation borehole is more problematic than sampling from a production well, for two reasons:

1. The observation borehole may have no permanent pump or provision for removal of water samples.
2. The observation borehole will usually contain a (possibly static) column of water whose quality may no longer be fully representative of that in the aquifer due to contact with the casing materials, lack of equilibrium with the aquifer environment or mixing of water from different horizons.

With a purpose-drilled observation or monitoring borehole, it is important to confirm that the methods and materials used for construction are compatible with the analytical parameters of concern – they should not leach or adsorb these parameters. It is also important to ensure that the materials will not be attacked by contaminants likely to be present; for example, PVC well screens may be attacked by organic solvents (Section 4.1.4; British Standards Institution, 2010b). Furthermore, the borehole should have been clearance-pumped adequately following construction to remove traces of drilling fluids and cuttings. Following clearance pumping, the borehole should have been allowed to stabilize for at least one week prior to sampling (British Standards Institution, 2010b). Thereafter, a scheme for purging the borehole prior to sampling should be selected (Box 8.6), and also a device that will extract a representative sample of water. Several issues will govern the choice of sampling equipment and sampling regime:

- Is the sampling device (and, indeed, the materials of which the well is constructed) chemically inert with respect to the parameters to be determined? Will it, for example, adsorb metals or organic compounds to its internal surfaces, or will it leach them to the water with which it is in contact? Stainless steel and glass are regarded as compatible with most organic determinands, and polyethene with most inorganic parameters. Stainless steel and fluoropolymers (polytetrafluoroethene or PTFE) are widely considered to be relatively chemically inert (though expensive) with respect to most parameters in most situations. Even stainless steel, however, can leach a range of metals under some circumstances (Hewitt, 1992) and the inertness of PTFE has also been questioned (Fetter, 1999).
- For analytes that are volatile or sensitive to atmospheric exposure, will the sampling method cause unacceptable suction, cavitation, turbulence or atmospheric contact? For volatile or redox-sensitive parameters, a bladder pump (Section 8.4.5), a good depth sampler (Section 8.4.2) or some types of submersible pump (Section 8.4.4) will probably be most suitable.
- Will the sampling device deliver an adequate flow to satisfactorily purge the borehole, or should the purging be performed by a different device (Box 8.6)?
- Can the sampling device be used repeatedly in a number of boreholes? For major ions analysis, the answer is often yes. For determining trace

amounts of some organic compounds, however, care needs to be taken to avoid cross-contamination of boreholes or samples. It may be that the bailers or pumps have to be thoroughly decontaminated, by distilled water, detergents or solvents in a dedicated cleaning area before they can be reused in another borehole. It may even be necessary to install a dedicated sampling device permanently in each borehole, or use disposable one-use bailers or submersible pumps.

Before sampling it is vital first of all to ascertain the construction details of the borehole. It is very important to know the origin of the water to be sampled – in other words, to know the location of the open sections of the borehole, the intervals of well screen, and the fractures or aquifer horizons which contribute groundwater flow to the borehole (and the proportion of the total discharge that each contributes). Some form of downhole flow logging may be very useful in this regard (Section 6.4). Furthermore, it is also important to know what the observation borehole is made of (will the construction materials influence water quality?) and what condition it is in (will the sampling equipment become entangled with debris?). With a newly constructed borehole, these details will hopefully be diligently recorded. For an older, existing well, a downhole CCTV inspection can be of great value.

In a borehole with a long well screen or open section, the sample of groundwater will typically be a mixture of water from a number of different horizons. In the example shown in Figure 8.5, it should not be assumed that, if the pump is placed in the open borehole at a depth of 40 m, the sampled water will be derived from an aquifer horizon corresponding to that depth. Section 8.4.6 suggests several means of obtaining depth-specific groundwater samples.

8.4.2   Bailers and depth samplers

Bailers and depth samplers are relatively cheap, easy to clean, easy to operate and portable (and cheap disposable versions are available). A bailer, in its simplest form, is a bucket! A modern bailer

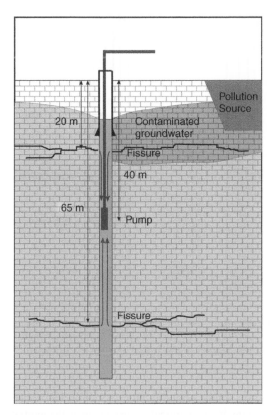

Figure 8.5 *A water sample may be a mixture of water from different flow horizons. In this diagram, a sample taken from 40 m depth in the well is not representative of groundwater from 40 m depth in the aquifer, but a mixture of contaminated water from a fissure at 20 m depth and clean water from 65 m depth*

Figure 8.6 *A transparent bailer has been used to extract fluid from a monitoring borehole at an ex-Soviet Army fuel dump in Riga, Latvia. The hydrogeologist, Arve Misund, has found a large thickness of fuel oil (LNAPL), floating on a layer of groundwater. Photo by David Banks*

for water sampling (Figure 8.6) consists of a narrow cylinder of chemically inert material suspended on a cable. It can be used to lift a volume of water from a borehole to the surface. It is often fitted with a foot-valve (ball or flap), such that it minimizes physical disruption of the water column on its entry to the borehole. The foot valve closes when the bailer is retrieved. Even so, with a standard open-topped bailer, the potential for mixing, volatilization and cross contamination is high and such equipment should not be used for sensitive sampling operations (for VOC, for example).

A depth-sampler is a modified bailer with valves at the top and bottom of an open sampling cylinder. When the cylinder has reached the correct depth in the open hole or screened section (*not* in the 'stagnant' water in the casing), some form of electrical or mechanical signal from the surface seals a volume of water in the bailer by closing the head and foot valves. Because of the low surface area to volume ratio of most depth samplers, they can be acceptable for sensitive sampling parameters (including VOC). However, degassing can occur when the sample is transferred from the bailer to the sample flask if great care is not taken. This can be avoided by transferring the sample from the cylinder to the sample flask using, for example, a closed PTFE sample line and a peristaltic pump.

Prior to sampling with a bailer (as with any sampling equipment), the observation borehole must be purged (Box 8.6). If the bailer is used for this task, it is difficult to monitor effectively the stability of parameters such as EC or pH and the borehole must usually simply be bailed until the requisite number of borehole volumes have been removed. Because of the turbulent nature of the bailing process, this can cause significant disruption to the water column in the borehole, and may even affect the contaminant distribution and hydrochemical environment in the aquifer immediately adjacent to the borehole. Thus, rather than using the bailer for purging, the borehole will ideally be purged by pumping at a low rate (to minimize disruption to the water column) prior to depth sampling (British Standards Institution, 2009).

8.4.3   Simple pumps

In principle, any of the pumps described in Section 3.7 could be used to extract samples of water from wells and boreholes. However, their pump and rising main materials may not be chemically inert. Also, turbulence, cavitation and suction within the pump system may alter the physical and chemical properties of the water, due to atmospheric mixing and degassing (e.g., of VOCs, including some fuel hydrocarbons, solvents, methyl tertiary butyl ether). For example, a standard surface-mounted suction pump is ideal for sampling shallow wells if the objective is to determine major ion chemistry. It is not suitable for collecting representative samples for VOC analysis, however.

In general, positive displacement pumps (e.g., helical screw pumps) will be more suited to sampling of hydrochemically sensitive environments than will centrifugal or surface-mounted motorized suction pumps.

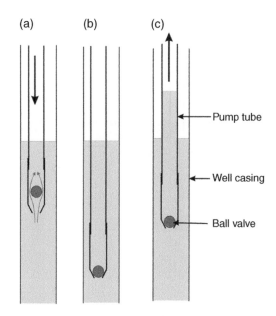

Figure 8.7 *An inertial lift pump for sampling groundwater. The diagram shows (a) the downstroke, (b) the bottom of the stroke and (c) the upstroke*

One simple sampling pump which is well-suited to many hydrochemical environments is the inertial lift pump: this device is essentially a single long tube or pipe, fitted with a foot valve that, by means of up-and-down motion (with a typical 'stroke' of 30–50 cm), progressively displaces a column of water to the surface (Figure 8.7). Such devices are portable and can be operated by hand or by a motor attachment. They are designed for boreholes as narrow as 50 mm, to depths of up to around 60 m (British Standards Institution, 2009) and the larger variants can displace in excess of 10 l min^{-1}.

8.4.4   Submersible pumps

Specialized stainless steel/PTFE submersible pumps are available for groundwater sampling in narrow diameter (50 mm) boreholes [Figure 8.8, 8.9(a)]. Small units can be powered by 12-volt batteries, while the larger pumps operate off electric generators and can provide flows of several litres per minute against heads of several tens of metres.

Figure 8.8 *A lightweight electrical submersible pump, used for groundwater sampling. Photo by Bjørn Frengstad. Reproduced by permission of Norges geologiske undersøkelse (Geological Survey of Norway)*

Tiny plastic submersible pumps that were originally designed for small scale water supply to caravans and holiday cottages, running off 12-volt car batteries, are available very cheaply (Figure 8.9b). These can be joined in series to offer significant lifts and have been found to be suitable for simple, low cost sampling of narrow (50 mm) diameter boreholes. As they are not designed specifically for groundwater sampling, and the leachability/adsorption characteristics of the plastic are unknown, they are probably only suitable for extraction of samples for straightforward inorganic chemical analysis. Their cost is so low that they can be purchased in bulk for dedicated installation in boreholes or even regarded as disposable items.

8.4.5 Other pumps

A range of other specific pump types is available for sampling, especially where parameters are volatile and sensitive to turbulence and atmospheric contact.

The *peristaltic pump* is essentially a positive displacement suction pump mounted at the wellhead. It delivers a flow of water by means of an eccentric rotating wheel squeezing water along a flexible tube (Figure 8.10). The main disadvantages are that (a) as a suction device, the pumping head is limited to 7–8 m and (b) the flow rate is so slow that it is not suitable for effective well-purging. Despite it being a suction device, the flow rate is generally low (several tens of ml min^{-1} to more than 1 l min^{-1}), such that once a continuous flow is established, it may be regarded as being acceptable for sampling volatile and redox-sensitive species. It can also be emplaced in very narrow diameter boreholes.

Gas drive pumps (Figure 8.11) are positive displacement pumps that allow entry of water via a strainer and foot valve to a pumping chamber. This volume of water is then expelled gently from the pump chamber by a slow stream of compressed gas. This may be compressed air or, if the sample is sensitive to oxygen, an unreactive gas such as nitrogen or helium. When sampling hydrocarbons, it is important to ensure that any compressor does not contaminate the air stream with oil. The gas drive mechanism may be connected directly to a well-point, which may be driven into the ground (in unconsolidated sediments) to the horizon of interest.

A *bladder pump* is a variant of the gas drive pump, where the water does not enter a pump chamber but a flexible bladder, preferably made of a relatively inert material such as PTFE. The repeated entry of compressed gas thus periodically squeezes the bladder and expels the water to the surface without any gas/water contact.

The advantages and disadvantages of the gas drive and bladder pumps are similar. On the one hand, they can fit inside narrow diameter (<25 mm) boreholes and can achieve high pumping heads. However, they are relatively expensive and require large gas volumes and cycle times in deep boreholes. Pumping rates, while greater than with peristaltic pumps, are lower compared to submersible or motorized suction pumps (Driscoll, 1986). The lack of any gas-water

(a)

Rising Main

Electric cable   Generator

Borehole

Frequency regulator

Spool of hose and cable

(b)

Figure 8.9 *(a) David Banks uses a narrow-diameter electric submersible pump to sample a monitoring borehole near a landfill at Trandum, Norway. The pump is powered by a small portable generator and its speed is controlled by a frequency regulator. (b) Bruce Misstear uses a tiny, low-cost submersible pump (of the kind employed as a bilge pump in boats) to sample an observation borehole near a landfill in southern England. The pump is powered from a 12 V car battery*

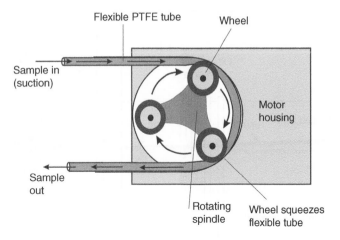

Flexible PTFE tube   Wheel

Sample in
(suction)

Motor
housing

Sample
out

Rotating
spindle

Wheel squeezes
flexible tube

Figure 8.10   *A simplified diagram of a peristaltic pump*

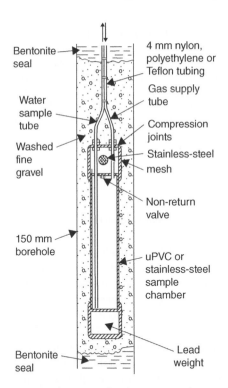

Bentonite
seal

4 mm nylon,
polyethylene or
Teflon tubing

Gas supply
tube

Water
sample
tube

Compression
joints

Washed
fine
gravel

Stainless-steel
mesh

Non-return
valve

150 mm
borehole

uPVC or
stainless-steel
sample
chamber

Bentonite
seal

Lead
weight

Figure 8.11   *The principle of operation of a gas drive pump. In this case, the "pump" is actually a sampler emplaced in a gravel-filled section of a borehole, although such a pump could, in principle, be suspended freely in a water filled well. Water enters the top chamber through a filter and then flows to the sample chamber via a non-return valve. An inert gas is pumped into the sample chamber displacing the volume of water up the water sample line to the surface (after Clark, 1988)*

contact renders the bladder pump suitable for sampling volatile or redox-sensitive compounds (and thus preferable to the gas-drive pump).

8.4.6   Sampling at specific depths

We have already established that taking groundwater samples from boreholes with long open sections can be problematic. The sample will typically represent some permeability-weighted average of the water quality from several different horizons or fractures. When sampling for contamination, this can lead to the erroneous conclusion that the contamination is more dilute than it actually is (Martin-Hayden and Robbins, 1997; Figure 8.12). The practice of purging (Box 8.6) large quantities of water from a borehole may result in substantial disruption of the depth profile and mixing of water from different levels. The technique of 'micro-purging', where a small volume of the well within a screened section or adjacent to a known flow horizon, is purged at a low rate, may allow more depth-specific samples to be taken (British Standards Institution, 2009). However, such low-flow purging may not overcome vertical flows in observation boreholes where strong vertical head gradients exist, and the sample provenance will be very sensitive to the pump position and to the duration and rate of pumping (McMillan *et al.*, 2014).

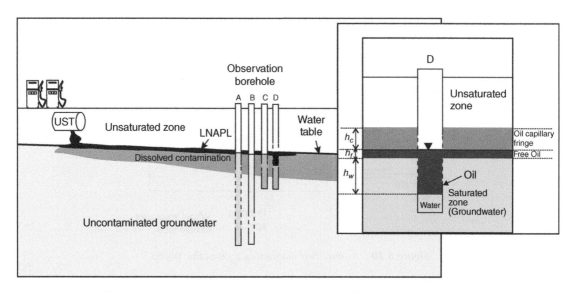

Figure 8.12 *Different ways in which observation borehole length and position can cause us to underestimate groundwater pollution. Borehole A will yield a mixture of water from uncontaminated and contaminated horizons, to the extent that the contaminant may be significantly diluted and underestimated as compared with a depth-specific piezometer C, emplaced within the contaminated zone. Piezometer B will not detect LNAPL phase contamination and may not detect dissolved phase contamination, as the well screen is open only at some distance below the water table. Borehole D's well screen straddles the water table and is positioned to detect the LNAPL but may overestimate the thickness of the LNAPL layer (see text and inset). Inset shows a close-up of Borehole D: h_f is thickness of free oil phase in the aquifer; h_w is depth of oil in the borehole below the base of free oil in the aquifer; h_c is thickness of the oil capillary fringe on top of the water table*

In many circumstances, it will be desirable to collect samples from specific horizons. To this end, packers can be used to isolate a section of open borehole for pumping and sampling (Figure 8.13(a)). Alternatively, Bishop *et al.* (1992) devised a system of pumping, using auxiliary scavenger pumps, to try to ensure that pumped water samples in open borehole sections are derived from the target aquifer horizons (Figure 8.13(b)). The application of Bishop *et al.*'s method is not straightforward, however, and requires a good knowledge of the borehole flow regime. The drawback with both packers and scavenger systems in open holes is that, prior to sampling, the open borehole may have allowed vertical exchange of groundwater between different horizons of the aquifer. The presence of the open borehole may thus have modified the

original water quality distribution in the aquifer system.

If the objective is to obtain samples of water from a particular depth or aquifer horizon, it is best to avoid long open sections of monitoring borehole. In some situations, cone penetrometer technology can be used, avoiding the need for constructing an observation borehole (see Section 8.2.2). For a permanent installation, an appropriate design is a *piezometer* – a narrow borehole with only a very limited section of well screen that is open to the aquifer (Section 3.5). This may be installed in a pre-drilled borehole (Figure 8.14) or be driven into shallow unconsolidated sediments as a drive point.

To sample multiple horizons at differing depths (or indeed to determine heads at different depths to understand the three-dimensional flow regime), a cluster or a nest of piezometers can be installed

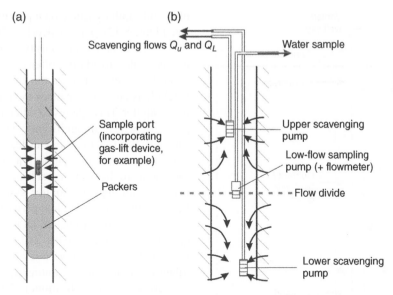

Figure 8.13 *Two possible means of obtaining depth-specific water quality samples from an open borehole in an aquifer system: (a) using packers, (b) using hydraulic scavenging pumps. Based on a concept by Bishop et al. (1992)*

(Figures 3.14, 8.15, 8.16). The nest is installed in a single borehole, such that individual piezometers must be hydraulically separated from each other by the emplacement of a low permeability bentonite or grout layer. The piezometer cluster is simply a closely-spaced array of single piezometers, each installed in its own borehole to a different depth. While the nest results in cost savings (single borehole), hydraulic isolation between intakes can be difficult to achieve and the cluster will usually result in greater integrity of sampling.

Other, more sophisticated, multilevel sampling devices have been devised. The variety in design and materials is large, but the conceptual design is similar in most cases. Several filter openings or ports are mounted on a central stem or core, which contains a number of channels or tubes connecting to the surface. The ports are separated hydraulically either by packers or by permanent bentonite seals in the borehole (British Standards Institution, 2010b). Sampling from such multilevel samplers is typically by gas-drive or bladder pump installations. Multilevel devices are available that can fit in 75 mm diameter boreholes.

8.4.7   Sampling for non-aqueous phase liquids

For sampling non-aqueous phase contaminants, such as oils, solvents and hydrocarbons, particular consideration must be given to the observation borehole design. The open section of the borehole must straddle the water table to allow floating hydrocarbons (light non-aqueous phase liquids – LNAPLs) to enter the borehole (Figure 8.12). LNAPLs in an observation borehole may be detected by several means:

- a transparent bailer, in which any column of oil should be visible floating on groundwater (Figure 8.6);
- an indicator paste that can be applied to a dipper tape and which changes colour where it encounters the LNAPL;
- a special electric dipper tape that detects the LNAPL-air and LNAPL-water interfaces;
- globules of LNAPL within a pumped water sample, or a film of LNAPL on the sample surface;
- the results of the sample analysis, where these show concentrations of NAPL compounds that are greater than the theoretical solubility of the NAPL in water.

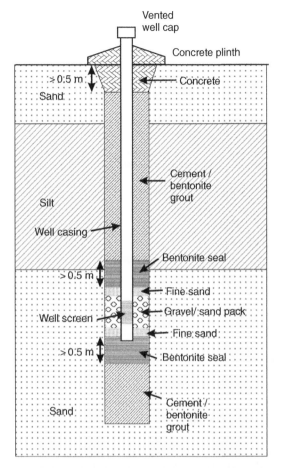

Figure 8.14 *Typical installation of a piezometer (an observation well with only a short section of open well screen) in a sandy aquifer, based on guidance provided by the Institution of Civil Engineers (1989), SFT (1991) and Fetter (2001). In some circumstances, it may be permissible for some of the grouted section to be substituted with other low permeability backfill material*

It is important to be aware that the height of LNAPL in the borehole is, due to buoyancy effects, not representative of the thickness of the LNAPL lens in the aquifer. Abdul *et al.* (1989) contend that:

$$\frac{h_w}{h_c} = \frac{\rho_o}{\left(\rho_w - \rho_o\right)} \qquad (8.9)$$

where h_w is the depth of oil in the borehole below the base of free oil in the aquifer, h_c is the thickness

of the oil capillary fringe on top of the water table (see Figure 8.12), ρ_o is the density of the hydrocarbon (LNAPL) and ρ_w the density of the groundwater. As the density of oil is often around 0.8 g cm^{-3}, it can be assumed that the thickness of oil in the borehole is around four times the thickness of oil in the aquifer. This ignores the thickness of free oil phase, h_f, which, if significant compared with h_c, can make this rule of thumb misleading.

Dense non-aqueous phase liquids (DNAPLs), such as many halogenated solvents or creosotes, may sink through the aquifer system into blind fractures or until a very low-permeability base is reached. Their occurrence in an aquifer is consequently difficult to predict and any monitoring strategy should aim to investigate contaminant distribution throughout the entire aquifer thickness (see British Standards Institution, 2009, 2010b).

Finally, it should be noted that purging a NAPL-contaminated well prior to sampling may simply spread the NAPL phase throughout the borehole column and may not be beneficial (British Standards Institution, 2009).

8.5 Sample filtration, preservation and packaging

As noted above, many water quality parameters have the potential to alter or degrade during the interval between sampling and analysis at the laboratory. In order to minimize this, it is important to:

i. Filter *some* types of sample to ensure that particulate matter does not enter the analytical process, causing overestimation of the dissolved concentrations of the element in question.
ii. Select appropriate flask types to ensure that packaging materials do not adversely affect sample quality. For most inorganic parameters, translucent polyethene flasks with uncoloured polyethene stoppers are adequate, and have been shown by Reimann *et al.* (1999b) not to present significant disadvantages compared with, for example, very expensive fluorocarbon

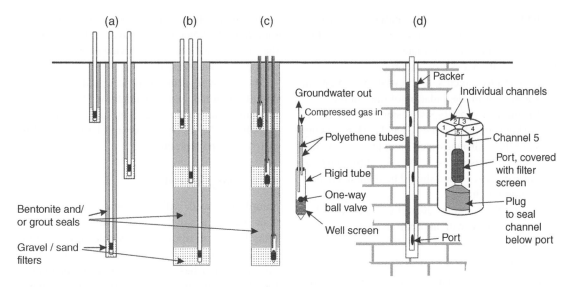

(a)   (b)   (c)   (d)

Packer

Individual channels

Groundwater out

Compressed gas in

Polyethene tubes

Channel 5

Rigid tube

Port, covered with filter screen

One-way ball valve

Plug to seal channel below port

Well screen

Port

Bentonite and/ or grout seals

Gravel / sand filters

Figure 8.15 *A selection of installations to allow sampling at different levels in an aquifer system: (a) a piezometer cluster; (b) a piezometer nest; (c) a multilevel device using buried gas-lift samplers, separated by bentonite seals; (d) a continuous multi-channel device with sampling horizons in a fractured aquifer separated by packers (each channel opens to a screened port, and can be sampled using a gas-lift device or a peristaltic pump)*

Figure 8.16 *Clusters of piezometers at Pollardstown Fen, Ireland. Photo by Bruce Misstear*

bottles. For most organic analyses, analytical-grade baked amber glass flasks, with PTFE-lined stoppers are adequate.

iii. Select, if necessary, appropriate chemical preservation agents to prevent precipitation, sorption or biodegradation of the sample.

iv. Ensure (for most types of sample) that flasks are kept cool (around 4 °C) and dark following sampling. Groundwater samples *can,* under some circumstances, be preserved by freezing, but specialist advice should be obtained before attempting this.

v. Ensure that samples are delivered to the laboratory rapidly, with appropriate quality control documentation, and that they are analysed promptly.

Unless known to have been adequately pre-cleaned, flasks (and their stoppers) for water samples for inorganic analyses should usually be rinsed three times with the groundwater to be sampled prior to sampling (unless the water is known to be grossly contaminated). For filtered samples, they should *then* be rinsed at least twice with the filtered water. For samples for organic analysis, rinsing of flasks should be avoided for some parameters (including hydrocarbons) and relevant protocols should be consulted prior to sampling.

Some laboratories will supply clean flasks pre-filled with appropriate quantities of a preservation agent. In this case, flasks should *not* be rinsed prior to sampling. They should be filled slowly and carefully, care being taken not to cause overflow resulting in loss of the pre-filled preservation agent.

Different parameters have differing sampling requirements. Figure 8.17 illustrates some of the factors which are most important for various analytes. Selection of appropriate flask types, filtration and preservation practices will, to some extent, depend on the treatment of samples at the laboratory and on the type of analytical equipment utilized. It is thus important to take guidance first and foremost from the laboratory on these issues. Indeed, many laboratories will provide pre-cleaned sampling bottles of appropriate materials, pre-filled

preservatives, filter units, labels, instructions for sampling, documents, cool boxes and ice packs to the sampler, and may also provide guaranteed courier delivery and pick-up.

The following sections reflect guidelines based on international recommendations. If laboratory recommendations deviate significantly from these guidelines, the laboratory should be requested to justify their own practice.

8.5.1   Sampling order

Sampling a well will often involve much more than filling a single flask with groundwater. Usually several aliquots of sample (for different analyses) will be required. The US Geological Survey (2015) has developed a recommendation for the optimum order of collection of these aliquots (Table 8.5). For example, as extensive pumping can result in well turbulence and disruption of the groundwater regime in the near aquifer, sensitive parameters such as volatile organic compounds (VOCs) are sampled first (following purging of the well).

8.5.2   Physicochemical parameters

Most physicochemical parameters are best determined on site using portable electrodes, meters and field kits (Section 8.2). It can, however, be useful to obtain confirmatory analyses of physico-chemical parameters in the laboratory (Figure 8.17).

For laboratory determination of pH, electrical conductivity and alkalinity, it is usually sufficient to collect a single large aliquot of groundwater (0.5 to 1 l) in a polyethene flask. The sample does not usually need to be filtered. No chemical preservation should be performed. If the sample contains a very high particulate content, filtration of a separate aliquot at 0.45 μm may be considered for the alkalinity determination.

For determinations of colour and turbidity, a similar flask of 0.5 to 1 l should be sufficient. Neither filtration nor chemical conservation should be performed.

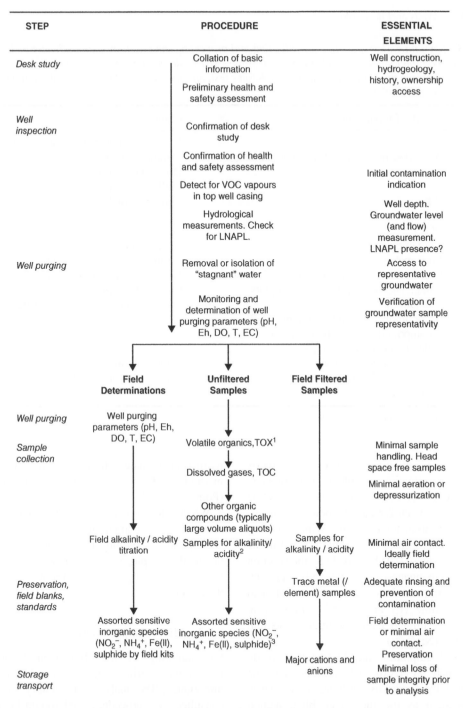

STEP	PROCEDURE	ESSENTIAL ELEMENTS

Desk study — Collation of basic information / Preliminary health and safety assessment — Well construction, hydrogeology, history, ownership access

Well inspection — Confirmation of desk study / Confirmation of health and safety assessment / Detect for VOC vapours in top well casing / Hydrological measurements. Check for LNAPL. — Initial contamination indication / Well depth. Groundwater level (and flow) measurement. LNAPL presence?

Well purging — Removal or isolation of "stagnant" water / Monitoring and determination of well purging parameters (pH, Eh, DO, T, EC) — Access to representative groundwater / Verification of groundwater sample representativity

Field Determinations — Unfiltered Samples — Field Filtered Samples

Well purging — Well purging parameters (pH, Eh, DO, T, EC)

Sample collection — Volatile organics, TOX[1] / Dissolved gases, TOC / Other organic compounds (typically large volume aliquots) — Minimal sample handling. Head space free samples / Minimal aeration or depressurization

Field alkalinity / acidity titration — Samples for alkalinity/acidity[2] — Samples for alkalinity / acidity — Minimal air contact. Ideally field determination

Trace metal (/ element) samples — Adequate rinsing and prevention of contamination

Preservation, field blanks, standards — Assorted sensitive inorganic species (NO_2^-, NH_4^+, Fe(II)), sulphide by field kits — Assorted sensitive inorganic species (NO_2^-, NH_4^+, Fe(II), sulphide)[3] — Field determination or minimal air contact. Preservation

Major cations and anions — Minimal loss of sample integrity prior to analysis

Storage transport

Figure 8.17 *Schematic diagram of groundwater sampling, identifying important factors to consider at each step (modified after Barcelona et al., 1985). [1]TOX are organic halogenated compounds. [2]Alkalinity and acidity determinations should ideally be performed in the field but can be performed in the laboratory. Such determinations are often performed on unfiltered groundwater samples. If significant turbidity or particulate matter is present, however, samples should be filtered. [3]Field kits are available for estimation of sensitive inorganic parameters in the field. Where determinations are made in the laboratory, samples are usually preserved. Samples are often unfiltered, but may be filtered depending on the particulate content of the sample and the parameter in question*

Table 8.5   *Recommended sequence for collecting and processing samples from a groundwater well (based on recommendations from US Geological Survey, 2015). Remember, in a contamination investigation, sample the least contaminated wells/sites first, and the most contaminated last, to minimize risk of cross-contamination*

Step	Samples for parameters	Comments
1	VOCs Pesticides, herbicides, PCBs, other organics	Unfiltered samples first, followed by filtered samples (if any) Do not field-rinse bottles Chill immediately
2	TOC (DOC) (SOC)	Chill immediately
3	Inorganic constituents: Trace metals Major cations, individually analysed elements (e.g., Hg, Se, As), if separate aliquots required) Major anions, alkalinity, nutrients (N, P compounds – ammonium, nitrate) Radiochemicals and isotopes	Filtered samples followed by unfiltered samples (for groundwater) Field rinse flasks as required Chill flasks (especially for anions/alkalinity/nutrient parameters) immediately
4a	(DOC, if capsule-filtered)	(if DOC is capsule-filtered)
4b	Radon and chlorofluorocarbons	Do not field-rinse flasks
5	Microorganisms	

DOC, dissolved organic carbon; SOC, suspended organic carbon; TOC, total organic carbon; VOCs, volatile organic compounds

Temperature, dissolved oxygen and redox potential determinations should be performed in the field. A method does exist for the preservation of dissolved oxygen for later laboratory analysis, but is seldom used today. In this Winkler (1888) method, any oxygen in the sample is fixed in the field as manganese (III) hydroxide by the addition of alkaline manganese sulphate and potassium iodide. Subsequent laboratory addition of acid generates iodine (I_2) in proportion to oxygen, which can be titrated with a solution of sodium thiosulphate ($Na_2S_2O_3$).

Dissolved sulphides and hydrogen sulphide (H_2S), which are redox sensitive, can also be determined by field kits, utilizing colorimetric reactions such as the methylene blue method (Fischer, 1883). Alternatively, alkaline zinc acetate solution can be added to the sample, which will precipitate any sulphides as relatively stable solid zinc sulphide (ZnS). In the laboratory, an iodine/thiosulphate titration or a colorimetric method can be applied to quantify the sulphide (Pomeroy, 1954).

When conducting a groundwater sampling exercise, the sampler should note any colour or odour at the point of sampling. The sampler should avoid tasting the sample, especially if any contamination is suspected, and exercise caution when assessing the odour.

8.5.3   Microbial parameters

For sampling of microbiological parameters, maintaining sterile conditions is of paramount importance. The analytical laboratory should be consulted over procedures and should be able to supply sterile (usually glass) flasks. The flasks should have been sealed after sterilization and should not have been opened until sampling takes place. Care should be taken not to contaminate flasks by handling during sampling; especially by touching flask necks. Clean, powder-free latex gloves can be used to minimize risk. Any nozzles, tubes or filters should be removed from any sample tap. Ideally the sample tap should be unthreaded and of an inert metal (e.g., brass). If the tap is metal, it is good practice to sterilize it carefully using a small gas-blowtorch or even a cigarette lighter. The sampling tap should then be run for at least three minutes before sampling. The sample, once taken, should be sealed, kept in a cool (4 °C) dark place and transported on the same day (and preferably within a few hours) to the laboratory. If this is not possible, the use of a portable laboratory should be considered (Section 8.2.3). The sample volume required will depend on the laboratory and also the type (likely degree of contamination) of the sample (Table 8.6). A sample volume of 500 ml should be adequate for analysis of standard faecal indicators. Samples taken for microbiological analysis should not be used for chemical analysis: it is best to collect separate sample aliquots for each type of analysis.

Note that if the sampled water has been disinfected by chlorination, a reducing agent, such as

sodium thiosulphate (at a rate of 0.1 ml of 1.8% sodium thiosulphate solution, per 100 ml sample, according to Bartram *et al.*, 1996), should ideally be added at the point of sampling, to consume any residual chlorine and thus halt any disinfecting effect during sample storage and incubation.

Analysis for other, more exotic, microbiological parameters (including protozoa) may require larger sample volumes and/or wellhead filtration techniques. It is advisable to consult the laboratory for specific advice.

8.5.4   Inorganic parameters: acidification and filtration

Analysis of metallic elements and cations poses some sampling and analysis problems. Such analyses are often undertaken by *inductively coupled plasma optical emission spectrometry* (ICP-OES), *inductively coupled plasma mass spectrometry* (ICP-MS) or *atomic absorption spectroscopy* (AAS) techniques (Section 8.8; Table 8.7). With all of these methods, the sample is usually acidified prior to analysis, either at the time of sampling or by the laboratory. If the sample is acidified, then in most cases it should also be filtered, for reasons explained below.

Acidification. Degassing of carbon dioxide from unacidified water samples during storage and transport can result in small increases in sample pH. This can cause some metals and cationic elements to precipitate as hydroxides or carbonates. These elements may also be susceptible to sorption on the walls of a sample flask. Acidification of the sample can prevent precipitation and sorption and keep the sample's total dissolved element load in solution during transport. Therefore, it is common practice to add ultra-pure concentrated nitric acid (HNO_3) to the sample in order to lower its pH to around or below a value of pH 2 (for most normal groundwater samples, between 0.5–2 ml acid per 100 ml should be sufficient). This should ideally be performed in the field immediately *after* sampling and filtration. However, in some locations it can be problematic to maintain purity of the acid in the field, and to legally arrange safe transport of the

Table 8.6 *Sample volumes (modified from recommendations by Oxfam/Delagua, 2000) for analysis of faecal coliforms using membrane filtration techniques*

Source	Amount
Lakes, ponds and other surface waters	10 ml
Protected groundwaters e.g. wells and springs	100 ml
Unprotected groundwaters e.g. open dug wells and springs	50 ml
Waters in treatment plants after partial treatment	50–100 ml
Waters in treatment plants after full treatment	100 ml
Reservoirs, distribution networks and household taps	100 ml

Reproduced by permission of the Robens Centre for Public and Environmental Health

Table 8.7 *A typical suite of inorganic chemical parameters offered by a laboratory, with associated detection limit and relative analytical uncertainty. Note that ICP equipment can be calibrated in different ways, offering differing detection limits for differing elements. This is one example only, tailored to a specific market*

Parameter	Measurement limit ($\mu g\ l^{-1}$)	Relative analytical uncertainty (%)	Parameter	Measurement limit ($\mu g\ l^{-1}$)	Relative analytical uncertainty (%)
	ICP-OES			ICP-MS	
Si	>20	±10	Al	>2.0	±10
Al	>20	±10	As	>0.05	±10
Fe	>2	±10	B	>5	±20
Ti	>1	±5	Be	>0.01	±10
Mg	>50	±10	Cd	>0.03	±20
Ca	>20	±5	Ce	>0.01	±5
Na	>50	±10	Co	>0.02	±10
K	>500	±20	Cr	>0.1	±10
Mn	>1	±5	La	>0.01	±5
P	>50	±10	Mo	>0.06	±5
Cu	>5	±5	Ni	>0.2	±10
Zn	>2	±5	Pb	>0.05	±5
Pb	>5	±10	Rb	>0.05	±10
Ni	>5	±10	Sb	>0.01	±10
Co	>1	±5	Se	>1	±20
V	>5	±10			
Mo	>5	±10			
Cd	>0.5	±10			
Cr	>2	±10			
Ba	>2	±5			
Sr	>1	±5			
Zr	>2	±10			
Ag	>5	±10			
B	>20	±10			
Be	>1	±5			
Li	>5	±10			
Sc	>1	±5			
Ce	>20	±5			
La	>5	±10			
Y	>1	±10			

Atomic absorption (graphite oven technique; cold vapour technique for Hg)			Ion chromatography (IC)		
Pb	>0.2	±10	F^-	>50	±10
Cd	>0.02	±10	Cl^-	>100	±10
As	>3.0	±10	Br^-	>100	±10
Se	>1.0	±10	NO_3^-	>50	±10
Sn	>2.0	±10	PO_4^{3-}	>200	±10
Sb	>2.0	±10	SO_4^{2-}	>100	±10
Hg	>0.01	±10			

Reproduced by permission of Norges geologiske undersøkelse (Geological Survey of Norway)

acid. Consequently, it may be necessary to request that the laboratory acidify the sample on receipt, in order to remobilize any sorbed or precipitated element content.

Filtration. If particles or colloids are present in the sample, there is a risk that these will be dissolved when the sample is acidified, thus leading to overestimation of *dissolved* concentrations of certain elements (metals, base cations). Thus, if the dissolved element content of a groundwater is to be determined (and the sample is to be acidified), it should also be field-filtered. The most commonly recommended filter size in international guidelines is 0.45 μm. This should exclude most particulate content but will allow some colloidal material to pass. The choice of filter size is somewhat arbitrary: some researchers have used 0.1 μm filters in order to exclude a greater proportion (but still not all) of colloidal matter (Reimann *et al*., 1999a). Filtration may be performed by:

- Mounting a filter unit directly on the sample line. Here, the line water pressure must be adequate to force water through the filter, but not enough to damage it. Yeskis and Zavala (2002) recommend that the first 500–1000 ml of filtered water should be discharged to waste.
- Use of a filtration unit powered by inert gas pressure.
- Drawing the groundwater into a clean inert plastic syringe, and then expelling it via a small filter capsule mounted on the tip of the syringe into the sample flask (Figure 8.18). The first few ml should be discarded as waste.
- Filtering through a filter paper mounted on a flask which can be depressurized by a simple hand pump to draw water through the filter. This method has the disadvantage of being more susceptible to contamination and to sample degassing than the other methods. It is *not* recommended, especially where volatile compounds are concerned.

In some circumstances, it may be required to determine total (as opposed to dissolved) element contents in a water sample. Some national drinking water standards may actually specify total

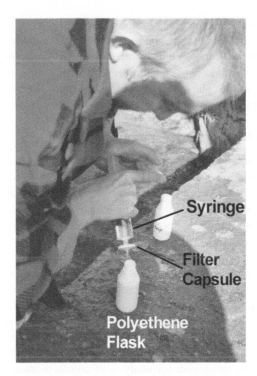

Figure 8.18 *A manual syringe-based filtration unit being used at Burojë, near Mitrovice, Kosova. Reproduced by permission of Kornelius Gustad*

contents, rather than dissolved contents (as any particulate or colloidal metals will still enter the acid digestion system that is the human stomach). Here, it *may* not be appropriate to filter the sample before acidification.

Some specific types of groundwaters (e.g., waters from abandoned mines – Banks *et al*., 1997) may have high loadings of metal species such as Al, Mn^{II} and Fe^{II}. These may start to oxidize, hydrolyse and precipitate in the mine or during filtration. In these situations, filtration of a sample may tend to exclude a proportion of the elements of interest. Therefore, for Fe–, Mn– or Al-rich waters, *two* aliquots of sample may be collected for metals analysis: one filtered and the second unfiltered (both of which should be acidified to solubilize the metals). The difference between the two analyses will be a measure of the dissolved or colloidal/particulate content of the elements in question at the time of sampling. Alternatively, for redox-sensitive parameters, such as Fe and Mn, on-line filtration may be attempted under ambient (anaerobic) conditions to prevent precipitation of oxidized phases on the filter medium (British Standards Institution, 2009).

8.5.5   Inorganic parameters: sampling

Aliquots of sample for inorganic analysis will be collected after the well has been purged (Box 8.6), and after the sample line has been run freely for at least three minutes. For anion analysis, a sample flask and its stopper should be rinsed thoroughly three times with the groundwater to be sampled. The flask is then filled and sealed, leaving a small headspace in the neck. For most groundwater samples that do not contain excessive particulate matter, filtration and chemical preservation are unnecessary for the analysis of the most common soluble anions (SO_4^{2-}, Cl^-, Br^-, F^-, NO_3^-, PO_4^{3-}). The sample volume will depend on the analytical technique employed: for analysis of common anions by ion chromatography, a sample volume of 50–100 ml in a polyethene flask should suffice.

For analysis of cations and metallic elements, a second polyethene sample flask should be used. The flask and stopper is rinsed three times with the groundwater to be sampled, and then at least twice with filtered sample water (assuming the sample is to be filtered). Then, the filtered sample can be taken, usually using a filter size of 0.45 μm (Section 8.5.4); the sample can then be acidified and the flask sealed leaving a small headspace in the neck. The sample volume will depend on the laboratory's requirements, but 10–50 ml is usually adequate for modern ICP methods.

Further aliquots (flasks) of sample may then be taken for specific parameters. Table 8.8 offers advice on recommended sampling and preservation procedures.

8.5.6   Organic parameters

The large number of organic compounds (many thousands), and the variety of analytical methods used to determine these, means that specific advice

on sampling for organic compounds should be sought from a reputable laboratory. The following guidelines are, of necessity, general in their scope.

Semi-volatile and non-volatile organic compounds are often sampled in analytical grade amber glass bottles, with a fluorocarbon-lined (PTFE) stopper. The amber glass hinders photochemical reaction or degradation. The issue of whether to filter samples for organic parameters is problematical, as filtration presents significant opportunities for volatilization and contamination. Indeed, Barcelona *et al.* (1985) recommend that filtration of samples for organic analysis should be avoided unless absolutely necessary. On the other hand, several contaminants that may be present in rather low quantities in the dissolved phase (including many pesticides and PAH) sorb strongly to particulate matter. The presence of contaminated particles in the sample can therefore lead to significant overestimation of dissolved concentrations. This is a particular concern for observation boreholes drilled into contaminated soils, where the well-screen does not adequately exclude particulate matter. Filtration of samples destined for organic analysis is possible, but specific advice should be sought from the analytical laboratory.

The US Geological Survey field guide (US Geological Survey, 2015) contains comprehensive advice on sampling. Preservation routines for samples for organic analysis vary and may be specified in national standards. In some standards, acidification of samples for mineral oils, hydrocarbons and total organic carbon is recommended, using sulphuric (or sometimes hydrochloric) acid to pH ≤2. According to US Geological Survey (2015) and SFT (1991), samples for analysis of phenols may be preserved by acidification with phosphoric acid to pH ≤4 and the addition of copper sulphate. Samples for organic analysis should be stored and transported in dark and cool (4 °C) conditions. They should be analysed promptly, preferably within 1–2 days for most parameters according to SFT (1991), although some laboratories claim longer sample lifetimes. Table 8.8 contains SFT (1991) and US Geological Survey (2015) recommendations for a variety of organic compounds.

Table 8.8 Sample preservation and filtration guidance for various chemical parameters (although specific guidance from analytical laboratories and national standards also should be sought). In all cases, samples should be stored and transported at 4 °C and analysed rapidly

Parameter	Bottle type and typical volume	Filtration/conservation	References
Ammonium (NH_4^+)	Translucent or brown polyethene or analytical grade glass (>125 ml)	Filtration may be advisable if visible particulate matter. Acidify with ultrapure conc. H_2SO_4 to pH≤2	Barcelona et al. (1985) SFT (1991) UNESCO/WHO/UNEP (1996) USGS (2015)
Anions (Cl^-, $SO_4^=$, F^-, NO_3^-, PO_4^{3-}, Br^-)	Polyethene (c. 100 ml for ion chromatography)	No filtration necessary (unless sample obviously turbid). No conservation (unless long delay before analysis, in which case, see below for nitrate)	SFT (1991) UNESCO/WHO/UNEP (1996) USGS (2015).
Cations, metallic elements	Polyethene (c. 100 ml for ICP techniques)	Filtration at 0.45 μm usually recommended. Acidify with ultrapure conc. HNO_3 to pH≤2. For mercury (Hg) add 6N HCl instead of HNO_3	SFT (1991) UNESCO/WHO/UNEP (1996) Mather (1997) USGS (2015)
Nitrate (NO_3^-)	As for ammonium	If sample kept cool and delivered rapidly, no special conservation required (can be sampled with other anions). Otherwise filtration/preservation as for ammonium. Some authorities recommend addition of a bactericidal agent such as thymol or chloroform to hinder biodegradation	As for ammonium Appelo & Postma (2005) UNESCO/WHO/UNEP (1996)
Oils and hydrocarbons	Amber glass with PTFE-sealed stopper. 1000 ml	Acidify with conc. HCl or H_2SO_4 to pH≤2. Extract promptly	SFT (1991) USGS (2015)
PAH (polycyclic aromatic hydrocarbons)	Amber glass with PTFE-sealed stopper 1000 ml	None	SFT (1991)
Pesticides, PCB, other "semi-volatile" organic compounds	Amber glass with PTFE-sealed stopper. 1000 ml (but several flasks may be needed for various analyses)	None. (Some pesticide compounds may require specific preservation)	SFT (1991) USGS (2015)

(Continued)

Table 8.8 (Continued)

Parameter	Bottle type and typical volume	Filtration/conservation	References
Phenols	Amber glass, with PTFE-sealed stopper. 500 to 1000 ml	Acidify with 2M phosphoric acid to pH≤4 (CuSO$_4$ added), or acidify with H$_2$SO$_4$ to pH≤2	SFT (1991) USGS (2015)
α, ß-Radioactivity, radium	1–2 l polyethene	Filtered at 0.45 μm. Acidify to pH≤2 with ultrapure conc. HNO$_3$	USGS (2015)
Radon	c. 20–25 ml glass scintillation vials	10 ml sample collected from flowing water in non-turbulent manner in syringe and injected beneath c. 10 ml of liquid scintillation gel pre-filled in vial	
Sulphide	Differing sources recommend glass or polyethene. c. 250 ml	Filtration may be desirable if sample visibly turbid. Preservation with alkaline zinc (or cadmium) acetate solution to precipitate ZnS (or CdS)	SFT (1991) UNESCO/WHO/UNEP (1996) Appelo & Postma (2005)
TOC (total organic carbon)	Amber analytical grade glass, preferably with PTFE-sealed stopper	Acidification to pH≤2 with analytical grade conc. H$_2$SO$_4$ or HCl may assist in preservation of TOC. Laboratory advice should be sought	SFT (1991) UNESCO/WHO/UNEP (1996) USGS (2015)
VOCs (volatile organic compounds)	40 ml amber glass vials with PTFE septum	Unfiltered. No headspace. No gas bubbles. Some protocols require acidification to pH≤2 with 1:1 HCl + H$_2$O (no more than 5 drops)	USGS (2015)

Volatile organic compounds (VOCs) will require particular care. The sample line should be run at a very low flow rate (not exceeding 100 ml min⁻¹). If a bailer is used, some form of PTFE flow line should be used to pump the sample from the bailer barrel to the sample flask. Filtration should not generally be attempted (Barcelona *et al.*, 1985). Typically, special 40 ml amber glass vials for volatiles are carefully and slowly filled with the sample water and sealed with a fluorocarbon septum. It is important that the sample contains no headspace and no gas bubbles. Some protocols for VOC sampling recommend preservation with two to five drops of 50% hydrochloric acid to acidify the sample to pH ≤2.

Alternatively, methods exist to sample groundwater for VOCs in a special headspace vial. Here, the 40 ml vial is partially (about 60%) filled with sample water and sealed. The air in the remaining headspace then acquires a VOC content in equilibrium with the concentration of VOC in solution. The headspace gas can thus be analysed directly to yield an estimate of dissolved VOC concentrations. The headspace technique is also commonly used to estimate VOC contents in soils.

At the laboratory, the organic content of aqueous samples will typically be extracted using some form of solvent, which is then injected into a spectrometer or chromatograph. It is theoretically possible to add the relevant extracting solvent (such as hexane) to the sample flask in the field. This may stabilize the sample content and extend its preanalysis lifetime, but should only be attempted by experienced samplers with the approval of the analytical laboratory, and only under conditions where the purity of the extractant can be guaranteed.

8.5.7   Stable isotopes

The most commonly analysed stable isotopes in groundwater (Geyh, 2000) are:

- oxygen-18 (^{18}O) and deuterium (^{2}H), which occur in the water molecule (H_2O) itself;
- carbon-13 (^{13}C), in the dissolved inorganic carbon ($CO_3^= - HCO_3^- - H_2CO_3 - CO_2$) system; the unstable ^{14}C isotope is also sometimes used to date water;

- sulphur-34 (^{34}S), which can occur in the sulphide phase, or the sulphate phase;
- oxygen-18 within the sulphate ion ($S^{18}O_4^=$).

Sampling methods are described by IAEA/UNESCO (2000), IAEA (2010), US Geological Survey (2015) and USGS Stable Isotope Laboratory (2015). For ^{2}H and ^{18}O, all that is required is a very small (several 10 ml should be more than adequate) sample of water (which can be filtered if there is a visible particulate content). The only major factor that can disturb ^{2}H and ^{18}O concentrations is evaporation. If the sample is to be analysed promptly, simply collecting a full sample in a screw-topped polyethene flask is usually sufficient. Wrapping a paraffin film seal around the top may further hinder opportunities for evaporation. However, as polyethene is gas-permeable, a glass vial, with a gas-type cap or seal is preferable (IAEA, 2010).

For ^{13}C and ^{14}C it has been common to precipitate the inorganic carbon in the field immediately after sampling (Figure 8.4), by adding carbonate-free sodium or potassium (or even ammonium) hydroxide to raise the pH to around 11, followed by the addition of strontium (or barium) chloride. The alkaline pH converts all carbon dioxide and bicarbonate to carbonate, allowing it to be precipitated as insoluble strontium or barium carbonate. Iron sulphate is sometimes also added to hasten precipitation (IAEA, 2010). In the laboratory, acid is added to the precipitate, regenerating carbon dioxide and allowing it to be analysed. Many laboratories are now beginning to accept water samples for analysis without the immobilization of carbon as a precipitate. A precondition for this is that CO_2 degassing is prevented. Thus, a fully-filled dark glass container (0.25 to 1 l), with a gas-tight seal, can be used. Biocides can be added to prevent biological transformations of carbon, if the sample will not be analysed immediately.

To analyse ^{34}S in sulphide, alkaline zinc or cadmium acetate is added to the water sample, to precipitate zinc or cadmium sulphide, which is relatively stable. Silver nitrate solution could also be used as a precipitating agent, to precipitate sulphide as silver sulphide (Ag_2S). An alternative method for collection of sulphide is the acidification of the water to convert all sulphide to H_2S gas, stripping the gas using high purity nitrogen, and passing the gas through silver nitrate solution, where the sulphide is precipitated as Ag_2S (Sacks, 1995; US Geological Survey, 2015).

To analyse ^{34}S and ^{18}O in sulphate, and provided the water contains only sulphate and no significant sulphide, the water sample can simply be delivered to the laboratory and processed there. Alternatively (after any sulphide has been stripped from the water), the sulphate can be precipitated as barium sulphate from the water sample in the field by addition of excess barium chloride. The sample is usually acidified to a pH of around 4 using hydrochloric acid, prior to barium chloride addition, to prevent the precipitation of barium carbonate (Sacks, 1995). The precipitate can be collected by filtration. If sulphate concentrations are low, anion exchange resins can be utilized to concentrate sulphate prior to precipitation. If the sample is to be stored for long periods, consideration can be given to the addition of biocides to prevent bacterial sulphate reduction.

8.5.8   Dissolved gases

Dissolved gases which are of interest to the hydrogeologist include:

- Carbon dioxide (CO_2), which can affect the aggressivity and corrosivity of the water and its suitability for treatment (McAllan *et al.*, 2009). It can also be toxic if allowed to degas into a confined space.
- Methane (CH_4), which can be explosive.
- Nitrogen (N_2), which is often found in groundwater in concentrations slightly above those representing atmospheric equilibrium ("excess air"). Concentrations of nitrogen and its various oxides can tell us about recharge conditions and temperature (Heaton and Vogel, 1981) and nitrification and denitrification reactions in the subsurface environment (Vogel *et al.*, 1981; Plummer *et al.*, 2004; Jahangir *et al.*, 2010, 2012).

- Oxygen (O_2), which controls the solubility of many elements and has a significant impact on the redox state of the water.
- Several inert gases [e.g., argon (Ar), helium (He)] and their isotopes (e.g., ^3He). Excess Ar concentrations, together with nitrogen, can yield information on recharge temperature (Heaton and Vogel, 1981; Plummer *et al.*, 2004).
- Chlorinated fluorocarbons (CFCs), which are man-made and have been released during recent decades to the atmosphere (and thence to rainfall recharge and groundwater). The concentrations, and relative proportions, of the CFCs can be used to date groundwater (Misstear *et al.*, 2008).
- Sulphur hexafluoride (SF_6), which is used similarly to CFCs.

Furthermore, when concentrations of these dissolved gases exceed atmospheric pressure, they can degas, forming bubbles and changing the redox or pH of the solution. As we saw in Chapter 4, the presence of bubbles can have major implications for clogging of reinjection wells, while sharp changes in pH or Eh can lead to precipitation of unwanted minerals (iron oxyhydroxides, calcite – see Section 9.1).

Some of these gases can be determined directly in groundwater in the field (see Section 8.1.2). Oxygen can be measured by a dissolved oxygen electrode (its solubility is around 10 mg l^{-1} in water at 15 °C). Carbon dioxide (both free and in the form of carbonic acid) can be determined by an acidity titration, assuming that no other sources of acidity are present, using addition of sodium hydroxide to a phenolphthalein endpoint at pH 8.2 (see McAllan *et al.*, 2009). It can also be determined by the difference between total dissolved inorganic carbon (TDIC) and total alkalinity. TDIC can be determined by precipitation of all inorganic carbon as strontium carbonate using alkaline strontium chloride solution.

However, dissolved gases can also be determined by direct sampling and analysis using a gas chromatograph (US Geological Survey, 2014a,b). There are several methods of collecting dissolved gas samples, all of which presuppose a pumping method that avoids degassing, excessive turbulence or admixture of air in the sampled water (low-stress sampling techniques such as bladder pumps may be employed: US Environmental Protection Agency, 2006):

1. A bucket is gently (i.e., from a slowly flowing submerged hose, to avoid mixing with air) filled with the water to be sampled. A glass bottle is then filled with sample water from a submerged hose (0.5 to 2 l min⁻¹) and allowed to overflow. While overflowing, it is placed in the bucket. When all bubbles have been displaced from the sides of the flask, a rubber-type stopper, pre-penetrated by a hypodermic needle is inserted into the bottle neck (the needle allows any trapped air to escape). While still underwater, the needle is withdrawn (Figure 8.19a; slightly varying proce-

dures are recommended for SF$_6$ and CFCs – US Geological Survey, 2014a).

2. Direct collection of water samples using low-stress sampling techniques into glass serum bottles with butyl rubber septa and sealed with an aluminium crimping cap (Jahangir *et al.*, 2010, 2012).

3. Capturing the sample in a glass throughflow 'Fisher vessel'. Groundwater is passed through the vessel to flush out any bubbles and sorbed gases. The taps are then closed, trapping the water sample and connecting it to a pre-evacuated side chamber, thus creating a vapour volume that will gradually be filled by the water's equilibrium dissolved gas content [Hobba *et al.*, 1976; Thorstenson *et al.*, 1979; Plummer *et al.*, 2004; Figure 8.19(b)].

Dissolved gas samples, once collected, should be kept cool and dark and can even be preserved

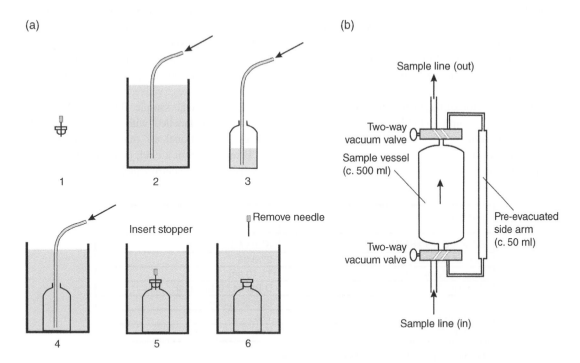

Figure 8.19 *Sampling for dissolved gases (a) by collecting the sample underwater, after a public domain diagram by US Geological Survey (2014a); (b) using a 'Fisher vessel', based on descriptions in Hobba et al. (1976) and Thorstenson et al. (1979)*

under water in a larger container of the sampled water. When analysing the sample, an inert gas is typically injected into the sample flask, displacing water and creating a headspace. The gases in the headspace eventually equilibrate with the water sample and can be extracted directly into a gas chromatograph.

8.6 Packing and labelling samples

Water samples should be labelled as soon as the sample is taken (or even before). The sample label will usually be laboratory-specific, but may contain the following fields (Figure 8.20):

1. Sample number, which is common to all sample flasks taken from the particular well at the particular time.
2. Aliquot qualifier: for example, an aliquot may be labelled *FA* for filtered and acidified, *FU* for filtered and unacidified, *UU* for unfiltered unacidified, and so forth. Additionally, the aliquot may be labelled with the type of analysis for which it is destined (e.g., ICP-MS or HgCVAA – mercury by cold vapour atomic absorption).
3. Name of person who has collected the sample, and identification of the firm or individual who is sending the sample to the laboratory.
4. Date of sampling.
5. Time of sampling.
6. Site name/location.

The sample labels should be water-resistant, and writing should be in a clear waterproof ink. Alternatively, many laboratories nowadays use bar-coded labels that can be electronically read and cross-referenced to a sample list.

Samples will typically be kept cool (4 °C) and dark for storage and transport, and a cool box or portable refrigerator is ideal for this. If ice packs are used to chill a cool box, they should avoid direct contact with the flasks, such that the samples do not freeze. A 'maximum-minimum' thermometer can be included in the cool box to verify that the sample temperature did not rise above a certain level during transit. Samples should ideally be packed upright and secure. They should be adequately padded and protected, especially if glass flasks are involved.

Transport of samples should be arranged prior to commencing the sampling project, and the laboratory should be informed about the arrival date and time of the samples (it is useful to check the opening times of the laboratory). When international transport is involved, any restrictions on exporting and importing samples should be ascertained, together with any restrictions the carrier might apply (especially if contaminated samples, or samples with toxic/flammable preservatives or extractants are involved). Individual sample flasks or shipments may need to be labelled with appropriate hazard warnings. International air transport may involve flasks being stored in a cold, depressurized aircraft hold, which could be detrimental to certain samples.

Kosova Project	Holymoor
SAMPLE NUMBER*	**KS03 *FA***
Date of Sampling*	27/3/05
Time of Sampling	11:10 am
Sampled by	D Banks / B Misstear
Location	Burojë Spring
Field filtered	Yes
Field acidified	Yes

FU means field filtered at 0.45 µm. Unacidified.
UU means unfiltered, unacidified.
FA means field filtered at 0.45 µm. Acidified in field with conc. nitric acid.

Figure 8.20 *A typical label for a groundwater sample flask*

A *request for analysis* and a *sample list* should be packed with each sample shipment and should be faxed or emailed ahead to the laboratory. The sample list should provide details of all samples and sample aliquots included in the shipment. Against each sample in the list, the following information will usually be recorded:

- sample number;
- number of aliquots for each sample and aliquot numbers/filtration/preservation details;
- type of sample (e.g., groundwater, surface water, spring);
- information on sample which may have relevance to selection of analytical method: for example, pH, electrical conductivity, existence of known contamination;
- any hazard warning (presence of contaminant or hazardous additive).

The organization carrying out the sampling should additionally record the following information on their copy of the sample list:

- UTM Grid Reference and location name;
- sample depth;
- name/nature of aquifer horizon;
- well construction/dimension;
- pumping status and rate at time of sampling;
- water level at time of sampling;
- method of sample collection;
- sample appearance and odour;
- field parameters corresponding with sample (pH, EC, temperature, etc.).

The request for analysis should specify:

- the analyses to be carried out on the individual aliquots of sample;
- field filtration, preservation or extraction already carried out on various aliquots;
- any further necessary filtration or acidification routines to be carried out *at the laboratory*.

8.7 Quality control and record keeping

Increasingly rigorous quality assurance requirements are being demanded both for drinking water quality and contamination investigation. Not only is there a need to generate accurate analyses of water quality, it is necessary to document how the samples have been taken and how they have been analysed in line with national and international standards.

Many laboratories will have achieved national or international quality assurance certification. This should cover documented analytical procedures at the laboratory but may also cover the entire chain of operations from sampling to reporting. In such cases, it will be wise to follow the laboratory's procedures for sampling and transport of samples (or even delegate these tasks to laboratory staff) to ensure an unbroken chain of quality assurance. If this is not possible, however, the following steps should be observed:

- Become familiar with laboratory recommendations and national standards, and follow these when sampling.
- Prepare clear method statements and protocols for field determinations and for collection, storage and transport of samples. These may be required (or specified) by regulators but, even if they are not, they will assist the sampler in achieving good practice and consistency between different sample rounds and sample sites.
- Document sampling procedures on pre-prepared forms in the field. For example, when purging a well, flow rates and values of field parameters (temperature, pH, conductivity, redox potential) should be recorded at given times, such that the number of well-volumes pumped and the stabilization of field parameters can be checked (Box 8.6).
- Use a *chain of custody* document when transporting samples to the laboratory, so that it is possible to prove who has been responsible for the package and how it has been transported.

Despite the current trend to quality-assure all aspects of human activity, a water sample sent to two different laboratories, both quality-assured according to relevant national standards, may yield two significantly diverging analytical results

Table 8.9 *Types of quality assurance sample that may be submitted to an analytical laboratory*

Type of quality assurance sample	Description and purpose
Replicate sample	These are simply two identical sets of samples taken from the same point at the same time. If a single sample quantum is split into two replicate aliquots, these can be used to check the reproducibility of analytical results from the laboratory. If, however, the replicate samples are two separate samples taken consecutively from the same source, then they will act as a check of the reproducibility both of the sampling technique and of the laboratory analysis
Blank sample	These are essentially flasks of deionized water submitted anonymously to the laboratory. Note that with the extreme sensitivity of today's ICP-MS and GC-MS techniques (Section 8.8), even deionized water is not 'pure' water, so that traces of analytes may be detected even in blank samples. Blanks can be of several types (see below)
• Field blank	These are samples of deionized water taken out to the well site and subjected to the same packing and transport procedures as the real samples. These confirm that samples have not been contaminated during packing and transport
• Equipment blank	These might consist of deionized water passed through the bailer or filtration apparatus used to obtain real samples. These confirm whether any contamination can be ascribed to the field equipment or filter (Reimann *et al.*, 1999a)
• Acid/reagent blank	These are deionized water aliquots where the relevant acid or other preservation agent has been added in the field. These confirm the purity of the acid used
• Laboratory blank	These are simply aliquots of deionized water submitted to the laboratory to ascertain whether false positives might be generated in the laboratory, either due to poor instrumental calibration or contamination in the analytical environment
Spiked sample	Here, a known concentration of one or more analytes is added either to deionized water or to a real sample. This checks the analytical precision and range of calibration of the laboratory

(Banks, 2004). This may be because of different sample transport and storage conditions, different analytical procedures or simply analytical error. It is therefore advisable to send in a selection of blank, replicate or spiked samples to the laboratory, and to send duplicate flasks of selected samples to a second analytical laboratory. In major sampling programmes it is not uncommon for at least 10% of the total samples taken to represent such quality control samples of the various types listed in Table 8.9.

A further, and often powerful, independent means of laboratory quality control is to check the ion balance error (Section 8.1.3) of the sample. For most uncontaminated potable groundwater samples, the error should not exceed ±5% (Box 8.7).

8.8   Sample chemical analysis

Many methods of water chemical analysis are available in the laboratory, ranging from traditional wet chemical titrations to modern ICP-MS multi-element determinations on small sample volumes. This section will discuss in outline the most common instrumental techniques found in modern automated laboratories. These are not necessarily better than traditional wet chemical techniques, but they are a lot faster.

Inductively coupled plasma (ICP) techniques introduce a small portion of sample into an inert carrier gas (such as argon) which is then heated to generate a plasma. In optical/atomic emission spectrometry (ICP-OES or ICP-AES), the excited

Box 8.7 Examining a groundwater analysis

The first three columns in the following table [Table B8.7(i)] present the result of an inorganic chemical analysis of groundwater from a well in the Iddefjord Granite at Råde, south-east Norway (published by Frengstad, 2002).

The concentrations of major anions and cations can be converted to meq l^{-1} (shown in the fourth column) using the conversion factors listed in Table 8.2. (There is no need to convert the alkalinity, as it is already cited in meq l^{-1}). pH can be approximately converted to an activity of hydrogen ions in meq l^{-1} using the algorithm:

$$\left[H^+\right] = 1000 \times 10^{-pH} \text{ in meq l}^{-1}$$

Table B8.7(i)   *Data from analysis of groundwater from a well in the Iddefjord Granite, Råde, Norway*

Parameter	Unit	Analytical data	meq l^{-1}	%
Field parameters				
pH		8.41	3.9E-06	2.0E-05
Temperature	°C	9.1		
Alkalinity	meq l^{-1}	6.3	6.30	33.0
Laboratory parameters				
Electrical Conductivity (EC)	µS cm^{-1}	820		
Colour[1]	mg l^{-1} Pt	23.5		
Turbidity	FTU	1.50		
Major anions				
Cl$^-$	mg l^{-1}	89.4	2.52	13.2
SO$_4^=$	mg l^{-1}	1.36	0.03	0.1
NO$_3^-$	mg l^{-1}	<0.05	0.00	0.0
F$^-$	mg l^{-1}	5.84	0.31	1.6
Major cations				
Na$^+$	mg l^{-1}	217	9.44	49.4
K$^+$	mg l^{-1}	4.41	0.11	0.6
Ca^{++}	mg l^{-1}	4.37	0.22	1.1
Mg^{++}	mg l^{-1}	2.00	0.16	0.9
Other parameters				
Si	mg l^{-1}	5.91		
Fe	µg l^{-1}	52.4		
Mn	µg l^{-1}	22.0		
Al	µg l^{-1}	4.5		
U	µg l^{-1}	16.3		
Rn	Bq l^{-1}	31900		
Calculated parameters				
Sum cations	meq l^{-1}		9.93	
Sum anions	meq l^{-1}		9.16	
Ion total	meq l^{-1}		19.09	
IBE	%		4.1	

Note: the Formazine Turbidity Unit (FTU) is identical to the Nephelometric Turbidity Unit (NTU).
1 True Colour Unit (TCU) = 1 mg l^{-1} Pt.

The sum of cations ($Na^+ + K^+ + Ca^{++} + Mg^{++}$) is 9.93 meq l^{-1}, while the sum of anions (alkalinity $+ Cl^- + SO_4^= + NO_3^- + F^-$) is 9.16 meq l^{-1}. This allows an ion balance error of 4.1% to be calculated (Section 8.1.3), which is less than 5% and indicates that the analysis is broadly acceptable. The proportions of each ion contributing to the total ionic content of the water can be calculated (right-hand column) and expressed as a pie diagram (Figure B8.7(i)). It will be seen that the water is dominated by sodium and (bi)carbonate alkalinity.

Equation (8.4) can be used to confirm that the sum of cations is approximately 1/100th of the electrical conductivity. The water hardness can be estimated as the sum of $Mg^{++} + Ca^{++} = 0.38$ meq l^{-1} (equivalent to 19.0 mg l^{-1} calcium carbonate).

In comparing the water chemistry to the World Health Organization drinking water guidelines (WHO, 2011; see Appendix 4), it can be seen that fluoride significantly exceeds the guideline of 1.5 mg l^{-1}. Uranium, while relatively high, is less than the provisional guideline of 30 μg l^{-1}. The sample has a relatively high colour: the WHO (2011) suggests that colouration above 15 TCU will be unacceptable to many consumers. The WHO (2011) sets no health-based guideline for turbidity, but states that turbidity should generally be below 5 NTU (4 NTU can be distinguished by the naked eye), although values of <1 NTU are advisable for effective disinfection and <0.2 NTU on average should be achievable in large well-run municipal supplies. Likewise, the World Health Organization (2011) suggests that sodium concentrations in excess of 200 mg l^{-1} may give rise to unacceptable taste. In addition to the World Health Organization guidelines, national drinking water standards may also apply. In this context it should be noted that the radon concentration hugely exceeds the Norwegian guideline value of 500 Bq l^{-1} for domestic water.

We can also assess the water quality for its suitability for other purposes, such as irrigation (see Section 2.7.4). The sodium adsorption ratio (SAR) is calculated from the meq l^{-1} concentrations of sodium, magnesium and calcium using Equation (2.5):

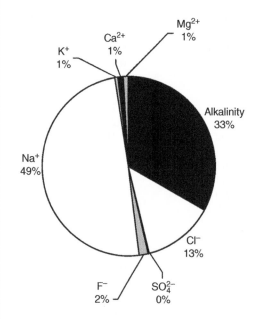

Figure B8.7(i) *Pie diagram illustrating ionic composition of water sample*

$$SAR = \frac{Na}{\sqrt{\dfrac{Ca + Mg}{2}}} = 9.44 / \sqrt{(0.38/2)} = 21.7$$

On a plot of SAR versus EC (Figure 2.27), the water falls in the field of "severe reduction in rate of infiltration", indicating a very high hazard of sodium accumulation in the soil, unless the soil is very well drained or unless calcium additives are employed.

The water contains no detectable nitrate and has an unusually low sulphate concentration. This may be indicative of rather reducing conditions in the aquifer, promoting sulphate reduction by organic matter (and generating alkalinity):

$$SO_4^= + 2CH_2O = 2HCO_3^- + H_2S$$

The high pH, high alkalinity, high Na^+ and low Ca^{++} are characteristic of a mature groundwater that has been resident for a long time in the granite. Hydrolysis of plagioclase feldspar has progressed to such an extent that the water has become saturated with respect to calcite and calcite has been precipitated, removing calcium from solution but allowing sodium to accumulate (Frengstad and Banks, 2000):

$$2NaCaAl_3Si_5O_{16} + 4CO_2 + 7H_2O$$
$$= 2CaCO_3 + 2Na^+ + 2HCO_3^- + 4SiO_2$$
$$+ 3Al_2Si_2O_5(OH)_4$$

plasma generates emission spectra of electromagnetic radiation. The wavelengths of the various lines in the emitted spectrum correspond to a given element, and the intensity of the line is related to the element's concentration. Thus, a large number of elements (typically around 30) can be analysed simultaneously on a single sample to relatively low detection limits (Table 8.7). The technique is also relatively robust. In mass spectrometry (ICP-MS) techniques, the charged particles of the plasma are accelerated in a curve through an electromagnetic field. According to the mass-to-charge ratio of the particles, they will strike a detector in different locations. The method is very sensitive and can analyse a large number of elements (or even individual isotopes of elements) to extremely low detection limits. The instrument can be calibrated in a number of different ways to suit differing element suites. It is, however, subject to interferences between some elements which generate particles of similar mass-to-charge ratios; for example, interferences between chlorine and arsenic, or between silicon and scandium.

Atomic absorption spectroscopy (AAS) techniques employ lamps emitting light of a specific frequency, corresponding to the absorption frequency of various elements. The sample is injected into a high temperature flame. The excitation of electrons in the element under consideration (such as sodium) in the sample absorbs light of the lamp's characteristic wavelength and the strength of the absorption (as measured by a detector) can be related to the concentration of the element. The method can be modified (graphite oven techniques, cold vapour techniques) to suit certain metals/metalloids (Se, As, Cd, Sb, Sn, Pb, Hg) that can be problematic to determine by other techniques.

Ion chromatography (IC) is often used to identify anions in water samples. Here, the water sample is injected into a stream of eluent fluid passing through a column of polymer-based ion exchange resin. The various ionic species are retarded to varying degrees in the resin and they are thereby separated and eluted from the column at different times. The individual peaks of eluted anions are detected, for example, by a conductivity cell. There is potential for interference here, where species have similar retardations within the resin; for example, interferences between organic (humic/fulvic) complexes and fluoride are common.

An *autoanalyser* is any piece of equipment which performs a sequence of analytical steps in a rapid and consistent manner on a large number of samples. It is usually applied, however, to automated colorimetric or spectrophotometric analysis of water samples, by measuring the intensity of colour (absorbance of specific light wavelengths) generated when a reagent is added to the water sample.

Gas chromatography (GC) is commonly used to separate and analyse for organic compounds. Here, the various organic compounds are extracted from the water sample using a suitable organic solvent. The solvent extraction is then injected into an inert gas stream which flows through a chromatographic tube or column. The various organic compounds are retarded to varying degrees in the column and separated, thus emerging from the column at different characteristic times (typically related to the length of their carbon chains or their functional groups). Some form of detector registers each peak of organic compound emerging from the column at its characteristic delay time. One common form of detector is a flame ionization detector (*GC-FID*) which measures changes in electrical conductivity as organic compounds are combusted and ionized in a flame in a mixture of hydrogen and air. Another type of detector is the mass spectrometer (*GC-MS*). Here, each fraction (compound) emerging from the chromatographic column is fragmented into numerous particles of varying charge and mass. They are detected as a mass-to-charge spectrum by the spectrometer. Each organic compound has a characteristic spectrum. Thus, rather than simply measuring the mass of organic compound being eluted (as with *GC-FID*), the *GC-MS* will identify and quantify individual eluted compounds, to a very low detection limit.

Gas chromatography can also be applied to dissolved gases (N_2, Ar, etc.), typically by injecting the headspace gases from a sampling flask (which should be in equilibrium with the water) into a chromatographic column.

Infra-red spectroscopy (IR) is a technique where an organic compound is subject to irradiation by infra-red radiation. This causes individual chemical bonds within the molecule to vibrate with characteristic frequencies and to absorb those frequencies of IR light. Each frequency can be related to a specific type of bond or functional group (such as C-H, C=O bonds). However, in a complex mixture of compounds (a water sample), the *IR* method is often used simply to measure the total number of C-H bonds in solution and this is taken (sometimes misleadingly) as a surrogate measure of total hydrocarbon content of the sample.

8.9 Hydrochemical databases

Modern water analyses can generate huge amounts of information and we need to think very carefully about data archiving and management. ICP-MS and GC-MS analysis can quantify concentrations of many tens of elements and potentially hundreds of organic compounds *on each sample*. Obviously, paper copies of analytical results should be retained for quality assurance purposes, but computer-based databases will be invaluable for:

- rapid data retrieval;
- statistical processing (Banks *et al.*, 1998);
- generation of time-series plots (change of concentration versus time);
- generation of hydrogeochemical maps (Banks *et al.*, 2001; Banks, 2014c).

Spreadsheet-type applications are attractive as each row can correspond to a given sample and each column to a different analytical parameter (Na, K, Ca, Mg, and so forth). Other more specialized geochemical software can be purchased that uses a spreadsheet-type databasing facility, but with additional options for:

- statistical processing by non-parametric techniques;
- preparation of maps;
- calculation of saturation indices and element speciation;
- presentation of specialized graphics (Durov diagrams, Piper diagrams).

When entering data to a database, there is potential for errors in typing large strings of numbers. It can thus be advantageous to obtain digital files of analytical results directly from the laboratory, preferably in spreadsheet format. When designing a database for water quality data, we should pay special attention to:

- *Metadata*. It is seldom sufficient to enter a sample number, a sample location and a number of concentrations of different analytes to a database. It will often be necessary to include other metadata, such as: date and time of sampling, analytical laboratory, grid reference, sample depth, well details, whether sample is filtered/chemically preserved, and analytical method.
- *Units*. Laboratories tend to report different parameter concentrations in different units – for example, mg l^{-1} for major ions and µg l^{-1} for trace

elements – so special care should be taken to enter the correct units in the database.

- *Selection of parameters.* Caution should be exercised against 'mixing' subtly different measurements in the same column or field in a database. For example, field and laboratory measurements of pH probably should not be mixed in a single column, but rather in two separate fields, labelled 'pH_field' and 'pH_lab'. Similarly, there is a need to distinguish between metals analyses carried out on field filtered and unfiltered samples; 'Fe_Filtered' and 'Fe_Unfiltered' would be different parameters (spreadsheet columns). It may also be appropriate to treat separate analytical methods as different parameters. For example, total petroleum hydrocarbons (TPH) may be measured by infra-red spectroscopy (IR) as the total concentration of C-H bonds in solution, or by gas chromatography (GC) as the sum of eluted carbon compounds within a given range of carbon chain lengths. Although the analyses may share the same name, the methods actually determine somewhat differing properties, suggesting they should not be mixed in a database, but rather assigned to separate fields 'TPH_IR' and 'TPH_GC'.

- *Non-detects.* An analytical laboratory should never return a '0' for an analysis of a chemical element, ion or molecule. There will always be a finite limit of detection or quantification and the laboratory should thus return a 'less than detection/quantification limit' value (for example: <0.001 µg l^{-1}). This is valuable information and should neither be discarded nor entered as a '0' in the database. In a computer database, analytical values below detection limit (b.d.l.) should be digitally flagged in some way.

When plotting maps or time series diagrams, or generating statistics from a large chemical dataset, it is necessary to decide how to handle b.d.l. values. Many researchers will choose to temporarily set the result to a value of half the detection limit for the purposes of statistical/graphical presentation (Banks *et al.,* 1998). This practice is fine for datasets from single laboratories with fixed detection limits. When detection limits for a single parameter are variable (due to use of differing laboratories or due to salinity-related detection limits), this practice can quickly cause problems. The use of non-parametric statistical techniques, rather than arithmetic means and standard deviations, will minimize the impact of the low end of the distribution (b.d.l.) on the overall statistical characteristics of a data-set. As a guide for those involved in the management and optimal analysis of a large and complex dataset by non-parametric techniques, the environmental 'Geochemical Atlas of the Kola Peninsula' by Reimann *et al.* (1998) is recommended.

9

Well Monitoring and Maintenance

Once a well has been constructed, tested and put into operation, it is 'out of sight and out of mind', and often receives little attention by way of monitoring or maintenance. However, like any other engineering structure, a water well does deteriorate over the years and it does need periodic maintenance. In order for maintenance to be effective, the causes of the deterioration in well performance must be established through monitoring and diagnosis. It is not only the well that must be considered, but also the entire system for withdrawing groundwater – this includes the aquifer, pumping plant and any water treatment and distribution system (here referred to collectively as the well system). As many as 40% of wells worldwide experience operational difficulties (Howsam *et al.*, 1995). In the Netherlands, it is estimated that about 66% of the well fields suffer from clogging (van Beek, 2001). In developing countries, many well schemes fail owing to breakdown of the pump and inadequate facilities for its repair or replacement, or inability to maintain the necessary power supplies (Foster *et al.*, 2000). There are significant costs associated with well deterioration: in the United States it is estimated that the direct costs (in terms of plant) of well deterioration are

more than one billion dollars annually (Smith and Comeskey, 2010).

In this chapter we will consider the well system from the viewpoint of: the factors that influence performance; the approaches available for monitoring performance; and the measures that can be applied for maintenance or rehabilitation. The chapter ends with a section on decommissioning water wells, since there are many instances where it may not be feasible or desirable to rehabilitate a well, and it is important that the well is decommissioned in a manner that does not cause long-term damage to the aquifer.

The focus of this chapter is on monitoring and maintenance of water supply wells. Some types of wells may be particularly prone to problems of clogging and declining performance. Wells used for aquifer recharge (Sections 4.9 and 4.10) are susceptible to clogging by suspended solids and chemical incrustation resulting from mixing of different water types, whilst wells used for aquifer clean-up operations are often exposed to very aggressive groundwaters. Reinjection wells in ground source heating/cooling schemes can be prone to clogging from chemical precipitates (especially if oxygen comes into contact with

Water Wells and Boreholes, Second Edition. Bruce Misstear, David Banks and Lewis Clark.
© 2017 John Wiley & Sons Ltd. Published 2017 by John Wiley & Sons Ltd.

iron-rich waters), particulates or even bubbles of exsolved or entrained gas. Observation wells may also experience a deterioration in performance since, unlike pumped wells, they are not flushed regularly.

We strongly recommend to all those responsible for well schemes that priority is given to training technical staff in the monitoring of the state of a well system and in the maintenance of all the components of that system. This perhaps applies especially to those countries in the arid or semi-arid areas of the world where there is a shortage of trained staff and where the breakdown of a water supply can be most serious.

9.1 Factors affecting well system performance

The factors affecting well performance can be categorized according to the nature of the processes involved: physical, chemical or microbiological (Table 9.1). Other important factors to consider are the design, construction and operation of the well

system. Although we will discuss these factors separately in turn below, it is important to note that they are all interrelated – clogging of a well screen, for example, may result from a combination of physical, chemical and microbiological processes, exacerbated by poor construction and operation practices.

9.1.1 Physical processes

The main physical processes are clogging and abrasion. Clogging of the aquifer close to the borehole wall can result from drilling fluid invasion during well construction, highlighting the need for proper drilling fluid control to minimize the formation damage (Section 5.2.2) and good well development procedures to repair the damage caused by drilling (Section 5.9). Clogging of this zone may also be due to mobilization of fine particles in the aquifer during well operation. In a study of a well field constructed in an alluvial aquifer in the Netherlands, Timmer *et al.* (2003) found that there was a 50% reduction in porosity close to (less than 50 mm) the borehole wall of the wells investigated. They attributed this to accumulation

Table 9.1 *Factors affecting well system performance*

Factors		Aquifer	Well		Pumping system		Wellhead works and distribution system
			Gravel pack/ borehole wall	Casing/ screen	Pump	Rising main	
Physical	Clogging	√	√	√	√	√	–
	Abrasion	–	–	√	√	√	√
Chemical	Clogging	√	√	√	√	√	√
	Corrosion	–	–	√	√	√	√
Microbial	Clogging	√	√	√	√	√	√
	Corrosion	–	–	√	√	√	√
Structural	Design/ construction	–	√	√	√	√	√
	Materials	–	√	√	√	√	√
Operational	Intermittent pumping	–	√	√	√	–	–
	Over-abstraction	√	–	√	√	–	–

Based on Table 2.1 of Howsam P, Misstear BDR and Jones CR (1995) 'Monitoring, maintenance and rehabilitation of water supply boreholes', CIRIA Report 137, London, by permission of the Construction Industry Research and Information Association; go to: www.ciria.org

of fine particles in the pore throats of the formation, these fines having been loosened from their original positions in the aquifer and moved towards the well due to high groundwater velocities during pumping. The gravel pack may become clogged as a result of mixing of formation particles with the gravel pack material, perhaps due to poor gravel pack design or emplacement procedures (Section 4.4). Migration of aquifer and/or gravel pack material may lead to clogging of the screen slots and infilling of the well. A sand-pumping well may also result in clogging of the pump impellers and accumulation of sediment in the distribution and treatment system.

In addition to clogging, the migration of particles into a well can lead to abrasion of the well, pump and distribution system. The problems are usually greatest in the zones of highest flow velocity – the screen slots and the pump impellers.

9.1.2   Chemical processes

The two main chemical processes affecting well system performance are incrustation and corrosion. They often occur together, and are also typically associated with microbiological activity (see Section 9.1.3).

Incrustation. Incrustation may be of chemical or microbiological (biofouling) origin. In this section, we will deal with chemical incrustation, resulting from the precipitation of mineral deposits. Common encrusting materials include iron oxyhydroxide, calcium carbonate and iron sulphide, but any incrustation is rarely composed of a single mineral. Ferric oxyhydroxide precipitate commonly contains some manganese and usually a fair percentage of carbonates, while the calcium carbonate scale usually contains a proportion of magnesium carbonate. The cause of incrustation is the change in physical and chemical conditions in the groundwater between the body of the aquifer and the well. In pumping wells, incrustation is mainly due to the coming together of different water types at the well, for example deeper reducing groundwater containing dissolved metals

such as iron mixing with oxygenated water (McLaughlan, 2002) or even air itself, if air is entrained in the water column by cascading inflow horizons above the dynamic water level. Another incrustation process involves degassing of carbon dioxide due to temperature and pressure changes at the pumping well. Degassing of carbon dioxide increases pH and disturbs carbonate equilibria in solution and may lead to precipitation of carbonate or hydroxide minerals.

Although chemical precipitates may be found anywhere in the well system, common locations for incrustation are at the top of the well screen (Figure 9.1) – as this is often a mixing zone containing well-oxygenated water and where screen-inflow velocities are high – and in the pump inlet and pump bowls (where degassing of CO_2 can occur).

An indicative phase-stability diagram for the solubility of iron minerals in water with respect to Eh and pH is shown in Figure 9.2. The natural conditions of most groundwaters are close to the boundary between the fields of soluble (Fe^{2+}) and insoluble iron [ferric hydroxide [$Fe(OH)_3$] or oxyhydroxide [$FeO(OH)$]]. Normally, the level of dissolved iron in groundwater is low – less than $1\,mg\,l^{-1}$ total iron – but slight changes in the water chemistry can increase the solubility of iron appreciably. A drop in pH will increase the solubility of both ferrous and, ultimately, ferric (Fe^{3+}) iron. Iron can be mobilized if reducing conditions (low redox potential, or Eh) prevail – for example, if the aquifer or the groundwater contains sufficient organic matter to deoxygenate the water – reductive dissolution of iron minerals in the aquifer may occur, mobilizing iron in ferrous (Fe^{2+}) form:

$$2Fe_2O_3 + 7H^+ + CH_2O = 4Fe^{2+} + 4H_2O + HCO_3^-$$
$$(9.1)$$

where CH_2O represents the organic matter. This type of reaction is seen in an extreme form with water polluted by organic-rich landfill leachates when the water becomes anoxic, has a negative Eh, sometimes a low pH, and can contain very high levels of iron; commonly several tens of $mg\ l^{-1}$. Iron (and some other metals and metalloids) can

Figure 9.1 *Encrusted well screens, Pakistan. The characteristic iron oxide 'tubercles' are visible on the surface of these steel screens. Photo reproduced by permission of Mott MacDonald Ltd*

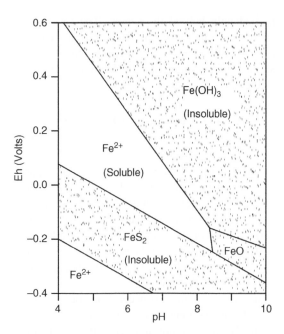

Figure 9.2 *Iron stability fields (for a particular temperature and pressure, dissolved iron, sulphur and carbon dioxide concentration). After Hem (1985)*

also be mobilized by the oxidation of sulphide minerals such as pyrite (FeS_2) – this is particularly characteristic of groundwaters in many mined voids. Finally, iron (and manganese) may also be mobilized by complexation with natural organic acids, such as humic and fulvic acids, possibly derived from peaty soils or bogs.

Groundwater in the unconfined zone of aquifers will normally be oxygenated through solution of atmospheric oxygen during recharge, and will have a positive Eh. If the water's pH is around neutral, the water will lie in the ferric hydroxide stability field, so that the concentration of dissolved iron will be low due to the insolubility of that mineral. The oxygen in the groundwater becomes depleted as the water infiltrates deeper or passes into the confined zone of the aquifer: the Eh falls, and the water gradually moves into the stability field of the soluble ferrous iron. In most confined aquifers, oxygenation is low and there is potential for the solution of iron in the ferrous state.

Wells represent localized points of access for oxygen into confined aquifers, and this local oxygenation, possibly combined with an increase in pH through the exsolution of CO_2, can move the groundwater quality back into the stability field of ferric hydroxide. This change of stability, if there is sufficient iron dissolved in the water, can lead to the precipitation of ferric oxyhydroxide where the oxygen meets native groundwater, that is, in the screen and in the gravel pack and formation immediately behind the screen. The reaction is shown in Equation (9.2) and is autocatalytic, in that ferrous iron oxidation takes place on the surfaces of the solid ferric oxyhydroxide phase as it forms.

$$4Fe^{2+} + O_2 + 6H_2O = 4FeO(OH)\downarrow + 8H^+$$
$$(9.2)$$

This means that the rate of incrustation in parts of the well system such as the gravel pack may increase as the ferric iron accumulates, until such time as the pore spaces become clogged and the surface area for the reaction is reduced (Houben, 2004). As we will see in the next section, there are also bacteria which can catalyse the precipitation of ferric oxyhydroxide.

Incrustation can also occur from precipitation of calcium carbonate minerals, although this is probably less common than clogging due to iron deposits. Groundwater has sufficient residence time in most carbonate aquifers, carbonate-cemented consolidated aquifers or even granular aquifers with a modest carbonate content, to become saturated with respect to calcite, through solution of calcium (and magnesium) carbonates and silicate minerals by weak carbonic acid. The carbonic acid is derived from carbon dioxide dissolved during the passage of recharge water through the soil zone, where a good deal of microbial respiration occurs. In groundwater, the carbonate system is the subject of a number of hydrochemical equilibria:

$$CO_2 + H_2O \leftrightarrow H_2CO_3 \leftrightarrow H^+ + HCO_3^-$$
$$\leftrightarrow 2H^+ + CO_3^{2-}$$
$$(9.3)$$

If carbon dioxide degasses from groundwater, as it may do in a pumped well (due to sudden pressure changes) or on exposure to the atmosphere, there will be a tendency towards an increase in pH. This pH increase can trigger the precipitation of calcium carbonate (whose solubility is strongly pH-dependent). In rather simplified form, we can consider the degassing of CO_2 forcing the following reaction to the right.

$$Ca^{2+} + 2HCO_3^- \leftrightarrow CO_2\uparrow + CaCO_3\downarrow + H_2O$$
$$(9.4)$$

The incrustation and cementation by both iron and carbonate minerals may entrap and incorporate fine materials moving out of the aquifer under the influence of pumping. This can make a bad situation worse.

The net result of incrustation, whether purely chemical or involving biofouling, is usually to impede groundwater flow in the aquifer in the vicinity of the well, in the gravel pack or across the well screen. This will increase the losses in head at the well (the well losses) and decrease the well's efficiency (Section 4.5). Furthermore, it may increase the head difference across the well screen, placing greater stress on the structure of the well screen. The clogging of well screen slots will also increase entrance velocities in those slots that still remain open, increasing the potential for sand pumping and abrasion. Furthermore, the incrustation layer can itself harbour anaerobic microenvironments beneath its surface, which favour the corrosion of metals: in other words, incrustation and corrosion can go hand-in-hand.

Corrosion. A water well can also deteriorate through the phenomenon of corrosion. Although corrosion might appear to be the opposite of incrustation, it can be intimately related. Damage to the well structure can be from solution of the metal casing and screen, or from incrustation by the by-products of corrosion.

The whole problem of chemical corrosion in groundwater is complex but, by and large, the main factors affecting corrosion are the nature of the metal corroded, the formation of corrosive microenvironments on the metal surface (or beneath biofilms or incrusting layers), and the physical and chemical condition of the water. Also, microbiological activity is increasingly recognized as having an important role in corrosion. Although corrosion is normally associated with metals, plastic casings and screens can 'corrode' through contact with certain organic chemicals (Section 4.1.4).

The simplest and most familiar form of corrosion is that of iron or steel exposed to oxygen and water, which forms a mixture of iron oxyhydroxides and oxides that we refer to as 'rust':

$$8Fe^0 + 6O_2 + 12H_2O \rightarrow 8Fe(OH)_3$$
$$= 8FeO(OH) + 8H_2O$$
$$= 4Fe_2O_3 + 12H_2O \qquad (9.5)$$

The fundamental corrosion reaction (for iron), however, is simply the oxidative dissolution of iron metal:

$$Fe^0 \leftrightarrow Fe^{2+} + 2e^- \qquad (9.6)$$

This is a redox reaction and is thus of an electrochemical nature, whereby anodic and cathodic areas are set up on the metal surface. Iron is oxidized to yield positive ions in solution at the metal/water interface. If the metal is steel and if dissolved oxygen is present (and the pH and Eh conditions are favourable), these ions will often reprecipitate as ferric oxide or oxyhydroxide [Equation (9.5), Figure 9.3]. Note that, in Figure 9.3, the cathodic reaction is, strictly speaking, the reduction of hydrogen ions to hydrogen gas. However, any such hydrogen gas will react quickly with dissolved oxygen present to produce water.

The formation of electrolytic cells is encouraged by electrochemical inhomogeneity. This may be due to:

- defects in the surface of a single metal;
- microenvironments on the metal surface; for example, reducing microenvironments *behind* a layer of incrustation or biofilm;
- the presence of two different metals in contact.

Important sites for chemical corrosion in a water well are at physical imperfections on pipes, for

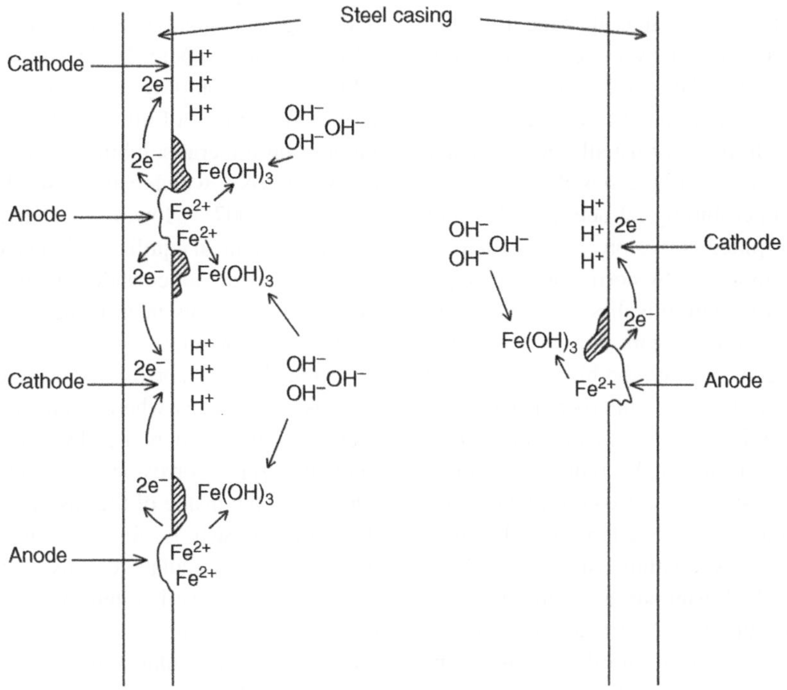

Figure 9.3 *Simplified diagram illustrating electrochemical corrosion and precipitation of iron (as ferric hydroxide)*

Table 9.2 *Electromotive series of metals used in well construction*

Active (anodic)	Magnesium
	Zinc
	Mild steel
	Brass
Noble (cathodic)	Stainless steel

example, the rough edges of screen slots or the screw threads of collars. Corrosion may be strongest where two different metals that are widely separated in the electromotive series (Table 9.2) are in contact with one another, resulting in a galvanic cell system. However, Smith and Comeskey (2010) note that only a small proportion of corrosion in drinking water wells is attributable to galvanic cell systems.

Two of the main factors in water chemistry that affect corrosion are the pH and Eh (Figure 9.2). The water in contact with steel has to be in the stability field of soluble Fe^{2+}, rather than the insoluble $Fe(OH)_3$, for corrosion to be very active. However, even where this is not the case for the bulk of the groundwater, microenvironments with the appropriate redox conditions will usually occur sooner or later on the steel surface, possibly aided by incrustation or microbiological activity. This will typically lead to pitted corrosion. Thus, in an aerated well, iron dissolved by corrosion is precipitated very close to where it is dissolved [Equation (9.2)]. In practice, this appears as a blister of rust on the pipe surface. Beneath the blister, protected from the aerated well water, a microenvironment is set up where conditions may be extremely reducing and corrosion enhanced; this reducing environment may be a site for corrosion induced by sulphate reducing bacteria (Section 9.1.3). Because the dissolved iron is continuously removed by precipitation on the blister surface, corrosion also is continuous (Figure 9.3). A similar situation may develop behind a layer of biofilm incrustation, where sulphate reducing microbes can thrive.

Other physicochemical factors that can strongly influence corrosion potential include: salinity,

temperature, dissolved gas content and the presence of organic acids (see Section 9.2.4 and Box 9.2).

There are a number of indices which are used to predict incrustation and corrosion potential. Three of the most common are the Langelier Saturation Index (LSI: Langelier, 1936), the Ryznar Stability Index (RSI: Ryznar, 1944) and the Larson-Skold Corrosion Index (LSCI: Larson and Skold, 1958). The LSI and RSI are based on the calcite saturation index and are calculated from pH, calcium and alkalinity measurements, with adjustments for temperature and salinity; they can be used to predict the likelihood of a calcium carbonate scale forming on the well components, which can inhibit corrosion. However, these two indices are based largely on the carbonate water chemistry and pH, and do not take account of other chemical parameters that may influence corrosion. The LSCI takes account of the water's salinity and is essentially the ratio of (chloride plus sulphate) ions to alkalinity (all as meq l^{-1}). For more nuanced assessments, a hydrogeochemical model such as PHREEQC (Parkhurst and Appelo, 1999; Appelo and Postma, 2005) can be employed to calculate saturation indices for a range of mineral phases, and also the impacts on mineral solubility of mixing of different water types, which can be useful (Houben and Treskatis, 2007). However, predictive methods based on chemical equilibria alone have the limitation that they do not consider the important role of microbial processes in influencing corrosion and incrustation potential.

As noted above, corrosion can be severe in metal casings and screens where contrasting metals are present, or where there are physical imperfections in the materials. Corrosion can affect the outside in addition to the inside of the casing, notably where the borehole seal is absent or incomplete, and water is able to circulate. Corrosion can also affect other parts of the well system. The 'splash zone' of the casing between the rest and pumping water levels may be particularly susceptible to corrosion, since this surface is regularly wetted with oxygenated water. The pump and rising main can also be severely impacted (Figure 9.4).

Figure 9.4 *Corroded pump rising main removed from a well in an alluvial aquifer, Oman. The rising main exhibits both pitted corrosion and iron oxide incrustation. Photo by Bruce Misstear*

Although stainless steel well components are often perceived as corrosion resistant, this is not strictly true for all varieties of stainless steel in all conditions. Stainless steel is a steel alloy with a chromium content typically between 16 and 20% (Sterrett, 2007). Its passivity relies on the formation of an inert, impermeable chromium oxide film on contact with oxygen, which protects the steel from attack. In highly reducing anaerobic conditions, this chromium (III) oxide (Cr_2O_3) film may not be able to form, leaving the steel vulnerable to corrosion. Corrosion may be enhanced in chloride- and sulphide-rich environments or in very acidic environments. It is important to remember that there is a range of grades of stainless steel, some of which are more suited to certain corrosive environments than others. For example, higher grades (Type 316 and better) of stainless steel are preferred in anaerobic environments typical of groundwater from deep aquifers or those with a high organic carbon content.

The main dangers that can result from corrosion are:

- Corrosion can perforate casing or cut through the casing joints, so weakening the structure and allowing potentially polluted water into the well. In severe cases, the structure may be so compromised that the well collapses.
- The screen can corrode sufficiently to widen the slots and allow the gravel pack or formation to

pass into the well, causing damage to the pumping plant.
- The products of corrosion, the precipitated ferric and other hydroxides, can clog both the screen and the formation in the same way as incrustation.
- Corrosion of the pump components may lead to pump failure. The joints on the rising main may be sufficiently weakened to allow the pump to fall down the well.
- Corrosion of the rising main can lead to perforation (Figure 9.4) and loss of pumped water back to the well. This results in apparent loss in pump capacity.

9.1.3 Microbiological processes

Microbes are involved in many of the processes that lead to clogging and corrosion of well systems. Indeed, Smith and Comesky (2010) suggest that "the number one contributor to reduced well performance in most regions across the globe is biofouling", whilst the American Society of Civil Engineers (2014) report that 80% of blockage in wells is due, at least in part, to biological growth. In production wells, biofouling can lead to clogging of the aquifer and gravel pack and hence to a decline in well yield. In observation boreholes, the presence of biomass may lead to adsorption of some of the more reactive chemical constituents

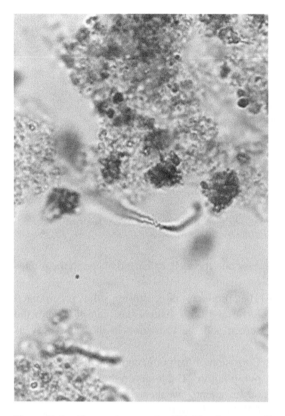

Figure 9.5 *Photomicrograph of biofilm from a well in Indiana, USA. Reproduced by permission of Stuart Smith, Smith-Comeskey Ground Water Science LLC*

from the groundwater and therefore adversely affect the results of a groundwater monitoring programme.

Although microbes may be introduced to the groundwater during drilling, there are many species of bacteria and other microbes that reside in aquifers. Some of the most common problems in well systems result from *iron-related biofouling* (IRB), where the deposition of iron and other metals is mediated by bacteria such as *Gallionella*, *Thiobacillus*, *Clonothrix* and *Leptothrix*. Many of the bacteria species are stalked or filamentous, and occur in colonies in which the bacteria are coated in slimy sheaths (biofilms; Figure 9.5). There is also a large number of non-filamentous rod-shaped precipitators

of iron, including a wide array of *Pseudomonas* strains. Some IRB bacteria may actually derive energy from oxidation reactions such as ferrous to ferric iron, whereas with some other species the polysaccharide slime merely provides a favourable locus for oxyhydroxide precipitation. In any case, the solid ferric hydroxide resulting from the oxidation of soluble ferrous iron is precipitated as flocs or granules in the biofilms. Similar reactions occur for manganese and other metals. The biofilm is thus a mixture of tangled bacterial filaments, polysaccharide slime, gelatinous metal hydroxides and particulate matter caught up in the whole. Where the biofilm accumulates sufficiently to cause problems of incrustation or corrosion in well systems, the process is usually known as biofouling.

The rate of biofilm development depends on (McLaughlan, 2002):

1. Bacterial activity, including nutrient availability and the production of the biofilm coating material known as extracellular polymers.
2. The availability of particles, which is governed by the nature of the formation, the volume of flow past the surface and the precipitation mechanisms (oxidation, degassing of CO_2).
3. Biofilm shear forces relating to flow rate and turbulence.

The early precipitate of iron hydroxide, both chemically and biologically precipitated, is soft and has an almost gelatinous consistency when wet, but is soft and powdery when dry. In biologically precipitated iron, carbon may be a significant percentage of the total incrustation. These deposits age with time, dehydrate, recrystallise and turn into a reddish brown, hard mixture of ferric oxyhydroxides and oxides. This aging process involves a decrease in both the reactivity and surface area of the iron deposit, which in turn means that older deposits are more difficult to remove by chemical treatment (Houben, 2003a,b).

Some biofouling reactions may result in the accumulation of biomass only, without metal precipitates. For example, biomass may be produced by methanotrophic bacteria from the mixing of methane-containing groundwater with oxygen-containing groundwater:

$$CH_4 + O_2 = CH_2O + H_2O \qquad (9.7)$$

where CH_2O represents the biomass.

Corrosion in anaerobic conditions may be accelerated through the actions of sulphate-reducing bacteria (SRB; Enning and Garrelfs, 2014). These bacteria are found in strongly reducing environments in the subsurface, often characterized by the absence of oxygen, the presence of organic matter and a supply of sulphate ions. SRB use the reduction of sulphate ions for their energy requirements (sulphate, rather than oxygen, is the electron receptor in their respiration), and the resulting hydrogen sulphide can enhance corrosion of iron and steel:

$$SO_4^{2-} + 2CH_2O \rightarrow H_2S + 2HCO_3^- \quad (9.8)$$

Anaerobic conditions favouring SRB-induced corrosion may also be found in an aerobic well within the microenvironment of a biofilm, as this can contain both anaerobic and aerobic layers. Thus, biologically-mediated corrosion may occur in a well where the groundwater chemistry indicates a low potential for corrosion.

For further information on biofilms and biofouling the reader is referred to McLaughlan (2002), Cullimore (2008) and Smith and Comeskey (2010).

9.1.4 Well design and construction

The design and construction of water wells has been covered in Chapters 3, 4 and 5, and the need to follow good practice in new wells should be clear so as to avoid, or at least minimize, problems with their future performance and reliability. With existing wells that have not been designed or constructed properly, problems commonly arise from:

i. Poor gravel pack design and/or emplacement leading to sand pumping, and clogging or abrasion of the screen, pump and headworks.
ii. Use of inappropriate materials in well design – many old wells were constructed with mild steel casings and screens that can deteriorate rapidly in corrosive groundwaters, potentially

resulting in entry of poor quality water to the well, or even well failure.
iii. Long screen sections may encourage mixing of different water types, leading to clogging. Whilst it may not be possible to avoid long screens in some production wells, observation boreholes are best completed with short lengths of screen (Section 3.5).
iv. Bad drilling fluid control during drilling, resulting in clogging of the aquifer.
v. Insufficient well development to repair the drilling damage.
vi. Poor installation of the casing string, with individual pipes not connected together correctly (cross-threading of joints), leading to enhanced corrosion and pitting of the pipe joints, sand pumping, entry of poor quality water, and possibly structural failure of the well.
vii. Poor grouting of the upper casing string and poorly-designed wellhead, allowing the ingress of polluted water into the well from the ground surface.
viii. Inappropriate choice of pump and rising main materials for use in aggressive groundwater.
ix. Incorrect setting of pump; for example, a pump set in the well screen will increase the risk of fouling of both the screen and the pump.

An understanding of the limitations of the original design and construction can help in diagnosing the causes of the performance problems in a well system.

9.1.5 Well system operation

It is generally considered that, for meeting a given demand, wells are best operated smoothly at relatively constant discharge rates, rather than pumped intermittently at higher discharges. However, this may not always be practicable, notably in the following cases:

• wells fitted with hand pumps;
• wells with motorized pumps where power/operator constraints limit the hours of pumping;
• small-scale automated wells which automatically cut in or cut out in response to pressure in a pressure tank or to water level in a reservoir.

Drawbacks associated with pumping a well intermittently include:

1. It creates an opportunity for particles in the aquifer or gravel pack to mobilize each time the pump starts up, as frequent stopping and starting of the pump leads to surging of the well.
2. The higher flow velocity associated with intermittent pumping may lead to greater particle migration than by pumping at a lower discharge for a longer period.
3. Intermittent pumping gives a higher drawdown than continuous pumping for the same daily discharge, and thus creates a longer 'splash zone' on the casing, where corrosion may be enhanced (Section 9.1.2).
4. Higher pumping rates increase the chance that water levels will fall close to the top of the screen, increasing the supply of oxygen and hence the risk of clogging from biofouling or chemical incrustation.
5. It requires a larger capacity pump and potentially a larger well than would be needed for a continuously (or nearly continuously) operating well, and hence involves greater capital costs.
6. The operating costs of pumping a well intermittently will also tend to be greater, since the well loss component of drawdown increases in relation to the square of the discharge rate (Section 4.5).

The well performance will also be influenced by any changes in the regional groundwater levels. Over-pumping and falling groundwater levels in an unconfined aquifer may result in a significant decrease in aquifer transmissivity and hence in well yield. Lower pumping water levels may also lead to enhanced corrosion and biofouling [in a similar manner to (3) and (4) above], as well as to increased pumping costs.

9.2   Monitoring well system performance

In order to be able to assess any decline in performance of a well system, and identify the causes behind this decline, the performance and condition of the well system must be monitored, together with the processes that influence the performance. Furthermore, the location and cause of any problem must be correctly diagnosed. If a well is underperforming, we need to know if this is related to:

1. *Aquifer-related factors*. Are regional groundwater levels (and thus, possibly, transmissivity) declining? If so, is this in response to temporarily low recharge or over-abstraction?
2. *Well-related factors*. Has the performance of the well declined in terms of specific capacity? Is this related to clogging or incrustation? If so, is this located in the aquifer near the well, the gravel pack or the well screen?
3. *Pump-related factors*. Is the pump performing efficiently? Has the rising main become incrusted, corroded, or perforated? Are the pump impellers worn? Is the flowmeter functioning correctly?

The main parameters that should be monitored to help diagnose the existence, location and cause of a potential problem are summarized in Table 9.3. As a very minimum, we would recommend that *every* abstraction well should be equipped with the following features, to allow the three most important parameters to be monitored:

1. *Water level*: access at the wellhead should be provided for manual water level dipping, preferably via an access tube, and for the pressure transducer and data logger systems that are desirable for continuous water level monitoring (see Section 7.2.4).
2. *Discharge*: some form of calibrated meter is required. This may be a combination of a real-time flowmeter, an integrator meter or a device for recording pump operating hours and power consumption.
3. *Water quality*: a sampling tap or line at the well head is required, suitable for water chemistry and bacteriological sampling and analysis.

The frequency of monitoring will depend to some extent on the use of the well and the monitoring facilities/capabilities available locally. Where automated systems for continuous monitoring of

Table 9.3 Well system monitoring

	Aquifer	Well	Pumping system	Wellhead works
Performance	Net abstraction/ Recharge	Discharge rate	Discharge rate	Flow meter and instrument accuracy
	Regional water level	Pumping water level	Discharge head	
	River base flows	Rest water level	Energy consumption	
	Regional water quality	Water quality		
		Specific capacity		
Condition	Net abstraction/ Recharge	Appearance	Pump appearance	Appearance
	Regional water level	Hydraulic efficiency	Noise and vibration	Leakage
	River base flows		Rising main appearance	
	Regional water quality		Earthing	
Process				
Physical	Formation grain size	Gravel pack grain size	Sand content	
	Flow rate/velocity	Gravel pack level		
		Flow rate/velocity		
Chemical	Water chemistry	Water chemistry	Water chemistry	Water chemistry
	Geochemistry	Materials	Materials	Materials
Microbial	Recharge water quality	Microbial activity	Microbial activity	Microbial activity
	Nutrient status	Nutrient status	Nutrient status	Nutrient status
		Flow rate/velocity	Flow rate/velocity	Flow rate/velocity
		Oxygenation	Oxygenation	Oxygenation
		Materials	Materials	Materials
Structural/ mechanical		Depth of infill or collapse	Failure	Failure
Operational	Aquifer status	Operating hours	Operating hours	

Based on Table 2.2 of Howsam P, Misstear BDR and Jones CR (1995) 'Monitoring, maintenance and rehabilitation of water supply boreholes', CIRIA Report 137, London, by permission of the Construction Industry Research and Information Association; go to: www.ciria.org

water level and discharge rate are not available, manual measurements should be taken at least weekly, and more frequently if possible. For preventative maintenance, water quality should be monitored at least monthly during the initial period of well operation after commissioning, and then the frequency can be reduced to quarterly if conditions do not appear to be changing after the first year (Smith and Comeskey, 2010).

Other important aids in monitoring and diagnosis include direct observation of the condition of the well and pumping plant, downhole CCTV and geophysical logging surveys, regular well pumping tests, pump efficiency measurements and the availability of an observation borehole at some distance from the well.

9.2.1 Monitoring well performance

The hydraulic performance of a well can be assessed by a step drawdown test. The test can be used to establish the components of well

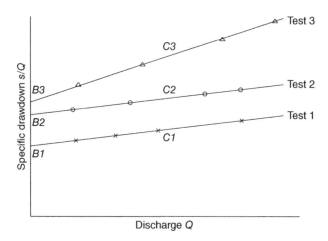

Figure 9.6   *Changes in well performance determined from step drawdown tests (see text for explanation)*

drawdown at different pumping rates attributable to aquifer loss and well loss (Section 4.5 and 7.4.3):

$$s_w = BQ + CQ^2 \qquad (4.2)$$

where s_w is the drawdown in the pumping well, Q the discharge rate, and B and C are the coefficients of linear and non-linear head loss, respectively. In practice, the term B is often dominated by aquifer loss and C by well loss. Well efficiency is sometimes determined from the proportion of total drawdown attributable to aquifer loss:

$$Efficiency = \frac{BQ}{BQ + CQ^2} \times 100\% \quad (7.10; 9.9)$$

Step tests can be carried out at regular intervals during the lifetime of the well, say annually, to determine if there has been a change in well performance. Figure 9.6 shows schematically a situation where Test 2 indicates an increase in coefficient B compared to the original Test 1. This may be ascribable to a decline in aquifer performance, perhaps as a result of a reduction in transmissivity owing to a fall in regional water levels. Test 3, on the other hand, also shows an increase in coefficient C compared to the original test, suggesting that a reduction in well condition has also occurred. However, we must take care not to rely

overmuch on step test data to determine the proportions of aquifer and well loss i.e., in determining the efficiency of a well. As explained in Section 4.5, the aquifer loss term BQ may include a component of laminar well loss, whilst the well loss term CQ^2 may include some turbulence effects in the aquifer. So, for example, the increase in coefficient B between Test 1 and Test 2 in Figure 9.6 might be due to clogging of the aquifer rather than a decline in regional water levels.

Some workers (including Helweg *et al.* 1983) recommend determining changes in well efficiency by monitoring the specific capacity of the well, and comparing these data with a set of normalized specific capacity results derived from the original step test. Changes in well efficiency can be assessed from:

$$Efficiency = \frac{SC_C}{SC_O} \times 100\% \qquad (9.10)$$

where SC_C is the current specific capacity of the well and SC_O the original specific capacity determined from the step drawdown test. Figure 9.7 shows a curve of specific capacity versus discharge rate for a set of original step test data. Two subsequent specific capacity measurements indicate a decline in hydraulic performance of the well. The change in efficiency for the most recent

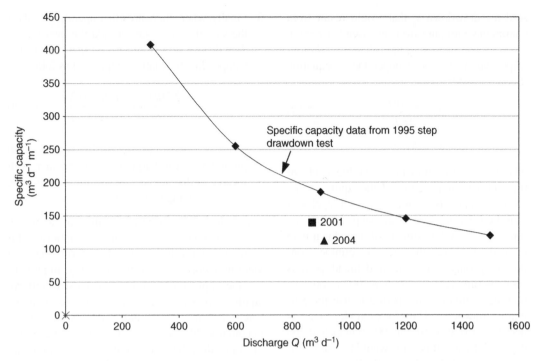

Figure 9.7 *Example of specific capacity versus discharge assessment*

measurement (2004) is determined from $SC_e/SC_o = 110/185\,\text{m}^3\,\text{d}^{-1}\,\text{m}^{-1}$ (at a discharge rate of approximately $900\,\text{m}^3\,\text{d}^{-1}$), representing a decline of around 41%. Note that the meaning of well efficiency in Equation (9.10) is not the same as that in the earlier example in Equation (9.9): in the earlier example, efficiency is being measured against an ideal well, whereas for the specific capacity change it is being measured against the original condition of the real well.

For comparisons to be realistic, the specific capacity measurements during well operation have to be performed under similar conditions to the original step test. This means that rest water levels should be approximately equivalent, the pumping rates should be similar and that specific capacity should be determined after the same period of pumping, since the BQ component of drawdown (the aquifer loss) in Equation (4.2) is time-dependent. Thus, for example, if the original step drawdown test comprised steps of 100 minutes, then the future specific capacity measurements should be taken

100 minutes after pump start-up, following a suitable period of shutdown that allowed water level recovery to static, or near static, conditions. In a well field with several production wells, the specific capacity tests should be made with the same wells pumping or resting as in the original test, so that well interference effects are roughly equivalent. It is also important to ensure that the all the discharge measurements are accurate.

The data from the original pumping tests can be used to predict the 'long-term' drawdown of the well for a range of pumping rates, against which operational pumping water level data can be compared (Misstear and Beeson, 2000). The approach involves three steps:

1. Determine the short-term drawdown for a range of discharge rates from the step test results; for example, the 100-minute drawdown if the test comprised 100-minute discharge steps. This component of drawdown includes the well loss, which is not time-dependent.

2. Add the drawdown s_a that would result from continuous pumping over an extended period. This can be done using the following relationships based on the Cooper-Jacob equation (Section 7.4.5):

$$\Delta s = \frac{2.30Q}{4\pi T} \qquad (7.27)$$

$$s_a = \Delta s \times t_n \qquad (9.11)$$

where Δs is the drawdown per log cycle of time, Q the pumping rate, T the aquifer transmissivity and t_n is the number of log cycles of time between the end of the step test and the time for which the estimate of drawdown is to be made. For wells that are to be pumped continuously only during high demand periods such the summer months, then 100 or 200 days may be a suitable time period to use for estimating drawdown. Thus, for example, the number of log cycles between 100 minutes and 200 days is 3.46, and this value would be used for t_n in Equation (9.11). For wells that are to be operated continuously throughout a year, then a longer time period may be appropriate when making predictions.

3. Finally, where there are other operating wells at the site, it is necessary to add the interference drawdowns, which can be estimated from the Cooper-Jacob equation, expressed as follows:

$$s = \frac{2.30Q}{4\pi T} \log_{10}\left(\frac{2.25Tt}{r^2 S}\right) \qquad (7.25)$$

where s is the interference drawdown due to pumping at discharge Q from another well located at distance r; t is the time of pumping and S the storage coefficient.

It is important to be aware that this approach makes a number of simplifying assumptions and if these are not fulfilled, the long-term predicted yield may not be achievable in reality. In particular, the approach assumes an extensive uniform aquifer whereas in reality the aquifer may be heterogeneous, with boundary effects limiting the potential yield of the well (Box 9.1). An example of some drawdown calculations for a pumping well is included in Table 9.4, and the resultant discharge-drawdown predictions are illustrated in Figure 9.8. This is for a well site where there are two production wells.

Box 9.1   Assessing the reliable yield of a water well

The simple method described in Section 9.2.1 for predicting the long-term yield of a well presupposes an extensive, thick, homogeneous aquifer. Essentially, it relies on the extrapolation of short-term pumping test results to a much longer pumping period under uniform conditions. It does not take into account, for example, any reduction in specific capacity that may result from the water level cone of depression spreading out during extended pumping to encounter a low permeability hydraulic barrier (Section 7.6.1). Nor does it recognize that long-term pumping in certain situations may lead to water quality problems such as saline water upconing in a coastal aquifer. In addition to possible limitations on well

yield that may result from the nature of the aquifer, there are also the constraints imposed by the engineering set-up and operation of the well system – the pumping plant, distribution system, and so on.

Approaches for estimating the reliable yields of production wells have been described in Beeson *et al.* (1997) and Misstear and Beeson (2000). These approaches allow the calculation of both the *potential yield* (the well yield achievable subject only to aquifer constraints) and the *deployable output* (the yield achievable also taking into account constraints such as the pump capacity, the well abstraction permit, etc.). In both cases it is necessary to define a deepest advisable pumping water level (DAPWL) for

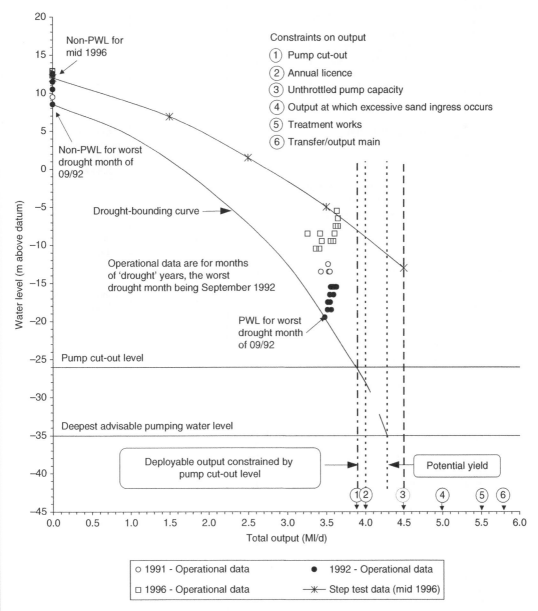

Figure B9.1(i) *Relationship between well water level and total output, showing the constraints on output. PWL = pumping water level. Based on Figure 1 in Misstear and Beeson (2000) 'Using operational data to estimate the reliable yields of water wells', Hydrogeology Journal 8:177–187. Reproduced by kind permission of Springer Science+Business Media*

the well, beyond which undesirable effects may occur. In a fractured consolidated aquifer, the DAPWL may correspond to a particular inflow zone, below which a rapid increase in drawdown is observed (see Figure 7.20). In an unconfined sand and gravel aquifer, the DAPWL may be defined at some distance above the top of the screen (which may be set against the lower third

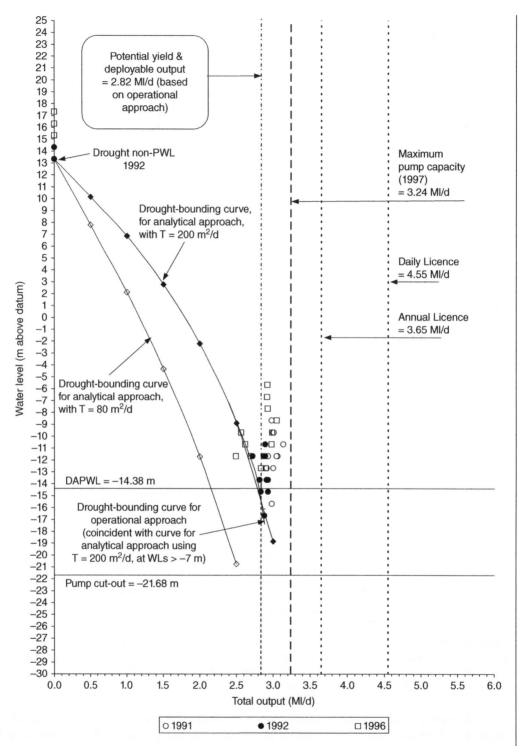

Figure B9.1(ii) *Water level output-graph for a single well. Based on Figure 2 in Misstear and Beeson (2000) 'Using operational data to estimate the reliable yields of water wells', Hydrogeology Journal 8:177–187. Reproduced by kind permission of Springer Science+Business Media*

of the aquifer (Section 3.1.4), or alternatively, could relate to the pumping water level corresponding to a discharge rate above which sand pumping is known to be a problem.

The methods used for estimating potential yield and deployable output were originally developed for aquifers in the UK, especially the heterogeneous fractured Chalk aquifer which is the main source of groundwater supplies in England. Where there is significant aquifer heterogeneity, as in the Chalk, the preferred approach for estimating the potential yield and deployable output is to rely on operational data rather than the more idealized analytical approach described in Section 9.2.1. The operational approach uses the pumping water levels and discharge rates measured during periods of extended drought; a drought-bounding curve is fitted through the maximum pumping and non-pumping water levels to indicate the performance of the well during the worst drought for which operational data are available. In the example in Figure B9.1(i), the potential yield is given by the intersection of the drought-bounding curve with the DAPWL, at around $4,300 \, m^3 \, d^{-1}$. The deployable output is defined by the intersection of the drought-bounding curve with the first of the operational constraints on well yield, in this case the pump cut-out level, giving a deployable output of $3,900 \, m^3 \, d^{-1}$.

A second example, also for a single well source, is included in Figure B9.1(ii). Here, drought-bounding curves are drawn from analytical results as well as from operational data. The analytical approach gives two curves based on two different values of aquifer transmissivity derived from different pumping tests. The higher value of transmissivity ($200 \, m^2 \, d^{-1}$) produces a drought-bounding curve almost identical to that defined by the operational data. The potential yield and deployable output are the same in this example ($2,800 \, m^3 \, d^{-1}$), since the potential yield is below the limits that would be imposed by any of the operational factors such as the maximum pump capacity or the abstraction licence quantities.

The methodology described briefly here can also be applied to intermittently-pumped wells and to well fields with a large number of pumping wells (Misstear and Beeson, 2000). Whatever the application, it is important to remember that well system performance will change over time and must be monitored. If continued monitoring of discharge rate and operating pumping water levels indicate a reduction in the hydraulic performance in the future, additional data collection (Section 9.2) will be required for diagnosing the cause of this decline. Depending on the outcome of the monitoring and diagnosis, work-over operations may be required on the well or pumping plant to restore the system performance, or the reliable yield estimates for the well may need to be reappraised if the problem is outside of the well operator's control, for example, a fall in regional water levels.

In wells that are pumped continuously, or nearly continuously, it may not be possible to shut down the pump for long enough to allow water levels to recover sufficiently to permit a proper estimate of specific capacity once the pump is switched back on, that is, drawdown can only be established if both the static and the pumping water levels can be measured. In such a situation it is even less likely that the well can be taken out of supply for a sufficiently long period to do a controlled step test, to compare performance with the original. In these circumstances much can be learnt from regular monitoring of pumping water levels and discharge rates, and comparing these with the predicted performance from the original tests. Alternatively, the change in yield for a given change in drawdown can, with some caution, be used as a surrogate for specific capacity. The operational data can also be used to establish the reliable yield of the well, as explained in Box 9.1.

Table 9.4   Example of long-term drawdown calculations

Discharge (m³ d⁻¹)	100-min drawdown[a] (m)	200-day drawdown[b] (m)	500-day drawdown[b] (m)	200-day drawdown, second well pumping[c] (m)	500-day drawdown, second well pumping[c] (m)
0	0	0	0	0	0
300	0.74	1.21	1.26	1.70	1.81
600	2.35	3.30	3.41	4.28	4.50
900	4.85	6.28	6.44	7.74	8.07
1200	8.23	10.13	10.35	12.09	12.52
1500	12.50	14.87	15.14	17.31	17.86

[a] Based on the step test results, $B = 0.00098$ d m⁻²; $C = 0.0000049$ d² m⁻⁵.
[b] The additional drawdown for 200 days pumping and 500 days pumping is calculated from Equation (9.11). Transmissivity $= 400$ m² d⁻¹.
[c] The interference drawdown is calculated using Equation (7.26). The second well is 50 m away, pumping at the same discharge rate. Storage coefficient $= 0.02$

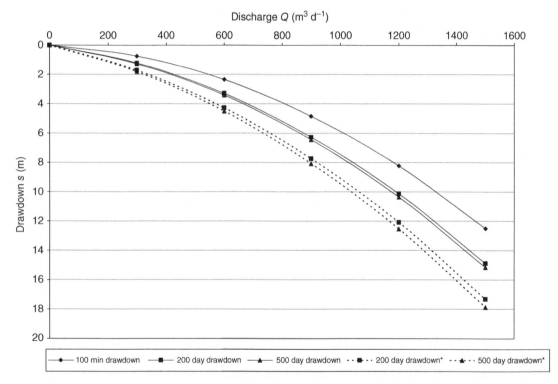

Figure 9.8   *Long-term drawdown predictions for a pumping well. The dashed lines show the 200-day and 500-day drawdowns when there is a second well pumping at the site (see Table 9.4)*

It is conventional to monitor water levels inside a pumping well and in neighbouring observation wells. In the pumping well, it can be helpful to also monitor water levels in the gravel pack between the borehole wall and the screen, as this will aid the diagnosis of the cause of any clogging problems. Figure 9.9 shows two wells with monitoring dip tubes installed both inside the well and in

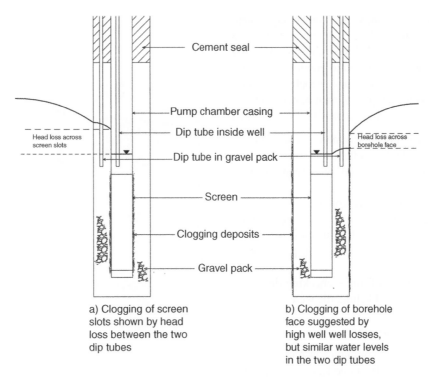

Cement seal

Pump chamber casing

Dip tube inside well

Head loss across screen slots

Dip tube in gravel pack

Head loss across borehole face

Screen

Clogging deposits

Gravel pack

a) Clogging of screen slots shown by head loss between the two dip tubes

b) Clogging of borehole face suggested by high well well losses, but similar water levels in the two dip tubes

Figure 9.9   *Location of well clogging indicated by monitoring of pumping water levels both inside the well and in the gravel pack*

the gravel pack. In these examples, performance monitoring, supplemented by water quality analyses, has indicated that the wells are experiencing clogging problems. In well (a), the operational water level monitoring data show a significant head difference between the pack and the inside of the well, indicating that the screen slots may be blocked. In well (b), the water levels in the well and the pack during pumping are similar, suggesting that clogging of the screen is not the problem, and that the clogging lies in the gravel pack and/or the aquifer close to the well. This diagnosis will influence our choice of well maintenance measures.

9.2.2   Well inspection tools

Down-hole camera (CCTV) surveys and other geophysical logging methods can provide very useful information on the condition of a well and on the processes that might be adversely affecting well performance. The logging techniques are described in Chapter 6. In many cases, it will be necessary to remove the pump in order to allow sufficient access for the logging tools. Great care must be taken to minimize the disturbance to the well, since removing the pumping equipment may dislodge pieces of biofilm and other deposits, and reduce the clarity of the water for the CCTV survey. In larger diameter wells, however, it may be possible to install a guide tube (75 to 100 mm diameter) to permit the logging tools to be lowered down the well past the pump. Such an arrangement will facilitate much more frequent logging inspections.

CCTV surveys are the most useful logging tool for well inspections. They can provide information on the condition of the casing and screen, identifying the presence of pitting corrosion, damage to the casing or screen joints, clogging of the screen slots, the occurrence of biofilms, significant infilling of the well bottom with formation debris or, in some cases, with old pumps and other equipment 'dropped' from the surface. It is important to keep a collection of the CCTV recordings so that new inspections can be compared to previous surveys.

Other potentially relevant geophysical logs (Table 6.4), with examples of their applications for diagnosing well condition and performance, include:

- caliper – checking enlargements in well diameter, for example due to failure of a casing joint; checking position of top of screen or open section of borehole;
- flow and temperature/conductivity logs – comparing the profile of groundwater inflows and outflows at the well with those originally recorded; determining changes in groundwater quality with depth;
- acoustic 'sonic bond' log – checking the integrity of the grout seal behind the well casing.

9.2.3   Pump performance

Pump problems are a common cause of failure of the well system. Much can be learnt from routine observations of the pump and well system in operation, for example by visually inspecting the headworks for signs of corrosion, incrustation or biofouling which, if present, might indicate that the pump is also experiencing such problems (Figure 9.10). Again, the operator should take note of vibrations or untoward noises coming from the pumping plant, such as might occur when a vertical turbine pump has not been installed properly and where there is wear on the pump shaft.

A monitoring programme for evaluating pump performance (for motorized pumps) should include: discharge rate versus total dynamic head, wire-to-water efficiency and the net positive suction head. The discharge rate and the total dynamic head should be monitored and compared with the data collected at the time of the original pumping tests. Note that the total dynamic head includes pipe friction and velocity head energy losses in addition to the total water lift from the pumping water level to the delivery point (plus any pressure head if the pump is operating against a system pressure), and these additional losses need to be

Figure 9.10 *Biofouled submersible pump removed from a carbonate aquifer well, Ohio, United States. Iron oxide deposits are visible on the pump bowls, while a black iron sulphide slime can be seen on the motor. Reproduced by permission of Stuart Smith, Smith-Comeskey Ground Water Science LLC*

taken account of when assessing the performance of the pumping system.

The efficiency of a pumping system was considered briefly as part of pump selection in Chapter 3 (Box 3.9). The wire-to-water efficiency (η: Lipták, 1999; Rishel, 2001) is the overall efficiency of the pump and combines both the pump efficiency (η_p – the mechanical-to-hydraulic efficiency) and the motor efficiency (η_m). The wire-to-water efficiency is simply the ratio of the hydraulic energy produced by the pumping system (the water power, P_{out}) to the prime (usually electrical) energy put into that system (P_{in}), and is given by:

$$\eta = \eta_m \times \eta_p = \frac{Qg\rho_w h}{kW} \qquad (9.12)$$

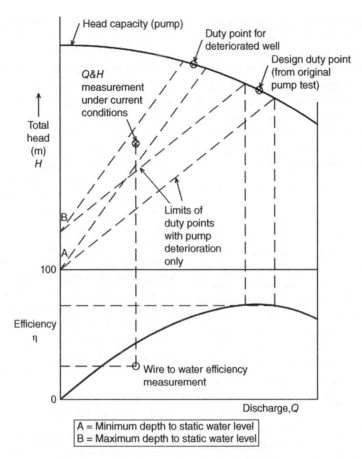

Figure 9.11 *Comparing well performance data with a pump curve (schematic). Based on Figure A1.2 of Howsam P, Misstear BDR and Jones CR (1995) 'Monitoring, maintenance and rehabilitation of water supply boreholes', CIRIA Report 137, London, by permission of the Construction Industry Research and Information Association; go to: www.ciria.org*

where Q is the discharge in m³ s⁻¹, g the acceleration due to gravity in m s⁻², ρ_w the water density in kg m⁻³, h the operating head in m, and kW the electrical power input to the motor (kW). For a mechanically powered pump, P_{in} or kW represents the mechanical power applied to the prime mover. The relationship between efficiency, head and discharge rate should be established initially at the same time as the original step drawdown test, as this will give benchmark data for future comparisons with operational performance (Helweg *et al.*, 1983). A decline in wire-to-water efficiency may result from a number of causes, including increased

friction losses due to clogging of the pump bowls and wear on the pump components. Figure 9.11 shows a schematic example of a situation where discharge, head and wire-to-water efficiency measurements in a well equipped with a variable displacement pump suggest a decline in performance of both the well and the pumping plant.

With centrifugal pumps such as submersible and vertical turbine pumps, insufficient operating pressures at the pump inlet can lead to the damaging process of *cavitation*. The pressure of water reduces with increasing velocity as water enters the pump. Cavitation can occur if the absolute pressure of water falls below its vapour pressure – the water is converted to vapour pockets, which can collapse violently when they encounter higher pressure areas in the pump. This can lead to severe damage to the pump impellers. To avoid cavitation, the pump inlet pressure must be greater than that specified for the particular pump by the pump manufacturer; that is, the *available* net positive suction head (NPSH) must exceed the *required* NPSH. The available NPSH is given by:

$$NPSH = H_a + H_s - H_{vp} - H_f \qquad (9.13)$$

where H_a is the atmospheric pressure (which varies according to the elevation of the well site); H_s is the positive inlet head (the head of water above the eye of the lowermost impeller – this value is negative for a suction lift pump); H_{vp} is the absolute vapour pressure of water; and H_f is the friction loss in the suction piping. If cavitation is occurring, the available NPSH can be increased by lowering the pump (which increases H_s) or by reducing the discharge rate and thereby raising the pumping water level.

When assessing pump performance, it is useful to remember that seemingly trivial factors can make a pump appear to under-perform. For example, a perforated rising main will mean that power is being expended simply to pump water through a hole in the rising main back into the well. A three-phase submersible pump that has been wrongly wired (i.e., two of the three connections 'switched') may still work, but the impeller will rotate the wrong way, resulting in highly inefficient pumping. Further guidance on monitoring pump performance can be found in Roscoe Moss (1990) and Smith and Comeskey (2010).

9.2.4 Water quality monitoring

Water quality parameters that may help us when diagnosing the physical, chemical and microbiological processes affecting well system performance are summarized in Box 9.2. Key chemical parameters to monitor include pH, electrical conductivity, Eh, iron (Fe^{2+}) and the dissolved gases carbon dioxide (CO_2), hydrogen sulphide and dissolved oxygen. It is essential that measurements are made at the wellhead. A flow-through cell should be used to avoid the exposure of the sample to air (Section 8.2.2). McLaughlan (2002) suggests that total iron (incorporating Fe^{3+} as well as Fe^{2+}) should be determined, as this will include any iron that has been oxidized during the sample collection. For more information on sampling and analysis procedures for the main physical and chemical determinands of water quality, the reader is referred to Chapter 8. Concerning bacteriological analysis, direct examination of a water sample using a microscope may just possibly show up some of the stalked or filamentous bacteria associated with biofouling, if these sessile bacteria have become detached from the biofilm, but more detailed and targeted bacteriological sampling and analysis will be required to gain a better understanding of the problem, as discussed in Section 9.2.5. Occasional, sporadically high values of total heterotrophic plate counts (THPC) *can* be an indicator of biofouling problems, the high counts being due to 'sloughing off' of pieces of biofilm from well surfaces.

9.2.5 Monitoring microbial processes

Samples for biofouling analysis can be collected in several ways (Howsam *et al.*, 1995; McLaughlan, 2002; Smith and Comeskey, 2010):

- Pieces of biofilm can be removed from well system components that are accessible.
- Sterilized glass slides or metal coupons can be immersed in the well for a period of time to collect samples of microbiological growth.
- Membrane filters and flow-through devices can be placed at the wellhead to collect bacterial samples – examples include the Robbins device (McCoy and Costerton, 1982), the Moncell (Howsam and Tyrrel, 1989) and the flow cell (Smith, 1992).

Samples can be analysed for their bacteriological composition by light microscopy and culturing techniques, preferably used in combination. Microscopy can reveal the stalked or filamentous species of iron-related and sulphate reducing

---

Box 9.2 Water quality indicators of clogging and corrosion problems

Colour	Colour of water samples can give clues as to the nature of incrustation and biofouling problems
Suspended solids/turbidity	High suspended solids indicates pumping of formation/gravel pack material and/or biofouling problems
pH and Eh	The solubility of metal casings is strongly influenced by the groundwater pH and Eh (see Figure 9.2)
Salinity	High water salinity (measured as EC or TDS) encourages corrosion
Chloride	High chloride levels increase corrosion potential
Temperature	High temperature encourages corrosion
Iron and manganese	High Fe^{2+} and Mn^{2+} levels may lead to chemical precipitates or clogging in contact with air or oxygenated water. They may also indicate corrosion of metallic well components. Erratic concentrations suggest biofouling
Hardness	High carbonate hardness increases incrustation potential
Nitrate	High nitrate levels may encourage bacterial growth
Oxygen	Anaerobic conditions cause low Eh and aid the growth of sulphate-reducing bacteria. Aerobic conditions provide the oxygen necessary for Equation (9.2) to proceed
Carbon dioxide	High CO_2 levels lower the pH and increase corrosion potential. High CO_2 levels may also indicate microbial activity
Hydrogen sulphide/sulphate	High H_2S levels from reduction of sulphate promote steel corrosion. They may also indicate a low Eh and other corrosion/biofouling problems
Organic acids	Acids from peat or pollution lower the pH and the oxygen level and increase corrosion potential
Bacteria	Water sample analyses may identify iron related bacteria or other bacteria that cause biofouling and/or biocorrosion

---

bacteria (Figure 9.5), and can also help identify the mineralogical components of a biofilm, whereas culturing techniques can detect the non-filamentous bacteria which are not easily identified during microscope examination, and which may contribute to biofouling (Smith, 2004).

There are several other analysis techniques that can be used to study the overall composition of a biofilm sample (McLaughlan, 2002). These include: the total organic carbon content of the sample, application of the loss on ignition method (for determining the ratio of organic to inorganic carbon present), x-ray diffraction (to identify the mineral elements present) and acid dissolution followed by ICP analysis (to identify elemental composition; Banks and Banks, 1993). As an alternative to laboratory-based analysis techniques, electrochemical in-line sensors have been developed to provide early warning of biofouling problems (Smith and Comeskey, 2010).

9.3 Well maintenance and rehabilitation measures

As noted in the introduction to this chapter, there are many thousands of production wells across the world which have fallen into disuse. This may be

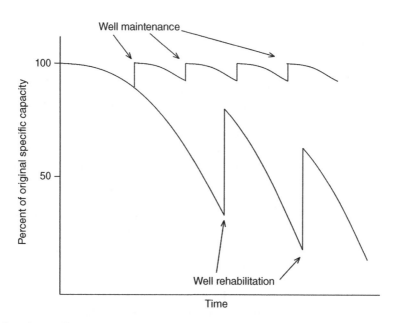

Figure 9.12   *The relative effectiveness of regular well maintenance compared to infrequent well rehabilitation, as shown by changes in specific capacity over time*

for economic reasons: for example, the failure of the community to support the economic burden that a potable water supply can represent or the inadequacy of a revenue collection system. It may be for technical reasons: for example, the breakdown of pumps or deterioration of the well structure through lack of maintenance. Maintenance is an essential part of the well system operation, and proper maintenance depends on having correct procedures in place for monitoring and diagnosis. It is also worth reiterating here that the most effective measures against clogging and corrosion in a well are those taken at the design stage (Section 9.1.4). If low grade steel materials, for example, are used in a well system where the groundwater is corrosive, then corrosion and clogging problems will arise in the well, pump and headworks and it will probably not be possibly to keep this well system performing at its original level, even if regular monitoring and maintenance are carried out. The problems will be exacerbated if the well is pumped intermittently with large fluctuations in water levels, since this increases the length of the vulnerable 'splash zone'

(Section 9.1.5). The choice of a screen with a large open area (Section 4.3), to enable well maintenance to be undertaken efficiently, is also an important aspect of the design process.

Preventative *maintenance* involves cleaning and other actions which are undertaken on the well system on a regular basis, with the aim of keeping the well system at, or close to, its original level of performance. *Rehabilitation*, on the other hand, is the process of trying to restore a well system to its original condition and performance after it has deteriorated significantly. Similar techniques are used in maintenance and rehabilitation; the difference is often only in the extent to which the techniques are applied (although there may be some differences in the choice of chemicals, for example). The important thing to appreciate is that regular maintenance is much more likely to be successful in preserving the original well performance than occasional rehabilitation. This is illustrated by the schematic plot of specific capacity versus time in Figure 9.12. The figure also illustrates that there is often a longer time interval between the well commissioning and the first rehabilitation exercise than between subsequent rehabilitation programmes (van Beek, 2001). Consistent with this, Houben (2004), in a study of iron oxide incrustation, found that partial unclogging of the pores in the gravel pack as a result of rehabilitation actually reactivates the iron precipitation reaction and hence leads to a relatively rapid decline in well performance after the initial improvements due to rehabilitation.

The problem with leaving a well until its performance has deteriorated appreciably is that by that time the screen and aquifer are nearly totally blocked and any chemical incrustation or other fouling deposits have had time to age. Well restoration depends on being able to get water into the clogged gravel pack and/or aquifer to remove the clogging material, and if blockage is total it may not be possible for water to enter the pack or formation material. Also if an incrustation has aged and recrystallized it will be much harder than in its early state, have a smaller surface area and lower chemical reactivity, and will be much more difficult to break up and remove.

The frequency with which preventative maintenance should be carried out depends on a number of factors and local experience. Factors to consider include (McLaughlan, 2002):

- hydraulic performance of the well system;
- water quality guidelines;
- structural integrity of the system;
- economic factors;
- social and environmental factors.

Taking a well out of supply can be costly and difficult, and maintenance operations themselves can be expensive, so it is necessary to strike some sort of balance between prevention and cure when developing a sensible preventative maintenance strategy. On hydraulic grounds, it is considered advisable to carry out cleaning operations in a well when the specific capacity has reduced to no less than 80-85% of the original value – the American Society of Civil Engineers (2014) notes that a large amount of development energy is required for well restoration when the specific capacity has fallen below about 75-85% of the original value.

However, other factors may influence the frequency of maintenance operations; for example, the local availability of the necessary expertise and equipment. The practical aspects of removing a well from supply for maintenance are also important. For a drinking water supply well, the best time is normally in winter, since this is usually the period of lowest demand. With irrigation wells, the non-irrigation season is obviously the most suitable. However, in practice, problems with the well system often only become readily apparent during the period of maximum pumping, which is the least suitable time to carry out well work-over operations.

The methods used for maintenance and rehabilitation are summarized in Table 9.5. The maintenance programme needs to encompass the full well system – the aquifer, pumping plant and headworks in addition to the well itself. The aim of the restoration programme should be based on the diagnosis of the problem causing the reduction in condition or performance. In the well, the location of the clogging problem will have a strong bearing on the techniques used and their likely effectiveness. In a screened well, clogging of the screen slots is more accessible, and therefore easier to deal with, than clogging of the gravel pack and especially the aquifer beyond the pack (Figure 9.9). The upper part of the screen often experiences the most incrustation and biofouling, and needs to be targeted in cleaning operations. Where the gravel pack and surrounding formation are badly clogged, injection of the appropriate chemicals through access tubes in the pack or into the formation via satellite boreholes around the pumping well, may be more effective than treatments applied inside the well (Howsam *et al.*, 1995). In an unscreened well in a consolidated or crystalline aquifer, maintenance operations need to target the main fracture inflow zones.

The methods used for well restoration after incrustation and biofouling are similar to those for well development, and include standard physical methods like surge pumping and jetting. However, in well restoration the clogging deposits have to be broken up before they can be removed and so

Table 9.5 *Maintenance and rehabilitation aims and methods*

Location	Aims	Methods
Aquifer	Water level recovery	• reduce local/regional abstraction • reduce interference between boreholes operating at the same time • induce additional recharge
	Water quality recovery	• reduce abstraction • induce fresh-water recharge • prevent further pollution • remove contaminant from aquifer or *in-situ* remediation • blend with other sources
Well	Removal of infilling deposits	• airlift pumping • bailing
	Removal of deposits from internal surface of casing or screen	• high pressure jetting • brushing or swabbing • explosive or ultrasonic treatment • pasteurization • chemical treatment
	Removal of deposits from screen slots	• high pressure jetting and pumping • surging and pumping • explosive or ultrasonic treatment • chemical treatment
	Removal of deposits from external surface of casing or screen	• chemical treatment • combined physical-chemical process • explosive or ultrasonic treatment
	Removal of material from gravel pack and adjacent formation	• surge-block surging or air-lift pumping • high pressure jetting and pumping • chemical treatment • combined physical-chemical process
	Removal of material from formation fissures	• high pressure jetting and pumping • airlift surging and pumping • hydrofracturing • chemical treatment
	Repair of ruptured or perforated casing or screen	• reline • *in-situ* repair • seal off section of borehole • retrieve and replace
	Correction to faulty design	• reline • *in-situ* repair • seal off section of borehole
	Corrosion reduction	• coatings • cathodic protection • reline
	Disinfection of parts	• use of chemical disinfectants or biocides • pasteurization • irradiation
Pump system (after retrieval)	Removal of deposits from external surfaces of pump and rising main	• high pressure jetting • brushing • chemical treatment
	Removal of deposits from internal surfaces of pump and rising main	• dismantle and high pressure jetting • dismantle and brushing or swabbing • dismantle and chemical treatment
	Rectify electrical fault	• dismantle and replace part
	Rectify physical/mechanical fault	• dismantle and replace part
	Corrosion reduction	• coatings • cathodic protection • replace materials
	Disinfection of parts	• use of chemical disinfectants/biocides • pasteurization

Table 9.5   *(Continued)*

Location	Aims	Methods
Wellhead works	Removal of deposits from internal surfaces of wellhead pipes and fittings	• dismantle and high-pressure jetting • dismantle and brushing or swabbing • dismantle and chemical treatment
	Repair of non-functioning part	• dismantle and repair or replace
	Repair of rupture/perforation to prevent leakage	• dismantle and repair or replace • reseal or patch
	Disinfection of parts	• use of chemical disinfectants or biocides • pasteurization

Based on Table 3.1 of Howsam P, Misstear BDR and Jones CR (1995) 'Monitoring, maintenance and rehabilitation of water supply boreholes', CIRIA Report 137, London, by permission of the Construction Industry Research and Information Association; go to: www.ciria.org

chemical treatments are normally also employed. Carbonate incrustation can be removed by acidization (usually sulphamic or hydrochloric acid, with the former being much easier and safer to handle), as with the development of limestone water wells. If the incrustation is heavy, the treatment will be improved by use of wall scratchers and very high-pressure jetting to break up the carbonate and allow the acid better access. Acids such as sulphamic acid, hydrochloric acid, oxalic acid and ascorbic acid can also dissolve iron oxides, but the solubility of these deposits is much less when they have recrystallized and hardened (Houben, 2003b). With hardened incrustation of ferric hydroxide, restoration has to depend on breaking up the cemented deposits by physical or hydraulic methods. Care must be taken when using vigorous methods, not to damage the well and make the situation worse. Very high-pressure jetting, for example, can cut through plastic casing. Other physical methods sometimes used for breaking up and dislodging incrusting materials include: shocking a well by explosives (or by some other less dramatic sonic/vibratory method), the introduction of deep-cold CO_2 into the well, or the application of fluid-pulse tools which inject pulses of gas or water under high pressure (Smith and Comeskey, 2010).

Where the main source of clogging is biological, biocides can be included in the treatment programme. The agent most commonly employed for well maintenance is chlorine, usually in the form of sodium or calcium hypochlorite (although use of the latter is not encouraged because of the risk of forming insoluble precipitates in the well; American Society of Civil Engineers, 2014), with hydrogen peroxide having some limited applications. Acetic or glycolic organic acid blends are also increasingly used for treating biofouling problems in wells in the United States.

Apart from acids and disinfectants, the other main class of chemicals used in maintenance and rehabilitation is the dispersing agent, to assist in the removal of 'fines', where its application is similar to that in well development for the removal of drilling mud. Traditionally, the dispersing agent of choice was some form of polyphosphate chemical. However, phosphate is a nutrient for bacteria in groundwater and any remaining in the well after treatment could stimulate further biofouling problems. For this reason, non-P alternatives such as polyacrylamide are now recommended (Smith and Comeskey, 2010; American Society of Civil Engineers, 2014).

Acids and other chemicals used in a well restoration programme can be harmful to people and the environment if they are not used properly. Local health and safety regulations and guidelines should be followed when handling these chemicals, and only experienced personnel equipped with the necessary protective clothing should be employed to do this type of work. A health and safety plan should be prepared for any well restoration programme (Appendix 3).

Many of the problems in a well system are connected to the pumping plant, and the performance of the pumping system should be monitored as explained in Section 9.2.3. Maintenance measures will obviously depend on the type of pump and the nature of the problem, but may include:

- pump retrieval, disassembly and inspection of the rising main, pump impellers, casing, shaft and other parts of the pump for indications of corrosion and clogging;
- lubrication of moving parts in the pump and prime mover;
- routine replacement of bushings, wear rings and other protective parts;
- replacement of other worn or damaged parts as necessary.

The restoration programme for the well will depend on the results of monitoring and diagnosis, but typically may involve the following stages:

1. Assessment of the current hydraulic performance of the well using operational data, specific capacity data or, preferably, a step drawdown test, and comparison with the performance data at time of commissioning.
2. Checking of water quality parameters, including EC, temperature, pH, suspended solids and bacteriological content.
3. Removal of the pump from the well for inspection and repair, as described above.
4. Down-hole CCTV and geophysical logging survey to identify location of zones of clogging and/or corrosion.
5. Cleaning of casing and screen using a wall scratcher or wire brush.
6. Removal of debris by bailing or air-lift pumping.
7. Acidization or other chemical treatment.
8. Surging or jetting the well to enhance effectiveness of chemical treatment.
9. Removal of spent acid or other chemicals by bailing or pumping.
10. Disinfection of the well, pump and rising main using sodium hypochlorite or other disinfectant.
11. Replacement of pump.
12. Step drawdown test or specific capacity test to assess extent of improvement in performance.
13. Measurement of water quality parameters, including EC, temperature, pH, suspended solids and bacteriological content, to confirm that all traces of acids, oxidizing agents and other chemicals have been removed to acceptable levels, and that the water quality is suitable for its intended purpose.

A number of the above stages can be repeated to try and improve the effectiveness of the well restoration programme. It is important to remember that well maintenance can involve acids and other hazardous chemicals, plus high-energy physical methods. Therefore, only experienced personnel should be used, and these personnel must follow the necessary health and safety procedures in all operations. Also, well restoration or maintenance can produce quantities of sludge or severely contaminated water, and measures must be taken for temporary storage of this material on-site, and its ultimate disposal to a safe waste management facility or sewage works, in accordance with the provisions of any permit required for the work.

Metal casing and screen can be protected from corrosion by cathodic protection, whereby sacrificial electrodes of a metal higher in the electromotive series than steel are provided; these electrodes are then corroded in preference to the casing. This system is used in some water wells to protect the pump installations, but is rarely used to protect the casing and screen string. More commonly in water wells, corrosion-resistant materials are chosen at the design stage. These materials may be plastic, fibreglass or stainless steel, or coated materials such as bitumen–coated steel casing (Section 4.2). The problem with coated materials is that if the coating is scratched during installation, as is likely, then the corrosion resistance may be lost.

A water well damaged by corrosion to such an extent that the casing is perforated or that it is pumping sand, can only be restored by partial or total relining. The necessary course of action may be decided after a geophysical logging programme (Sections 6.4 and 9.2.2). Holes in the casing can be revealed by features on the differential temperature and conductivity logs, by flow logs or by caliper logs. The nature and extent of the perforations can be shown by CCTV inspection, provided that the water column is clear.

The casing used for relining should be chosen to avoid a repetition of the problem, that is, it should be corrosion resistant. Seals between the new casing and the corroded section must be sound to avoid seepage of polluting water between the casings. The seals could be plastic rings or malleable metal swedge rings. If the casing being re-lined is the pump chamber of a well, then the size of pump which can be installed will be reduced. Corroded screen should not be relined if possible, because concentric screens can induce turbulence and physical abrasion in the annulus between the two screens. The corroded screen should be removed, if possible, and replaced with corrosion-resistant screen.

The discussion above has focused on maintenance and rehabilitation of drilled wells. Hand-dug wells also need to be kept in good repair: they are prone to damage from floods, vandalism, civil unrest and other causes. The steps for rehabilitating and cleaning hand-dug wells include (Godfrey and Reed, 2013): removal, inspection and repair of the hand-pump; cleaning out the well; repair and relining of damaged sections of the well lining; construction of new headworks, including a concrete apron, drainage channel and well cover (see Section 3.2). The rehabilitated well should be disinfected using a hypochlorite solution before it is returned to supply.

9.4   Well decommissioning

Relining or other repairs may not be feasible or cost-effective for some wells, which are then decommissioned or abandoned. If a well is not decommissioned properly it can lead to a number of problems:

- It may pose a physical danger to people and animals. This is especially so with the open well top of a hand-dug well or a large diameter drilled well.

- It may allow polluted surface water to enter the aquifer system.
- It may allow poor quality groundwater from one aquifer to contaminate another aquifer.
- It may result in a depletion of head and of the groundwater resource in an aquifer by enabling groundwater from this aquifer to leak into another aquifer. If the well is artesian, continued uncontrolled discharge may represent a serious wastage of aquifer resources.

To avoid such problems, the well should be decommissioned in an engineered manner so as to prevent the vertical movement of water from the surface to an aquifer or from one aquifer to another. The well structure also must be stabilized so that it does not pose a risk to people or animals; the wellhead in particular must be sealed.

The first step in a well decommissioning programme is to find out as much as possible about the construction of the well: how deep it is, how it was lined, whether the casing was grouted, whether there is an artificial gravel pack around the screen, and so forth. This information should be available in the drilling records – if these can be found. The second step is to inspect the well. Visual inspection from the well top may give some clues as to the condition of well, and lowering a plumb-bob or similar device can indicate if there are obstructions present that will need to be removed. However, a CCTV and logging survey will be necessary to identify the nature of any obstruction and to establish the condition of the casing and screen (Section 9.2.2). It is essential that the presence of a poorly grouted casing does not compromise the effectiveness of the proposed well sealing operation. If there is a risk that the annulus between the casing string and the borehole wall will allow water to move vertically, then the casing string should be removed – if possible. It is normally easier to jack-out steel casing and screen than plastic linings owing to their greater strength. If it is not practicable to remove a steel casing, then the casing at the interval where the seal is defective should be ripped or punctured before a low-permeability backfill material is inserted. With

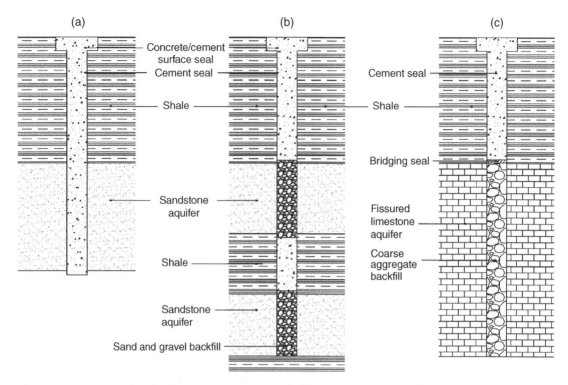

Figure 9.13 *Approaches for decommissioning (a) a shallow well, (b) a deep well and (c) a large diameter, deep well in a fissured aquifer*

plastic casing strings, it is sometimes necessary to break these up by drilling them out.

The approaches adopted for decommissioning a well will depend on the type and condition of the well and on the hydrogeology. A small diameter, shallow production well or exploration borehole is often backfilled with a low permeability material – often a cement grout, but sometimes concrete, bentonite clay or a bentonite-cement mix – throughout its entire length [Figure 9.13(a)], care being taken that the grout is emplaced from the bottom of the hole upwards so as to avoid bridging. A larger diameter and deeper drilled well can be decommissioned by backfilling the aquifer sections of the hole with sand and gravel and by placing low permeability seals across the aquitards that separate the individual aquifer horizons [Figure 9.13(b)]. It is essential that the coarse material used for the backfill is clean and non-reactive. If the well being decommissioned is in a

sensitive area, for example in the vicinity of an operating production well, it is advisable to disinfect any granular backfill material using a solution of hypochlorite before placing the backfill down the well being decommissioned. The backfill should be poured or pumped into the well using a tremie tube (Section 4.4). In some situations the amount of backfill material required could be very large, for example for infilling a large diameter and deep well, or a deep well constructed in a highly fissured aquifer where significant quantities of sand and gravel backfill could be lost into the formation. In these situations a very coarse aggregate can be used as backfill instead of sand and gravel, or the lowermost aquifer section can be left as open hole, with a bridging seal placed above this [Figure 9.13(c)]. Specialized cement-based devices are available for bridging seals (National Ground Water Association, 1998).

Flowing artesian wells pose particular challenges for decommissioning, since the flow must be controlled before the hole can be backfilled and grouted. The artesian flow can be controlled by:

- raising the wellhead above the level of the potentiometric surface (this is only practicable for small artesian heads);
- lowering the head – for example, by pumping neighbouring wells;
- placing a heavy drilling mud in the well (Section 5.2.2);
- installing a pre-formed plug in the well at the top of the artesian aquifer.

With all decommissioning programmes, the relevant regulatory authority should be contacted beforehand to approve the proposed methodology, including the proposed sealing intervals and materials.

Further discussion on well decommissioning methods can be found in publications by the Environment Agency (2012b), American Society of Civil Engineers (2014) and Australian Drilling Industry Training Committee (2015), whilst standards that cover well decommissioning include ANSI/AWWA A100-06 (American Water Works Association, 2006), ASTM D5299-99 (American Society for Testing and Materials, 2012c) and ANSI/NGWA-01-14 (National Ground Water Association, 2014).

10

Well and Borehole Records

There is a need to manage the information generated during the construction, testing and routine operation of water wells. Well and borehole records essentially fall into two categories:

1. An archived record, containing construction, location, geological and hydrogeological data gleaned from the construction and testing of a well or borehole. The most important elements of such a record will often be public domain.
2. An operational database, containing the data generated during the routine running of a well source and enabling the owner or operator to diagnose potential problems with the source at an early stage.

10.1 Well archives

When constructing a new well, we are generating geological and hydrogeological information about a portion of an aquifer that may never have been investigated before. This new information needs to be archived systematically so that other hydrogeologists and engineers can access it in the future. Most countries operate some form of well and borehole archive, containing all the details of wells reported by drillers, owners and operators.

The archive may be in the form of paper records (for example, a set of maps with a corresponding card index), but it will usually be computerized. It may be lodged with a geological survey, or with a water or environment agency or ministry. Increasingly, such databases are available via the internet, often as a web-based geographical information system (Web-GIS). The well and borehole data may be integrated with other environmental data, such as geological maps, topographical or hydrological data. Excellent examples of such Web-GIS systems are the British "GeoIndex", the Danish "JUPITER" and the Norwegian "GRANADA" databases – currently available at http://www.bgs.ac.uk/geoindex/, http://data.geus.dk/geusmap/?mapname=jupiter and http://geo.ngu.no/kart/granada/, respectively.

There may be legislation *requiring* the driller or owner of any new well or borehole to submit details to the archive. Even if no legislation is in place, it is in the interest of the driller and well owner to archive the well records, since the archive will:

1. Accumulate a body of data that will lead to a better understanding of the geology and hydrogeology of the area. An improved conceptual

Water Wells and Boreholes, Second Edition. Bruce Misstear, David Banks and Lewis Clark.
© 2017 John Wiley & Sons Ltd. Published 2017 by John Wiley & Sons Ltd.

model of the aquifer system will benefit the owner or operator of a well field, as it will enable more effective management of the groundwater sources and resources. An improved understanding of the geology will benefit the well driller by potentially increasing his drilling success rate in the future.

2. Provide a tool for planners and licensing authorities which will help protect the well against new developments that may adversely impact on groundwater flow or quality. If the well is recorded in a national well archive, and there is subsequently a proposal to license a new groundwater abstraction nearby, the regulatory authority can assess whether the existing well will be affected and, if so, whether the new abstraction should be allowed to proceed.

3. Assist the operator of a well field to manage data. In the absence of a good well archive, we have observed that even rather professional water companies can lose track of which well is which, or indeed of the number of wells in a large well field.

Figure 10.1 shows an example of a card index for a well record of the type used in England prior to the computerization of databases. The well is allocated an index number, the first part of which indicates a 1:25 000 scale UTM map grid square (TF 54) and the latter part of which is a sequential well number within that map square.

Box 10.1 demonstrates the value of a comprehensive well database. The operator of any national or regional well archive faces a difficult task in persuading some drillers to invest time in reporting (especially reporting of 'dry' or abandoned wells, which, though possibly of little further direct interest to the driller, may be hugely important for the hydrogeologist). Figure 10.2 shows a map and corresponding record generated by the Geological Survey of Norway's computerized database. The fact that the database is available online has been a major incentive for drillers to submit data to the database. In addition to being a useful resource for them, it acts as a form of 'showroom' for their work.

The fields that should be included in a well database will depend to some extent on the geological context of the region considered. Nevertheless, some of the most likely fields are shown in Table 10.1. The database will contain information on the well location; its operation, construction details and geological log; water level and salinity; basic hydrogeological parameters; a summary of pumping test results; and it may also contain date-related hydrochemical data. The archiving of hydrochemical data is considered in Section 8.9: key issues include:

- how to include 'below detection limit' data;
- how to include 'metadata' (information *about* the data, including conditions of sampling, analytical laboratory and method of analysis);
- how to define different hydrochemical parameters: for example, should analysis of fluoride by an ion-selective electrode be treated as the same parameter as analysis of fluoride by ion chromatography?

10.2   Operational well databases

In previous sections of this book, we have emphasized the need for equipping every well with the following:

- a means of access for water level measurement;
- a means of determining instantaneous discharge rates (flows) from the well, cumulative abstracted quantities and duration of pump operation;
- a means of collecting water samples at or near the wellhead.

Every well operator should use these facilities to make regular observations of rest water level, pumping water level and corresponding pumping rates. Water samples should also be taken at intervals depending on the size of the abstraction, its purpose and the population served. It is also good practice to carry out a step drawdown test (Section 7.3.2) on the well at regular intervals, to measure the drawdown for a range of pumped yields and to calculate the specific capacity (ratio

SLEEPYVILLE PARVA No. 5 WELL					TF 54 / 13E		

| Owner Sleepyville Water Company | | Licence No. T13/54/12 | | Nat. Grid Ref. TF 5312 4969 | | | |
| Occupier Sleepyville Water Company | | IGS Ref. No. | | Status Public water supply | | | |

Ground Level	32.195 m OD	105.63 ft. OD	Aquifer UPPER CHALK		
Level of Well Top	32.515 m OD	106.69 ft. OD			

Rest Water Level	15.24 m bwt	50.00 ft. bwt	Summary of Geological Section	Thickness	Depth
(Date 29/7/85)	17.28 m OD	56.69 ft. OD	Alluvial sands & gravels	4 m	4 m
Construction Borehole			Lower London Tertiary Strata	34 m	38 m

Depth bwt	Dia.	Linings (below well top)				Chalk with flints	97 m	135 m
		From	To	Dia.	Type	Very hard Chalk	3 m	138 m
158 m	610mm	0 m	45 m	610 mm	Plain steel	Chalk with flints	20 m	158 m
		45 m	65 m	610 mm	Slotted steel			
						Data from drillers' log		

Abstraction Rates	Type of Pump 250 mm elec. sub.
340 m³ hr⁻¹ on test	Chem./Bact. Anal. (YES) NO
gpd	Well Driller Acme Drilling May 1985

If insufficient space has been allowed, continue in 'Notes' overleaf.	Geophysical logs: Temp./ EC / pH, caliper, guard resistivity, natural gamma

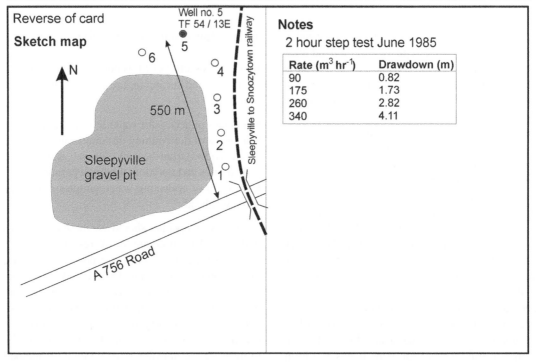

Reverse of card

Sketch map

Well no. 5
TF 54 / 13E

550 m

Sleepyville gravel pit

A 756 Road

Sleepyville to Snoozytown railway

Notes

2 hour step test June 1985

Rate (m³ hr⁻¹)	Drawdown (m)
90	0.82
175	1.73
260	2.82
340	4.11

Figure 10.1 *A typical card index database entry for a (fictional) well. Based on the format of records held by the Environment Agency of England & Wales (Thames Region). Public domain information, provided by and reproduced with the permission of the Environment Agency of England & Wales (Thames Region)*

Box 10.1   The value of well databases in crystalline rock aquifers

Hydrogeological databases are of enormous value in any situation. Historical well data allow us to identify systematic changes in well yield, water chemistry and aquifer properties with distance or with depth. They also assist us in building conceptual models of aquifer systems.

Many fractured crystalline rock aquifers, however, are highly anisotropic, heterogeneous and discontinuous. For practical purposes, many hydrogeologists argue that it is impossible to *predict* parameters such as well yield and water chemistry for a proposed well in such aquifers. Instead, they can, with a good well database, describe the *probability* of a new well achieving a given yield (Banks *et al.*, 2005).

In Sweden, a database exists of some 59 000 wells in Precambrian and Palaeozoic crystalline bedrock. The median yield is found to be 600l hr^{-1}. The arithmetic average is 1643l hr^{-1}, although this has little meaning for single wells, as the dataset is not normally distributed, but is highly skewed (Gustafson, 2002). The cumulative frequency distribution in Figure B10.1(i) shows that there is an 80% probability of achieving a yield of some 200l hr^{-1}, ample for a domestic or small farm supply (this explains the apparent success of *water witches* or *dowsers*, who typically work for such clients, and also the ability of some drillers to offer 'water guarantees', absorbing the 20% risk of failure into their pricing structures).

A similar database also exists in Norway of some 30 000 wells. Intriguingly, the median yield here is also exactly 600l hr^{-1} (n = 12 757 quality controlled wells; Morland, 1997). The database can be broken down into lithological subsets, and some minor differences in yield distributions are observed between these various lithologies [Figure B10.1(ii)].

Banks *et al.* (2010) have noted that Norway, Sweden and Finland have remarkably similar yield distribution statistics for crystalline rock aquifers. While one might argue that these Fennoscandian nations have similar (though by no means identical) climate, geology and weathering history, one might expect the yield distribution curves for hard-rock aquifers in Africa to be very different. However, well data in crystalline rock terrain of eastern Chad suggests that the well 'success rate' is only 30–40%, and of

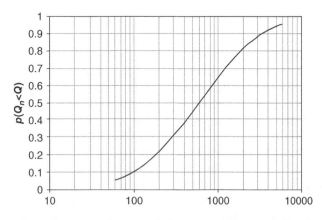

Figure B10.1(i)   *Cumulative frequency diagram of all 59 000 wells in crystalline bedrock in the Swedish well archive (y-axis shows probability, x-axis shows yield in l hr⁻¹). After Gustafson (2002), reproduced with permission of Norges geologiske undersøkelse (Geological Survey of Norway)*

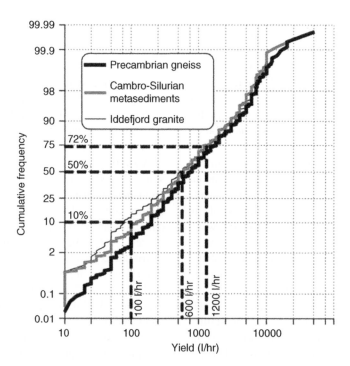

Figure B10.1(ii) *Cumulative frequency diagram showing yield distribution curves for Norwegian water boreholes in the Precambrian Iddefjord Granite, Cambro-Silurian metasediments of the Norwegian Caledonian terrain, and Precambrian gneisses. A guideline shows the approximate 10% yield for most lithologies (indicating that 90% of wells yield better than this value). It will be seen that 90% of all wells yield more than 100 l hr^{-1} in most lithologies. The 50% guideline indicates the median yield. After Banks and Robins (2002) and based on data from Morland (1997). Reproduced by permission of Norges geologiske undersøkelse (Geological Survey of Norway)*

the successful wells, the median yield is only 1500–2000 l hr^{-1} (P. Hansbury, *personal communication*). This is also intriguingly suggestive of an overall median yield of some several hundred litres per hour, an observation that is tentatively supported by data from other "crystalline rock" nations in Banks *et al.* (2010).

The yield statistics obtained from such databases depend on the quality of the input data, and how representative these data are of all the wells in an area. For example, it is common practice for drillers not to report failed or 'dry' boreholes, such that that the low-yield end of a yield-distribution curve may often be under-represented. It should also be noted that databases might contain only the short-term yields reported by drillers on completion of the well: the long-term sustainable yield may be considerably lower.

It is often found that many well yield distributions in fractured rock aquifers are approximately log-normal. This is also true of distributions of certain hydrochemical parameters (especially those lacking any solubility control, such as radon and uranium – see Box 2.9). This may be related to the observation by rock geomechanicists, that fracture lengths and apertures are typically log-normally distributed.

Bedrock well no. 24550

		Location	
Total well depth:	150.00 metre	**Country:**	Sør-Trøndelag
Depth to rockhead:	1.00 metre	**Municipality:**	Trondheim (1601)
Water Yield:		**Property no.:**	
Drilling date:	21.05.2003	**Site no.:**	
Use:	Test well	**UTM zone:**	32 V
		Easting:	572030.00
Waterworks:		**Northing:**	7037029.00
Drilled diameter:		**Map sheet (1:50 000)**	Trondheim (1621-4)
Casing material:	Steel	**Location method:**	GPS after May 2000
Casing length:	3.00 m		
Deviation:	Vertical	**Location accuracy:**	1000 cm

Drilling firm:	Båsum Boring Trøndelag AS
Drillers name:	Olav/ Ola
Other remarks:	Well no. 2. Borehole used for calibration of geophysical instruments.

Contact details:

Well address: Geological Survey of Norway, Leiv Eirikssonsvei 39, Lade, 7040 Trondheim

Formations (bedrock well):

Depth from surface(meter)

From	To	Water ingress (l hr^{-1}) Colour Rock type Other remarks
0.00	100.00	50–500
100.00	150.00	>1000

Figure 10.2   *An extract from the Norwegian internet-based well and borehole database. The map shows all registered wells and boreholes in a section of Mid-Norway around Trondheim (GSH is ground source heat; 'Drift well' is a well in superficial Quaternary deposits). The lower portions show the registered data available for a single borehole. Reproduced by permission of Norges geologiske undersøkelse (Geological Survey of Norway)*

Table 10.1   *Fields that will often be included in a well database*

Field	Comments
Identifier	
Unique well index number	-
Well name	-
Location	
Grid reference	UTM coordinates or latitude/longitude, with means of measurement (GPS, map)
Address	-
Map sketch	-
Management	
Well owner	Contact details
Well operator	Contact details
Purpose of well	Abstraction for public water supply, domestic water supply, irrigation, industry, ground-source heat, environmental investigation, or other purpose
Licence number	-
Licensed quantities	-
Construction	
Driller/well constructor	Contact details. Also any consultant involved
Year of construction	-
Total depth	-
Deviation from vertical	-
Casing/lining details	Type, length, position, diameter
Well screen/well intake details	Type, length, position, diameter
Pump installed	-
Yield enhancement technique	Hydrofraccing, acidization (if performed)
Geology	
Main aquifer(s)	Name(s), details (unconsolidated, consolidated, crystalline)
Geological column/sequence	Depth to top and base of each unit
Hydrogeology	
Measurement datum	Top of casing or ground level, usually
Elevation of datum	Elevation in m above sea level
Rest water level/date	Several measurements could be included
Step test results	Date, duration of step, pumping rate, pumping water level at end of step
Constant rate test results	Digital file may be appended, or a hyperlink to the pumping test data
Aquifer parameters	Derived values of transmissivity/storage
Maximum sustainable yield	Assessed on basis of test pumping
Water Quality	
Date of measurement	A sequence of dates could be included
Parameter	Likely to be a large number of parameters
Digital flag(s)	Digital flag(s) for 'below detection limit'
Metadata	Sampling conditions, filtration, preservation, analytical laboratory, analytical method
Links	
Cross references or hyperlinks	To other data, reports, digital files, photographs etc

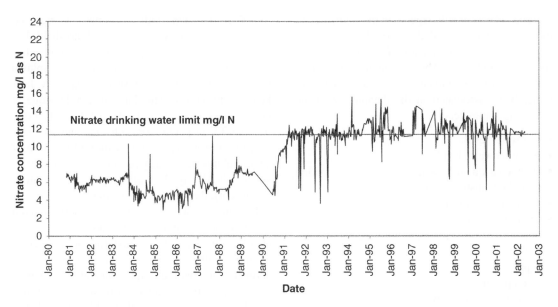

Figure 10.3   *A time series plot of nitrate concentrations in groundwater abstracted from a single well in the Sherwood Sandstone of Yorkshire, England, from 1980 to 2003. A sharp increase in nitrate concentration, such as that evident in 1990, should trigger further investigation. It might have been due to a change in analytical laboratory or to a change in data reporting practice (from mg l^{-1} N to mg l^{-1} NO$_3^-$, for example). In this case, however, the sharp increase in nitrate concentrations proved to be real, and was related to the closure of a neighbouring well that had pumped high-nitrate water. Reproduced from M.F. Knapp (2005), 'Diffuse pollution threats to groundwater...', in Quarterly Journal of Engineering Geology and Hydrogeology, 38, p39-51 by permission of the Geological Society (London) and Yorkshire Water Services*

of yield to drawdown) and the well efficiency (Section 9.2.1). Regular operational monitoring of this nature will enable the early recognition and diagnosis of problems such as:

- *Over-abstraction* of the aquifer relative to available recharge. This may be indicated by continuously declining rest and pumping water levels in both pumping and observation wells.
- *Decrease in well performance*, possibly due to clogging or biofouling (Section 9.1). This may be indicated by increasing drawdowns for a given pumping rate, or declining yield for a given drawdown (i.e., gradually decreasing specific capacity).
- *Increasing demand.* This will be indicated by increased hours of pumping, for a given pumping rate.

- *Deterioration in water quality*. Regular monitoring of water quality may indicate progressive incursion of a front of contaminated water or saline intrusion. It may indicate increasing nitrate contamination (as is the case in many wells in the UK, due to changes in agricultural intensity over periods of decades; Figure 10.3), or increasing frequency of episodes of faecally-related microbiological contamination.

Of course, it is not enough simply to measure water levels, pumping rates and water quality. It is also necessary to manage the data collected efficiently and to interrogate the data at regular intervals in order to recognize potential trends with time, and thus initiate appropriate maintenance operations (Section 9.3).

An operational database will typically comprise a sequence of time-related data. The time interval for data collection will depend on the frequency of monitoring, but the availability of electronic data-logging renders the collection of data on an hourly (or even more frequent) basis feasible for the larger well abstractions. Given dates and times will typically be associated in the database with the following parameters:

- rest water level;
- pumping water level;
- abstracted quantity within a given time period (e.g. daily abstracted quantity);
- instantaneous abstraction rate;
- calculated specific capacity;
- operational water quality monitoring parameters, including:
  - pH,
  - temperature,
  - electrical conductivity,
  - dissolved oxygen,
  - faecal coliforms;
- results of water quality analyses;
- records of any well maintenance or rehabilitation actions carried out.

The pitfalls of archiving water quality data have been noted above and in Section 8.9. For operational monitoring of water quality, it can be advantageous to use a single, accredited, analytical laboratory, with proven and consistent laboratory methods and limits of detection. It is not unheard of for changes in laboratory or in analytical method to result in apparent 'steps' in water quality trends: these do not necessarily reflect real changes in water quality; rather, they can be due to inconsistencies in laboratory analytical procedure. Nevertheless, it is useful to carry out periodic checks on the performance of the selected laboratory by sending duplicate samples to another laboratory, and then to thoroughly investigate reasons for any differences in the results obtained.

An operational database need not simply be an archive, but should be a management tool. For larger abstractions, data can be telemetered to a central operations room and software can be used to generate graphs showing time-related trends in real time. Such a database can be equipped with

'red flags' (warning indicators) that are triggered when a parameter exceeds a critical threshold. For example, if the pumping water level approaches the top of a well screen, the database can display a warning signal. Again, if specific capacity falls below a previously stipulated threshold value, a prompt may appear to indicate that the well is due for a CCTV inspection or even a maintenance service to remove incrustation or biofouling. Also, if a particular water quality parameter reaches a trigger threshold (say, 90% of the national regulatory maximum), the well operator can be alerted.

In summary, a well is a relatively sophisticated piece of equipment designed to abstract water efficiently from an aquifer. A modern motor vehicle is equipped with onboard monitoring – speedometer, odometer, oil pressure gauge, temperature warning – and is typically scheduled for routine inspections, checks and services. A well should be regarded in a similar way: its performance, longevity and fluid characteristics should likewise be observed and a 'log-book' kept. Warning signals should be heeded and the well should be scheduled for regular inspection and maintenance. A well represents a substantial investment for any community or utility. Sensible operational monitoring and maintenance will serve to protect that investment. As noted in the previous chapter, regular maintenance of a well based on the collection and interpretation of good operational data is likely to be much more effective in prolonging the life of that well compared to a strategy which relies on the application of rehabilitation measures when the well supply fails.

10.3 An example of a hydrogeological database – Afghanistan

For several decades, Afghanistan has suffered successive periods of civil strife and foreign military interventions. Hydrogeological science had enjoyed a brief flowering in the 1960s and 1970s, with some excellent hydrogeological maps being published in collaboration with Soviet specialists (Mishkin, 1968; Marinova, 1974; Abdullah and

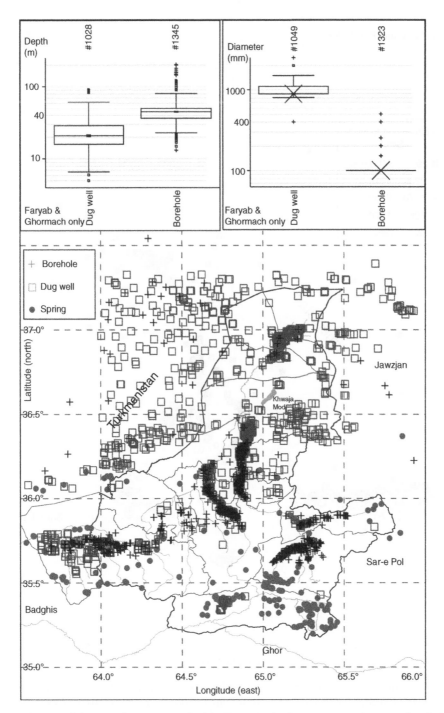

Figure 10.4 *A map of the Faryab province of Afghanistan and adjacent areas, showing the distribution of registered hand-dug wells, springs and boreholes (drilled wells) (N=3140). It will be seen that springs are typical of the mountainous south, with its karstic limestones; boreholes are typical of the long river valley corridors, while hand-dug wells are typical of the northern semi-desert areas. The insets are boxplots showing the distribution of depths and diameters of registered hand-dug wells and boreholes in Faryab. In boxplots, the 'box' shows the interquartile range, the 'whiskers' the extraquartile range and a central line indicates the median. Squares and crosses indicate near and far outliers, respectively. The hand-dug wells are typically constructed of 900 mm diameter concrete rings and most have depths in the range 7-60 m; the boreholes are typically 25–80 m deep and of 100 mm completed diameter. After Banks (2014c). Reproduced by permission of Asplan VIAK AS, trading as NORPLAN*

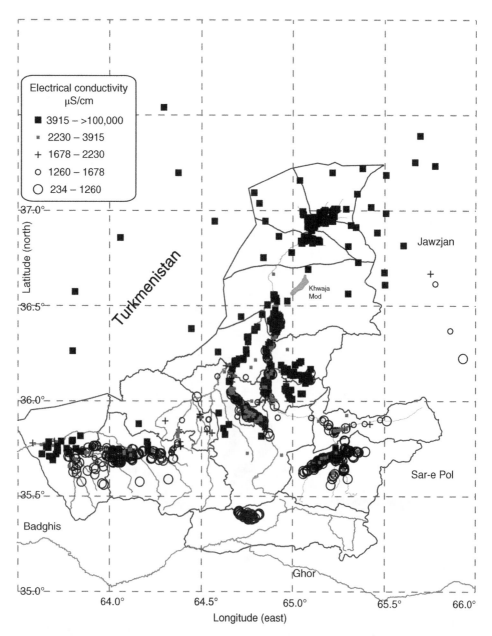

Figure 10.5 *A map of the Faryab province of Afghanistan and adjacent areas, showing the electrical conductivity of groundwater from hand-dug wells, springs and boreholes (drilled wells) (N = 2354), increasing dramatically from the mountainous south to the semi-desert north. After Banks (2014c). Reproduced by permission of Asplan VIAK AS, trading as NORPLAN*

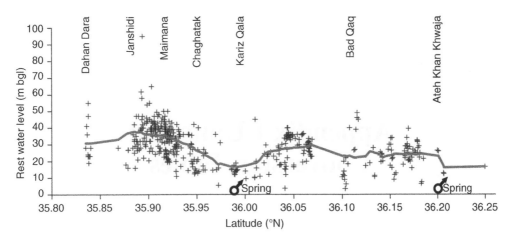

Figure 10.6 *Registered groundwater levels (crosses) in m below ground level (m bgl) in hand-dug wells and boreholes (drilled wells) along the south-north corridor of the Maimana River valley, Faryab, Afghanistan. The line shows a moving average through the data, and suggests that the water level approaches the surface in locations where major springs occur. After Banks (2014c). Reproduced by permission of Asplan VIAK AS, trading as NORPLAN*

Chmyriov, 2008). However, by the time of the Taliban government at the end of the millennium, many of the groundwater specialists had left Afghanistan and only a handful of hydrogeologists remained in the country; their attention, together with that of international non-governmental organizations, was mainly directed towards very basic rural water supply projects and trying to ensure that unmanaged well drilling did not damage groundwater resources or supplies from traditional springs, shallow hand-dug wells or *karezes* (horizontal water tunnels, also known as *qanats* or *aflaj* – Banks and Soldal, 2002; see Box 3.8).

Following the fall of the Taliban government, the science of hydrogeology in Afghanistan has been slowly renewed, such that regional groundwater assessments and modelling projects have now been completed – for example, in the Kabul basin (Tünnermeier *et al.*, 2005/2006) and in the northern province of Faryab (Banks, 2014c). Such projects owe a debt of gratitude to one particular non-governmental organization – the Danish Committee for Aid to Afghan Refugees (DACAAR) – for their

determination in carrying out hydrogeological science in Afghanistan throughout a prolonged period of civil strife. DACAAR has maintained a hydrogeological database of hand-dug and drilled wells, with geological logs and water analyses (and also an operational database, documenting routine maintenance activities), which has proved an invaluable tool to other non-governmental organizations and government ministries.

The recent hydrogeological mapping project in Faryab (Banks, 2014c) relied heavily on the well and borehole data archived by DACAAR. These data were supplemented by relatively sparse data from other organizations and by a new campaign of field registration and sampling to result in a Hydrogeological Atlas (Banks, 2014c) and a WebGIS database. Figures 10.4 to 10.6 show various methods of presenting data from the Faryab dataset, typically using techniques of non-parametric statistical analysis (i.e., relying on percentiles and medians, rather than means and standard deviations, and using presentations such as the boxplot: Banks *et al.*, 2005).

Appendix 1 Units and Conversion Tables

Length (SI unit, metre, m)

	m	ft	in
1 m	1.000	3.281	39.37
1 ft	0.3048	1.000	12.00
1 in	2.540×10^{-2}	8.333×10^{-2}	1.000

Area (SI unit, square metre, m²)

	m²	ft²	acre	hectare
1 m²	1.000	10.76	2.471×10^{-4}	1.0×10^{-4}
1 ft²	9.29×10^{-2}	1.000	2.29×10^{-5}	9.29×10^{-6}
1 acre	4.047×10^{3}	4.356×10^{4}	1.000	4.047×10^{-1}
1 hectare	1.0×10^{4}	1.076×10^{5}	2.471	1.000

Volume (SI unit, cubic metre, m³)

	m³	l	Imp. gal	US gal	ft³
1 m³	1.000	1.000×10^{3}	2.200×10^{2}	2.642×10^{2}	35.32
1 l	1.000×10^{-3}	1.000	0.2200	0.2642	3.532×10^{-2}
1 Imp. gal	4.546×10^{-3}	4.546	1.000	1.201	0.1605
1 US gal	3.785×10^{-3}	3.785	0.8327	1.000	0.1337
1 ft³	2.832×10^{-2}	28.32	6.229	7.480	1.000

Water Wells and Boreholes, Second Edition. Bruce Misstear, David Banks and Lewis Clark.
© 2017 John Wiley & Sons Ltd. Published 2017 by John Wiley & Sons Ltd.

Time (SI unit, second, s)

	s	min	h	d
1 s	1.000	1.667×10^{-2}	2.777×10^{-4}	1.157×10^{-5}
1 min	60.00	1.000	1.667×10^{-2}	6.944×10^{-4}
1 h	3.600×10^3	60.00	1.000	4.167×10^{-2}
1 d	8.640×10^4	1.440×10^3	24.00	1.000

Discharge rate (SI unit, cubic metre per second, $m^3 s^{-1}$)

	$m^3 s^{-1}$	$m^3 d^{-1}$	$l s^{-1}$	Imp. gal d^{-1}	US gal d^{-1}	$ft^3 s^{-1}$
$1 m^3 s^{-1}$	1.000	8.640×10^4	1.000×10^3	1.901×10^7	2.282×10^7	35.315
$1 m^3 d^{-1}$	1.157×10^{-5}	1.000	1.157×10^{-2}	2.200×10^2	2.642×10^2	4.087×10^{-4}
$1 l s^{-1}$	1.000×10^{-3}	86.40	1.000	1.901×10^4	2.282×10^4	3.531×10^{-2}
1 Imp.gal d^{-1}	5.262×10^{-5}	4.546×10^{-3}	5.262×10^{-5}	1.000	1.201	1.858×10^{-6}
1 US gal d^{-1}	4.381×10^{-8}	3.785×10^{-3}	4.381×10^{-5}	0.8327	1.000	1.547×10^{-6}
$1 ft^3 s^{-1}$	2.832×10^{-2}	2.447×10^3	28.32	5.382×10^5	6.463×10^5	1.000

Hydraulic conductivity (SI unit, cubic metre per second per square metre, $m^3 s^{-1} m^{-2}$ or $m s^{-1}$)

	$m s^{-1}$	$m d^{-1}$	Imp.gal $d^{-1} ft^{-2}$	US gal $d^{-1} ft^{-2}$	$ft s^{-1}$
$1 m s^{-1}$	1.000	8.640×10^4	1.766×10^6	2.12×10^6	3.281
$1 m d^{-1}$	1.157×10^{-5}	1.000	20.44	24.54	3.797×10^{-5}
1 Imp.gal $d^{-1} ft^{-2}$	5.663×10^{-7}	4.893×10^{-2}	1.000	1.201	1.858×10^{-6}
1 US gal $d^{-1} ft^{-2}$	4.716×10^{-7}	4.075×10^{-2}	0.8327	1.000	1.547×10^{-6}
$1 ft s^{-1}$	0.3048	2.633×10^4	5.382×10^5	6.463×10^5	1.000

Transmissivity (SI unit, cubic metre per second per metre, $m^3 s^{-1} m^{-1}$ or $m^2 s^{-1}$)

	$m^2 s^{-1}$	$m^2 d^{-1}$	Imp.gal $d^{-1} ft^{-1}$	US gal $d^{-1} ft^{-1}$	$ft^2 s^{-1}$
$1 m^2 s^{-1}$	1.000	8.640×10^4	5.793×10^6	6.957×10^6	10.76
$1 m^2 d^{-1}$	1.157×10^{-5}	1.000	67.05	80.52	1.246×10^{-4}
1 Imp.gal $d^{-1} ft^{-1}$	1.726×10^{-7}	1.491×10^{-2}	1.000	1.201	1.858×10^{-6}
1 US gal $d^{-1} ft^{-1}$	1.437×10^{-7}	1.242×10^{-2}	0.8327	1.000	1.547×10^{-6}
$1 ft^2 s^{-1}$	9.29×10^{-2}	8.027×10^3	5.382×10^5	6.463×10^5	1.000

Mass (SI unit, kilogram, kg)
1 kg = 2.205 pounds mass.

Force (SI unit, Newton, N)
1 N = 0.2248 pounds force.

Pressure (SI unit, pascal, $Pa = 1 Nm^{-2}$)
1 megapascal (Mpa) = 145 pounds force per square inch (psi)
101 325 pascals = 1 standard atmosphere (atm) = 1.01325 bar
9806.65 pascals = 1 m head of water = 96.78×10^{-3} atm
2989 pascals = 1 ft head of water.

Appendix 2 Hydraulic Equations for Groundwater Engineers

A2.1 Energy requirements

Input power $\dot{E}$ (W) required to pump a quantity Q (m^3 s^{-1}) of fluid against a pressure difference ΔP (Pa):

$$\dot{E} = Q \cdot \Delta P \cdot \eta = Q \cdot \Delta h_{fl} \cdot \rho \cdot g \cdot \eta$$

where ρ = fluid density, in kg m^{-3};
  η = pump efficiency as a fraction (e.g. 0.6 = 60%);
  g = acceleration due to gravity = 9.81 m s^{-2};
  Δh_{fl} = total head difference being overcome (sum of elevation and pressure head difference and frictional head loss), expressed in terms of the fluid being considered, in m.

A2.2 Turbulence

A2.2.1 Reynolds number (*Re*) for flow in pipes

$$Re = \frac{\rho \cdot \bar{v} \cdot d_{hyd}}{\mu} = \frac{\bar{v} \cdot d_{hyd}}{\nu}$$

where $\bar{v}$ is the average flow velocity (volumetric flow rate divided by pipe cross-section).

For a circular pipe $\bar{v} = 4 \cdot Q / (\pi \cdot d^2)$ and thus:

$$Re = \frac{4 \cdot \rho \cdot Q}{\pi \cdot \mu \cdot d} = \frac{2 \cdot \rho \cdot Q}{\pi \cdot \mu \cdot r}$$

where d_{hyd} = internal hydraulic diameter (m) = pipe internal diameter d for circular pipes (see below);
  r = hydraulic internal radius;
  μ = fluid dynamic viscosity (kg m^{-1} s^{-1});
  ν = fluid kinematic viscosity (m^2 s^{-1}) = μ/ρ.

For Reynolds numbers less than 2000, the flow is regarded as laminar. For $Re > 4000$, the flow is regarded as fully turbulent. For $2000 < Re < 4000$, the flow is said to be transitional turbulent. These thresholds depend somewhat on pipe surface roughness - the rougher the surface, the earlier the onset of turbulence. For relatively smooth pipes, the onset of transitional turbulent flow is often taken to be around $Re = 2300$ (Holman, 2010).

A2.2.2 Hydraulic diameter of a pipe

For a circular pipe, the hydraulic diameter d_{hyd} is equal to the internal pipe diameter d.

Water Wells and Boreholes, Second Edition. Bruce Misstear, David Banks and Lewis Clark.
© 2017 John Wiley & Sons Ltd. Published 2017 by John Wiley & Sons Ltd.

For a rectangular or square pipe, where all four sides are of similar length:

$$d_{hyd} = \frac{4 \cdot A}{Pm}$$

where A is cross-sectional area (m²) and Pm = wetted perimeter of pipe (m).

For an annulus (e.g. the casing of a well, partially occupied by an internal pipe such as a rising main):

$$d_{hyd} = d_o - d_i$$

where d_o is the inner diameter of the outside pipe and d_i is the outer diameter of the inner pipe.

A2.2.3   Turbulent flow in porous media

Smith and Sayre (1964) calculated the theoretical critical flow velocities (average advective inter-granular velocity) for water in an idealized porous medium (smooth open-packed array of uniform spheres, with a porosity of 48%), where turbulent flow commences. They derived the diagram in Figure A2.1.

It will be seen that for spheres of the size 'fine gravel' and smaller, the critical velocity is of such a magnitude that it would require a hydraulic head gradient in excess of 0.1 to achieve, which is seldom encountered in natural sub-horizontally flowing saturated groundwater systems.

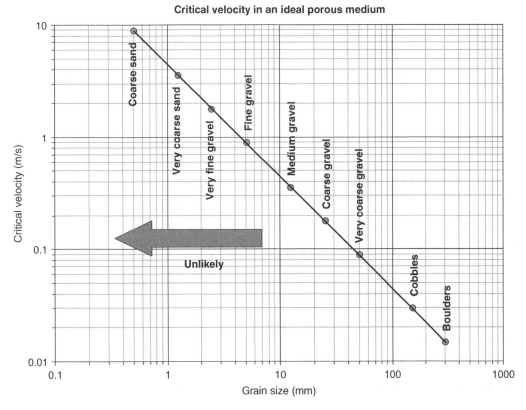

Figure A2.1   *Critical flow velocities for onset of turbulent flow in idealized porous media of differing grain sizes. The field marked 'unlikely' shows where rather high velocities would be required in fine-grained media, which are unlikely to be achieved in natural conditions. Derived from the principles of Smith and Sayre (1964)*

A2.3   Pressure Loss in Pipes

A2.3.1   Pressure loss by height gain in pipes

This is given by:

$$\Delta P = \Delta h \cdot \rho \cdot g$$

where Δh=change in elevation (m) and $\Delta P=$ pressure loss (Pa).

A2.3.2   Pressure loss by component resistance in pipes

Each component of a pipe network (valves, bends, restrictions) will have a hydraulic resistance coefficient ξ:

$$\Delta P = \frac{\xi \cdot \rho \cdot \bar{v}^2}{2}$$

where $\bar{v}$ is the average flow velocity.

A2.3.3   Pressure loss by friction in pipes

The *Darcy-Weisbach equation* (for steady, incompressible flow) states that:

$$\Delta P = \frac{\rho \cdot \lambda \cdot L \cdot \bar{v}^2}{2 \cdot d_{hyd}}$$

where L is the length of pipe (m);

d_{hyd} is the pipe hydraulic diameter=internal diameter in circular pipes (m);

ρ=fluid density, in kg m^{-3};

λ is a friction coefficient, which depends on the nature of the flow and the pipe surface - formally, λ is a function of Re (the Reynolds Number) and k_r/d_{hyd} (the roughness ratio);

k_r=absolute roughness (m) - which is often taken as being synonymous with the equivalent sand roughness k_s - although the real relationship may be somewhat more complex (Marriott and Jayaratne, 2010; McGovern, 2011; Adams *et al.*, 2012).

A2.3.4   Pressure loss by friction in rough pipes

Different internal pipe surfaces, of varying materials, will have different surface roughness. The greater the roughness ratio (k_r/d_{hyd}), the greater the frictional

Table A2.1   Absolute roughness of various pipe materials (after Chaurette, 2003; Beck and Collins, 2008 and Engineering Toolbox, 2014)

Pipe surface material	Absolute roughness (m x 10^{-6})
New copper	1 to 2
Plastic/PVC	1.5 to 7
Fibreglass	c. 5
Stainless steel	15 to 45
Commercial/welded steel	45 to 90
Corroded/rusted steel	150 to 4000
Ordinary concrete	250 to 3000

pressure loss. The equations below are for smooth pipes (roughness coefficient=c. 0), but can be modified by the incorporation of roughness into the friction coefficient λ (see Engineering Toolbox, 2014, for the relevant equations).

Table A2.1 documents some suggested absolute roughness coefficients.

A2.3.5   Pressure loss by friction in smooth circular pipes

For a circular pipe:

$$\bar{v} = 4 \cdot Q / (\pi \cdot d^2)$$

and thus

$$\Delta P = \frac{8 \cdot \rho \cdot \lambda \cdot L \cdot Q^2}{\pi^2 \cdot d^5} \text{ for a circular pipe}$$

For laminar flow:

$$\lambda = 64 / Re$$

thus

$$\Delta P = \frac{512 \cdot \rho \cdot L \cdot Q^2}{Re \cdot \pi^2 \cdot d^5} = \frac{128 \cdot \mu \cdot L \cdot Q}{\pi \cdot d^4} = \frac{8 \cdot \mu \cdot L \cdot Q}{\pi \cdot r^4}$$

For turbulent flow:

$$\frac{1}{\sqrt{\lambda}} = 2 \cdot \log_{10}\left(\frac{Re \cdot \sqrt{\lambda}}{2.51}\right)$$

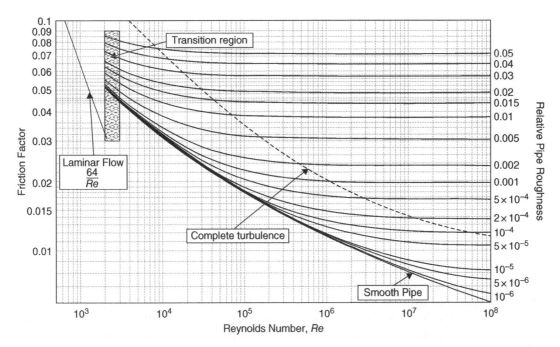

Figure A2.2 *A Moody diagram, plotting friction factor and pipe roughness against Reynolds Number. Created by Beck and Collins (2008) using the Swamee and Jain (1976) approximation. Diagram modified and reproduced under the terms of the Creative Commons Attribution-Share Alike 3.0 Unported license. Attribution: Donebythesecondlaw at the English language Wikipedia*

This is the *Prandtl* or *Colebrook-White* (Colebrook, 1939) *equation*, which cannot be solved explicitly, but can be solved diagrammatically [see the Moody (1944) diagram in Figure A2.2].

Alternatively, for turbulent flow:

$$\lambda = 0.3164 / \sqrt[4]{Re}$$

Thus:

$$\Delta P = \frac{2.5312 \cdot \rho \cdot L \cdot Q^2}{\sqrt[4]{Re} \cdot \pi^2 \cdot d^5}$$

which is known as the *Blasius approximation*. Or

$$\lambda = 0.25 \left(\log_{10} \left(\frac{5.74}{Re^{0.9}} \right) \right)^{-2}$$

This is the *Swamee-Jain* (1976) *approximation*.

A2.3.6   The Hazen-Williams equation

The Hazen-Williams equation also estimates head loss (Δh) along a pipe length L, according to an empirically derived relationship. It is only valid for water and does not take into account changes in viscosity or temperature. For a full-flowing circular pipe:

$$\frac{\Delta h}{L} = \frac{10.67 Q^{1.85}}{C^{1.85} d^{4.87}}$$

where d = internal pipe diameter (m);
  Q = flow rate ($\mathrm{m^3\,s^{-1}}$);
  C is a roughness coefficient, which is around 140 for polyethene pipes, 150 for PVC, 90–150 for steel and 60–130 for cast iron (decreasing with age and degree of rusting - Engineering Toolbox, 2014).

Appendix 3 Health and Safety Plans

Health and safety at the workplace is becoming an increasingly significant issue for all occupations, including those related to well construction. Hydrogeologists, engineers and other people involved in the siting, construction, testing or rehabilitation of water wells need to be aware of the potential risks to themselves and to members of the public from their activities, and how best to manage these risks. Although contaminated sites often pose the greatest hazards, we must recognize that all well construction projects involve some aspect of risk. We should also be aware that health and safety is a major source of legal liability for both the employee and the employer.

As we explained in the preface, this book is not intended as a well construction manual. Nor are the authors experts in health and safety policies and practices (indeed, it is clear that many of the photos included in the book predate the current requirements for the wearing of hard hats, protective boots and high visibility clothing). Whilst the importance of adopting good safety practice and using experienced personnel has been emphasized at stages throughout the text, the aim of this appendix is to provide some guidance on the need for, and the content of, health and safety plans for water well projects. The hydrogeologist or engineer should always follow the relevant legislation, regulations and codes in their own countries, as well as the guidance that may be available from their own organizations. Textbooks that contain useful sections on health and safety in relation to water wells include those by the Australian Drilling Industry Training Committee (2015) and Smith and Comeskey (2010), the latter relating more specifically to dealing with hazards surrounding well maintenance and rehabilitation.

A3.1 Scope of a health and safety plan for a water well project

The plan should set out the nature of the project, including its scope, location, timescale and main working practices. There should be a full discussion of the potential hazards and of the measures proposed to manage these hazards (Section A3.2). The plan is often developed in phases, with a preliminary plan prepared at the outset of the project

Water Wells and Boreholes, Second Edition. Bruce Misstear, David Banks and Lewis Clark.
© 2017 John Wiley & Sons Ltd. Published 2017 by John Wiley & Sons Ltd.

before the hydrogeological investigations for well siting are carried out, followed by a more detailed plan prepared when the well sites, well designs and construction methods have been finalized.

General information required in most health safety plans for water well projects will include:

1. An outline of the relevant legislation, regulations and codes of practice that apply in the country where the project is being carried out.
2. Information on the personnel who will be carrying out the work, identifying their individual responsibilities for the project and their relevant training in health and safety; for example, in the use of equipment and handling of chemicals. Contact details for staff should be included.
3. Details of project timescale, including any restrictions on working hours.
4. The scope of the project, including a brief description of the hydrogeological investigations for locating the well sites (with, for example, details of the proposed geophysical surveys); the likely numbers and location of wells; the proposed well design; the methods of well construction, testing and sampling.
5. Details of potential hazards associated with the project, including an indication of their likelihood and relative severity, followed by the measures proposed to deal with these hazards safely.
6. Procedures for safely managing drilling fluids, well test discharges and chemicals used on site.
7. Emergency response procedures in the event of a serious incident occurring, including information on first aid treatments and personnel decontamination facilities available on site, the locations and contact details for the nearest hospitals, the procedures to follow in the event of a fire, and so forth.

Good record keeping is essential for the successful operation of a health and safety plan, and therefore the plan should require the contractor (or other responsible agency) to record details of any accidents or other problems that occur during the project.

A3.2   Risk assessment

There are many hazards associated with the construction or rehabilitation of water wells, including (but not limited to):

- operating the drilling rig, which contains many moving parts that can trap clothing or fingers, or lead to other types of accidents such as eye injuries from flying debris;
- unsafe lifting and handling of heavy equipment;
- exposure to fumes and noise from the drilling rig, compressor or other equipment;
- erecting the mast of the drilling rig near high voltage power lines (this should be avoided);
- accidental contact with, and damage to, underground services (gas or water pipes, cable ducts, etc.) during drilling, or even risks from encountering unexploded ordinance (UXO);
- working on crowded sites in confined spaces; this is especially a problem in contamination investigations;
- working in areas where the public have access;
- working close to main roads, railways, rivers;
- exposure to gases or vapours from volatile chemicals, and to hazardous materials in soils such as asbestos fibres;
- transporting, storing, handling and mixing of hazardous chemicals used in well development and maintenance operations;
- exposure to vermin and disease, for example, Weil's disease from contact with rats;
- chemical and microbiological testing of water samples.

As an example of an identifiable risk during the hydrogeological investigation stage of a well project, the work programme may require river flows to be monitored as part of the groundwater recharge assessment (Section 2.6). The main hazards here are injury or drowning in the river. The risk will be lower during periods of low river flows and will increase as the flow increases. The severity of the hazard could be high (drowning). The risk management procures might include: a) there should be two people involved, one of whom should remain on the river bank and have an emergency rescue line available; b) the person doing the in-stream flow gauging should wear a life jacket; c) he/she should only wade into the river when it is safe to do so (when the river stage is below a particular level).

A second example of risk is where a well project requires entering a confined space, such as a below-ground wellhead chamber. One of the major dangers here is exposure to gases, especially carbon dioxide (an asphyxiant), hydrogen sulphide (toxic) or methane (potentially explosive if mixed in air at certain percentages). The severity of the risk is high – sadly, there have been many fatalities over the years from exposure to these gases. Therefore, entering a confined space should be avoided if possible. Where it is necessary, then only personnel with relevant health and safety training for working in confined spaces should be involved. Measures that might be recommended to manage the risk include: a) working in pairs, with only one person entering the chamber and the second person staying above ground; b) the person entering the chamber should use a gas monitor to warn of dangerous gases and wear a safety jacket with a line to facilitate rescue; c) gas masks/respirators should be available for both people if required.

Well development (Section 5.9) and, especially, well rehabilitation projects (Section 9.3) often require the use of hazardous chemicals such as acids or oxidizing agents. The risk of injury from improper transport, handling, storage or mixing of chemicals can be high, ranging from skin burns and eye injuries, to fires or explosions. The key risk management advice is that such work should only be performed by people with the necessary qualifications and experience. Specific risk management measures might include: a) always follow the in-country regulations and codes regarding the storage and handling of these chemicals (for example, corrosive chemicals should not be stored in metal drums); b) choose less hazardous chemical alternatives where available (for example, sulphamic acid is safer to handle than hydrochloric acid); c) take care in mixing different chemical solutions (add acid to water, not vice versa); d) wear protective clothing (for acidization procedures, respirators will be required); e) have wash-down facilities available for personnel on the site. The use of hazardous chemicals also presents risks to the environment and to other groundwater users, and these risks need to be considered as part of the overall project planning.

Appendix 4 World Health Organization Drinking Water Guidelines

The following tables present the most recent (2011) World Health Organization drinking water guidelines (for those parameters for which guideline values have been established). The guidelines are updated periodically by the WHO, and the reader should always consult the most up to date set of guidelines.

Table A4.1 *Guideline values for verification of microbial quality*[a]

Organisms	Guideline value
All water directly intended for drinking	
E. coli or thermotolerant coliform bacteria[b,c]	Must not be detectable in any 100-ml sample
Treated water entering the distribution system	
E. coli or thermotolerant coliform bacteria[b]	Must not be detectable in any 100-ml sample
Treated water in the distribution system	
E. coli or thermotolerant coliform bacteria[b]	Must not be detectable in any 100-ml sample

From World Health Organization (2011), reproduced by permission of the World Health Organization.
[a] Immediate investigative action must be taken if *E. coli* are detected.
[b] Although *E. coli* is the more precise indicator of faecal pollution, the count of thermotolerant coliform bacteria is an acceptable alternative. If necessary, proper confirmatory tests must be carried out. Total coliform bacteria are not acceptable as an indicator of the sanitary quality of water supplies, particularly in tropical areas, where many bacteria of no sanitary significance occur in almost all untreated supplies.
[c] It is recognized that in the great majority of rural water supplies, especially in developing countries, faecal contamination is widespread. Especially under these conditions, medium-term targets for the progressive improvement of water supplies should be set.

Water Wells and Boreholes, Second Edition. Bruce Misstear, David Banks and Lewis Clark.
© 2017 John Wiley & Sons Ltd. Published 2017 by John Wiley & Sons Ltd.

Table A4.2 *Guideline values for naturally occurring chemicals that are of health significance in drinking water*

Chemical	Guideline value[a] (mg l^{-1})	Remarks
Inorganic		
Arsenic	0.01 (A, T)	
Barium	0.7	
Boron	2.4	
Chromium	0.05 (P)	For total chromium
Fluoride	1.5	Volume of water consumed and intake from other sources should be considered when setting national standards
Selenium	0.04 (P)	
Uranium	0.03 (P)	Only chemical aspects of uranium addressed
Organic		
Microcystin-LR	0.001 (P)	For total microcystin-LR (free plus cell-bound)

From World Health Organization (2011), reproduced by permission of the World Health Organization.
[a] A, provisional guideline value because calculated guideline value is below the achievable quantification level; P, provisional guideline value because of uncertainties in the health database; T, provisional guideline value because calculated guideline value is below the level that can be achieved through practical treatment methods, source protection, and so on.

Table A4.3 *Guideline values for chemicals from industrial sources and human dwellings that are of health significance in drinking water*

Inorganics	Guideline value (mg l^{-1})	Remarks
Cadmium	0.003	
Mercury	0.006	For inorganic mercury

Organics	Guideline value[a] (µg l^{-1})	Remarks
Benzene	10[b]	
Carbon tetrachloride	4	
1,2-Dichlorobenzene	1000 (C)	
1,4-Dichlorobenzene	300 (C)	
1,2-Dichloroethane	30[b]	
1,2-Dichloroethene	50	
Dichloromethane	20	
Di(2-ethylhexyl)phthalate	8	
1,4-Dioxane	50[b]	Derived using Tolerable Daily Intake approach as well as linear multistage modelling
Edetic acid	600	Applies to the free acid
Ethylbenzene	300 (C)	
Hexachlorobutadiene	0.6	
Nitrilotriacetic acid (NTA)	200	
Pentachlorophenol	9[b] (P)	
Styrene	20 (C)	
Tetrachloroethene	40	
Toluene	700 (C)	
Trichloroethene	20 (P)	
Xylenes	500 (C)	

From World Health Organization (2011), reproduced by permission of the World Health Organization.
[a] C, concentrations of the substance at or below the health-based guideline value may affect the appearance, taste or odour of the water, leading to consumer complaints; P, provisional guideline value because of uncertainties in the health database.
[b] For non-threshold substances, the guideline value is the concentration in drinking-water associated with an upper-bound excess lifetime cancer risk of 10^{-5} (one additional cancer per 100 000 of the population ingesting drinking-water containing the substance at the guideline value for 70 years). Concentrations associated with estimated upper-bound excess lifetime cancer risks of 10^{-4} and 10^{-6} can be calculated by multiplying and dividing, respectively, the guideline value by 10.

Table A4.4 *Guideline values for chemicals from agricultural activities that are of health significance in drinking water*

Non-pesticides	Guideline value[a] (mg l^{-1})	Remarks
Nitrate (as NO$_3^-$)	50	Short-term exposure
Nitrite (as NO$_2^-$)	3	Short-term exposure; a provisional guideline value for chronic effects of nitrite that was in the 3rd edition of the WHO guidelines has been suspended and is under review owing to significant uncertainty surrounding the endogenous formation of nitrite and concentrations in human saliva.

Pesticides used in agriculture	Guideline value[a] (µg l^{-1})	Remarks
Alachlor	20[b]	
Aldicarb	10	Applies to aldicarb sulfoxide and aldicarb sulfone
Aldrin and dieldrin	0.03	For combined aldrin plus dieldrin
Atrazine and its chloro-s-triazine metabolites	100	
Carbofuran	7	
Chlordane	0.2	
Chlorotoluron	30	
Chlorpyrifos	30	
Cyanazine	0.6	
2,4-D (2,4-dichlorophenoxyacetic acid)	30	Applies to free acid
2,4-DB (2,4-Dichlorophenoxybutyric acid)	90	
1,2-Dibromo-3-chloropropane	1[b]	
1,2-Dibromoethane	0.4[b] (P)	
1,2-Dichloropropane	40 (P)	
1,3-Dichloropropene	20[b]	
Dichlorprop	100	
Dimethoate	6	
Endrin	0.6	
Fenoprop	9	
Hydroxyatrazine	200	Atrazine metabolite
Isoproturon	9	
Lindane	2	
MCPA [4-(2-Methyl-4-chlorophenoxy)acetic acid]	2	
Mecoprop	10	
Methoxychlor	20	
Metolachlor	10	
Molinate	6	
Pendimethalin	20	
Simazine	2	
2,4,5-T (2,4,5-Ttrichlorophenoxyacetic acid)	9	
Terbuthylazine	7	
Trifluralin	20	

From World Health Organization (2011), reproduced by permission of the World Health Organization.

[a] P, provisional guideline value because of uncertainties in the health database.

[b] For substances that are considered to be carcinogenic, the guideline value is the concentration in drinking-water associated with an upper-bound excess lifetime cancer risk of 10^{-5} (one additional cancer per 100 000 of the population ingesting drinking-water containing the substance at the guideline value for 70 years). Concentrations associated with estimated upper-bound excess lifetime cancer risks of 10^{-4} and 10^{-6} can be calculated by multiplying and dividing, respectively, the guideline value by 10.

Table A4.5 *Guideline values for chemicals used in water treatment or materials in contact with drinking water that are of health significance in drinking water*

Disinfectants	Guideline value[a] (mg l^{-1})	Remarks
Chlorine	5 (C)	For effective disinfection, there should be a residual concentration of free chlorine of ≥0.5 mg l^{-1} after at least 30 min contact time at pH <8.0. A chlorine residual should be maintained throughout the distribution system. At the point of delivery, the minimum residual concentration of free chlorine should be 0.2 mg l^{-1}.
Monochloramine	3	
Sodium dichloroisocyanurate	50	As sodium dichloroisocyanurate
	40	As cyanuric acid

Disinfection by-products	Guideline value[a] (µg l^{-1})	Remarks
Bromate	10[b] (A,T)	
Bromodichloromethane	60[b]	
Bromoform	100	
Chlorate	700 (D)	
Chlorite	700 (D)	
Chloroform	300	
Dibromoacetonitrile	70	
Dibromochloromethane	100	
Dichloroacetate	50[b] (D)	
Dichloroacetonitrile	20 (P)	
Monochloroacetate	20	
N-Nitrosodimethylamine	0.1	
Trichloroacetate	200	
2,4,6-Trichlorophenol	200[b] (C)	
Trihalomethanes		The sum of the ratio of the concentration of each to its respective guideline value should not exceed 1

Contaminants from treatment chemicals	Guideline value[a] (µg l^{-1})	Remarks
Acrylamide	0.5[b]	
Epichlorohydrin	0.4 (P)	

Contaminants from pipes and fittings	Guideline value[a] (µg l^{-1})	Remarks
Antimony	20	
Benzo[a]pyrene	0.7[b]	
Copper	2000	Staining of laundry and sanitary ware may occur below guideline value
Lead	10 (A, T)	
Nickel	70	
Vinyl chloride	0.3[b]	

From World Health Organization (2011), reproduced by permission of the World Health Organization.
[a]A, provisional guideline value because calculated guideline value is below the achievable quantification level; C, concentrations of the substance at or below the health-based guideline value may affect the appearance, taste or odour of the water, leading to consumer complaints; D, provisional guideline value because disinfection is likely to result in the guideline value being exceeded; P, provisional guideline value because of uncertainties in the health database; T, provisional guideline value because calculated guideline value is below the level that can be achieved through practical treatment methods, source control, etc.
[b]For substances that are considered to be carcinogenic, the guideline value is the concentration in drinking-water associated with an upper-bound excess lifetime cancer risk of 10^{-5} (one additional cancer per 100 000 of the population ingesting drinking-water containing the substance at the guideline value for 70 years). Concentrations associated with estimated upper-bound excess lifetime cancer risks of 10^{-4} and 10^{-6} can be calculated by multiplying and dividing, respectively, the guideline value by 10.

Table A4.6 *Guideline values for pesticides that were previously used for public health purposes and are of health significance in drinking water*

Pesticides used in water for public health purposes	Guideline value ($\mu g\, l^{-1}$)
DDT and metabolites	1

From World Health Organization (2011), reproduced by permission of the World Health Organization.

Table A4.7 *Guidance levels for radionuclides in drinking water*

Radionuclides	Guidance level ($Bq\, l^{-1}$)[a]	Radionuclides	Guidance level ($Bq\, l^{-1}$)[a]	Radionuclides	Guidance level ($Bq\, l^{-1}$)[a]
^{3}H	10 000	^{93}Mo	100	^{140}La	100
^{7}Be	10 000	^{99}Mo	100	^{139}Ce	1 000
^{14}C	100	^{96}Tc	100	^{141}Ce	100
^{22}Na	100	^{97}Tc	1 000	^{143}Ce	100
^{32}P	100	^{97m}Tc	100	^{144}Ce	10
^{33}P	1 000	^{99}Tc	100	^{143}Pr	100
^{35}S	100	^{97}Ru	1 000	^{147}Nd	100
^{36}Cl	100	^{103}Ru	100	^{147}Pm	1000
^{45}Ca	100	^{106}Ru	10	^{149}Pm	100
^{47}Ca	100	^{105}Rh	1 000	^{151}Sm	1 000
^{46}Sc	100	^{103}Pd	1 000	^{153}Sm	100
^{47}Sc	100	^{105}Ag	100	^{152}Eu	100
^{48}Sc	100	^{110m}Ag	100	^{154}Eu	100
^{48}V	100	^{111}Ag	100	^{155}Eu	1 000
^{51}Cr	10 000	^{109}Cd	100	^{153}Gd	1 000
^{52}Mn	100	^{115}Cd	100	^{160}Tb	100
^{53}Mn	10 000	^{115m}Cd	100	^{169}Er	1 000
^{54}Mn	100	^{111}In	1 000	^{171}Tm	1 000
^{55}Fe	1 000	^{114m}In	100	^{175}Yb	1 000
^{59}Fe	100	^{113}Sn	100	^{182}Ta	100
^{56}Co	100	^{125}Sn	100	^{181}W	1 000
^{57}Co	1 000	^{122}Sb	100	^{185}W	1 000
^{58}Co	100	^{124}Sb	100	^{186}Re	100
^{60}Co	100	^{125}Sb	100	^{185}Os	100
^{59}Ni	1 000	^{123m}Te	100	^{191}Os	100
^{63}Ni	1 000	^{127}Te	1 000	^{193}Os	100
^{65}Zn	100	^{127m}Te	100	^{190}Ir	100
^{71}Ge	10 000	^{129}Te	1 000	^{192}Ir	100
^{73}As	1 000	^{129m}Te	100	^{191}Pt	1 000
^{74}As	100	^{131}Te	1 000	^{193m}Pt	1 000
^{76}As	100	^{131m}Te	100	^{198}Au	100
^{77}As	1 000	^{132}Te	100	^{199}Au	1 000
^{75}Se	100	^{125}I	10	^{197}Hg	1 000
^{82}Br	100	^{126}I	10	^{203}Hg	100
^{86}Rb	100	^{129}I	1	^{200}Tl	1 000

(Continued)

Table A4.7   (*Continued*)

Radionuclides	Guidance level (Bq l^{-1})^a	Radionuclides	Guidance level (Bq l^{-1})^a	Radionuclides	Guidance level (Bq l^{-1})^a
^{85}Sr	100	^{131}I	10	^{201}Tl	1 000
^{89}Sr	100	^{129}Cs	1 000	^{202}Tl	1 000
^{90}Sr	10	^{131}Cs	1 000	^{204}Tl	100
^{90}Y	100	^{132}Cs	100	^{203}Pb	1 000
^{91}Y	100	^{134}Cs	10	^{210}Pb^b	0.1
^{93}Zr	100	^{135}Cs	100	^{206}Bi	100
^{95}Zr	100	^{136}Cs	100	^{207}Bi	100
^{93m}Nb	1 000	^{137}Cs	10	^{210}Bi^b	100
^{94}Nb	100	^{131}Ba	1 000	^{210}Po^b	0.1
^{95}Nb	100	^{140}Ba	100	^{223}Ra^b	1
^{224}Ra^b	1	^{235}U^b	1	^{242}Cm	10
^{225}Ra	1	^{236}U^b	1	^{243}Cm	1
^{226}Ra^b	1	^{237}U	100	^{244}Cm	1
^{228}Ra^b	0.1	^{238}U^{b,c}	10	^{245}Cm	1
^{227}Th^b	10	^{237}Np	1	^{246}Cm	1
^{228}Th^b	1	^{239}Np	100	^{247}Cm	1
^{229}Th	0.1	^{236}Pu	1	^{248}Cm	0.1
^{230}Th^b	1	^{237}Pu	1 000	^{249}Bk	100
^{231}Th^b	1 000	^{238}Pu	1	^{246}Cf	100
^{232}Th^b	1	^{239}Pu	1	^{248}Cf	10
^{234}Th^b	100	^{240}Pu	1	^{249}Cf	1
^{230}Pa	100	^{241}Pu	10	^{250}Cf	1
^{231}Pa^b	0.1	^{242}Pu	1	^{251}Cf	1
^{233}Pa	100	^{244}Pu	1	^{252}Cf	1
^{230}U	1	^{241}Am	1	^{253}Cf	100
^{231}U	1 000	^{242}Am	1 000	^{254}Cf	1
^{232}U	1	^{242m}Am	1	^{253}Es	10
^{233}U	1	^{243}Am	1	^{254}Es	10
^{234}U^b	1			^{254m}Es	100

From World Health Organization (2011), reproduced by permission of the World Health Organization.
^{a}Guidance levels are rounded according to averaging the log scale values (to 10^n if the calculated value was below 3×10^n and above $3 \times 10^{n-1}$).
^{b}Natural radionuclides.
^{c}The provisional guideline value for uranium in drinking-water is 30 µg l^{-1} based on its chemical toxicity for the kidney.

Appendix 5 FAO Irrigation Water Quality Guidelines

Table A5.1   *Guidelines for interpretation of water quality for irrigation*

Potential irrigation problems	Units	Degree of restriction on use		
		None	Slight to moderate	Severe
Salinity (affects crop water availability)[a]				
EC_w	dS m^{-1}	<0.7	0.7–3.0	>3.0
or TDS	mg l^{-1}	<450	450–2000	>2000
Infiltration (affects infiltration rate of water into the soil; evaluate using EC_w and SAR together)[b]				
SAR = 0–3 and EC_w =		>0.7	0.7–0.2	<0.2
SAR = 3–6 and EC_w =		>1.2	1.2–0.3	<0.3
SAR = 6–12 and EC_w =		>1.9	1.9–0.5	<0.5
SAR = 12–20 and EC_w =		>2.9	2.9–1.3	<1.3
SAR = 20–40 and EC_w =		>5.0	5.0–2.9	<2.9
Specific ion toxicity (affects sensitive crops)[c]				
Sodium (Na)				
- surface irrigation	SAR	<3	3–9	>9
- sprinkler irrigation	meq l^{-1}	<3	>3	
Chloride (Cl)				
- surface irrigation	meq l^{-1}	<4	4–10	>10
- sprinkler irrigation	meq l^{-1}	<3	>3	
Boron (B)	mg l^{-1}	<0.7	0.7–3.0	>3.0
Miscellaneous effects (on susceptible crops)				
Nitrate (NO$_3$-N)[d]	mg l^{-1}	<5	5–30	>30
Bicarbonate (HCO$_3$) (overhead sprinkling only)	meq l^{-1}	<1.5	1.5–8.5	>8.5
pH			Normal range 6.5–8.4	

Adapted from Ayers and Westcot (1985). Reproduced by permission of the *Food and Agriculture Organization of the United Nations*.

Notes:
[a] EC_w, Electrical conductivity of water, recorded at 25° C; TDS, Total dissolved solids content.
[b] SAR, sodium adsorption ratio (see Chapter 2, Section 2.7.4).
[c] See Ayers and Wescot (1985) for further information on sodium and chloride tolerances of sensitive crops, and also for information concerning trace elements other than boron.
[d] Ammonia and organic nitrogen should be included when wastewater is used for irrigation.

References

Abbott DW (2013) Wells and words: Tools in the hydrogeologist's field kit—The Imhoff Settling Cone. *Hydrovisions (Newsletter of the Groundwater Resources Association of California)* **22**(4): 15–16.

Abdul AS, Kia SF and Gibson TL (1989) Limitations of monitoring wells for the detection and quantification of petroleum products in soils and aquifers. *Ground Water Monitoring Review* Spring **1989**: 90–99.

Abdullah SH and Chmyriov VM (eds.) (2008) *Geology and Mineral Resources of Afghanistan. Volume 2: Mineral Resources of Afghanistan.* Ministry of Mines and Industries of the Democratic Republic of Afghanistan/Afghanistan Geological Survey. British Geological Survey Occasional Publication No. 15.

Acworth I (2001) The electrical image method compared with resistivity sounding and electromagnetic profiling for investigation in areas of complex geology: A case study from groundwater investigation in a weathered crystalline rock environment. *Exploration Geophysics* **32**: 119–128.

Acworth RI and Brain T (2008) Calculation of barometric efficiency in shallow piezometers using water levels, atmospheric and earth tide data. *Hydrogeology Journal* **16**: 1469–1481.

Adams B and Foster SSD (1992) Land surface zoning for groundwater protection. *Journal of the Chartered Institution of Water and Environmental Management* **6**: 312–320.

Adams T, Grant C and Watson H (2012) A simple algorithm to relate measured surface roughness to equivalent sand-grain roughness. *International Journal of Mechanical Engineering and Mechatronics* **1**: 66–71.

Ahmed MF (2001) An overview of arsenic removal technologies in Bangladesh and India. In: Ahmed MF, Ali MA and Adeel Z (eds) Proc. International Workshop on *'Technologies for Arsenic Removal from Drinking Water'*, Bangladesh University of Engineering and Technology, Dhaka and The United Nations University, Tokyo, Japan, pp. 251–269.

Akindunni FF and Gillham RW (1992) Unsaturated and saturated flow in response to pumping of an unconfined aquifer: numerical investigation of delayed drainage. *Ground Water* **30**(6): 873–884.

Ali S and Bandyopadhyay R (2013) Use of ultrasound attenuation spectroscopy to determine the size distribution of clay tactoids in aqueous suspensions. *Langmuir* **29**: 12663–12669.

Aller LT, Bennett T, Lehr JH, Petty RJ and Hackett G (1987) *DRASTIC: A Standardized System for Evaluating Ground Water Pollution Potential Using Hydrogeologic Settings.* US Environmental Protection Agency EPA/600/2-87/035.

Aller L, Bennett TW, Hackett G, Petty RJ, Lehr JH, Sidoris H, Nielsen DM and Denne J (1991) *Handbook of Suggested Practices for the Design and Installation of Ground-Water Monitoring Wells.* US Environmental Protection Agency EPA160014-891034.

American Petroleum Institute (2012) *API Specification for Line Pipe*, 45th edn. API Specification 5L.

American Society of Civil Engineers (1996) *Operation and Maintenance of Ground Water Facilities*. ASCE Manuals and Reports on Engineering Practice 86, New York, USA.

American Society of Civil Engineers (2014) *Hydraulics of Wells: Design, Construction, Testing, and Maintenance of Water Well Systems*. ASCE, Reston, USA.

American Society for Testing and Materials (2008) *Standard test method for determining unsaturated and saturated hydraulic conductivity in porous media by steady-state centrifugation*. ASTM D6527-00.

American Society for Testing and Materials (2009a) *Standard Specification for Welded Unannealed Austenitic Stainless Steel Tubular Products*. ASTM A778-01 (2009).

American Society for Testing and Materials (2009b) *Practice for Description and Identification of Soils (Visual-Manual Procedure)*. ASTM D2488-09a.

American Society for Testing and Materials (2010a) *Standard Specification for Electric-Fusion (Arc)-Welded Steel Pipe (NPS 4 and Over)*. ASTM A139-04 (2010).

American Society for Testing and Materials (2010b) *Standard Guide for Selection of Aquifer Test Method in Determining Hydraulic Properties by Well Techniques*. ASTM D4043-96(2010)E1.

American Society for Testing and Materials (2010c) *Standard Practice for Design and Installation of Groundwater Monitoring Wells*. ASTM D5092-04(2010)E1.

American Society for Testing and Materials (2011) *Standard Test Method for Penetration Test and Split-Barrel Sampling of Soils*. ASTM D1586-11.

American Society for Testing and Materials (2012a) *Standard Specification for Pipe, Steel, Black and Hot-Dipped, Zinc-Coated, Welded and Seamless*. ASTM A53-12.

American Society for Testing and Materials (2012b) *Standard Specification for Seamless and Welded Carbon Steel Water-Well Pipe*. ASTM A589-06 (2012).

American Society for Testing and Materials (2012c). *Standard Guide for Decommissioning of Groundwater Wells, Vadose Zone Monitoring Devices, Boreholes, and Other Devices for Environmental Activities*. ASTM D5299-99(2012)e1.

American Society for Testing and Materials (2014a) *Standard Specification for Thermoplastic Well Casing Pipe and Couplings made in Standard Dimension Ratios (SDR), SCH40 and SCH80*. ASTM F480-14.

American Society for Testing and Materials (2014b). *Standard Guide for Conceptualization and Characterization of Groundwater Systems*. ASTM D5979-96(2014).

American Society for Testing and Materials (2015) *Standard Specification for Seamless, Welded and Heavily Cold Worked Austenitic Stainless Steel Tubular Products*. ASTM A312-15.

American Water Works Association (1998) *AWWA Standard for Water Wells*. ANSI/AWWA A100-97, AWWA, Denver.

American Water Works Association (2006) *AWWA Standard for Water Wells*. ANSI/AWWA A100-06, AWWA, Denver.

American Water Works Association (2011) *Field Welding of Steel Water Pipe*. AWWA C206-11, AWWA, Denver.

American Water Works Association (2012) *Steel Water Pipe: 6 in and Larger*. AWWA C200-12, AWWA, Denver.

Anderson MP (2005) Heat as a ground water tracer. *Ground Water* **43**(6): 951–968.

Anderson MP and Woessner WW (1992) *Applied Groundwater Modeling: Simulation of Flow and Advective Transport*. Academic Press Inc, San Diego.

Anderson M, Dewhurst R, Jones M and Baxter K (2005) Characterisation of turbidity and well clogging processes in a double porosity Chalk aquifer during the South London Artificial Recharge Scheme trials. In: '*Recharge systems for protecting and enhancing groundwater resources*'. Proceedings of the 5th International Symposium on Management of Aquifer Recharge (ISMAR5), Berlin, Germany, 11-16 June 2005. *UNESCO IHP-VI Series on Groundwater* 13: 593–598.

Andersson O (2009) The ATES project at Stockholm Arlanda Airport – technical design and environmental assessment. *Proceedings Effstock 2009 – The 11th International Conference on Thermal Energy Storage for Efficiency and Sustainability, Stockholm, Sweden*, June 14–17 2009, Session 6.3, paper 56.

Andreo B, Ravbar N and Vias JM (2009) Source vulnerability mapping in carbonate (karst) aquifers by extension of the COP method: application to pilot sites. *Hydrogeology Journal* **17**: 749–758.

Appelo CAJ and Postma D (2005) *Geochemistry, Groundwater and Pollution*, 2nd edition. AA Balkema, Leiden, Netherlands.

Archie GE (1942) The electrical resistivity log as an aid in determining some reservoir characteristics. *Transactions of the American Institute of Mining, Metallurgical and Petroleum Engineers (AIMME)* **146**: 54–62.

Asaba RB, Fagan GH, and Kabonesa C (2015) Women's access to safe water and participation in community management of supply. In: Fagan GH, Linnane S, McGuigan KG and Rugumayo AI (eds) '*Water is Life: Progress to secure safe water provision in rural Uganda*', Practical Action Publishing, Rugby, United Kingdom, pp. 15–29.

Asikainen M and Kahlos H (1979) Anomalously high concentrations of uranium, radium and radon in water from drilled wells in the Helsinki region. *Geochimica et Cosmochimica Acta* **43**: 1681–1686.

Attwa M, Basokur AT and Akca I (2014) Hydraulic conductivity estimation using direct current (DC) sounding data: a case study in East Nile Delta, Egypt. *Hydrogeology Journal* **22**: 1163–1178.

Australian Drilling Industry Training Committee Limited (2015) *The Drilling Manual*, 5th edn. CRC Press, Taylor & Francis Group, Florida.

Australian Standards – see Standards Australia International.

Ayers RS and Westcot DW (1985) *Water Quality for Agriculture*. UN-FAO Irrigation and Drainage Paper 29, Rome.

Bakiewicz W, Milne DM and Noori M (1982) Hydrogeology of the Umm Er Radhuma aquifer, Saudi Arabia, with reference to fossil gradients. *Quarterly Journal of Engineering Geology* **15**: 105–126.

Bakiewicz W, Milne DM and Pattle AD (1985) Development of public tubewell designs in Pakistan. *Quarterly Journal of Engineering Geology* **18**: 63–77.

Ball DF and Herbert R (1992) The use and performance of collector wells within the regolith aquifer of Sri Lanka. *Ground Water* **30**(5): 683–689.

Banks DC (1972) In-situ measurements of permeability in granite. In: Proc. Symposium on '*Percolation through Fissured Rock*', Section TI-A, International Society for Rock Mechanics, Stuttgart, Germany.

Banks D (1984) *An investigation into the suitability of glacial sands, clays and gravels for use as basal liners of a landfill site*. Unpublished MSc dissertation, University of Birmingham, UK.

Banks D (1989) Appropriate Geology in Africa. *British Geologist* **15**(2): 65–70.

Banks D (1992a) Optimal orientation of water-supply boreholes in fractured aquifers. *Ground Water* **30**: 895–900.

Banks D (1992b) Estimation of apparent transmissivity from capacity-testing of boreholes in bedrock aquifers. *Applied Hydrogeology* **4**: 5–19.

Banks D (2004) Monitoring of fresh and brackish water resources. In: '*Encyclopaedia of Life Support Systems (EOLSS)*', Topic Level Contribution E4.18.03. UNESCO, Paris.

Banks D (2009a) An introduction to "thermogeology" and the exploitation of ground source heat. *Quarterly Journal of Engineering Geology and Hydrogeology* **42**: 283–293.

Banks D (2009b) Thermogeological assessment of open-loop well-doublet schemes: a review and synthesis of analytical approaches. *Hydrogeology Journal* **17**: 1149–1155.

Banks D (2011) The application of analytical solutions to the thermal plume from a well doublet ground source heating or cooling scheme. *Quarterly Journal of Engineering Geology and Hydrogeology* **44**: 191–197.

Banks D (2012a) *An Introduction to Thermogeology: Ground Source Heating and Cooling.* 2nd Edition. John Wiley & Sons, Chichester, UK.

Banks D (2012b) From Fourier to Darcy, from Carslaw to Theis: the analogies between the subsurface behaviour of water and heat [Da Fourier a Darcy, da Carslaw a Theis: le analogie del comportamento delle acque e del calore nel sottosuolo]. *Acque Sotterranee (Italian Journal of Groundwater)*, **1**(3): 9–19.

Banks D (2014a) William Thomson – father of thermogeology. Published online in *Scottish Journal of Geology*. doi: 10.1144/sjg2013-017

Banks D (2014b) Horatio Scott Carslaw and the origins of the well function and line source heat function. Published online in *Scottish Journal of Geology*. doi: 10.1144/sjg2014-021

Banks D (2014c) *A Hydrogeological Atlas of Faryab Province, Northern Afghanistan.* NORPLAN report for Afghan Ministry of Rural Rehabilitation and Development, funded by NORAD. Published November 2014 by Asplan VIAK AS, Kristiansand, Norway and available at www.holymoor.co.uk/Faryab.html or www.norplan.af.

Banks D (2015) Dr T.G.N. "Graeme" Haldane – Scottish heat pump pioneer. *The International Journal for the History of Engineering and Technology* **85**(2): 168–176.

Banks D and Mauring E (1993) Grunnvannsundersøkelser i Flatanger kommune. Oppfølging av GiN-prosjektet i Nord-Trøndelag fylke *[Groundwater investigations in Flatanger municipality: follow-up of the GiN Project in Nord-Trøndelag county – in Norwegian]*. Norges geologiske undersøkelse Rapport 93.034.

Banks D and Less C (1999) Chapter 3.3 'Well Construction'. In: Lloyd JW (ed) *'Water Resources of Hard Rock Aquifers in Arid and Semi-Arid Zones'*, UNESCO Studies and Reports in Hydrology 58, 103–127. UNESCO, Paris.

Banks D and Robins N (2002) *An introduction to groundwater in crystalline bedrock.* Norges geologiske undersøkelse, Trondheim.

Banks D and Soldal O (2002) Towards a policy for sustainable use of groundwater by non-governmental organisations in Afghanistan. *Hydrogeology Journal* **10**: 377–392.

Banks D, Solbjørg M-L and Rohr-Torp E (1992a) Permeability of fracture zones in a Precambrian granite. *Quarterly Journal of Engineering Geology* **25**: 377–388.

Banks D, Rohr-Torp E and Skarphagen H (1992b) An integrated study of a Precambrian granite aquifer, Hvaler, Southeastern Norway. *Norges geologiske undersøkelse Bulletin* **422**: 47–66.

Banks D, Cosgrove T, Harker D, Howsam P and Thatcher JP (1993a) Acidisation: borehole development and rehabilitation. *Quarterly Journal of Engineering Geology* **26**: 109–125.

Banks D, Rohr-Torp E and Skarphagen H (1993b) Groundwater chemistry in a Precambrian granite island aquifer, Hvaler, Southeastern Norway. In: Banks SB and Banks D (eds) *'Hydrogeology of Hard Rocks'*, Memoir 24th Congress of International Association of Hydrogeologists, 28 June–2 July 1993, Ås (Oslo), Norway, pp 395–406.

Banks D, Rohr-Torp E and Skarphagen H (1993c) Groundwater resources in hard rock; experiences from the Hvaler study, Southeastern Norway. In: Banks SB and Banks D (eds) *'Hydrogeology of Hard Rocks'*, Memoir 24th Congress of International Association of Hydrogeologists, 28 June–2 July 1993, Ås (Oslo), Norway, pp 39–51. Republished in *Applied Hydrogeology* 1994, 2: 33–42.

Banks D, Davies C and Davies W (1995) The Chalk as a karstic aquifer: the evidence from a tracer test at Stanford Dingley, Berkshire. *Quarterly Journal of Engineering Geology* **28**: S31–S38.

Banks D, Odling NE, Skarphagen H and Rohr-Torp E (1996) Permeability and stress in crystalline rocks. *Terra Nova* **8**: 223–235.

Banks D, Burke SP and Gray CG (1997) Hydrogeochemistry of coal mine drainage and other ferruginous waters in north Derbyshire and south Yorkshire, UK. *Quarterly Journal of Engineering Geology* **30**: 257–280.

Banks D, Frengstad B, Midtgård Aa K, Krog JR and Strand T (1998) The chemistry of Norwegian Groundwaters: I. The distribution of radon, major and minor elements in 1604 crystalline bedrock groundwaters. *The Science of the Total Environment* **222**: 71–91.

Banks D, Sæther OM, Ryghaug P and Reimann C (2001) Hydrochemical distribution patterns in stream waters, Trøndelag, Central Norway. *The Science of the Total Environment*, **267**: 1–21.

Banks D, Markland H, Smith PV, Mendez C, Rodriguez J, Huerta A and Sæther OM (2004) Distribution, salinity and pH-dependence of elements in surface waters of the catchment areas of the Salars of Coipasa and Uyuni, Bolivian Altiplano. *Journal of Geochemical Exploration* **84**: 141–166.

Banks D, Morland,G and Frengstad B (2005) Use of non-parametric statistics as a tool for the hydraulic and hydrogeochemical characterization of hard rock aquifers. *Scottish Journal of Geology*, **41**(1): 69–79.

Banks D, Fraga Pumar A and Watson I (2009) The operational performance of Scottish minewater-based ground source heat pump systems. *Quarterly Journal of Engineering Geology and Hydrogeology* **42**: 347–357.

Banks D, Gundersen P, Gustafson G, Mäkelä J and Morland G (2010). Regional similarities in the distributions of well yield from crystalline rocks in Fennoscandia. *Norges geologiske undersøkelse Bulletin* **450**: 33–47.

Banks SB and Banks D (1993) Groundwater microbiology in Norwegian hard-rock aquifers. In: Banks SB and Banks D (eds) '*Hydrogeology of Hard Rocks*', Memoir 24th Congress of International Association of Hydrogeologists, 28 June–2 July 1993, Ås (Oslo), Norway, pp 407–418.

Barcelona MJ, Gibb JP, Helfrich JA and Garske EE (1985) *Practical guide for ground-water sampling*. Illinois State Water Survey, ISWS Contract Report 374.

Barker JA (1985) Generalized well function evaluation for homogeneous and fissured aquifers. *Journal of Hydrology* **76**: 143–154.

Barker JA (1988) A generalised radial flow model for hydraulic tests in fractured rock. *Water Resources Research* **24**: 1796–1804.

Barker JA (2012) Discussion on 'The application of analytical solutions to the thermal plume from a well doublet ground source heating/cooling scheme' *Quarterly Journal of Engineering Geology and Hydrogeology* **45**: 123.

Barker JA and Black JH (1983) Slug tests in fissured aquifers. *Water Resources Research* **19**: 1558–1564.

Barker JA and Herbert R (1992a) Hydraulic tests on well screens. *Applied Hydrogeology* **0**/92: 7–19.

Barker JA and Herbert R (1992b) A simple theory for estimating well losses: with application to test wells in Bangladesh. *Applied Hydrogeology* **0/92**: 20–31.

Barker JA, Downing RA, Gray DA, Findlay J, Kellaway GA, Parker RH and Rollin KE (2000) Hydrogeothermal studies in the United Kingdom. *Quarterly Journal of Engineering Geology and Hydrogeology* **33**: 41–58.

Barker RD (1981) The offset system of electrical resistivity sounding and its use with a multicore cable. *Geophysical Prospecting* **29**: 128–143.

Barker RD (1999) Surface and borehole geophysics. Section 3.2 in: Lloyd JW (ed) '*Water resources of hard rock aquifers in arid and semi-arid zones*', UNESCO Studies and reports in hydrology 58, UNESCO, Paris, pp 72–101.

Barrash W, Clemo T, Fox JJ and Johnson TC (2006) Field, laboratory, and modelling investigation of the skin effect at wells with slotted casing, Boise Hydrogeophysical Research Site. *Journal of Hydrology* **326**: 1818–198.

Bartram J and Ballance R (1996) '*Water Quality Monitoring – A Practical Guide to the Design and Implementation of Freshwater Quality Studies and Monitoring Programme*', United Nations Environment Programme (UNEP) and the World Health Organization (WHO).

Bartram J, Mäkelä A and Mälkki E (1996) Field work and sampling. Chapter 5 *in* Bartram J and Ballance R (eds.) '*Water Quality Monitoring – A Practical Guide to the Design and Implementation of Freshwater Quality Studies and Monitoring Programme*', United Nations Environment Programme (UNEP) and the World Health Organization (WHO).

Baumann E and Furey SG (2013) *How three hand-pumps revolutionized rural water supplies: A brief history of the India Mark II/III, Afridev and Zimbabwe Bush Pump*. Rural Water Supply Network Field Note No 2013-1, SKAT Foundation, St Gallen Switzerland

Bawden GW, Sneed M, Stork SV and Galloway DL (2003) Measuring human-induced land subsidence from space. *US Geological Survey Factsheet* 069-03.

Bear J (1972) *Dynamics of Fluids in Porous Media.* Dover, New York, USA.

Beck S and Collins R (2008). *Moody diagram. Lines created using Swami and Jaine formula. Plot created on Matlab.* English Language Wikipedia website http://commons.wikimedia.org/wiki/File:Moody_diagram.jpg, accessed November 2014.

Beeson S, Misstear BDR and van Wonderen JJ (1997) Assessing the reliable outputs of groundwater sources. *Journal of the Chartered Institution of Water and Environmental Management* 11: 295–304.

Begemann HKS (1974) The Delft continuous soil sampler. *Bulletin of the International Association of Engineering Geology* 10: 35–37.

Berg SJ, Hsieh PA and Illman WA (2011) Estimating hydraulic parameters when poro-elastic effects are significant. *Ground Water* 49(6): 815–829.

Berthold S (2010) Synthetic convection log – characterization of vertical transport processes in fluid-filled boreholes. *Journal of Applied Geophysics* 72: 20–27.

BGS and DPHE (2001) *Arsenic contamination of groundwater in Bangladesh.* Kinniburgh DG and Smedley PL (eds), British Geological Survey Report WC/00/19, Keyworth, UK.

Bierschenk WH (1963) Determining well efficiency by multiple step-drawdown tests. *International Association of Scientific Hydrology Publication* 64: 493–507.

Birsoy YK and Summers WK (1980) Determination of aquifer parameters from step tests and intermittent pumping data. *Ground Water* 18: 137–146.

Bishop AW (1948) A new sampling tool for use in cohesionless sands below groundwater level. *Géotechnique* 1: 125–136.

Bishop PK, Gosk E, Burston MW and Lerner DN (1992) Level-determined groundwater sampling from open boreholes. *Quarterly Journal of Engineering Geology* 25: 145–157.

Black JH (1994) Hydrogeology of fractured rocks – a question of uncertainty about geometry. *Applied Hydrogeology* 3: 56–70.

Blasius PRH (1913) Das Aehnlichkeitsgesetz bei Reibungsvorgängen in Flüssigkeiten. *Mitteilungen über Forschungsarbeiten auf dem Gebiete des Ingenieurwesens* 131: 1–41. Verein Deutscher Ingenieure (VDI) Verlag, Berlin.

Bliss JC and Rushton KR (1984) The reliability of packer tests for estimating the hydraulic conductivity of aquifers. *Quarterly Journal of Engineering Geology and Hydrogeology* 17: 81–91.

Boak RA and Packman MJ (2001) A methodology for the assessment of risk of *Cryptosporidium* contamination of groundwater. *Quarterly Journal of Engineering Geology and Hydrogeology* 34: 187–194.

Bloetscher F, Muniz A and Witt GM (2005) *Groundwater injection: modeling, risks, and regulations.* McGraw Hill, New York, USA.

Bobeck P (2006) Henry Darcy in his own words. *Hydrogeology Journal* 14: 998–1004.

Bones J, Thomas KV and Paull B (2007) Using environmental analytical data to estimate levels of community consumption of illicit drugs and abused pharmaceuticals. *Journal of Environmental Monitoring* 9: 701–707.

Boniface ES (1959) Some experiments in artificial recharge in the lower Lee Valley. *Proceedings of the Institution of Civil Engineers* 14(4): 325–338.

Bos MG (1989). *Discharge measurement structures*, 3rd edn. International Institute for Land Reclamation and Improvement, Publication 20. ILRI, Wageningen, Netherlands.

Boulton NS (1963) Analysis of data from non-equilibrium pumping tests allowing for delayed yield from storage. *Proc. Institute of Civil Engineers* 26: 469–482.

Bourdet D, Whittle TM, Douglas AA and Pirard YM (1983) A new set of type curves simplifies well test analysis. *World Oil* May 1983: 95–106.

Bourdet D, Ayoub JA and Pirard YM (1989) Use of pressure derivative in well-test interpretation.

Society of Petroleum Engineers Formation Evaluation June 1989: 293–302.

Bouwer H (1989). The Bouwer and Rice slug test--an update. *Ground Water* **27**: 304–309.

Bouwer H and Rice RC (1976) A slug test method for determining hydraulic conductivity of unconfined aquifers with completely or partially penetrating wells. *Water Resources Research* **12**: 423–428.

Bradbury KR and Muldoon MA (1994) Effects of fracture density and anisotropy on delineation of wellhead-protection areas in fractured-rock aquifers. *Applied Hydrogeology* **2**(3): 23–30.

Brassington R (2007) *Field Hydrogeology*, 3rd edn. John Wiley & Sons Ltd, Chichester, UK.

Brassington FC and Walthall S (1985) Field techniques using borehole packers in hydrogeological investigations. *Quarterly Journal of Engineering Geology and Hydrogeology* **18**: 181–193.

Bredehoeft JD (1967) Response of well-aquifer systems to earth tides. *Journal of Geophysical Research* **72**: 3075–3087.

Bredehoeft JD (2011) Monitoring regional groundwater extraction: the problem. *Ground Water* **49**(6): 808–814.

Bredehoeft JD, Papadopulos SS and Cooper HH (1982) Groundwater: the water-budget myth. Chapter 4 in: '*Studies in Geophysics: Scientific Basis of Water Resource Management*'. National Academy Press, Washington DC, pp 51–57.

Bredenkamp DB, Botha LJ, van Tonder GJ and van Rensburg HJ (1995) *Manual on Quantitative Estimation of Groundwater Recharge and Aquifer Storativity*. Water Research Commission, Pretoria.

British Standards Institution (1983) *Code of practice for test pumping of water wells*. British Standard BS 6316:1983. [Now superseded by British Standards Institution (2003)].

British Standards Institution (1985) *Water well casing, Part 1. Specification for steel tubes for casing*. British Standard BS 879: Part 1: 1985.

British Standards Institution (1988) *Water well casing, Part 2. Specification for thermoplastic tubes for casing and slotted casing*. British Standard BS 879: Part 2: 1988.

British Standards Institution (2003) *Hydrometric determinations – Pumping tests for water wells – Considerations and guidelines for design, performance and use*. British Standard BS ISO 14686: 2003.

British Standards Institution (2005a) *Pore size distribution and porosity of solid materials by mercury porosimetry and gas adsorption – Part 1: Mercury porosimetry*. British Standard BS ISO 15901-1:2005.

British Standards Institution (2005b) *Manual methods for the measurement of a groundwater level in a well*. British Standard BS ISO 21413: 2005.

British Standards Institution (2006) *Geotechnical investigation and testing. Sampling methods and groundwater measurements. Technical principles for execution*. British Standard BS EN ISO 22475-1:2006.

British Standards Institution (2007) *Eurocode 7. Geotechnical design. Ground investigation and testing*. British Standard BS EN 1997-2:2007.

British Standards Institution (2008) *Water quality. Sampling. Guidance on sampling of bulk suspended solids*. British Standard BS ISO 5667-17:2008.

British Standards Institution (2009) *Water quality – Part 6. Sampling – Section 6.11 Guidance on sampling of groundwaters*. British Standard BS ISO 5667-11:2009/BS 6068-6.11:2009.

British Standards Institution (2010a) *Code of practice for site investigations*. British Standard BS 5930: 1999 + A2: 2010.

British Standards Institution (2010b) *Water quality – Sampling Part 22: Guidance on the design and installation of groundwater monitoring points*. British Standard BS ISO 5667-22:2010.

British Standards Institution (2012a) *Geotechnical investigation and testing – Geohydraulic testing. Part 4: Pumping tests*. British Standard BS EN ISO 22282-4:2012.

British Standards Institution (2012b) *Water quality. Sampling. Preservation and handling of water samples*. British Standard BS EN ISO 5667-3:2012.

British Standards Institution (2014a) *Soil quality – determination of dry bulk density*. British Standard BS EN ISO 11272:2014.

British Standards Institution (2014b) *Water quality. Sampling. Guidance on sampling of rivers and streams*. British Standard BS ISO 5667-6:2014.

Brown GO (2002) Henry Darcy and the making of a law. *Water Resources Research* **38**(7): 11–1 to 11–12.

Building Research Establishment (1976) *Soils and Foundations: 2*. Building Research Establishment, Digest 64, Watford, UK.

Burbey TJ and Zhang M (2010) Assessing hydrofracing success from earth tide and barometric response. *Ground Water* **48**: 825–835.

Burgess WG, Dottridge J and Symington RM (1998) Methyl tertiary butyl ether (MTBE): a groundwater contaminant of growing concern. In: Mather J, Banks D, Dumpleton S and Fermor M. (eds) '*Groundwater Contaminants and their Migration*', Geological Society, London, Special Publication **128**, pp 29–34.

Butler JJ, Garnett EJ and Healey JM (2003) Analysis of slug tests in formations of high hydraulic conductivity. *Ground Water* **41**: 620–630.

Cairncross S and Feachem R (1993) *Environmental Health Engineering in the Tropics, An Introductory Text*, 2nd edn. J Wiley & Sons, Chichester, UK.

Campbell MD and Lehr JH (1973) *Water Well Technology*. McGraw-Hill, New York.

Carlsson L and Carlstedt A (1977) Estimation of transmissivity and permeability in Swedish bedrock. *Nordic Hydrology* **8**: 103–116.

Carslaw HS (1921) *Introduction to the Mathematical Theory of the Conduction of Heat in Solids*. Macmillan, London.

Carter RC (2005) *Human-Powered Drilling Technologies: An overview of human-powered drilling technologies for shallow small diameter well construction, for domestic and agricultural water supply*. Cranfield University, UK.

Carter R, Chilton, PJ, Danert K and Olschewski A (2010) *Siting of Drilled Water Wells: A Guide for Project Managers*. Rural Water Supply Network Field Note 2010-5, St Gallen, Switzerland.

Chakraborty R (2012) Centuries old aflaj irrigation system in Oman. *ICE Proceedings, Engineering History and Heritage*, **165** (EH2): 73–80.

Chapman D (ed. 1996) *Water quality assessments – a guide to use of biota, sediments and water in environmental monitoring, 2nd edn*. Published on behalf of the United Nations Educational, Scientific and Cultural Organization (UNESCO), the World Health Organization (WHO) and the United Nations Environment Programme (UNEP) by E & F Spon, London.

Chaurette J (2003) Pipe roughness values. Available at website http://www.pumpfundamentals.com/download-free/pipe_rough_values.pdf, accessed November 2014.

Chenaf D and Chapuis RP (2007). Seepage face height, water table position, and well efficiency at steady state. *Ground Water* **45**: 168–177.

Clark J and Page R (2011) Inexpensive geophysical instruments supporting groundwater exploration in developing nations. *Journal of Water Resources and Protection* **3**: 768–780.

Clark L (1977) The analysis and planning of step drawdown tests. *Quarterly Journal of Engineering Geology* **10**: 125–143.

Clark L (1985) Groundwater abstraction from Basement Complex areas of Africa. *Quarterly Journal of Engineering Geology* **18**: 25–34.

Clark L (1988) *The Field Guide to Water Wells and Boreholes*. Geological Society of London Professional Handbook Series, John Wiley & Sons Ltd, Chichester, UK.

Clark L and Turner PA (1983) Experiments to test the hydraulic efficiency of well screens. *Ground Water* **21**(3): 270–281.

Cla-val (2012) *Hydraulic pipe valve (HPV): Aquifer thermal energy storage applications*. Information brochure HPV001DE, Cla-val Europe, Romanel/Lausanne: 4 pp.

Clayton CRI, Matthews MC and Simon NE (1995) *Site Investigation. 2nd edition*. Wiley-Blackwell, Oxford.

Clyde CG and Madabhushi GV (1983) Spacing of wells for heat pumps. *Journal of Water Resources Planning & Management* **109**(3): 203–212.

Clyn, J (2007) *The annals of Ireland*. Edited and translated by B Williams, Four Courts Press.

Colebrook CF (1939) Turbulent flow in pipes, with particular reference to the transition region between smooth and rough pipe laws. *Journal of the Institution of Civil Engineers* **11**(4): 133–156.

Collins S (2000) *Hand-Dug Shallow Wells*. SKAT, Swiss Centre for Development Cooperation in Technology and Management, St Gallen, Switzerland.

Comte J-C, Cassidy R, Nitsche J, Ofterdinger U, Pilatova K, Flynn R (2012) The typology of Irish hard-rock aquifers based on an integrated hydrogeological and geophysical approach. *Hydrogeology Journal* **20**: 1569–1588.

Connorton BJ (1988) Water management and the lithosphere: Water resources management and rising groundwater levels in the London Basin. *Revista Águas Subterrâneas (São Paulo, Brasil)* **12**(1): 10–27.

Cook PG and Solomon DK (1997) Recent advances in dating young groundwater: chlorofluorocarbons, ^{3}H/^{3}He and ^{85}Kr. *Journal of Hydrology* **191**: 245–265.

Cooper HH and Jacob CE (1946) A generalised graphical method for evaluating formation constants and summarizing well field history. *Transactions of the American Geophysical Union* **27**: 526–534.

Cooper HH, Bredehoeft JD and Papadopulos SS (1967) Response of a finite-diameter well to an instantaneous charge of water. *Water Resources Research* **3**: 263–269.

Coxon C and Drew D (2000) Interdependence of groundwater and surface water in lowland karst areas of western Ireland: management issues arising from water and contaminant transfers. In: Robins NS and Misstear BDR (eds) *'Groundwater in the Celtic Regions: Studies in Hard Rock and Quaternary Hydrogeology'*, Geological Society, London, Special Publication 182, pp 81–88.

Cronin AA and Pedley S (2002) *Microorganisms in groundwater: tracers and troublemakers*. Environment Agency R&D Technical Report P2-290/TR, Bristol, UK.

Cullimore DR (2008) *Practical Manual of Groundwater Microbiology*, 2nd edn. CRC Press, Taylor & Francis Group, London.

Cussler EL (2009) *Diffusion: Mass Transfer in Fluid Systems*. Cambridge University Press, Cambridge, UK.

Daily W, Ramirez A, Binley A and LaBrecque D (2004) Electrical resistance tomography. *The Leading Edge* **23**(5): 438–442.

Daly D and Warren WP (1998) Mapping groundwater vulnerability: the Irish perspective. In: Robins NS (ed.) *'Groundwater Pollution, Aquifer Recharge and Vulnerability'*, Geological Society, London, Special Publication 130, pp179–190.

Danert K (2015) *Manual Drilling Compendium 2015*. Rural Water Supply Network Publication 2015-2, Skat Foundation, St Gallen, Switzerland.

Danert K, Carter RC, Adekile D, MacDonald A (2008) Cost-effective boreholes in sub-Saharan Africa. In Proc 33rd WEDC International Conference, "*Access to Sanitation and Safe Water: Global Partnerships and Local Actions*", Accra, Ghana, 2008.

Daniel DE, Anderson DC and Boynton SS (1985) Fixed-wall versus flexible-wall permeameters. *In*: Johnson AI, Frobel RK, Cavalli NJ and Pettersson CB (eds) *Hydraulic barriers in soil and rock*. American Society for Testing and Materials STP 874, Philadelphia, USA, 107–126.

Darcy H (1856) *Les fontaines publiques de la Ville de Dijon*. Dalmont, Paris. English translation by Patricia Bobeck, Kendall/Hunt Publishing Co., 2004.

Darcy H (1857) *Recherches expérimentales relatives au mouvement de l'eau dans les tuyaux*. Mallet-Bachelier, Paris.

Darcy H and Bazin H (1865) *Recherches Hydrauliques, Enterprises par MH Darcy*. Imprimerie Nationale, Paris.

Darling WG and Talbot JC (2003) The O and H stable isotope composition of freshwaters in the British Isles. 1 rainfall. *Hydrology and Earth System Sciences* **7**: 163–181.

Darling WG, Bath AH and Talbot JC (2003) The O and H stable isotope composition of freshwaters in the British Isles. 2 surface waters and groundwater. *Hydrology and Earth System Sciences* **7**: 183–195.

Darling WG, Gooddy DC, MacDonald AM and Morris BL (2012) The practicalities of using CFCs and SF$_6$ for groundwater dating and tracing. *Applied Geochemistry* **27**: 1688–1697.

Darold AP, Holland AA, Morris JK and Gibson AR (2015) Oklahoma earthquake summary report 2014. *Oklahoma Geological Survey Open-File Report* OF1-2015.

Davis J and Annan AP (1989) Ground-penetrating radar for high-resolution mapping of soil and rock stratigraphy. *Geophysical Prospecting* **37**: 531–551.

Davis J and Lambert R (2002) *Engineering in emergencies*, 2nd edn. ITDG Publishing, London.

Deakin J (2000) Groundwater Protection Zone delineation at a large karst spring in western Ireland. In: Robins NS and Misstear BDR (eds) *'Groundwater in the Celtic Regions: Studies in Hard Rock and Quaternary Hydrogeology'*, Geological Society, London, Special Publication 182, pp 89–98.

De Glee GJ (1930) *Over grondwaterstromingen bij wateronttrekking door middel van putten*. Thesis, J Waltman, Delft, Netherlands.

De Glee GJ (1951) Berekningsmethoden voor de winning van grondwater. In: *'Drinkwatervoorziening'*, *Vacantie cursus* 38–80. Moorman's periodieke pers, Den Haag (Netherlands).

De Lange SS and van Tonder GJ (2000). Delineation of borehole protection zones in fractured rock aquifers. In: Sililo O *et al*. (eds) *'Groundwater: Past Achievements and Future Challenges'*, Balkema, Rotterdam, pp 741–746.

Denny SC, Allen DM and Journeay JM (2007) DRASTIC-fm: a modified vulnerability mapping method for structurally controlled aquifers in the southern Gulf Islands, British Columbia. *Hydrogeology Journal* **15**: 483–493.

Department of the Environment and Local Government, Environmental Protection Agency and Geological Survey of Ireland (1999) *A scheme for the Protection of Groundwater*. Geological Survey of Ireland, Dublin.

De Simone LA, Hamilton LC and Gilliom RJ (2009) *Quality of Water From Domestic Wells in Principal Aquifers of the United States, 1991–2004: Overview of Major Findings*. United States Geological Survey Circular 1332.

Detay M (1997) *Water Wells: Implementation, Maintenance and Restoration*. J Wiley & Sons Ltd, Masson, France.

De Vries JJ and Simmers I (2002) Groundwater recharge: an overview of processes and challenges. *Hydrogeology Journal* **10**: 5–17.

De Zwart AH (2007) *Investigation of clogging processes in unconsolidated aquifers near water supply wells*. Doctoral thesis, Department of Civil Engineering and Geosciences, Delft University of Technology, Netherlands.

DHV (2002) *National Water Supply and Sanitation Masterplan – Interim Masterplan, Volume 1: Main Report*. DHV Consultants BV Report (Environmental Support Project, Component 3) for Ministry of Water Resources, Federal Democratic Republic of Ethiopia, April 2002.

Digby A (undated) *A general guide to current geophysical borehole logging (for non-oil industry users)*. Privately published by the author and distributed by BEL Geophysical, Chesterfield, UK.

Dillon P, Pavelic P, Page D, Beringen H and Ward J (2009) Managed aquifer recharge: an introduction. *Waterlines Report Series* 13. Australian Government, National Water Commission, Canberra, Australia.

Doerfliger N, Jeannin P-Y and Zwahlen F (1999) Water vulnerability assessment in karst environments: a new method of defining protection areas using a multi-attribute approach and GIS tools (EPIK method). *Environmental Geology* **39**: 165–176.

Driscoll FG (1986) *Ground Water and Wells*, 2nd edn. Johnson Filtration Systems, St. Paul, Minnesota.

Dudeney B, Demin O and Tarasova I (2003) Control of ochreous deposits in minewater treatment. *Land Contamination & Reclamation* **11**(2): 259–266.

Dupuit J (1863) *Études théoretiques et pratiques sur le mouvement des eaux dans les canaux découverts et à travers les terrains perméables*, 2nd edn. Dunod, Paris.

Earlougher RC (1977) Advances in well test analysis. *SPE Monograph* **5**: 90–103. Society of Petroleum Engineers, New York.

Eden RN and Hazel CP (1973) Computer and graphical analysis of variable discharge pumping test of wells. *Institution of Engineers of Australia, Civil Engineering Transactions* **15**: 5–10.

Edmunds WM and Gaye CB (1994) Estimating the spatial variability of groundwater recharge in the Sahel using chloride. *Journal of Hydrology* **156**: 47–59.

Edmunds WM and Smedley PL (1996). Groundwater geochemistry and health: an overview. In: Appleton JD, Fuge R and McCall GJH (eds) *'Environmental Geochemistry and Health'*, Geological Society, London, Special Publication **113**, pp 91–105.

Ehlig-Economides CA, Hegeman P and Vik S (1994) Guidelines simplify well test interpretation. *Oil and Gas Journal* **92**(29): 33–40.

Ekelund Jr RB and Hebert RF (1999) *Secret Origins of Modern Microeconomics: Dupuit and the Engineers*. Chicago UP, USA.

Elci A, Molz III FJ and Waldrop WR (2001) Implications of observed and simulated ambient flow in monitoring wells. *Ground Water* **39**(6): 853–862.

Ellsworth WL (2013) Injection-induced earthquakes. *Science* **341**(6142): 1225942.

Elson B (1994) Low technology drilling. In: Proc. 20th WEDC Conference *'Affordable Water Supply and Sanitation'*, Colombo, Sri Lanka.

Engineering Toolbox (2014) Major loss in ducts, tubes and pipes. http://www.engineeringtoolbox.com/major-loss-ducts-tubes-d_459.html, accessed November 2014.

Enning D and Garrelfs J (2014) Corrosion of iron by sulfate-reducing bacteria: new views of an old problem. *Applied Environmental Microbiology* **80**(4): 1226–1236.

Environment Agency (2012a) *Guidance on the discharge of small quantities of substances for scientific purposes. Additional guidance for groundwater tracer tests and substances used as part of specified remediation schemes.*

Environment Agency Horizontal guidance H1 – Annex J 2. Environment Agency, Bristol, UK.

Environment Agency (2012b) *Good practice for decommissioning redundant boreholes and wells*. Environment Agency, Bristol, UK.

Environment Agency (2013) *Groundwater protection: Principles and Practice*. The Environment Agency, Bristol, UK.

Environmental Protection Agency (2009) *Code of Practice: Wastewater treatment and disposal systems serving single houses (p.e. ≤10)*. Environmental Protection Agency, Wexford, Ireland.

European Commission (1980) Directive 80/777/EEC: Council Directive of 15 July 1980 on the approximation of the laws of the Member States relating to the exploitation and marketing of natural mineral waters. *Official Journal of the European Communities* **L 229**: 1.

European Commission (1996) Directive 96/70/EC: Directive of the European Parliament and of the Council of 28 October 1996 amending Council Directive 80/777/EEC on the approximation of the laws of the Member States relating to the exploitation and marketing of natural mineral waters. *Official Journal of the European Communities* **L299**: 26–28.

European Commission (1998) Directive 98/83/EC: Council Directive of 3 November 1998 on the quality of water intended for human consumption. *Official Journal of the European Communities* **L330**: 32–54.

European Commission (2003) Directive 2003/40/EC: Council Directive of 16 May 2003 establishing the list, concentration limits and labelling requirements for the constituents of natural mineral waters and the conditions for using ozone-enriched air for the treatment of natural mineral waters and spring waters. *Official Journal of the European Communities* **L126**: 34–39.

European Commission (2013) Directive 2013/51/EURATOM: Council Directive of 22 October 2013 laying down requirements for the protection of the health of the general public with regard to radioactive substances in water intended for human consumption. *Official Journal of the European Communities* **L296**: 12–21.

Fagan B (2011) *Elixir: A human history of water.* Bloomsbury. London.

Fagerlind T (1979) Högtryckspumpning, sprängning och gradborrning. *Information från Brunnsarkivet* **2**(1979): 4–8. Sveriges geologiska undersökning, Uppsala, Sweden.

Fagerlind T (1982) Tryckning av brunnar. *Information från Brunnsarkivet* **2**(1982): 16–17. Sveriges geologiska undersökning, Uppsala, Sweden.

Faraldo Sánchez M (2007) *Thermal balance of mine-water treatment lagoons and wetlands.* MSc Dissertation in Environmental Engineering, School of Civil Engineering and Geosciences, University of Newcastle, August 2007.

Ferris JG, Knowles DB, Brown RH and Stallman RW (1962) Theory of aquifer tests. *U.S. Geological Survey Water Supply Paper* 1536-E: 174 pp.

Fetter CW (1999) *Contaminant Hydrogeology*, 2nd edn. Prentice-Hall, New Jersey.

Fetter CW (2001) *Applied Hydrogeology*, 4th edn. Pearson Education, Prentice-Hall, New Jersey.

Fewtrell L (2004) Drinking-water Nitrate, Methemoglobinemia, and Global Burden of Disease: A Discussion. *Environmental Health Perspectives* **112**: 1371–1374.

Finger J and Blankenship D (2010) *Handbook of Best Practices for Geothermal Drilling.* Sandia Report No. SAND2010-6048. Sandia National Laboratories, Albuquerque, New Mexico, USA.

Fischer E (1883) Bildung von Methylenblau als Reaktion auf Schwefelwasserstoff. *Berichte der deutschen chemischen Gesellschaft* **16**(2): 2234–2236.

Fitzsimons VP and Misstear BDR (2006) Estimating groundwater recharge through tills: a sensitivity analysis of soil moisture budgets and till properties in Ireland. *Hydrogeology Journal* **14**: 548–561.

Flanagan SV, Johnston RB and Zheng Y (2012) Arsenic in tube well water in Bangladesh: health and economic impacts and implications for arsenic mitigation. *Bulletin of the World Health Organization* **90**: 839–846.

Flavin RJ and Hawnt RJE (1979) Artificial recharge for resource storage by the Thames Water Authority in the Lee Valley, London, England. *Proc. International Symposium on Artificial Groundwater Recharge, Dortmund, Germany, 14th–18th May 1979.* Paper no. 14.

Food and Agriculture Organization (1998) *Crop Evapotranspiration: Guidelines for Computing Crop Water Requirements.* FAO Irrigation and Drainage Paper 56, Rome, Italy.

Forbes JD (1846) Some experiments on the temperature of the Earth at different depths and in different soils near Edinburgh. *Transactions of the Royal Society of Edinburgh* **16**(2): 189–236.

Forchheimer P (1930) *Hydraulik.* Teubner Verlagsgesellschaft, Stuttgart.

Förster A, Schrötter J, Merriam DF and Blackwell DD (1997) Application of optical-fiber temperature logging – an example in a sedimentary environment. *Geophysics* **62**: 1107–1113.

Foster SSD (2012) Hard-rock aquifers in tropical regions: using science to inform development and management policy. *Hydrogeology Journal* **20**: 659–672.

Foster SSD, Geake AK, Lawrence AR and Parker JM (1985) Diffuse groundwater pollution: lessons of the British experience. In: '*Hydrogeology in the Service of Man*', Memoir 18th Congress of International Association of Hydrogeologists, Cambridge, England, pp 168–177.

Foster SSD, Chilton J, Moench M, Cardy F and Schiffler M (2000) *Groundwater in Rural Development: Facing the Challenges of Supply and Resource Sustainability.* World Bank Technical Paper 463, World Bank, Washington DC.

Foster S, Hirata R and Andreo B (2013) The aquifer pollution vulnerability concept: aid or impediment in promoting groundwater protection. *Hydrogeology Journal* **21**: 1389–1392.

Freeze A (1985) Historical correspondence between C. V. Theis and C. I. Lubin. *Eos* **66**(20): 442.

Freeze RA and Cherry JA (1979) *Groundwater.* Prentice Hall, New Jersey.

Frengstad BS (2002) *Groundwater quality of crystalline bedrock aquifers in Norway.* NTNU

Trondheim (Norwegian University of Science and Technology), Department of Geology and Mineral Resources Engineering, Dr.Ing. Thesis 2002: 53.

Frengstad B and Banks D (2000) Evolution of high-pH Na-HCO₃ groundwaters in anorthosites: silicate weathering or cation exchange? In: Sililo O *et al.* (eds) *'Groundwater: Past Achievements and Future Challenges'*, Proc. XXXIInd Congress of the International Association of Hydrogeologists, Cape Town, South Africa. Balkema, Rotterdam, pp 493–498.

Frengstad B and Banks D (2014) Uranium distribution in groundwater from fractured crystalline aquifers in Norway. In: Sharp JM (ed.) *'Fractured Rock Hydrogeology'*, International Association of Hydrogeologists Selected Papers, 20, 257–276, CRC Press/Taylor & Francis, London.

Frengstad B, Midtgård Skrede Aa K, Banks D, Krog JR and Siewers U (2000) The chemistry of Norwegian groundwaters: III. The distribution of trace elements in 476 crystalline bedrock groundwaters, as analysed by ICP-MS techniques. *The Science of the Total Environment* **246**: 21–40.

Fritz SJ (1994) A survey of charge-balance errors on published analyses of potable ground and surface waters. *Ground Water* **32**: 539–546.

Furey SG (2014) *Handpumps: where now? A synthesis of online discussions (2012–2014).* Rural Water Supply Network, SKAT Foundation, St Gallen, Switzerland.

Gaciri SJ and Davies TC (1993) The occurrence and geochemistry of fluoride in some natural waters of Kenya. *Journal of Hydrology* **143**: 395–412.

Gandy CJ, Clarke L, Banks D and Younger PL (2010) Predictive modelling of groundwater abstraction and artificial recharge of cooling water. *Quarterly Journal of Engineering Geology and Hydrogeology* **43**: 279–288.

Gehlin S (2002) *Thermal response test: method development and evaluation.* Luleå University of Technology (Department of Environmental Engineering) Doctoral Thesis 2002.39.

GeoDrilling International (2002) The development of sonic drilling technology – a historical review from its origins in the late 1940s. *GeoDrilling International*, December 2002: 7–10.

Geyh MA (2000) *Environmental isotopes in the hydrological cycle. Principles and applications. Volume 4: Groundwater, saturated and unsaturated zone.* International Atomic Energy Agency (Water Resources Programme) and UNESCO.

Ghassemi F, Jakeman AJ and Nix HA (1995) *Salinisation of Land and Water Resources: Human causes, extent, management and case studies.* CAB International, Australia.

Godfrey S and Reed R (2013) *Cleaning and rehabilitating hand-dug wells.* Technical Note 1 on Drinking-Water, Sanitation and Hygiene in Emergencies, prepared for the World Health Organization by the Water, Engineering and Development Centre, Loughborough University. WHO.

Goodfellow F, Ouki SK and Murray V (2002) Permeation of organic chemicals through plastic water supply pipes. *Water and Environment Journal* **16**: 85–89.

Gretener PE (1967) On the thermal instability of large diameter wells – an observational report. *Geophysics* **32**: 727–738.

Grindley, J. (1970) Estimation and mapping of evaporation. In: *International Association of Hydrological Sciences Publication* 1, pp 200–213.

Grombach P, Haberer K, Merkl G and Trüeb EU (2000) *Handbuch der Wasserversorgungsteknik.* 3rd edn. Oldenbourg Industrieverlag, Munich, Germany.

Guérin FPM and Billaux DM (1994) On the relationship between connectivity and the continuum approximation in fracture flow and transport modelling. *Applied Hydrogeology* **2**(3): 24–31.

Gustafson G (2002) Strategies for groundwater prospecting in hard rocks: probabilistic approach. *Norges geologiske undersøkelse Bulletin* **439**: 21–25.

Haldane TGN (1930) The heat pump – an economic method of producing low-grade heat from electricity. *Journal of the Institution of Electrical Engineers* **68**/402: 666–675.

Handa BK (1975) Geochemistry and genesis of fluoride-containing ground waters in India. *Ground Water* **13**: 275–281.

Hantush MS (1956) Analysis of data from pumping tests in leaky aquifers. *American Geophysical Union Transactions* **37**: 702–714.

Hantush MS (1964) Hydraulics of wells. In: Chow VT (ed) '*Advances in Hydroscience, Volume 1*', Academic Press, New York, pp 281–432.

Hare PW and Morse RE (1997) Water-level fluctuations due to barometric pressure changes in an isolated portion of an unconfined aquifer. *Ground Water* **35**: 667–671.

Harker D (1990) Effect of acidisation on Chalk boreholes. In: Howsam P (ed) (1990) '*Water Wells: Monitoring, Maintenance, Rehabilitation*', Proc. of the International Groundwater Engineering Conference, Cranfield University, 6–8 September 1990, E & FN Spon, London, pp 158–167.

Harris SJ, Adams M and Jones MA (2005) NLARS: Evolution of an artificial recharge scheme. In: '*Recharge systems for protecting and enhancing groundwater resources*'. Proceedings of the 5th International Symposium on Management of Aquifer Recharge (ISMAR5), Berlin, Germany, 11-16 June 2005. *UNESCO IHP-VI Series on Groundwater* 13: 103–108.

Hazell JRT, Cratchley CR and Preston AM (1988) The location of aquifers in crystalline rocks and alluvium in Northern Nigeria using combined electromagnetic and resistivity techniques. *Quarterly Journal of Engineering Geology* **21**: 159–175.

Healy E (2001) *In Search of Ireland's Holy Wells*. Wolfhound Press, Dublin.

Healy RW (2010) *Estimating Groundwater Recharge*. Cambridge University Press, Cambridge, UK.

Healy RW and Cook PG (2002) Using groundwater levels to estimate recharge. *Hydrogeology Journal* **10**: 91–109.

Hearst JR, Nelson PH and Paillet FL (2000) *Well Logging for Physical Properties*. John Wiley & Sons Ltd, Chichester, UK.

Heath RC (1983) *Basic Ground-Water Hydrology*. US Geological Survey Water-Supply Paper 2220, Washington, DC, USA.

Heaton THC and Vogel JC (1981) "Excess air" in groundwater. *Journal of Hydrology* **50**: 201–216.

Heigold PC, Gilkeson RH, Cartwright K and Reed PC (1979) Aquifer transmissivity from surficial electrical methods. *Ground Water* **17**(4): 338–345.

Helweg OJ, Scott VH and Scalmanini JC (1983) *Improving Well and Pump Efficiency*. American Water Works Association, Denver.

Hem JD (1985) *Study and Interpretation of the Chemical Characteristics of Natural Water*, 3rd edn. US Geological Survey Water-Supply Paper 2254, Washington DC.

Hewitt AD (1992) Potential of common well casing materials to influence aqueous metal concentrations. *Ground Water Monitoring Review* Spring 1992: 131–136.

Hilmo BO, Sæther OM and Tvedten S (1992) Groundwater chemistry during test-pumping at Sundby, Verdal, Mid-Norway. *Norges geologiske undersøkelse Bulletin* **422**: 27–35.

Hiscock KM and Bense VF (2014) *Hydrogeology: Principles and Practice*, 2nd edn. J Wiley & Sons, UK.

Hiscock KM, Rivett MO and Davison RM (2002) Sustainable groundwater development. In: Hiscock KM, Rivett MO and Davison RM (eds) '*Sustainable Groundwater Development*', Geological Society, London, Special Publication 193, pp 1–14.

Hix GL (1995) Sand content measurements. *Water Well Journal* January 1995: 121–123.

Hobba WA, Chemerys JC, Fisher DW and Pearson FJ (1976) Geochemical and hydrologic data for wells and springs in thermal-spring areas of the Appalachians. *US Geological Survey Open File Report* 76–550, 34 pp.

Hodson JS (1837) *Repertory of Patent Inventions and Other Discoveries and Improvements in Arts, Manufactures and Agriculture*. New series, Volume 7. JS Hodson, London, UK.

Holman JP (2010) *Heat transfer*. 10th Edition. McGraw-Hill, Boston.

Hoopes JA and Harleman DR (1967) Wastewater recharge and dispersion in porous media. *Journal of the Hydraulics Division of the ASCE* **93**(5): 51–71.

Horner DR (1951) Pressure build-up in wells. *Proc. 3rd World Petroleum Congress,* Section II, 503–523.

Houben GJ (2003a) Iron oxide incrustations in wells. Part 1: genesis, mineralogy and geochemistry. *Applied Geochemistry* **18**(2003): 927–939.

Houben GJ (2003b) Iron oxide incrustations in wells. Part 2: chemical dissolution and modeling. *Applied Geochemistry* **18**(2003): 941–954.

Houben GJ (2004) Modeling the buildup of iron oxide encrustations in wells. *Ground Water* **42**(1): 78–82.

Houben GJ (2015a) Review: Hydraulics of water wells – flow laws and influence of geometry. *Hydrogeology Journal* **23**: 1633–1657.

Houben GJ (2015b) Review: Hydraulics of water wells – head losses of individual components. *Hydrogeology Journal* **23**: 1659–1675.

Houben GJ and Treskatis C (2007) *Water Well Rehabilitation and Reconstruction.* McGraw Hill, New York.

Houlsby AC (1976) Routine interpretation of the Lugeon water-test. *Quarterly Journal of Engineering Geology and Hydrogeology* **9**: 303–313.

Howard KWF and Lloyd JW (1979) The sensitivity of parameters in the Penman evaporation equations and direct recharge balance. *Journal of Hydrology* **41**: 329–344.

Howsam P (ed) (1990) *Water Wells: Monitoring, Maintenance, Rehabilitation.* Proc. of the International Groundwater Engineering Conference, Cranfield University, 6-8 September 1990, E & FN Spon, London.

Howsam P and Tyrrel S (1989) Diagnosis and monitoring of biofouling in enclosed flow systems – experience in groundwater systems. *Biofouling* **1**: 343–351.

Howsam P, Misstear BDR and Jones CRC (1995) *Monitoring, maintenance and rehabilitation of water supply boreholes.* Report 137 of the Construction Industry Research and Information Association, London.

Hrudey SE, Huck PM, Payment P, Gilham RW and Hrudey EJ (2002) Walkerton: Lessons learned in comparison with waterborne outbreaks in the developed world. *Journal of Engineering Environmental Science* **1**: 397–407.

Hsieh PA (1996) Deformation-induced changes in hydraulic head during ground-water withdrawal. *Ground Water* **34**(6): 1082–1089.

Huisman L (1972) *Groundwater Recovery.* Macmillan, London.

Hulme P, Rushton KR and Fletcher S (2001) Estimating recharge in UK catchments. In: '*Impact of Human Activity on Groundwater Dynamics*', International Association of Hydrological Sciences 269, pp 33–42.

Hunt H (2003) Leo Ranney: Industry inventor and innovator. *Water Well Journal* **57**(10): 60.

Hunter-Blair A (1970) *Well screens and gravel packs.* Water Research Association Technical Paper TP64, Bucks, UK.

Hunter Williams N, Misstear BDR, Daly D and Lee M (2013) Development of a national groundwater recharge map for the Republic of Ireland. *Quarterly Journal of Engineering Geology & Hydrogeology* **46**: 493–506.

Huntley D, Nommensen R and Steffey D (1992) The use of specific capacity to assess transmissivity in fractured rock aquifers. *Ground Water* **30**: 396–402.

Hurst A, Griffiths CM and Worthington PF (eds) (1992) *Geological applications of wireline logs II.* Geological Society, London, Special Publication 65, London.

Hurtig E, Grosswig S and Kuhn K (1995) Distributed fiber optic temperature sensing: a new tool for long-term and short-term temperature monitoring in boreholes. *Proc. World Geothermal Congress 1995*, Section 5 Drilling and Completion Technology, Paper 2-Hurtig: 1495–1498.

Hurtig E, Grosswig S and Kuhn K (1996) Fibre optic temperature sensing: application for subsurface and ground temperature measurements. *Tectonophysics* **257**: 101–109.

Hurtig E, Grosswig S and Kuhn K (1997) Distributed fibre optic temperature sensing: a new tool for long-term and short-term temperature monitoring in boreholes. *Energy Sources* **19**: 55–62.

Hussein MEA, Odling NE and Clark RA (2013), Borehole water level response to barometric pressure as an indicator of aquifer vulnerability. *Water Resources Research* **49**: 7102–7119.

Hvorslev MJ (1951) Time lag and soil permeability in groundwater observations. *US Army Corps of Engineers Waterway Experimentation Station Bulletin* **36**.

Hyder Z, Butler JJ, McElwee CD and Liu W (1994) Slug tests in partially penetrating wells. *Water Resources Research* **30**: 2945–2957.

Hynds P., Misstear B.D.R. and Gill L.W. (2012). Development of a microbial contamination susceptibility model for private domestic groundwater sources, *Water Resources Research*, **48**, *W12504*, *doi:*10.1029/2012WR012492

Hynds P, Misstear BDR and Gill LW (2013) Unregulated private wells in the Republic of Ireland: Consumer awareness, source susceptibility and protective actions. *Journal of Environmental Management* **127C**: 278–288.

IAEA/UNESCO (2000) *Environmental isotopes in the hydrological cycle. Principles and applications. Volume 1: Introduction: Theory, Methods, Review*. International Atomic Energy Agency (Water Resources Programme) and UNESCO.

IAEA (2010) *Sampling Procedures for Isotope Hydrology*. International Atomic Energy Agency, Water Resources Programme.

Institute of Geologists of Ireland (2007) *Guidelines on water well construction*. IGI, Dublin.

Institute of Hydrology (1980) *Low Flow Studies*. Research Report 3, Wallingford, UK.

Institute of Hydrology (1989) *Flow Regimes from Experimental and Network Data (FREND)*. Natural Environment Research Council, Wallingford, UK.

Institution of Civil Engineers (1989) *Specification for ground investigation with bill of quantities*. Institution of Civil Engineers, Thomas Telford, London.

Institution of Water Engineers and Scientists (1986) *Groundwater: Occurrence, Development and Protection*. IWES, London.

Inter-agency Standing Committee (2005) *Guidelines for gender-based violence interventions in humanitarian settings*. Inter-agency Standing Committee (IASC), September 2005, Geneva.

Inter-agency Standing Committee (2006) *Women, girls, boys and men: Different needs – equal opportunities*. Inter-agency Standing Committee (IASC) Gender Handbook in Humanitarian Action, December 2006, Geneva.

International Standards Organization (2010) *Water quality – Sampling – Part 22: Guidance on the design and installation of groundwater monitoring points*. ISO 5667-22:2010.

Jacob CE (1946) Radial flow in a leaky artesian aquifer. *Transactions American Geophysical Union* **27**: 198–208.

Jahangir MMR, Johnston P, Khalil MI and Richards KG (2010) Comparison of groundwater sampling methods for dissolved groundwater gas analysis. *Proc. Environ 2010, 20th Irish Environmental Researchers' Colloquium, Limerick Institute of Technology, 17th–19th February 2010, Paper PA-10*.

Jahangir MMR, Johnston P, Khalil MI, Grant J, Somers C and Richards KG (2012) Evaluation of headspace equilibration methods for quantifying greenhouse gases in groundwater. *Journal of Environmental Management* **111**: 208–212.

Javandel I and Tsang CF (1986) Capture-zone type curves: a tool for aquifer cleanup. *Ground Water* **24**: 616–625.

Jetel J and Kràsny J. (1968) Approximate aquifer characteristics in a regional hydrogeological study. *Vestnik Ustredniho ustavu geologickeho (Prague)* **43**: 459–461.

Johnson AI (1965) *Piezometers for pore-pressure measurements in fine-textured soils*. US Geological Survey Open File Report 65–83, USGS Water Resources Division, Denver.

Johnson Division (1966) *Ground Water and Wells*, 1st edn. Johnson Division, UOP, Saint Paul, USA.

Johnson RC, Kurt CE and Dunham GF (1980) Experimental determination of thermoplastic casing collapse pressures. *Ground Water* 18(4): 346–350.

Johnston K, Martin M and Higginson S (2013) Case study: Recharge of potable and tertiary-treated wastewater into a deep, confined, sandstone aquifer in Perth, Western Australia. In: Martin R (ed) *Clogging issues associated with managed aquifer recharge methods.* IAH Commission on Managing Aquifer Recharge, Australia, 174–183.

Johnston R and Heijnen H (2001) Safe water technology for arsenic removal. In: Ahmed MF, Ali MA and Adeel Z (eds) Proc. International Workshop on *'Technologies for Arsenic Removal from Drinking Water'*. Bangladesh University of Engineering and Technology, Dhaka and The United Nations University, Tokyo, Japan, pp 1–22.

Jones CRC and Singleton AJ (2000) Public water supplies from alluvial and glacial deposits in northern Scotland. In: Robins NS and Misstear BDR (eds) *'Groundwater in the Celtic Regions: Studies in Hard Rock and Quaternary Hydrogeology'*, Geological Society, London, Special Publication 182, pp 133–139.

Karami GH and Younger PL (2002) Analysing step-drawdown tests in heterogeneous aquifers. *Quarterly Journal of Engineering Geology and Hydrogeology* 35: 295–303.

Kazmann RG and Whitehead WR (1980) The spacing of heat pump supply and discharge wells. *Ground Water Heat Pump Journal* 1(2): 28–31.

Keary P, Brooks BM and Hill I (2002) *An Introduction to Geophysical Exploration.* Blackwell Science Ltd, Oxford.

Kelly BFJ, Timms WA, Andersen MS, McCallum AM, Blakers RS, Smith R, Rau GC, Badenhop A, Ludowici K, and Acworth RI (2013) Aquifer heterogeneity and response time: the challenge for groundwater management. *Crop and Pasture Science* 64: 1141–1154.

Keranen KM, Savage HM, Abers GA and Cochran ES (2013) Potentially induced earthquakes in Oklahoma, USA: Links between wastewater injection and the 2011 Mw 5.7 earthquake sequence. *Geology* 41(6): 699–702.

Kim B-W (2014) Effect of filter designs on hydraulic properties and well efficiency. *Groundwater* 52: 175–185.

King FH (1892) Observations and experiments on the fluctuations in the level and rate of movement of ground water on the Wisconsin agricultural experiment station farm, and at Whitewater, Wisconsin. *US Weather Bureau Bulletin* 5.

Kirjukhin VA, Korotkov AN and Shvartsev SL (1993) *Гидрогеохимия (Hydrogeochemistry – in Russian).* Nedra, Moscow.

Kirschner AK, Atteneder M, Schmidhuber A, Knetsch S, Farnleitner AH and Sommer R (2012) Holy springs and holy water: underestimated sources of illness? *Journal of Water and Health* 10(3): 349–357.

Klönne FW (1880) Die periodischen Schwankungen des Wasserspiegels in den inundirten Kohlenschächten von Dux in der Periode vom 8. April bis 15 September 1879. *Sitzungsberichte der Kaiserlichen Akademie der Wissenschaften mathematisch-naturwissenschaftliche (Wien)* 81: 101–116.

Knapp MF (2005) Diffuse pollution threats to groundwater: a UK water company perspective. *Quarterly Journal of Engineering Geology and Hydrogeology* 38: 39–51.

Knêžek M and Kubala P (1994) Experience with artificial groundwater recharge in Káraný. In: Reeve C and Watts J (eds) *'Groundwater – Drought, Pollution & Management'*, Balkema, Rotterdam.

Knobeloch L, Salna B, Hogan A, Postle J and Anderson H (2000) Blue Babies and Nitrate-Contaminated Well Water. *Environmental Health Perspectives* 108: 675–678.

Kràsny J (1975) Variation in transmissivity of crystalline rocks in southern Bohemia. *Vestnik Ustredniho ustavu geologickeho (Prague)* 50: 207–216.

Kresic N (2014) Experience of well drilling and testing in karst. In: Proc. International Association of Hydrogeologists (Irish Group) 21st Annual

Seminar '*Water Resources Management: The Role of Hydrogeology*', Tullamore, 15-16 April 2014, I-1–I-8.

Krige LJ (1939) Borehole temperatures in the Transvaal and Orange Free State. *Proceedings of the Royal Society of London A* **173**: 450–474.

Kristensen AH, Jensen ED, Karlsen E, Karlby H, Sørensen I, Nielsen JO, Brinck K, Ellemose K, Ramsay L, Jørgensen M, Jacobsen P, Roslev P, Schmidt S and Bai W (2014) *Vandforsyning (Water supply – in Danish)*. Nyt Teknisk Forlag. 784 pp.

Kroening DE, Sipes DS, Brame SE, Hodges RA, Price V and Temples TT (1996) The rehabilitation of monitoring wells clogged by calcite precipitation and drilling mud. *Ground Water Monitoring and Remediation* **6**: 114–123.

Kruseman G (1997) Recharge from intermittent flow. In: Simmers I (ed) '*Recharge of Phreatic Aquifers in (Semi-) Arid Areas*', International Association of Hydrogeologists 19, Balkema, Rotterdam.

Kruseman GP and De Ridder NA, with Verweij JM (1990) *Analysis and evaluation of pumping test data*, 2nd edn. International Institute for Land Reclamation and Improvement, Wageningen, Netherlands.

Kurt CE (1979) Collapse pressure of thermoplastic water well casings. *Ground Water* **17**(6): 550–555.

Lancaster-Jones PFF (1975) The interpretation of the Lugeon water-test. *Quarterly Journal of Engineering Geology and Hydrogeology* **8**: 151–154.

Langelier WF (1936) The analytical control of anti-corrosion water treatment. *Journal of the American Water Works Association* **28**: 1500–1521.

Langguth HR and Treskatis C (1989) Reverse water level fluctuations in semiconfined aquifer systems – 'Rhade effect'. *Journal of Hydrology* **109**: 79–93.

Larson TE and Skold RV (1958) Laboratory studies relating mineral quality of water to corrosion of steel and cast iron. *Corrosion* **14**(6): 285–288. Also published as Illinois State Water Survey, Circular 71, Urbana, USA.

Lawrence AR and Foster SSD (1986) Denitrification in a limestone aquifer in relation to the security of low-nitrate groundwater supplies. *Journal of the Institution of Water Engineers and Scientists* **40**(2): 159–171.

Leaf AT, Hart DJ and Bahr JM (2012) Active thermal tracer tests for improved hydrostratigraphic characterization. *Ground Water* **50**: 726–735.

Lebbe L (1995) Land subsidence due to groundwater withdrawal from the semi-confined aquifers of southwestern Flanders. In: Barends FB, Brouwer FJ, Schröder FH (eds.) 'Land Subsidence', *IAHS Publication* **234**: 47–54. International Association of Hydrological Sciences, Wallingford, UK: (IAHS).

Lerner DN (1997) Too much or too little: Recharge in urban areas. In: Chilton J *et al.* (eds) '*Groundwater in the Urban Environment, Volume 1: Problems, Processes and Management*', Balkema, Rotterdam, pp 41–47.

Lerner DN, Issar AS and Simmers I (1990) *Groundwater Recharge: A Guide to Understanding and Estimating Natural Recharge*. IAH: International Contributions to Hydrogeology 8, Heise, Germany.

Less C and Andersen N (1994) Hydrofracture – state of the art in South Africa. *Applied Hydrogeology* **2**(2): 59–63.

Lewis DC, Kriz GJ and Burgy RH (1966) Tracer dilution sampling technique to determine hydraulic conductivity of fractured rock. *Water Resources Research* **2**: 533–542.

Lgotin V and Makushin Y (1998) Groundwater monitoring to assess the influence of injection of liquid radioactive waste on the Tomsk public groundwater supply, Western Siberia, Russia. *Geological Society, London, Special Publication* **128**: 255–264.

Lightfoot DR (2000) The origin and diffusion of qanats in Arabia: new evidence from the northern and southern Peninsula. *The Geographical Journal* **166**(3): 215–226.

Limaye SD (2013) A brief history of Indian hydrogeology. In: Howden N and Mather J (eds) '*History of Hydrogeology*', International Contributions to Hydrogeology 28, International

Association of Hydrogeologists. CRC Press, Taylor & Francis Group, London, UK, pp. 127–133.

Lippmann MJ and Tsang CF (1980) Groundwater use for cooling: associated aquifer temperature changes. *Ground Water* **18**: 452–458.

Lipták BG (1999) *Instrument Engineers' Handbook*, 3rd Edition. Process Control. CRC Press, Boca Raton, USA. pp. 688–689.

Lloyd JW (ed) *Water resources of hard rock aquifers in arid and semi-arid zones*. UNESCO Studies and reports in hydrology 58, UNESCO, Paris.

Logan J (1964) Estimating transmissibility from routine production tests of water wells. *Ground Water* **2**: 35–37.

Logan P (1980) *The Holy Wells of Ireland*. Colin Smythe Ltd, Gerrards Cross, UK.

Loke MH and Barker RD (1996) Rapid least squares inversion of apparent resistivity pseudosections by a quasi-Newton method. *Geophysical Prospecting* **44**: 131–152.

Long JCS, Remer JS, Wilson CR and Witherspoon PA (1982) Porous media equivalents for networks of discontinuous fractures. *Water Resources Research* **18**(3): 645–658.

Lubwama M, Corcoran B, Kirabira JB, Sebbit A and Sayers K (2015) Improving reliability and functional sustainability of groundwater hand-pumps by coating the rubber piston seals with diamond like-carbon. In: Fagan GH, Linnane S, McGuigan KG and Rugumayo AI (eds) '*Water is Life: Progress to secure safe water provision in rural Uganda*', Practical Action Publishing, Rugby, United Kingdom, pp. 125–137.

Luo J and Kitanidis PK (2004) Fluid residence times within a recirculation zone created by an extraction–injection well pair. *Journal of Hydrology* **295**: 149–162.

MacArthur J (2015) *Handpump standardization in Sub-Saharan Africa: seeking a champion*. Rural Water Supply Network Publication 2015-1, SKAT Foundation, St Gallen Switzerland.

MacDonald AM (2003) Groundwater and rural water supply in sub-Saharan Africa. In: Proc. International Association of Hydrogeologists (Irish Group) 23rd Annual Seminar, '*Groundwater:*

Its Stakeholders', Tullamore, Ireland, 29-30 April 2003.

MacDonald AM, Burleigh J and Burgess WG (1999) Estimating transmissivity from resistivity soundings: an example from the Thames gravels. *Quarterly Journal of Engineering Geology* **32**(2): 199–206.

MacDonald AM, Davies J and Ó Dochartaigh BÉ (2002) *Simple methods for assessing groundwater resources in low permeability areas of Africa*. British Geological Survey Technical Report CR/01/168.

MacDonald AM, Davies J, Calow R and Chilton J (2005) *Developing Groundwater: A Guide for Rural Water Supply*. ITDG Publishing, Bourton-on-Dunsmore, UK.

Macfarlane AP, Förster A, Merriam DF, Schrötter J and Healey JM (2002) Monitoring artificially stimulated fluid movement in the Cretaceous Dakota aquifer, western Kansas. *Hydrogeology Journal* **10**: 662–673.

Macler BA and Merkle JC (2000) Current knowledge on groundwater microbial pathogens. *Hydrogeology Journal* **8**: 29–40.

MacMillan GJ (2009) An analytical method to determine ground water supply well network designs. *Ground Water* **47**: 822–827.

Margat J and van der Gun J (2013) *Groundwater around the World: A Geographic Synopsis*. CRC Press, Taylor & Francis Group, London, UK.

Margat J, Pennequin D and Roux J-C (2013) History of French hydrogeology. In: Howden N and Mather J (eds) '*History of Hydrogeology*', International Contributions to Hydrogeology 28, International Association of Hydrogeologists. CRC Press, Taylor & Francis Group, London, UK, pp. 59–99.

Marinova NA [Маринова НА] (ed.) (1974) Гидрогеология Азии (*Hydrogeology of Asia*). Nedra, Moscow, 576 pp.

Marriott MJ and Jayaratne R (2010) Hydraulic roughness – links between Manning's coefficient, Nikuradse's equivalent sand roughness and bed grain size. In *Proceedings of Advances in Computing and Technology*, (AC&T). The School of Computing and Technology 5th Annual Conference, University of East London: 27–32.

Martin R (ed) (2013) *Clogging issues associated with managed aquifer recharge methods*. IAH Commission on Managing Aquifer Recharge, Australia.

Martin-Hayden JM and Robbins GA (1997) Plume distortion and apparent attenuation due to concentration averaging in monitoring wells. *Ground Water* **35**: 339–346.

Mather J (1997) Collection and analysis of groundwater samples. In: Sæther OM and de Caritat P (eds) '*Geochemical processes, weathering and groundwater recharge in catchments*', Balkema, Rotterdam.

Mathias SA, Butler AP, Peach DW and Williams AT (2007) Recovering tracer test input functions from fluid electrical conductivity logging in fractured porous rocks. *Water Resources Research* **43**: W07443.

Maurice L, Barker JA, Atkinson TC, Williams AT and Smart PL (2011) A tracer methodology for identifying ambient flows in boreholes. *Ground Water* **49**: 227–238.

Maurice LD, Atkinson TC, Barker JA, Williams AT and Gallagher AJ (2012) The nature and distribution of flowing features in a weakly karstified porous limestone aquifer. *Journal of Hydrology* **438-439**: 3–15.

Mauring E, Koziel J, Lauritsen T, Rønning JS and Tønnesen JF (1994) *Målinger med georadar: Teori, anvendelse, teknikker og eksempler på opptak [Measurements with georadar: Theory, application, techniques and examples of acquired data – in Norwegian]*. Norges geologiske undersøkelse Rapport 94.024.

McAllan J, Banks D, Beyer N and Watson I (2009) Alkalinity, temporary (CO_2) and permanent acidity: an empirical assessment of the significance of field and laboratory determinations on mine waters. *Geochemistry: Exploration, Environment, Analysis* **9**: 299–312.

McClements DJ (2006) Ultrasonic measurements in particle size analysis. *In* Meyers RA (ed) *Encyclopedia of Analytical Chemistry*, John Wiley & Sons Ltd, Chichester, UK.

McCorry M and Jones GL (eds) (2011) *Geotrainet Training Manual for Designers of Shallow*

Geothermal Systems. GEOTRAINET/European Federation of Geologists, Brussels, Belgium.

McCoy WF and Costerton JW (1982) Fouling biofilm development in tubular flow systems. *Developments in Industrial Microbiology* **23**: 551–557.

McGovern J (2011) *Technical note: Friction factor diagrams for pipe flow*. Dublin Institute of Technology.

McLaughlan RG (2002) *Managing Water Well Deterioration*. IAH-publication: International Contribution to Hydrogeology 22, AA Balkema, Netherlands.

McMillan LA, Rivett MO, Tellam JH, Dumble P and Sharp H (2014) Influence of vertical flows in wells on groundwater sampling. *Journal of Contaminant Hydrology* **169**: 50–61.

Meier P, Bethmann F, Ollinger D, Zingg O, Guinot F and Ladner F (2015) Future multi-stage EGS projects in Switzerland. *Proc. 1st Schatzalp Workshop on Induced Seismicity, Davos Schatzalp, Switzerland, 10-13 March 2015*. Swiss Seismological Service/ETH Zürich.

Meijerink AMJ, Bannert D, Batelaan O, Lubczynski MW and Pointet T (2007) *Remote Sensing Applications to Groundwater*. IHP-VI Series on Groundwater No 16, UNESCO, Paris, France.

Michalski A (1989) Application of temperature and electrical-conductivity logging in ground water monitoring. *Ground Water Monitoring and Remediation* **9**(3): 112–118.

Milvy P and Cothern CR (1990) Scientific background for the development of regulations for radionuclides in drinking water. In: Cothern CR and Rebers P (eds) '*Radon, radium and uranium in drinking water*', Lewis Publishers, Chelsea, Michigan, USA, pp 1–16.

Minzdrav (2007) Предельно допустимые концентрации (ПДК) химических веществ в воде водных объектов хозяйственно-питьевого и культурно-бытового водопользования. Гигиенические нормативы ГН 2.1.5.1315-03 *[The maximum permissible concentration (MPC) of chemicals in waters for drinking and cultural/community use. Hygienic standards GN*

2.1.5.1315-03 – in Russian]. With amendments dated 28th September 2007. Минздрав России (*Ministry of Health of Russia*), Moscow, 2003.

Mishkin LP [Мышкин ЛП] (1968) Схематическая карта гидроизогипси минерализации подземных вод четвертичных отложений центральной части Северного Афганистана. Масштаб 1:500,000 (*Schematic map if hydroisohypses of groundwater mineralization in the Quaternary deposits of the central part of northern Afghanistan. Scale 1:500,000*).

Misstear BDR (2000) Groundwater recharge assessment: a key component of river basin management. In: Proc. National Hydrology Seminar on '*River Basin Management*', Irish National Committees of the International Hydrology Programme and the International Committee for Irrigation and Drainage, Tullamore, 21 November 2000, pp 52–59.

Misstear BDR (2001) The value of simple equilibrium approximations for analysing pumping test data. *Hydrogeology Journal* **9**: 125–126.

Misstear BDR (2012) Some key issues in the design of water wells in unconsolidated and fractured rock aquifers, *Acque Sotterranee, Italian Journal of Groundwater*, AS02006: 9–17.

Misstear BDR and Beeson S (2000) Using operational data to estimate the reliable yields of water wells. *Hydrogeology Journal* **8**: 177–187.

Misstear BDR and Daly D (2000) Groundwater protection in a Celtic Region: the Irish example. In: Robins NS and Misstear BDR (eds) '*Groundwater in the Celtic Regions: Studies in Hard Rock and Quaternary Hydrogeology*', Geological Society, London, Special Publication **182**, pp 53–65.

Misstear BDR and Fitzsimons V (2007) Estimating groundwater recharge in fractured bedrock aquifers in Ireland. Chapter 16 in: Krasny J, Sharp (eds) '*Groundwater in Fractured Rocks*', Special Publication 9 of the International Association of Hydrogeologists, Taylor and Francis, 243–257.

Misstear BDR, Daly EP, Daly D and Lloyd JW (1980) *The groundwater resources of the Castlecomer Plateau.* Geological Survey of Ireland Report RS 80/3.

Misstear BDR, White ME, Bishop PK and Anderson G (1996) *Reliability of sewers in environmentally vulnerable areas.* Project Report 44 of the Construction Industry Research and Information Association, London.

Misstear BDR, Ashley RP and Lawrence AR (1998a) Groundwater pollution by chlorinated solvents: the landmark Cambridge Water Company case. In: Mather J, Banks D, Dumpleton S and Fermor M (eds) '*Groundwater Contaminants and their Migration*', Geological Society, London, Special Publication **128**, pp 201–215.

Misstear BDR, Rippon PW and Ashley RP (1998b) Detection of point sources of contamination by chlorinated solvents: a case study from the Chalk aquifer of eastern England. In: Mather J, Banks D, Dumpleton S and Fermor M (eds) '*Groundwater Contaminants and their Migration*', Geological Society, London, Special Publication **128**, pp 217–228.

Misstear BDR, Brown L and Hunter Williams N (2008) Groundwater recharge to a fractured limestone aquifer overlain by glacial till in County Monaghan, Ireland, *Quarterly Journal of Engineering Geology and Hydrogeology*, 2008, **41**(4): 465–476.

Misstear BDR, Brown L and Daly D (2009a) A methodology for making initial estimates of groundwater recharge from groundwater vulnerability mapping, *Hydrogeology Journal*, **17**: 275–285.

Misstear BDR, Brown L and Johnston PM (2009b) Estimation of groundwater recharge in a major sand and gravel aquifer in Ireland using multiple approaches. *Hydrogeology Journal* **17**: 693–706.

Moench AF (1997) Flow to a well of finite diameter in a homogeneous, isotropic water table aquifer. *Water Resources Research* **33**: 1397–1407.

Moench AF and Hsieh PA (1985) Slug testing in wells with finite-thickness skin. *Proc. 10th Workshop on Geothermal Reservoir Engineering, Stanford University, California, USA, 22-24*

Jan. 1985. Stanford Geothermal Program Technical Report SGP-TR-84: 169–175.

Molz FJ and Kurt CE (1979) Grout-induced temperature rise surrounding wells. *Ground Water* **17**(3): 264–269.

Moody LF (1944) Friction factors for pipe flow. *Transactions of the ASME* **66** (8): 671–684.

Moore C, Nokes C, Loe B, Close M, Pang L, Smith V and Osbaldiston S (2010) *Guidelines for separation distances based on virus transport between on-site domestic wastewater treatment systems and wells*. Environmental Science and Research Ltd, Porirua, New Zealand.

Moore R, Kelson V, Wittman J and Rash V (2012) A modelling framework for the design of collector wells. *Ground Water* **50**: 355–366.

Moran KH and Kunz KS (1962) Basic theory of induction logging and application to study of two-coil sondes. *Geophysics* **27**: 829–858.

Morland G (1997) *Petrology, lithology, bedrock structures, glaciation and sea level. Important factors for groundwater yield and composition of Norwegian bedrock boreholes*. Norges geologiske undersøkelse Rapport 97.122 (2 volumes), Trondheim, Norway.

Mose DG, Mushrush GW and Chrosniak C (1990) Radioactive hazard of potable water in Virginia and Maryland. *Bulletin of Environmental Contamination and Toxicology* **44**: 508–513.

Moye DG (1967) Diamond drilling for foundation exploration. *Civil Engineering Transactions, Institute of Engineers (Australia)* April 1967: 95–100.

Müllern CF and Eriksson A (1977) Kapacitetsökning hos bergborrade brunnar genom högtrckspumpar. *Vannet i Norden* 2. (Referred to by Fagerlind, 1979).

Muñoz JF and Fernández B (2001) Estimate of the sustainable extraction rate for the Santiago valley aquifer. In: Seiler KP and Wohnlich S (eds) *'New approaches: characterising groundwater flow'*, Proc. XXXIst Congress International Association of Hydrogeologists, Munich, 10-14 September 2001, Balkema, Lisse, Netherlands, pp 601–605.

Murray EC (1997) *Guidelines for assessing single borehole yields in secondary aquifers*. Unpublished M.Sc. thesis, Rhodes University, South Africa.

Narasimhan TN (1998) Hydraulic characterization of aquifers, reservoir rocks, and soils: A history of ideas. *Water Resources Research* **34**: 33–46.

Nash H and Agius DA (2011) The use of stars in agriculture in Oman. *Journal of Semitic Studies* LVI/1: 167–182.

Nathan RJ and MacMahon TA (1990) Evaluation of automated techniques for baseflow and recession analysis. *Water Resources Research* **26**: 1465–1473.

National Ground Water Association (1998) *Manual of Water Well Construction Practices*, 2nd edn. NGWA, Ohio.

National Ground Water Association/ANSI (2014) *Water well construction standard*. National Ground Water Association/American National Standards Institute standard ANSI/NGWA-01-14. NGWA, Westerville, Ohio, USA.

National Research Council (2008). *Prospects for managed underground storage of recoverable water*. National Research Council, Washington, D.C.

National Uniform Drillers Licensing Committee (2012) *Minimum Construction Requirements for Water Bores in Australia*, 3rd edn. National Water Commission, Australian Government.

Nawlakhe WG, Kulkarni DN, Pathak BN and Bulusu KR (1975) Defluoridation of water by Nalgonda technique. *Indian Journal of Environmental Health* 17.1: 26–65.

Neuman SP (1972) Theory of flow in unconfined aquifers considering delayed response of the watertable. *Water Resources Research* **8**: 1031–1045.

Neuman SP (1975) Analysis of pumping test data from anisotropic unconfined aquifers considering delayed gravity response. *Water Resources Research* **11**: 329–342.

Neuman SP and Witherspoon P (1969) Theory of flow in a confined two aquifer system. *Water Resources Research* **5**: 803–816.

Neuman SP and Witherspoon P (1972) Field deter-mination of the hydraulic properties of a leaky multiple aquifer system. *Water Resources Research* **8**: 1284–1298.

Nimmer RE, Osiensky JL, Binley AM, Sprenke KF and Williams BC (2007) Electrical resistiv-ity imaging of conductive plume dilution in frac-tured rock. *Hydrogeology Journal* **15**: 877–890.

Nold JF (1980) *The Nold Well Screen Book*. JF Nold & Co, Stockstadt am Rhein, Germany.

NSF/ANSI (2012) *Drinking water treatment chemicals – health effects*. NSF International/ American National Standards Institute standard NSF/ANSI-60-2013. NSF International, Ann Arbor, Michigan, USA.

Nwankwor GI, Gillham RW, van der Kamp G and Akindunni FF (1992) Unsaturated and saturated flow in response to pumping of an unconfined aquifer: field evidence of delayed drainage. *Ground Water* **30**: 690–700.

O'Brien R, Misstear BDR, Gill LW, Deakin J and Flynn R (2013) Developing an integrated hydro-graph separation and lumped modelling approach to quantifying hydrological pathways in Irish river catchments. *Journal of Hydrology* **486**: 259–270.

O'Brien R, Misstear BDR, Gill LW, Johnston PM and Flynn R (2014) Quantifying flows along hydrological pathways by applying a new filter-ing algorithm in conjunction with master reces-sion curve analysis. *Hydrological Processes*, **28** (26): 6211–6221.

Olofsson B (2002) Estimating groundwater resources in hardrock areas – a water balance approach. *Norges geologiske undersøkelse Bulletin* **439**: 15–20.

Oosterbaan RJ and Nijland NJ (1994) Determining the saturated hydraulic conductvity. Chapter 12 *in* H.P. Ritzema (ed.) *'Drainage principles and appli-cations.'* ILRI Publication 16, 2nd revised edition. International Institute for Land Reclamation and Improvement. Wageningen, Netherlands.

Open University (1995) *Water Resources*. S268, Block 3, Physical Resources and Environment Science: a second level course, Open University, Milton Keynes, UK.

O'Shea M and Sage R (1999) Aquifer recharge: an operational drought-management strategy in North London. *Water and Environment Journal* **13**: 400–405.

O'Shea MJ, Baxter KM and Charalambous AN (1995) The hydrogeology of the Enfield-Haringey artificial recharge scheme, north London. *Quarterly Journal of Engineering Geology and Hydrogeology* **28**(Supplement 2): S115–S129.

Oxfam/Delagua (2000) *Oxfam-Delagua Portable Water Testing Kit Users Manual*, 4th edn. Robens Centre for Public and Environmental Health, Guildford, UK.

Panagopoulos GP, Antonakos AK and Lambrakis NJ (2005) Optimization of the DRASTIC method for groundwater vulnerability assess-ment via the use of simple statistical methods and GIS. *Hydrogeology Journal* **14**: 894–911.

Papadopulos IS and Cooper HH (1967) Drawdown in a well of large diameter. *Water Resources Research* **3**: 241–244.

Paradis D, Martel R, Karanta G, Lefebre R and Michaud Y (2007) Comparative study of meth-ods for WHPA delineation. *Ground Water* **45**: 158–167.

Parkhurst DL and Appelo CAJ (1999) *User's guide to PHREEQC (version 2) – a computer program for speciation, batch-reaction, one dimensional transport, and inverse geochemical calcula-tions*. US Geological Survey Water Resources Investigation Report 99–4259, 312 pp.

Parsons SB (1994) A re-evaluation of well design procedures. *Quarterly Journal of Engineering Geology* **27**: S31–S40.

Pitrak M, Mares S and Kobr M (2007) A simple borehole dilution technique in measuring hori-zontal ground water flow. *Ground Water* **45**: 89–92.

Plummer LN and Friedman LC (1999) Tracing and dating young groundwater. *US Geological Survey Factsheet* 134–199.

Plummer LN, Busenberg E and Widman PK (2004) Applications of dissolved N_2 and Ar in ground-water. *Geological Society of America, Annual Meeting & Exposition, Denver, Colorado,*

November 7-10, 2004, Abstracts with Programs **36**(5): 468.

Pomeroy R (1954) Auxiliary pretreatment by zinc acetate in sulfide analyses *Analytical Chemistry* **26**(3): 571–572.

Poulsen BK (1996) *Evaluation and optimization of the Nalgonda technique in Ngurdoto, Tanzania.* Centre for Developing Countries, Technical University of Denmark, Student report C 908767.

Puls RW and Barcelona MJ (1996) *Low-flow (minimal drawdown) ground-water sampling procedures.* Groundwater Issue Paper, US Environmental Protection Agency, EPA/540/S-95/504.

Purdin W (1980) Using non-metallic casing for geothermal wells. *Water Well Journal* **34**(4): 90–91.

Pyne RDG (2005) *Aquifer Storage Recovery: A Guide to Groundwater Recharge through Wells* 2nd edn. ASR Systems, Florida, USA.

Rackard A and O'Callaghan L (2001) *Fish, Stone, Water: Holy Wells of Ireland*. ATRIUM, Cork.

Rasmussen TC and Crawford LA (1997) Identifying and removing barometric pressure effects in confined and unconfined aquifers. *Ground Water* **35**: 502–511.

Ravenscroft P, Burgess WG, Ahmed KM, Burren M and Perrin J (2005) Arsenic in groundwater of the Bengal Basin, Bangladesh: Distribution, field relations, and hydrogeological setting. *Hydrogeology Journal* **13**: 727–751.

Ravenscroft P, Brammer H and Richards K (2009) *Arsenic pollution: a global synthesis*. Oxford: John Wiley & Sons.

Read T, Bour O, Bense V, Le Borgne T, Goderniaux P, Klepikova MV, Hochreutener R, Lavenant N and Boschero V (2013) Characterizing groundwater flow and heat transport in fractured rock using fiber-optic distributed temperature sensing. *Geophysical Research Letters* **40**: 2055–2059.

Read T, Bour, O, Selker JS, Bense VF, Le Borgne T, Hochreutener R and Lavenant N (2014) Active-distributed temperature sensing to continuously quantify vertical flow in boreholes. *Water Resources Research* **50**: 3706–3713.

Rebouças AC (1993) Groundwater development in the Precambrian shield of South America and West Side Africa. In: Banks SB and Banks D '*Hydrogeology of Hard Rocks*'. Memoir 24th Congress International Association of Hydrogeologists, 28 June–2 July 1993, Ås, Norway, Geological Survey of Norway, pp 1101–1114.

Reimann C and Banks D (2004) Setting action levels for drinking water: are we protecting our health or our economy (or our backs!)? *Science of the Total Environment* **332**: 13–21.

Reimann C, Äryäs M, Chekushin VA, Bogatyrev IV, Boyd R, de Caritat P, Dutter R, Finne TE, Halleraker JH, Jæger Ø, Kashulina G, Niskavaara H, Lehto O, Pavlov VA, Räisänen ML, Strand T and Volden T (1998) *Environmental Geochemical Atlas of the Central Barents Region*. Norges geologiske undersøkelse, Trondheim, Norway.

Reimann C, Siewers U, Skarphagen H and Banks D (1999a) Influence of filtration on concentrations of 62 elements analysed on crystalline bedrock groundwater samples by ICP-MS. *The Science of the Total Environment* **234**: 155–173.

Reimann C, Siewers U, Skarphagen H and Banks D (1999b) Do bottle type and acid washing influence trace element analyses by ICP-MS on water samples? – A test covering 62 elements and four bottle types (high density polyethene (HDPE), polypropene (PP), fluorinated ethene propene copolymer (FEP) and perfluoralkoxy polymer (PFA)). *The Science of the Total Environment* **239**: 111–130.

Renard P, Glenz D and Mejias M (2009) Understanding diagnostic plots for well-test interpretation. *Hydrogeology Journal* **17**: 589–600.

Reynolds JM (2011) *An Introduction to Applied and Environmental Geophysics*, 2nd edn. Wiley-Blackwell, Chichester, UK.

Reynolds KA, Mena KD and Gerba CP (2008) Risk of waterborne illness via drinking water in the United States. *Reviews of Environmental Contamination and Toxicology* **192**: 117–158.

Rishel JB (2001) Wire-to-water efficiency of pumping systems. *ASHRAE Journal,* April **2001**: 40–46.

Rivera A, Ledoux E and de Marsily G (1991) Nonlinear modeling of groundwater flow and total subsidence of the Mexico City aquifer-aquitard system. In: Johnson AI (ed.) *'Land Subsidence'*, Proceedings of the 4ᵗʰ International Symposium on Land Subsidence, May 1991. IAHS Publication 200: 45-58. International Association of Hydrological Sciences, Wallingford, UK: (IAHS).

Robins NS (1990) *Hydrogeology of Scotland.* British Geological Survey, Keyworth, UK.

Robins NS and Misstear BDR (2000) Groundwater in the Celtic Regions. In: Robins NS and Misstear BDR (eds) *'Groundwater in the Celtic Regions: Studies in Hard Rock and Quaternary Hydrogeology'*, Geological Society, London, Special Publication **182**, pp 5–17.

Robinson VK and Oliver D (1981) Geophysical logging of water wells. In: Lloyd JW (ed) *'Case studies in groundwater resources evaluation'*, Oxford Science Publications, UK, pp 45–64.

Robinson VK, Banks D and Morgan-Jones M (1987) *Hydrogeological investigation of the Gatehampton groundwater source 1984 to 1986.* Thames Water, Regulation and Monitoring Internal Report IR 147, UK.

Rodrigues JD (1983) The Noordbergum effect and the characterization of aquitards at the Rio Maion mining project. *Ground Water* **21**(2): 200–207.

Rorabaugh MI (1953) Graphical and theoretical analysis of step-drawdown test of artesian well. *Transactions American Society Civil Engineers* **79**, 362, 23 pp.

Roscoe Moss Company (1990) *Handbook of Ground Water Development.* John Wiley & Sons Inc, USA.

Rosén B, Gabrielsson A, Fallsvik J, Hellström G, Nilsson G (2001) System för värme och kyla ur mark – en nulägesbeskrivning *[Heating and cooling systems in the ground – a state of the art report – in Swedish].* Varia 511, Statens geotekniska institut (SGI), Linköping, Sweden.

Rossum JR (1954) Control of sand in water systems. *Journal of the American Water Works Association* **46**/2: 123–132.

Rural Water Supply Network (2009) *Myths of the rural water supply sector.* RWSN Perspectives No 4, July 2009

Rural Water Supply Network (2010) *Code of practice for cost effective boreholes.* RWSN, June 2010, British English version.

Rushton KR (2003) *Groundwater Hydrology: Conceptual and Computational Models.* John Wiley & Sons Ltd, Chichester, UK.

Rushton KR (2006) Significance of a seepage face on flows to wells in unconfined aquifers. *Quarterly Journal of Engineering Geology and Hydrogeology* **39**: 323–331.

Rutagarama U (2012) *The role of well testing in geothermal resource assessment.* MSc thesis. Faculty of Earth Sciences, University of Iceland.

Ryznar JW (1944) A new index for determining amount of calcium carbonate scale formed by water. *Journal of the American Water Works Association* **36**: 472–483.

Sacks LA (1995) Geochemical and isotopic composition of ground water with emphasis on sources of sulfate in the Upper Floridan aquifer in parts of Marion, Sumter, and Citrus counties, Florida. *United States Geological Survey Water-Resources Investigations Report* 95-4251.

Salonen L (1994) ²³⁸U series radionuclides as a source of increased radioactivity in groundwater originating from Finnish bedrock. In: Proc. IAHS Helsinki Conference *'Future Groundwater Resources at Risk'*, International Association of Hydrological Sciences Publication **222**, pp 71–84.

Sammel EA (1968) Convective flow and its effect on temperature logging in small-diameter wells. *Geophysics* **33**: 1004–1012.

Sander P (1999) Mapping methods. Section 3.1 in: Lloyd JW (ed) *'Water resources of hard rock aquifers in arid and semi-arid zones'*, UNESCO Studies and reports in hydrology 58, UNESCO, Paris, pp 41–71.

Sander P (2007) Lineaments in groundwater exploration: a review of applications and limitations. *Hydrogeology Journal* **15**: 71–74.

Sanford W (2002) Recharge and groundwater models: an overview. *Hydrogeology Journal* **10**: 110–120.

Sansom K and Koestler L (2009) *African hand-pump market mapping study*. Summary Report prepared by Delta Partnership for UNICEF WASH Section and Supply Division.

Scanlon BR, Healy RW and Cook PG (2002) Choosing appropriate techniques for quantifying groundwater recharge. *Hydrogeology Journal* **10**: 18–39.

Schlumberger (1989) *Log Interpretation Principles and Applications*. Schlumberger Ltd, Houston, USA.

Schlumberger (2014) *Diver manual*. Schlumberger Water Services, November 2014. Delft, Netherlands.

Schwartz MO (2006) Numerical modelling of groundwater vulnerability: the example Namibia. *Environmental Geology* **50**: 237–249.

Scott Keys W (1997) *A Practical Guide to Borehole Geophysics in Environmental Investigations*. CRC Press, Boca Raton, USA.

Selmer-Olsen R (1980) *Ingeniørgeologi. Del 1: Generell geologi*. Tapir, Trondheim.

SFT (1991) Veiledning for miljøtekniske grunn-undersøkelser *[Guidelines for environmental geological site investigations – in Norwegian]*. Statens forurensningstilsyn Guidance Document 91.01, Oslo.

Shakya SK, Singh SR, Anjaneyulu B and Vashisht AK (2009) Design of a low-cost bamboo well. *Ground Water* **47**: 310–313.

Shan C (1999) An analytical solution for the capture zone of two arbitrarily located wells. *Journal of Hydrology* **222**: 123–128.

Shaw E, Beven KJ, Chappell NA and Lamb R (2011) *Hydrology in Practice*, 4th edn. Spon, UK.

Shuter E and Teasdale WE (1989) Application of drilling, coring and sampling techniques to test holes and wells. Chapter F1 in '*Techniques of Water-Resource Investigations of the United States Geological Survey, Book 2: Collection of Environmental Data*'. United States Geological Survey, Denver, CO, USA.

Shuval H and Dweik H (eds.) (2007) *Water Resources in the Middle East: Israel-Palestinian Water Issues – From Conflict to Cooperation*. Springer, Berlin/Heidelberg, 375 pp.

Simmons CT (2008) Henry Darcy (1803-1858): Immortalised by his scientific legacy. *Hydrogeology Journal* **16**: 1023–1038.

Sing KSW (1975) The characterization of porous solids by gas adsorption. *Berichte der Bunsengesellschaft für physikalische Chemie* **79**(9), 724–730.

Singha K, Day-Lewis FD, Johnson T and Slater LD (2014) Advances in interpretation of subsurface processes with time-lapse electrical imaging. *Hydrological Processes*, Wiley online DOI: 10.1002/hyp.10280

Sloot R (2010) *Assessment of groundwater investigations and borehole drilling capacity in Uganda*. Report prepared by UNICEF for Ministry of Water and Environment, Uganda.

Smith B, Siegel D, Neslund C and Carter C (2014) Organic contaminants in Portland cements used in monitoring well construction. *Groundwater Monitoring & Remediation* **34**(4): 102–111.

Smith SA (1992) *Methods for Monitoring Iron and Manganese Biofouling in Water Supply Wells*. American Water Works Association, Denver.

Smith SA (2004) Microbial ecology of wells and you. *Water Well Journal* March 2004: 31–32.

Smith SA and Comeskey AE (2010) *Sustainable Wells: Maintenance, Problem Prevention, and Rehabilitation*. CRC Press, Taylor & Francis Group, Boca Raton, Florida.

Smith WO and Sayre AN (1964) Turbulence and groundwater flow. *US Geological Survey Professional Paper* 402-E.

Snow DT (1969) Anisotropic permeability of fractured media. *Water Resources Research* **5**(6): 1273–1289.

Solodov IN (1998) The retardation and attenuation of liquid radioactive wastes due to the geochemical properties of the zone of injection. *Geological Society, London, Special Publication* **128**: 265–280.

Soupios PM, Kouli M, Vallianatos F, Vafidis A and Stavroulakis G (2007) Estimation of aquifer hydraulic parameters from surficial geophysical methods: A case study of Keritis Basin in Chania (Crete-Greece). *Journal of Hydrology* **338**: 122–131.

Spane FA and Wurstner SK (1993) DERIV: A computer program for calculating pressure derivatives for use in hydraulic test analysis. *Ground Water* **31**(5): 814–822.

SPHERE (2011) *Humanitarian Charter and Minimum Standards in Disaster Response*, 2011 edn. The SPHERE Project, Practical Action Publishing, Rugby, UK.

Standards Australia (1990) *Test pumping of water wells*. AS 2368–1990.

Standards Australia (2000) *Steel Water Bore Casing*. AS 1396–2000.

Standards Australia (2001) *Arc-Welded Steel Pipes for Water and Wastewater*. AS 1579–2001.

Standards Australia International (2006) *Pressure Piping*. AS 4041–2006.

Standards Australia/Standards New Zealand (2006) *PVC Pipes and Fittings for Pressure Applications*. AS/NZS 1477–2006.

Standards Australia/Standards New Zealand (2011) *Solvent Cements and Priming Fluids for PVC (PVC-U and PVC-M) and ABS and ASA Pipes and Fittings*. AS/NZS 3879: 2011.

Standards Australia International/Standards New Zealand (2013) *Acrylonitrile Butadiene Styrene (ABS) Compounds, Pipes and Fittings for Pressure Applications*. AS/NZS 3518.1: 2013.

Statens strålskyddsinstitut (1996). *Radon i vatten*. Swedish Radiation Protection Institute Report 96–03.

Sterrett RJ (ed) (2007) *Ground Water and Wells*, 3rd edn. Johnson Screens, New Brighton, Minnesota.

Stoner RF, Milne DM and Lund PJ (1979) Economic design of wells. *Quarterly Journal of Engineering Geology* **12**: 63–78.

Stonestrom DA and Constantz J (2003) Heat as a tool for studying the movement of ground water near streams. *US Geological Survey Circular* 1260. Reston, Virginia, USA.

Strauss, M.F., Story, S.L. & Mehlhorn, N.E. (1989) Applications of dual-wall reverse-circulation drilling in ground water exploration and monitoring. *Ground Water Monitoring and Remediation* **9**(2), 63–71.

Suarez DL (1981) Relation between pHc and Sodium Adsorption Ratio (SAR) and an alternate method of estimating SAR of soil or drainage waters. *Soil Science Society of America Journal* **45**: 469–475.

Swamee PK and Jain AK (1976) Explicit equations for pipe-flow problems. *Journal of the Hydraulics Division (ASCE)* **102**(5): 657–664.

Swamee PK, Tyagi A and Shandilya VK (1999) Optimal configuration of a wellfield. *Ground Water* **37**(3): 382–386.

Tarbé de St-Hardouin FPH (1884) *Darcy: Inspector General, 2nd Class*. Notices Biographiques sur les Ingénieurs des Ponts et Chaussées depuis la création du Corps, en 1716, pp 224–226, Baudry and Cie, Paris. http:// biosystems.okstate.edu/darcy/English/ St-HardouinEnglish.htm.

Tedd KM, Misstear BDR, Coxon CE, Daly D and Hunter Williams N (2012) Hydrogeological insights from groundwater level hydrographs in southeast Ireland. *Quarterly Journal of Engineering Geology & Hydrogeology* **45**: 19–30

Terzaghi K and Peck RB (1948) *Soil Mechanics in Engineering Practice*. J Wiley & Sons, New York.

Theis CV (1935) The relation between the lowering of the piezometric surface and the rate and duration of discharge of a well using groundwater storage. *Transactions of the American Geophysical Union* **16**: 519–524.

Theis CV (1940) The source of water derived from wells: essential factors controlling the response of an aquifer to development. *Civil Engineering* **10**(5): 277–280.

Thiem A (1870) Die Ergiebigkeit artesischer Bohrlöcher, Schachtbrunnen und Filtergallerien. *Journal für Gasbeleuchtung und Wasserversorgung* **14**, 450–467.

Thiem A (1887) Verfahren zür Messung natürliche Grundwassergeschwindigkeiten. *Polytechnisches Notizblatt* **42**: 229–232.

Thiem GA (1906) *Hydrologische Methoden*. Gebhardt, Leipzig.

Thomsen R and Søndergaard V (2007) Dense hydrogeological mapping as a basis for establishing groundwater vulnerability maps in Denmark.

In: Witkowski AJ, Kowalczyk A and Vrba J (eds) *'Groundwater Vulnerability Assessment and Mapping'*, Taylor & Francis/Balkema, AK Leiden, The Netherlands, pp 33–43.

Thomsen R, Søndergaard VH and Sørensen KI (2004) Hydrogeological mapping as a basis for establishing site-specific groundwater protection zones in Denmark. *Hydrogeology Journal* **12**: 550–562.

Thomson W (1852) On the economy of heating or cooling of buildings by means of currents of air. *Proceedings of the Royal Philosophical Society Glasgow* **3**: 269–272.

Thomson W (1884) Compendium of the Fourier mathematics for the conduction of heat in solids, and the mathematically allied physical studies of diffusion of fluids and transmission of electric signals through submarine cables (based on articles prepared for the Encyclopaedia Britannica in 1880 and the Quarterly Journal of Mathematics in 1856). Article 72 in *Mathematical and Physical Papers by Sir William Thomson, Vol. II*, 41–60. Cambridge University Press.

Thorstenson DC, Fisher DW and Croft MG (1979) The geochemistry of the Fox Hills-Basal Hell Creek aquifer in Southwestern North Dakota and Northwestern South Dakota. *Water Resources Research* **15**(6): 1479–1498.

Timmer H, Verdel J-D and Jongmans AG (2003) Well clogging by particles in Dutch well fields. *Journal of the American Water Works Association* August **2003**: 112–118.

Todd DK (1980) *Groundwater Hydrology*, 2nd edn. John Wiley & Sons Inc, USA.

Todd DK and Mays LW (2005) *Groundwater Hydrology*, 3rd edn. John Wiley & Sons Inc, USA.

Todd F and Banks D (2009) Modelling of a thermal plume in the Sherwood Sandstone: a case study in North Yorkshire, UK. *Proceedings Effstock 2009 – The 11th International Conference on Thermal Energy Storage for Efficiency and Sustainability*, Stockholm, Sweden, June 14–17 2009.

Tucker ME (1996) *Sedimentary Rocks in the Field*, 2nd edn. The Geological Field Guide Series, John Wiley & Sons Ltd, Chichester, UK.

Tünnermeier T, Houben G and Niard N; edited by Himmelsbach T (2005/2006) *Hydrogeology of the Kabul Basin* (3 parts). Bundesanstalt für Geowissenschaften und Rohstoffe (BGR), Hannover, Germany.

Turner IL, Coates BP and Acworth RI (1996) The effects of tides and waves on water-table elevations in coastal zones. Hydrogeology Journal **4**(2): 51–69.

UNESCO/WHO/UNEP (1996) *Water quality assessments – a guide to use of biota, sediments and water in environmental monitoring* (2nd edn). Published on behalf of the World Health Organisation by F & FN Spon.

UNICEF/WHO (2012) *Progress on Drinking Water and Sanitation, 2012 Update*. WHO/UNICEF Joint Monitoring Programme for Water Supply and Sanitation.

US Environmental Protection Agency (1987) *Handbook: Guidelines for Delineation of Wellhead Protection Areas*. US Environmental Protection Agency Document EPA/440/6-87-010.

US Environmental Protection Agency (1994) *Handbook: Ground Water and Wellhead Protection*. US Environmental Protection Agency Document EPA/625/R-94/001.

US Environmental Protection Agency (1999a) *Method 200.2, Revision 2.8: Sample Preparation Procedure for Spectrochemical Determination of Total Recoverable Elements*. US Environmental Protection Agency Document EPA-821-R-99-018.

US Environmental Protection Agency (1999b) *The Class V underground injection control study. Volume 19 heat pump and air conditioning return flow wells*. United States Environmental Protection Agency report EPA/816-R-99-014 s.

US Environmental Protection Agency (2004) *Field sampling guidance document. Document #1220 Groundwater well sampling. Revision 1, Sept. 2004*. US Environmental Protection Agency, Region 9 Laboratory, Richmond, California, USA.

US Environmental Protection Agency (2005) *Groundwater Sampling and Monitoring with Direct Push Technologies*. US Environmental Protection Agency Document

US Environmental Protection Agency (2006) *Low-stress (low-flow) purging and sampling procedure for the collection of ground water samples from monitoring wells*. Low Flow – US EPA Region 1, SOP No. GW 0001, Revision 2. July 30, 1996.

US Geological Survey (1999) *Land subsidence in the United States*. USGS Circular 1182.

US Geological Survey (2011) GWPD-17. Conducting an instantaneous change in head (slug) test with a mechanical slug and submersible pressure transducer. Version 2010.1. *In* Cunningham WL and Schalk CW (eds.) *Groundwater technical procedures of the US Geological Survey: US Geological Survey Techniques and Methods 1–A1*, 145–151.

US Geological Survey (2014a) *Dissolved Gas N_2/Ar and ^4He Sampling*. United States Geological Survey website. http://water.usgs.gov/lab/dissolved-gas/sampling/. Last updated 12/9/14, accessed 31/1/15.

US Geological Survey (2014b) *Analytical procedures for dissolved gasses N_2/Ar*. United States Geological Survey website. http://water.usgs.gov/lab/dissolved-gas/lab/analytical_procedures/. Last updated 26/9/14, accessed 31/1/15.

US Geological Survey (2015) *National field manual for the collection of water-quality data: US Geological Survey Techniques of Water-Resources Investigations, book 9, chapters. A1-A9*. United States Geological Survey, continuously updated on Internet. Last accessed May 2015.

US Geological Survey Stable Isotope Laboratory (2015) *Instructions for collecting samples*. United States Geological Survey, Reston Stable Isotope Laboratory website. http://isotopes.usgs.gov/lab/instructions.html Last updated 4/8/15, accessed 15/8/15.

Utom AU, Odoh BI and Okoro AU (2012) Estimation of aquifer transmissivity using Dar Zarouk parameters derived from surface resistivity measurements: A case history from parts of Enugu town (Nigeria). *Journal of Water Resources and Protection* **4**: 993–1000.

van Beek CGEM (2001) Production well operation in gravel. In: Proc. International Association of Hydrogeologists (Irish Group) 21st Annual Seminar '*Gravel Aquifers: Investigation, Development and Protection*', Tullamore, 16-17 October 2001, pp 7.

Van der Elst NJ, Savage HM, Keranen KM and Abers GA (2013) Enhanced Remote Earthquake Triggering at Fluid-Injection Sites in the Midwestern United States. *Science* **341**(6142): 164–167.

Van der Wal (2010) *Understanding groundwater and wells in manual drilling: Instruction handbook for manual drilling teams on hydro-geology for well drilling, well installation and well development*, 2nd edn. PRACTICA foundation, Oosteind, The Netherlands.

Van Tonder G and Xu Y (2000) *A guide for the estimation of groundwater recharge in South Africa*. Report by the Institute for Groundwater Studies, Bloemfontein, for the Department of Water Affairs and Forestry (DWAF), June 2000, South Africa.

Van Tonder G, Kunstmann H and Xu Y (1998) *Estimation of the sustainable yield of a borehole including boundary information, drawdown derivatives and uncertainty propagation*. Institute for Groundwater Studies, Bloemfontein, South Africa.

Varner GR (2009) *Sacred Wells*, 2nd edtion. Algora Publishing, New York.

Verruijt A (2013) *Theory and problems of poro-elasticity*. Delft University of Technology, Netherlands.

Vogel JC, Talma AS and Heaton THC (1981) Gaseous nitrogen as evidence for denitrification in groundwater. *Journal of Hydrology* **50**: 191–200.

Volker A and Henry JC (1988) Side effects of water resources management: overviews and case studies. A contribution to IHP Project 11.1.a, prepared by a working group for IHP-III. *IAHS Publication* 172.

Vroblesky DA, Casey CC and Lowery MA (2006) Influence of in-well convection on well sampling. *US Geological Survey Scientific Investigations Report* 2006–5247. Reston, Virginia, USA.

Wall G (1986) *Exergy – a useful concept*. 3rd Edition. Doctoral Thesis, Chalmers Technical University, Göteborg, Sweden.

Walsh A (2013) *Walking Through History: Oman's World Heritage Sites*. Al Roya Press & Publishing House, Muscat, Sultanate of Oman.

Walsh JB (1981) Effect of pore pressure and confining pressure on fracture permeability. *International Journal of Rock Mechanics and Mining Sciences & Geomechanics Abstracts* **18**: 429–435.

Walton WC (1960) *Leaky aquifer conditions in Illinois*. Illinois State Water Survey (Urbana) Report of Investigation 39, USA.

Walton WC (1962) *Selected analytical methods for well and aquifer evaluation*. Illinois State Water Survey (Urbana) Bulletin 49, USA.

Walton WC (1970) *Groundwater Resource Evaluation*. McGraw-Hill, New York.

Wang HF (2000) *Theory of linear poroelasticity with applications to geomechanics and hydrogeology*. Princeton University Press, USA.

WaterAid Tanzania (2009) *Management for sustainability: Practical lessons from three studies the management of rural water supply schemes*. WaterAid Tanzania, dar es Salaam.

Waters A and Banks D (1997) The Chalk as a karstified aquifer: closed circuit television images of macrobiota. *Quarterly Journal of Engineering Geology* **30**: 143–146.

Waters P, Greenbaum D, Smart PL and Osmaston H (1990) Applications of remote sensing to groundwater hydrology. *Remote Sensing Reviews* **4**(2): 223–264.

Watt SB and Wood WE (1977) *Hand Dug Wells and their Construction*. ITDG Publishing, London.

Wendling G, Chapuis RP, Gill DE (1997) Quantifying the effects of well development in unconsolidated material. *Ground Water* **35**(3): 387–393.

Westaway R and Younger PL (2014) Quantification of potential macroseismic effects of the induced seismicity that might result from hydraulic fracturing for shale gas exploitation in the UK. *Quarterly Journal of Engineering Geology and Hydrogeology* **47**: 333–350.

Westaway R, Scotney PM, Younger PL, Boyce AJ (2015) Subsurface absorption of anthropogenic warming of the land surface: the case of the world's largest brickworks (Stewartby, Bedfordshire, UK). *The Science of the Total Environment* **508**: 585–603.

Weyrauch R (1914) Hydrologische Vorarbeiten. *In:* Lueger O (ed.) *Lexikon der gesamten Technik und ihrer Hilfswissenschaften, Bind 9*. Stuttgart/Leipzig, 378–382. Available at http://www.zeno.org/nid/20006159303

White JS and Mathes MV (2006) Dissolved-gas concentrations in ground water in West Virginia, 1997–2005. *US Geological Survey Data Series* **156**, 8 p.

White RR and Clebsch A (1994) CV Theis: The Man and His Contributions to Hydrogeology. In: Clebsch A (ed.) '*Selected Contributions to Ground-Water Hydrology by C.V. Theis, and a Review of His Life and Work*', US Geological Survey Water-Supply Paper 2415. http://www.olemiss.edu/sciencenet/saltnet/theisbio.html.

Whiteside G and Trace S (1993) The use of sludging and well-pointing techniques to sink small diameter tube-wells. *Waterlines* **11**(3).

Whittaker A, Holliday DW and Penn IE (1985) *Geophysical Logs in British Stratigraphy*. Geological Society, London, Special Report 18, London.

Wilde FD (2011) Water-quality sampling by the US Geological Survey – Standard protocols and procedures. *US Geological Survey Fact Sheet* 2010-3121.

Wilkinson JC (1977) *Water and Tribal Settlement in South-East Arabia: A Study of the Aflaj of Oman*. Clarendon Press, Oxford.

Williams DE (1985) Modern techniques in well design. *Journal of the American Water Works Association* September 1985: 68–74.

Williams EB (1981) Fundamental concepts of well design. *Ground Water* **19**(5): 527–542.

Williams GM, Hooker PJ, Noy DJ and Ross CAM (1998) Mechanisms for [85]Sr migration through glacial sand determined by laboratory and in situ tracer tests. *Geological Society, London, Special Publication* **128**: 35–48.

Winkler L (1888) Die Bestimmung des im Wasser Gelösten Sauerstoffes. *Berichte der Deutschen Chemischen Gesellschaft* **21**(2): 2843–2855.

World Bank (2005) *Towards a More Effective Operation Response: Arsenic Contamination of Groundwater in South and East Asian Countries.* Report 31303.

World Health Organization (1997) *Guidelines for Drinking Water Quality, Volume 3: Survellance and control of community supplies*, 2nd edn. World Health Organization, Geneva.

World Health Organization (2011) *Guidelines for Drinking-water Quality*, 4th edn. World Health Organization, Geneva.

World Health Organization (2014) *Household Fuel Combustion*. World Health Organization, Geneva.

World Health Organization/UNICEF (2012) *Rapid assessment of drinking-water quality: a handbook for implementation*. World Health Organization, Geneva.

Yamano M and Goto S (2005) Long-term monitoring of the temperature profile in a deep borehole: temperature variations associated with water injection experiments and natural groundwater discharge. *Physics of the Earth and Planetary Interiors* **152**: 326–334.

Yang YJ and Gates TM (1997) Wellbore skin effect in slug-test data analysis for low-permeability geologic materials. *Ground Water* **35**, 931–937.

Yeskis D and Zavala B (2002) *Ground-water sampling guidelines for superfund and RCRA project managers*. US Environmental Protection Agency, Office of Solid Waste and Emergency Response, Ground Water Forum Issue Paper EPA 542-S-02-001, May 2002. http://www.epa.gov/tio/tsp/download/gw_sampling_guide.pdf

Young ME, de Bruijn RGM and bin Salim Al-Ismaily A (1998) Exploration of an alluvial aquifer in Oman by time-domain electromagnetic sounding. *Hydrogeology Journal* **6**: 383–393.

Younger PL (1989) Devensian periglacial influences on the development of spatially variable permeability in the Chalk of southeast England. *Quarterly Journal of Engineering Geology* **22**: 343–354.

Younger PL (1992) The hydrogeological use of thin sections: inexpensive estimates of groundwater flow and transport parameters. *Quarterly Journal of Engineering Geology and Hydrogeology* **25**: 159–164.

Younger PL (2012) Real source of the Gaza crisis runs very deep. *The Scotsman*, 18th December 2012.

Zang A, Majer E and Bruhn D (2014a) Preface. *Geothermics* **52**: 1–5.

Zang A, Oye V, Jousset P, Deichmann N, Gritto R, McGarr A, Majer E, Bruhn D (2014b) Analysis of induced seismicity in geothermal reservoirs – an overview. *Geothermics* **52**: 6–21.

Zhou Y, Zwahlen F and Wang Y (2011) The ancient Chinese notes on hydrogeology. *Hydrogeology Journal* **19**: 1103–1114.

Index

Water Wells and Boreholes, Second Edition. Bruce Misstear, David Banks and Lewis Clark.
© 2017 John Wiley & Sons Ltd. Published 2017 by John Wiley & Sons Ltd.

Printed and bound by CPI Group (UK) Ltd, Croydon, CR0 4YY

27/10/2024

14580309-0003